Micropaleontology

Application in Stratigraphy and Paleoceanography

Micropaleontology

Application in Stratigraphy and Paleoceanography

Editor
Devesh K. Sinha

Narosa Publishing House
New Delhi Chennai Mumbai Kolkata

Editor
Devesh K. Sinha
Department of Geology
Banaras Hindu University
Varanasi, India

NAROSA PUBLISHING HOUSE PVT. LTD.

22 Daryaganj, Delhi Medical Association Road, New Delhi 110 002
35-36 Greams Road, Thousand Lights, Chennai 600 006
306 Shiv Centre, D.B.C. Sector 17, K.U. Bazar P.O., Navi Mumbai 400 703
2F-2G Shivam Chambers, 53 Syed Amir Ali Avenue, Kolkata 700 019

www.narosa.com

Printed from the camera-ready copy provided by the Editor.

ISBN-13: 978-81-7319-768-0
ISBN-10: 81-7319-768-7

Published by N.K. Mehra for Narosa Publishing House Pvt. Ltd.,
22 Daryaganj, Delhi Medical Association Road, New Delhi 110 002
and printed at Rajkamal Electrical Press, New Delhi 110 033, India

Dedicated to
My Revered Teacher

Prof. M. S. Srinivasan, FTWAS, FNA, FASc, FNASc

Preface

This volume entitled "Micropaleontology: Application in Stratigraphy and Paleoceanography" contains invited papers from eminent scientists from India and abroad on diverse application of microfossils.

Recent years have witnessed tremendous progress in the science of microfossils including their application in paleoceanographic and paleoclimatic studies besides their traditional biostratigraphic relevance. Major age revisions of the so called unfossiliferous strata were made with the discovery of microfossils. The systematic micropaleontology has gradually given way to interpretative micropaleontology and has completely changed our perception of this science.

Two most significant areas where micropaleontology has played a vital role are Stratigraphy and Paleoceanography. With the advent of the Deep Sea Drilling Project (DSDP) in 1968 and its successor the Ocean Drilling Programme (ODP), the science of micropaleontology reached new heights as excellently preserved undisturbed cores provided a unique opportunity to study the well preserved microfossils at very high stratigraphic resolution. Interesting correlations were observed between microfossil evolution, faunal turnover and paleomagnetic reversals. Recently, the extraterrestrial causes for extinctions were questioned based on discovery of microtektites in the deep sea sections where exhaustive records of microfaunal evolution already exist. Stable isotope measurement of foraminiferal tests offered brilliant means to reconstruct paleotemperature and ice volume records which in turn were used for paleoceanographic and paleoclimatic studies including the history of Asian Monsoon.

Though there have been a number of books on application of microfossils in stratigraphy and paleoceanography in foreign countries, there is a serious lack of a concise volume providing a comprehensive treatment to the application of microfossils in these two fields in our country. Keeping this in view, eminent and active workers from India and abroad were invited to contribute to this volume on diverse application of microfossils. All the papers were peer reviewed by reputed experts in the concerned field. It gives me immense gratification to dedicate this volume to my revered teacher Prof. M.S. Srinivasan, FNA, a legendary figure in the field of Micropaleontology, Stratigraphy and Paleoceanography.

I am thankful to all the authors who have contributed to this volume. I feel indebted to my students Ashutosh, Amit and Vikram for their help and moral support. I thank Nupur and Devanshu for their boundless patience during completion of this volume.

I am sure the book will be of immense use to young and active micropaleontologists who wish to utilize microfossils in solving multifarious geological problems related to stratigraphy, paleoceanography, paleoenvironment and paleoclimate.

Devesh K. Sinha

Contents

List of Contributors

A.D. Singh
Department of Geology, Banaras Hindu University, Varanasi 221 005, India

Alfonso Pardo
Fundación San Valero, c/ Violeta Parra 9, 50015 Zaragoza, Spain

Ajai K. Rai
Department of Earth & Planetary Sciences, Nehru Science Center, Allahabad University, Allahabad 211 002, India

Amit K. Ghosh
Birbal Sahni Institute of Palaeobotany, 53 University Road, Lucknow 226 007, India

Ashutosh Kumar Singh
Department of Geology, Banaras Hindu University, Varanasi 221005, India

Benjamin P. Flower
College of Marine Science, University of South Florida, St. Petersburg, FL 33701, USA

Bijai Prasad
KDM Institute of Petroleum Exploration, Oil and Natural Gas Corporation Limited, 9 Kaulagarh Road, Dehradun 248 195, India

B.N. Tiwari
Wadia Institute of Himalayan Geology, 33 GMS Road, Dehra Dun 248001, India

B. R. Rao
OpEC-II, MW, GSI, Visakhapatanam 530017, India

C. Rajshekhar
Geology & Palaeontology Group, Agharkar Research Institute, Pune 411 004, India

D. Bhattacharjee
OpEC-I, MW, GSI, Kolkata 700 032, India

Deepak Joshi
Wadia Institute of Himalayan Geology, 33 GMS Road, Dehra Dun 248 001, India

Devesh K. Sinha
Department of Geology, Banaras Hindu University, Varanasi 221005, India

Gerta Keller
Department of Geosciences, Princeton University, Princeton, NJ 08544, USA

H.N. Sinha
Department of Geology, Vinoba Bhave University, Hazaribagh 825301, India

H. Singh
Department of Geology, HNB Garhwal University, Srinagar 246 174, India

James C. Zachos
Earth Sciences Department, University of California, Santa Cruz CA, 95064, U.S.A.

Jayasri Banerji
Birbal Sahni Institute of Palaeobotany, 53 University Road, Lucknow 226 007, India

J. Daneshian
Department of Geology, Teachers' Training University of Tehran, Mofateh Avenue, Tehran, Iran

John A. Barron
U.S. Geological Survey, MS 910, Menlo Park, CA 94025, USA

K. Ayyasami
Palaeontology Division, Geological Survey of India, Bandlaguda, Hyderabad 500068, India

K. Kumar
Wadia Institute of Himalayan Geology, 33 General Mahadeo Singh Road, Dehradun 248 001, India

Kshitij Mohan
Department of Geography, Banaras Hindu University, Varanasi 221 005, India

Kezhakielie Whiso
Department of Geology, Nagaland University, Kohima 797002, India

Manish Tiwari
Physical Research Laboratory, Navrangpura, Ahmedabad 380009, India

Masako Ibaraki
Department of Biology and Geosciences, Faculty of Science, Shizuoka University, Shizuoka 422-8529, Japan

M.N. Joshi
Department of Geology, D.B.S. (P.G.) College, Dehra Dun 248 001, India

M.S. Srinivasan
Department of Geology, Banaras Hindu University, Varanasi 221005, India

N.N. Dogra
Department of Earth Sciences, Kurukshetra University, Kuruksehtra 136 119, India

Om Prakash Singh
Department of Geology, Banaras Hindu University, Varanasi 221 005, India

Paul N. Pearson
Department of Earth Sciences, University of Bristol, Bristol BS8 1RJ, UK

P. Kundal
Postgraduate Department of Geology, Nagpur University, Law College Campus, Nagpur 440 001, India

P.P. Reddy
Geology & Palaeontology Group, Agharkar Research Institute, Pune 411 004, India

P. Ramesh
Paleontology Laboratories KDMIPE, ONGC, Dehra Dun 248 195, India

Pratul Kumar Saraswati
Department of Earth Sciences, Indian Institute of Technology, Bombay, Mumbai 400 076, India

P. Vijoy Kumar
OpEC-II, MW, GSI, Visakhapatanam 530017, India

Rahul Mohan
National Centre for Antarctic and Ocean Research, Goa, India

Ram Awatar
Birbal Sahni Institute of Paleobotany, 53, University Road, Lucknow 226 007, India

Ramson Asher
KDM Institute of Petroleum Exploration, Oil and Natural Gas Corporation Limited, 9 Kaulagarh Road, Dehradun 248 195, India.

Ratan Kar
Birbal Sahni Institute of Paleobotany, Lucknow 226007, India

R. J. Azmi
Wadia Institute of Himalayan Geology, 33 GMS Road, Dehra Dun 248 001, India

R.P. Kachhara
Department of Geology, Nagaland University, Kohima 797002, India

R.K. Sukhtankar
Geology & Palaeontology Group, Agharkar Research Institute, Pune 411 004, India

R. Ramesh
Physical Research Laboratory, Navrangpura, Ahmedabad 380009, India

R. S. Rana
Department of Geology, HNB Garhwal University, Srinagar 246 174 UA, India

R.S. Singh
Birbal Sahni Institute of Paleobotany, Lucknow 226007, India

Ryoji Wani
Department of Geology, National Science Museum, 3-23-1, Hyakunincho, Shinjuku-ku, Tokyo 169-0073, Japan

R.Y. Singh
Department of Geology, Punjab University, Chandigarh-160 014, India

S. Anantharaman
Palaeontology Division, Geological Survey of India, Bandlaguda, Hyderabad 500068, India

S.N. Uniyal
KDM Institute of Petroleum Exploration, Oil and Natural Gas Corporation Limited, 9 Kaulagarh Road, Dehradun 248 195, India.

S.S. Srivastava
Department of Earth Sciences, IIT Roorkee, Roorkee 247 667, India

Sumedh K. Humane
Postgraduate Department of Geology, Nagpur University, Law College Campus, Nagpur 440 001, India

Virendra B. Singh
Department of Earth & Planetary Sciences, Nehru Science Center, Allahabad University, Allahabad 211 002, India.

V. Sharma
Department of Geology, University of Delhi, Delhi 110 007, India

Y. Raghumani Singh
P. G. Department of Geology, University of Jammu, Jammu 180006, India

Meso-Neoproterozoic organic-walled microfossils from the Vindhyan sediments of Son Valley, Madhya Pradesh, India

Bijai Prasad, S.N. Uniyal and Ramson Asher

KDM Institute of Petroleum Exploration, Oil and Natural Gas Corporation Limited, 9 Kaulagarh Road, Dehradun 248 195, India.

Introduction

The Vindhyan Basin represents an undisturbed Proterozoic sedimentary succession in Central India (Figs. 1a, b), which is classed as the Vindhyan Supergroup. The age of Vindhyan Supergroup has long remained a subject of controversy due to scantiness of fossil records. Diverse ages have been assigned to the Vindhyan succession mainly based on micro-and macrofossil evidences and varied radiometric datings that gave its age ranging from Late Paleoproterozoic to Early Paleozoic. The age of the Vindhyan Supergroup still remains a debatable issue in the light of contradictory nature of age assignments based on a number of fossil evidences and radiometric datings.

The evidences available on the Vindhyan Supergroup leading to its diverse age assignments are the organic-walled microfossils, stromatolites, macrofossils, small shelly microfossils (SSMs) and radiometric datings.

Organic-walled microfossils

Studies on acritarchs and other organic-walled microfossils from Vindhyan sediments have been made by a number of workers (Maithy and Shukla, 1977; Salujha *et al.*, 1971; Venkatachala *et al.*, 1990; Maithy and Babu, 1988), which were later reviewed and summarized by Maithy (1992) and Venkatachala *et al.* (1996). These studies broadly suggested an Early Mesoproterozoic to Terminal Proterozoic (Vendian) age-range to the Vindhyan Supergroup. However, Salujha (1973), based on the records of small acanthomorphic acritarchs (*Micrhystridium*) and herkomorphs (*Cymatiosphaera*) from Rohtas Limestone and Bijaigarh Shale (Salujha *et al.,* 1971), opined that Vindhyan Supergroup ranges in age from Late Precambrian to Early Silurian. Anabarasu (2001) recorded doubtful herkomorphitic acritarchs referable to *Cymatiosphaeroides kullingii* of Late Neoproterozoic aspect from the basal Semri Group. Recently, Prasad and Asher (2001) suggested a Middle Mesoproterozoic (Ectasian) to Terminal Proterozoic (Vendian) age-range for the Vindhyan Supergroup, based on the record of rich organic-walled microfossil assemblages from DMH-A well, drilled in the Vindhyan Basin.

Stromatolites

A variety of stromatolitic structures were recorded from different limestone bearing formations (Kajrahat, Salkhan and Nagod limestones) of Vindhyan Supergroup, which were later systematically studied and summarized by Kumar (1984). Kumar (1982) suggested an age of Early to Middle Riphean for the stromatolites of Semri Group and Late Riphaen to those from the Bhander Group.

Macro-and megafossils

The macro-and megafossil records from the Vindhyan sediments are the carbonaceous macrofossils referable to *Chuaria-Tawuia*. These forms were recorded from the Lower as well as Upper Vindhyan successions (Kumar and Srivastava, 1997; Kumar, 2001) and are broadly indicative of Precambrian age. A doubtful specimen of *Spriggina,* an Ediacaran fossil, is also reported from Pulkoa Shale (Rohtas Subgroup) by Kathal *et al.* (2000).

Small shelly microfossils and brachiopods

Azmi (1998) recorded small shelly microfossils and brachiopods from the Rohtas Limestone. Based on the above findings, Azmi (1998) suggested the Precambrian-Cambrian boundary within Rohtas Limestone, and regarded Kaimur, Rewa and Bhander groups (Upper Vindhyan) as the Lower Paleozoic representatives in the Vindhyan Basin. However, this finding has been under debate.

Geochronology

The radiometric datings of Vindhyan sediments at different stratigraphic levels also gave varied ages to a number of lithounits. Banerjee and Frank (1999) dated the Deonar Formation ca. 1073-920±30 Ma at high temperature and ca. 617±3.5 Ma at low temperature through Ar^{40}/Ar^{39} method. On the other hand, Ray *et al.* (2002) estimated the absolute age of this lithounit as 1631±5 Ma through U-Pb Zircon method. The glauconitic sandstone of Rampur Formation is also variously dated as ca.1110±60 Ma (Vinogradov *et al.,* 1964), ca.1080±40 Ma (Kreuzer *et al.,* 1977) and even ca.1599-1628 Ma (Rasmussen *et al.*, 2002). Very recently, Sarangi *et al.* (2004) dated the Rohtas Limestone as ca. 1599±48 Ma. Vinogradov *et al.* (1964) estimated the abstracted age of 940±30 Ma for Lower Kaimur, and 910±30 Ma for the upper Kaimur rocks; whereas, Crawford and Compston (1970) dated the Kimberlite pipes of Kaimur rocks at Majhgawan as ca.1140±12 Ma. However, Kreuzer *et al.* (1977) revised the Vinogradov *et al.* (1964) dating on Kaimur Sandstone as ca. 890±40 Ma on the basis of new recommended constants. Srivatava and Rajagopalan (1988) dated the Upper Rewa Sandstone as 700±100 Ma by Fission-Track method which are suggestive of Late Neoproterozoic age to the Rewa Group. Friedman *et al.* (1996) and Friedman and Chakraborty (1997) suggested the Precambrian-Cambrian boundary in Bhander Group on the basis of stable $\delta^{13}C$ isotope values.

A perusal of the above age data based on fossil as well as radiometric indicates that there are no conclusive evidences to ascertain the age of Vindhyan Supergroup, and the every new findings have added to its age controversies. It appears that the diverse ages were assigned to this sedimentary succession because a number of studies were limited to a particular or some lithounits, and the complete Vindhyan succession was not taken into account in many studies. To overcome the above age controversies, a detailed organic-walled microfossils study is made on the Vindhyan Supergroup in outcrop as well as subsurface sections, covering the complete succession. A number of samples recorded rich organic-walled microfossils that suggested a more authentic age to the Vindhyan Supergroup. An account of organic-walled microfossils, recovered from different lithounits of Vindhyan Supergroup, is presented here with discussion on their age alongwith its total age-range.

Geological setting

The Vindhyan Basin is an arcuate shaped large intracratonic basin in central India, covering parts of eastern Rajasthan, Madhya Pradesh, southern Uttar Pradesh and western Bihar (Figs. 1a, b). Biswas *et al.* (1993) considered Vindhyan Basin as an Interior Sag basin. However, Ram *et al.* (1996) are of the opinion that it is a pericratonic basin having pre-rift, syn-rift and post-rift phases of development. Detailed accounts on the geology of Vindhyan Basin are available by the classic works of Mallet (1869), Heron (1932) and Auden (1933). Later on, Prasad (1976,1984) and Sastri and Moitra (1984) have added further on the geology, structure and sedimentary history of the Vindhyan Basin.

A number of lithostratigraphic classifications were attempted on the Vindhyan Supergroup; however, Auden's (1933) classification is widely accepted. This classification was modified later by Sastri and Moitra (1984) considering the latest Code of Stratigraphic Nomenclature, followed in the present work (Fig. 2a). The entire Vindhyan succession is divided into four groups, *viz.* Semri, Kaimur, Rewa and Bhander (Auden, 1933; Sastri and Moitra, 1984). The Semri Group represents the Lower Vindhyan succession; whereas, Kaimur, Rewa and Bhander groups characterise the Upper Vindhyan succession. Sastri and Moitra (1984) further subdivided the Semri Group of Son Valley into four subgroups, *viz.* Mirzapur, Porcellanite, Kheinjua and Rohtas subgroups, each one having a number of formations (Fig. 2a).

Material and methods

The outcrop samples for microfossil studies were collected from Maihar-Govindhgarh area of Son Valley in northern Madhya Pradesh in two sections. These are the Barhi-Badanpur-Maihar section (section I) and Jhal-Bihara section (section II) [Figs. 1b, c]. The stratigraphic position of studied samples is shown in the lithocolumn of respective sections (Figs. 2b, c). Vindhyan sediments from well DMH-A (surface - 3500m $_{DD}$), recently drilled near Damoh in Madhya Pradesh (Fig. 1b) and representing almost complete Vindhyan succession (Fig 4), are also studied for organic-walled microfossils to suppliment the outcrop microfossil data.

Microfossil assemblages and biostratigraphy

The recovered organic-walled microfossils from Vindhyan sediments include acritarchs, coccoid and filamentous forms, and vase-shaped microfossils. The microfossil assemblages, recorded from different lithounits of Vindhyan Supergroup, are briefly outlined below with discussion on their age.

Semri Group (Lower Vindhyan)

Kajrahat Limestone

The Kajrahat Limestone from outcrops (Figs. 2b, c), and depth intervals from +3500$_{(DD)}$ - 3147m in DMH-A well (Fig. 4) representing this formation, yielded abundant filamentous microfossils along with frequent occurrence of simple unornamented sphaeromorphic acritarchs. The filamentous forms include *Siphonophycus* spp. (*S. septatum, S. robustum, S. kestron*), *Karamia segmentata, Polythrichoides lineatus, Tortunema pseudoseptata, Arctacellularia ellipsoidea* and abundant dark opaque filamentous sheaths (Pl. 1, Figs. 4-12). *Leiosphaeridia asperata, L. tenuissima, L. crassa* and *Kildinosphaera chagrinata* represent the unonamented sphaeromorphic acritarchs (Pl. 1, Figs. 1-3).

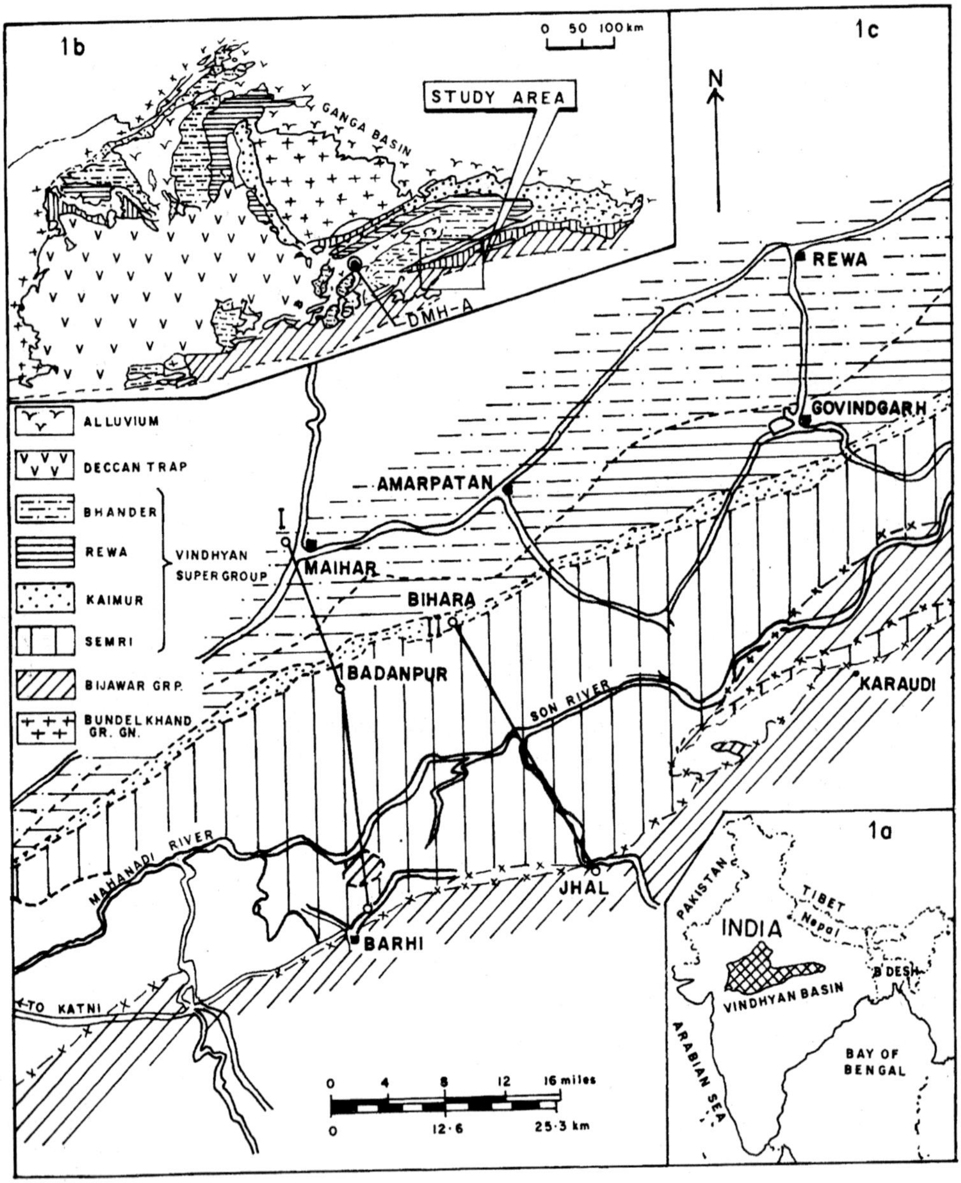

Fig. 1. **(a)** Index map of India showing location and settings of Vindhyan Basin. **(b)** An outline map of Vindhyan Basin (*after* Ahmad, 1962), showing present study area, and location of DMH-A well. **(c)** Detailed geological map of western Son Valley in East Madhya Pradesh (*after* Dutta *et al.*, 1998), showing different formations of Vindhyan Supergroup and studied sections of Barhi - Badanpur - Maihar (section-I) and Jhal- Bihara (section-II).

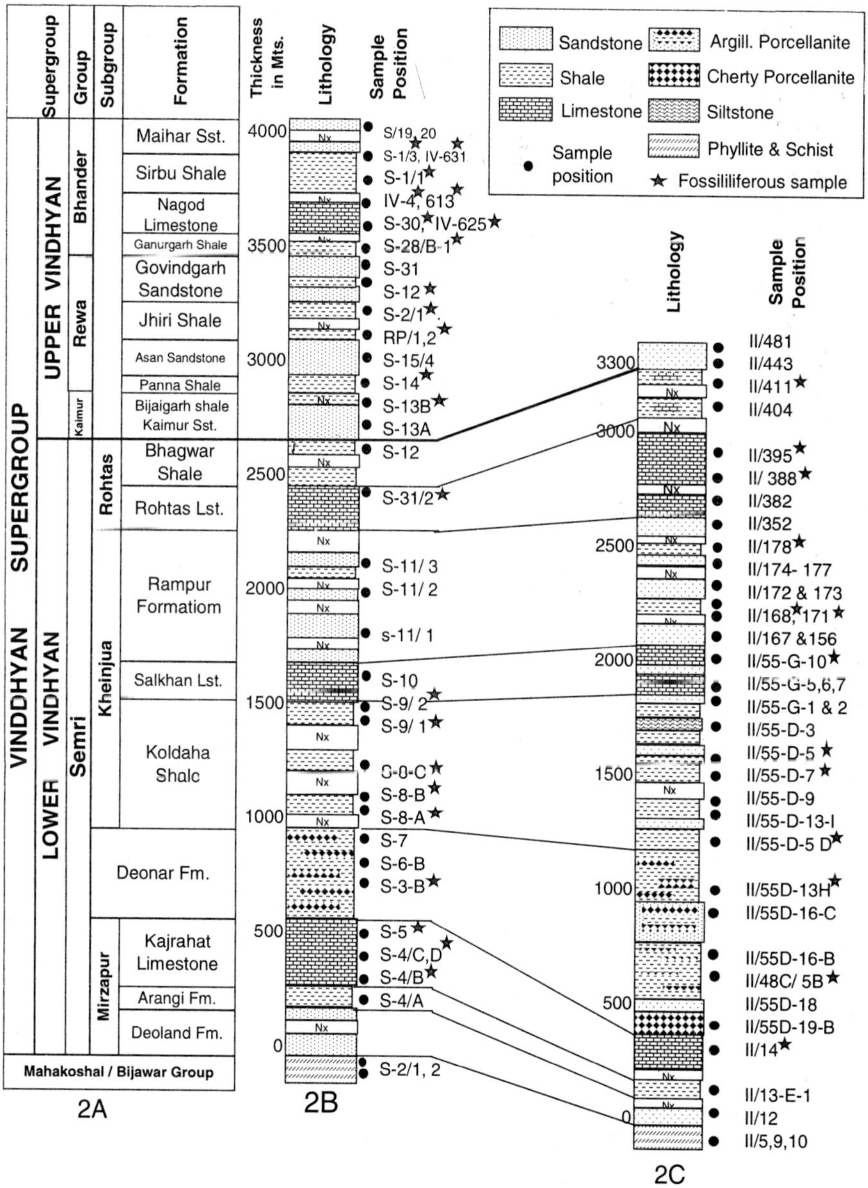

Fig. 2. **(a)** Generalised stratigraphic succession of Vindhyan Supergroup in western Son Valley, northern Madhya Pradesh (*after* Sastri and Moitra, 1984). **(a,b)** Generalised lithological succession of Vindhyan Supergroup around Maihar - Govindgarh area (*after* Dutta *et al.*, 1998), showing stratigraphic position of palynological samples in Barhi-Badanpur-Maihar section **(b)**, and Jhal - Bihara section **(c)**

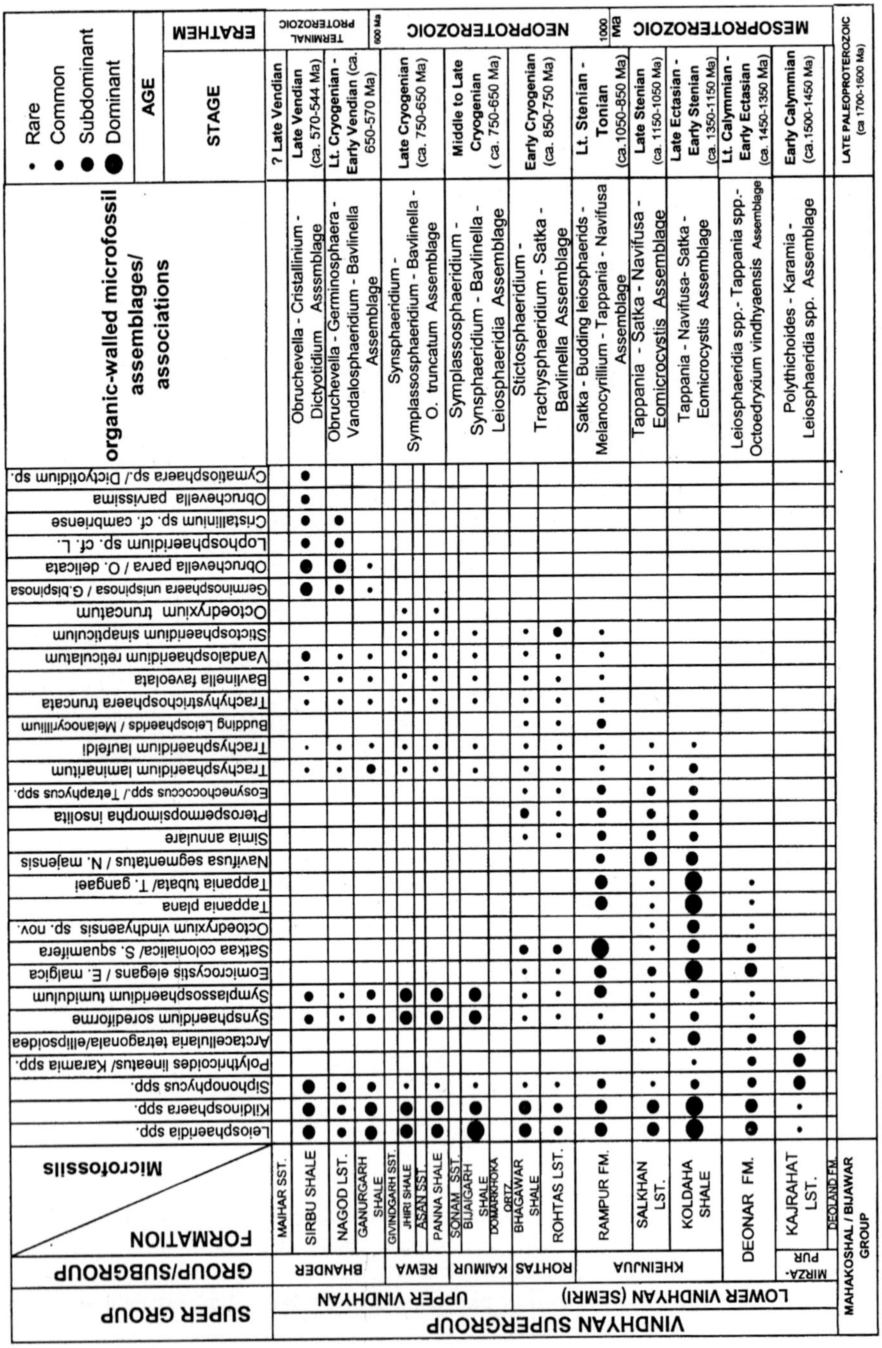

Fig. 3. Stratigraphic distribution of selected Meso-Neoproterozoic organic-walled microfossils in Vindhyan sediments of western Son Valley, northern Madhya Pradesh.

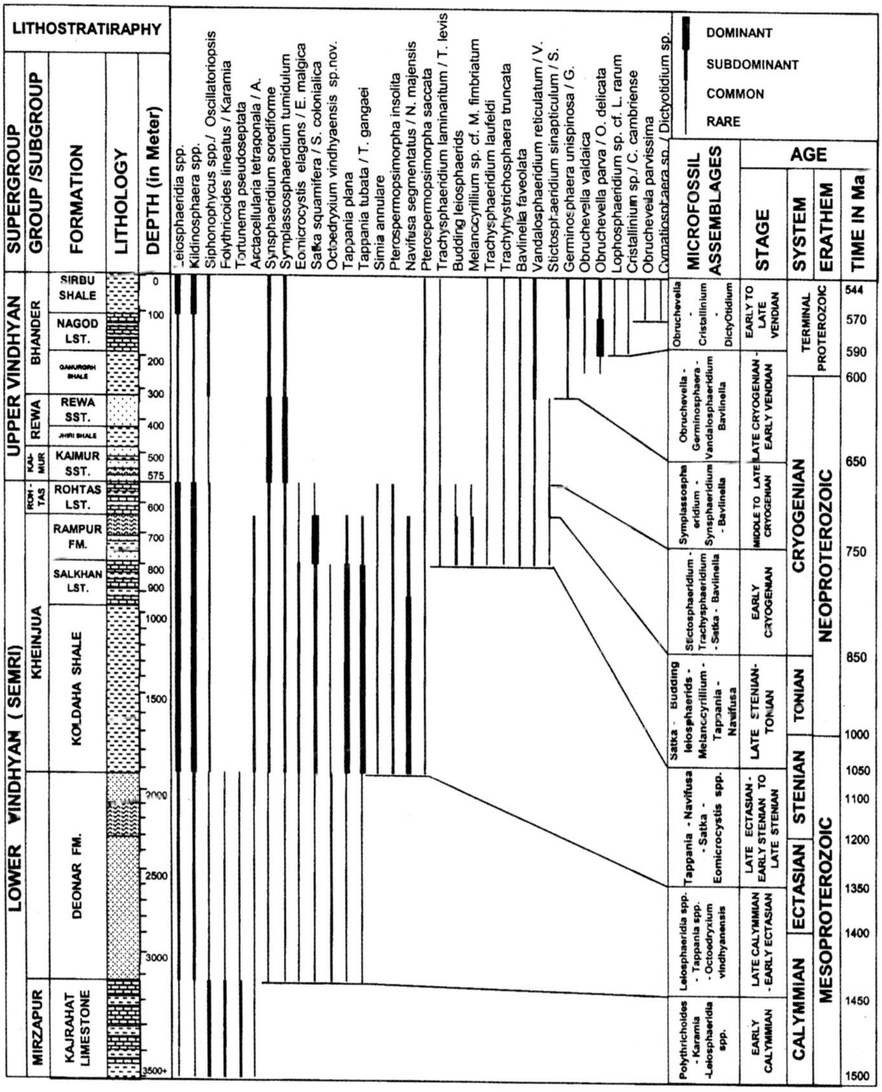

Fig. 4. Stratigraphic distribution of selected Meso-Neoproterozoic organic-walled microfossils in Vindhyan sediments of DMH - A well, central Madhya Pradesh.

The Kajrahat assemblage, by and large, compares with the Early Mesoproterozoic microfossil assemblage of Buryan Formation (Billyakh Group) from Southern Urals (Golovenok and Belova, 1984; Sergeev *et al.,* 1995), of which age is roughly estimated as ca.1600-1300Ma. Acritarch assemblages from Kajrahat Limestone and Billyakh Group are of low diversity and represented by simple unornamented sphaeromorphs (*Leiosphaeridia* spp., *Kildinosphaera chagrinata*) only, and are associated with abundant filamentous microfossils. The Kajrahat assemblage also broadly compares with the microfossil assemblage of Late Paleoproterozoic (ca. 1800 Ma) Chuanlingguo Formation of northern China (Hofmann and Chene, 1981; Zhang, 1986) as simple sphaeromorphic acritarchs and filamentous forms are common in both the assemblages. However, the Kajrahat assemblage shows younger age of Early Mesoproterozoic as it also includes abundant coccoid and filamentous microfossils (Venkatachala *et al.*, 1990). Venkatachala *et al.* (1990) recorded a rich assemblage of coccoid and filamentous microfossils from Kajrahat Limestone of Rohtasgarh area (Bihar) which is mainly represented by the various species of *Eoentophysalis, Tetraphycus, Glenobotrydion* and *Biocatenoides*. Though, Venkatachala *et al.* (1990) suggested a Middle Mesoproterozoic age to this assemblage, it shows close resemblence with Early Mesoproterozoic microfossil assemblage of Balbirini Dolomite (McArthur Group) of northern Australia, described by Oehler (1978). The age of Balbirini Dolomite is estimated approximately 1500 Ma (Oehler, 1978), overlain by the Roper Group having well constrained absolute age of 1450-1400 Ma (Kralik, 1982; Jackson *et al.*, 1999). The microfossil assemblage of Roper Group (ca. 1450-1400 Ma), northern Australia (Peat *et al.,* 1978; Javaux *et al.*, 2001) also broadly correlates with the Kajrahat assemblage as abundant filamentous forms and leiosphaerids are common in both the assemblages. However, the Kajrahat assemblage lacks the presence of much developed acritarchs, such as *Tappania*, *Valaria* and *Dictyosphaera*, which are

---

Plate 1.

Organic-walled microfossils from Kajrahat Limestone and Deonar Formation.

Kajrahat Limestone (specimens in Figs. 1-2 from outcrops; 3-12 from DMH-A well, CC-10: 3479-8m; magnifications approx. 600x, otherwise mentioned)

1 *Leiosphaeridia tenuissima* Eisenack. S-5/1; EFR H48/3. **2** *Leiosphaeridia crassa* (Naumova) Jankauskas. S-5/1; EFR L50. **3** *Leiosphaeridia asperata* (Naumova) Lindgren. DMH-A, CC-10, Seg.1/5/2; EFR F25/4. **4** Opaque filament. DMH-A, CC-10, Seg. 1/5/1; EFR U54; 200x. **5, 12** *Arctacellularia ellipsoidea* German. DMH-A, CC-10, Seg.1/5/1; Fig. 5, EFR Q36/2; 200x. Fig. 12, EFR G28. **6** *Tortunema pseudoseptata* Butterfield, Knoll and Swett. DMH-A, CC-10, Seg.1/5/2; EFR H68/1; 200x. **7** *Karamia segmentata* Jankauskas, Mikhaylova and German. DMH-A, CC-10, Seg.1/5/2; EFR P64; 200x. **8, 10** *Polythrichoides lineatus* Knoll, Swett and Mark. DMH-A, CC-10, Seg.1/5/1; Fig. 8, EFR G42/3; 450x. Fig. 10, EFR U59/2; 400x. **9** *Siphonophycus kestron* Schopf. DMH-A, CC-10, Seg.1/5/2; EFR F68/1. **11** *Siphonophycus robustum* Knoll, Swett & Mark. DMH-A CC-10, Seg.1/5/2; EFR G67/1; 350x.

Deonar Formation (all specimens from II/48/5/B; magnifications approx. 1000x, otherwise mentioned)

13-15 *Octoedryxium* sp. Fig. 13, EFR W64/2; Fig. 14, EFR T65/3; Fig.15, EFR T67/3. **16** *Satka squamifera* Pyatiletov. EFR T58. **17** *Eomicrocystis elegans* Golovenok & Belova. EFR G50/3; 750x. **18, 25** *Kildinosphaera granulata* Vidal. Fig.18, EFR K30. Fig. 25, EFR Q47/3. **19** *Leiosphaeridia asperata* (Naumova) Lindgren. EFR S37. **20** *Tappania plana* Yin. EFR N67; 800x. **21** *Tappania tubata* Yin. EFR H40/1. **22** *Synsphaeridium sorediforme* Eisenack. EFR N64/3. **23** *Siphonophycus septatum* Knoll, Swett and Mark. EFR S58. **24** *Leiosphaeridia tenuissima* Eisenack. EFR T34/3.

Plate 1.

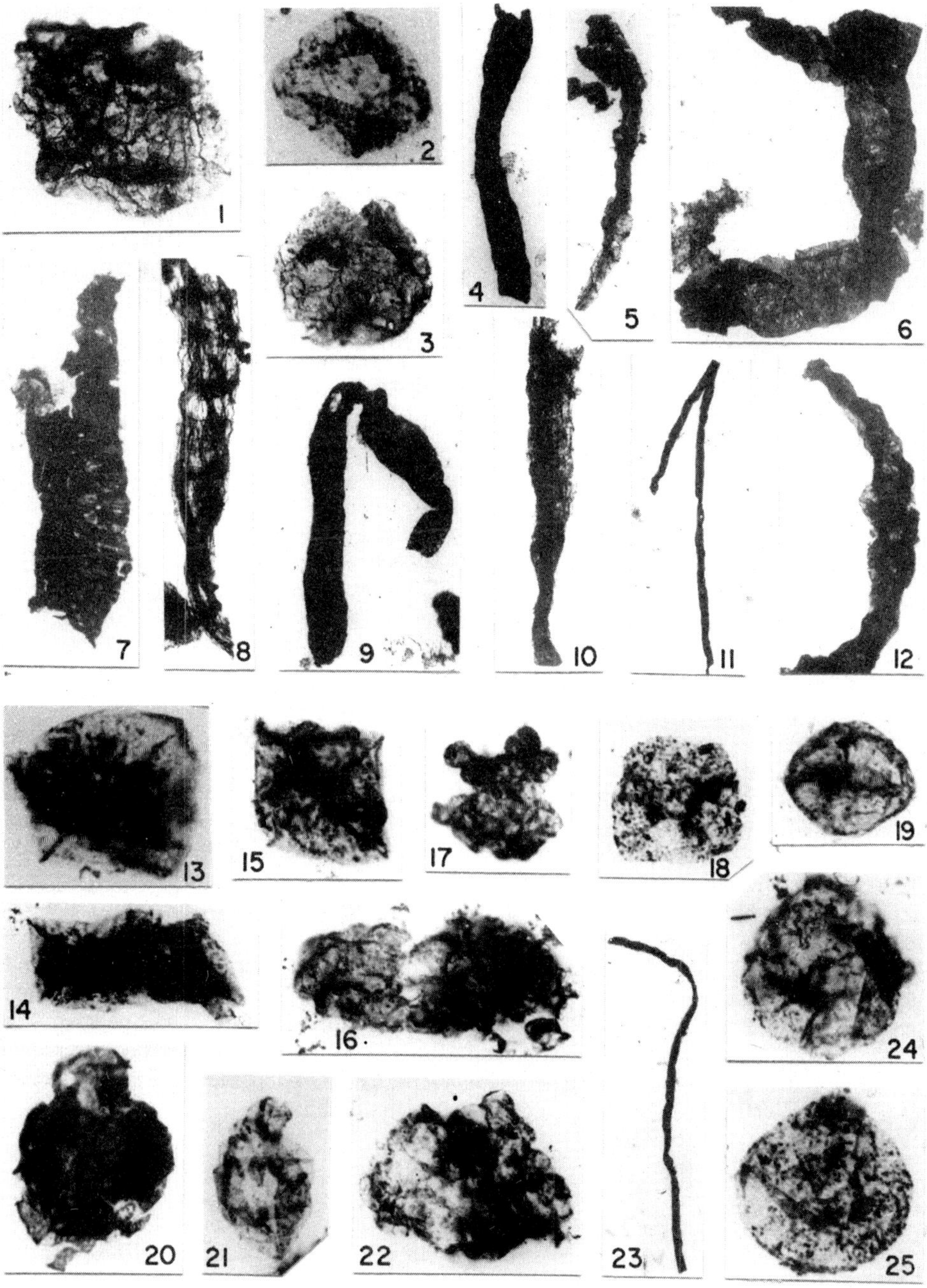

present in the Roper Group assemblage. It is worth mentioning that the succeeding Deonar Formation in Vindhyan Basin also recorded microfossil assemblage having the characteristic occurrence of *Tappania,* and associated with *Satka* and *Eomicrocystis* which is comparable with the microfossil assemblage of Roper Group that has been dated ca. 1450-1400 Ma.

It appears that the Kajrahat Limestone is little older than the Roper Group, and represents an Early Mesoproterozoic (Late Calymmian; ca.1500-1450 Ma) microfossil assemblage in the Vindhyan Basin.

Deonar Formation

Three samples from Jhal-Bihara section and one from Barhi-Badanpur-Maihar section (Fig. 2b, c), and porcellanite succession from DMH-A well (1765-3130m) yielded a moderate, but significant organic-walled microfossil assemblage.

Simple unornamented sphaeromorphic acritarchs are the main constituents of Deonar assemblage which include *Leiosphaeridia asperata, L. ternata, L. tenuissima, Kildinosphara chagrinata* and *K. granulata* (Pl. 1, Figs. 18-19, 24-25). Coccoid microfossils are represented by *Eomicrocystis* spp., *Synsphaeridium sorediforme* and *Satka squamifera,* and show their appearance within this formation. Filamentous forms, *viz. Siphonophycus* spp., *Polythrichoides lineatus* and *Karamia segmentata* that occur in the Kajrahat Limestone, also persist in this formation. However, the most distinctive feature of Deonar assemblage is the appearance of acanthomorphic acritarch *Tappania (T. plana, T. tubata)* in rarity (Pl. 1, Figs. 20-21) and several polygonomorphic acritarchs looking similar to the *Octoedryxium* Vidal (Pl. 1, Figs. 13-15) which are referred to as the *Octoedryxium* sp..

Deonar microfossil assemblage resembles the Early Mesoproterozoic assemblages, recorded from Jalboi, Corcoran and McMinn formations (Peat *et al.,* 1978; Javaux *et al.*, 2001) of the Roper Group (northern Australia), which have been dated ca.1450-1400Ma (Kralik, 1982; Jackson *et al.*, 1999). Species of *Tappania, Satka, Leiosphaeridia* and *Siphonophycus,* and coccoids (*Synsphaeridium* spp., *Eomicrocystis* spp.) are common in both the assemblages. The appearance of *Tappania* in Deonar Formation is very significant for dating of this unit. *Tappania* is an acanthomorphic acritarch with its oldest record from the Middle Mesoproterozoic Jalboi Formation (ca. 1450Ma) of Roper Group in northern Australia (Javaux *et al.*, 2001). The appearance of *Tappania, Satka* and *Eomicrocystis* in Deonar Formation, and presence of only the smooth sphaeromorphs, favours an age similar to the Roper Group (ca.1450-1400Ma) to it. This assemblage also compares with the Early to Middle Mesoproterozoic (ca. 1400Ma) Chamberlain Shale (Belt Supergroup) assemblage of Montana, U.S.A. (Horodyski, 1980) as abundant unornamented sphaeromorphs (*Leiosphaeridia* spp.), filamentous (*Siphonophycus* spp.) and coccoid (*Satka* spp.) forms are common in both the assemblages. However, the occurrence of polygonomorphic acritarchs (*Octoedryxium*) in Deonar Formation is perplexing. Its oldest record is from Late Neoproterozoic Upper Visingso Beds of Sweden (Vidal, 1976), and is very common in Terminal Proterozoic sediments (Vidal and Knoll, 1983). Its occurrence in this lithounit favours a little younger age to the Deonar assemblage than the Roper Group. But, in any case, this assemblage appears undoubtably older than the Middle to Late Mesoproterozoic *Tappania* dominated assemblages of Ruyang Group (ca. 1256 -1140 Ma), Eastern China (Yin, 1997) and Bahraich Group (ca. 1350 -1250 Ma) of Northern India (Prasad and Asher, 2001), as *Tappania* is rarely represented in this assemblage. It appears that the Deonar Formation represents a Middle Mesoproterozoic (ca.1450-1350 Ma) organic-walled microfossil assemblage in the Vindhyan Basin.

Koldaha Shale

This formation represents the lowermost lithounit of Kheinjua Subgroup and comprises thick succession olive green shale. Samples from outcrop as well as subsurface (DMH-A, 1760-940m) sections, representing Koldaha Shale succession (Fig. 2b, c & 4), recorded a very rich and diversified organic-walled microfossil assemblage.

The abundant occurrence of acanthomorphic acritarchs, *viz. Tappania plana* and *T. tubata* (Pl. 2, Figs. 3-4), netromorphic acritarchs, such as *Navifusa segmentatus* and *N. majensis* (Pl. 2, Figs. 11, 13-14) and filamentous forms, *viz. Arctacellularia tetragonala* and *A. ellipsoidea* (Pl. 2, Fig. 15) are the most distinctive features of the Koldaha Shale assemblage. The above species of *Navifusa* alongwith *Pterospermopsimorpha insolita, P. saccata* and *Simia annulare* (Pl. 2, Figs. 9, 16) make their first appearance in this formation; whereas, the species of *Tappania, Arctacellularia, Satka* and *Eomicrocystis* persist from the underlying Deonar Formation (Figs. 3, 4). Other coccoid forms, such as *Eosynechococcus moorei, Tetraphycus diminutivus* and *Eoentophysalis* sp. (Pl. 2, Figs. 1, 2, 5) also appear in Koldaha Shale. Various species of *Leiosphaeridia, Kildinosphaera* and *Siphonophycus* are also very common, and persist from the underlying Deonar Formation.

The Koldaha Shale microfossil assemblage resembles Middle to Late Mesoproterozoic assemblages of Beidajian Formation (ca. 1256-1140Ma) of Ruyang Group, Eastern China (Yin, 1997; Yin and Yuan, 2002) and Bahraich Group, northern India (Prasad and Asher, 2001), as the various species of *Tappania* occur abundantly in the above three assemblages. Above assemblages are also associated with the common occurrence of *Navifusa* spp. and *Arctacellularia* spp. along with a large number of sphaeromorphic acritarchs, filamentous and coccoid microfossils. The Ruyang microfossil assemblage was assigned a Late Mesoproterozoic age (Yin, 1997; Yin and Yuan, 2002), as the K-Ar dating of Beidajian Formation indicated an age of 1256-1140 Ma (Ma *et al.*, 1980). Prasad and Asher (2001) also recorded a rich acritarch assemblage, mainly dominated by *T. plana, T. tubata* and *N. segmentatus* from the subsurface metasedimentary succession (Bahraich Group) of Ganga Basin in northern India. Prasad and Asher (2001) assigned a Middle to Late Mesoproterozoic (Late Ectasian-Early Stenian; ca.1350-1150Ma) age to the *Tappania* dominated Bahraich assemblage, as the absolute age of Bahraich Group was estimated ca. 1320 ± 30Ma (Chug *et al.*, 1989), and its assemblage was equated with the Ruyang Group assemblage which was dated ca. 1256-1140 Ma.

Thus, the closed comparable relationship of Koldaha Shale assemblage with the known Middle to Late Mesoproterozoic organic-walled microfossil assemblages suggests a Late Ectasian - Early Stenian (ca. 1350 -1150 Ma) age of Middle to Late Mesoproterozoic to the Koldaha Shale.

Salkhan Limestone

One sample from Jhal-Bihara section and depth intervals from 940-780m in DMH-A well, representing Salkhan Limestone, recorded moderate occurrence of organic-walled microfossils. Sphaeromorphic acritarchs are the main constituents of the assemblage and represented by the various species of *Kildinosphaera* and *Leiosphaeridia*. Important forms of underlying Koldaha Shale, *viz. Tappania plana, T. tubata, Navifusa majensis, N. segmentatus, Pterospermopsimorpha insolita, P. saccata, Simia annulare, Satka squamifera, Eomicrocystis* spp. and *Tetraphycus diminituves* persist here. *Siphonophycus* spp. and *Oscillatoriopsis psilata* represent the filamentous forms.

The palynofossil assemblage of Salkhan Limestone is very much similar to that from the underlying Koldaha Shale, except that individual taxa are represented quantitatively less in number. It is also represented by the age marker taxa, *viz. Tappania plana, T.tubata* and *Navifusa segmentatus,* which are suggestive of Middle to Late Mesoproterozoic age, similar to the Koldaha Shale. However, the appearance of *Leiosphaeridia kulgunica* (Pl. 2, Fig. 17) in Salkhan Limestone, which, till now, are recorded from Late Mesoproterozoic and younger sediments, suggest a Little younger age of Late Mesoproterozoic (Late Stenian; ca. 1150 -1050 Ma) to this assemblage.

Rampur Formation

This formation includes glauconitic sandstone succession with intercalated siltstone. Siltstone samples from Jhal-Bihara section and well DMH-A (780-620m), representing this unit (Figs. 3, 4), recorded a diversified organic-walled microfossil assemblage (Pl. 2, Figs. 18-30).

The microfossil assemblage of Rampur Formation includes almost all the taxa recorded from the Koldaha Shale and Salkhan Limestone. However, the species of *Satka, Simia, Navifusa* and *Pterospermopsimorpha* are in quite abundance. The age potential taxa, such as *Tappania, T. tubata* (Pl. 2, Figs. 29-30), *Navifusa segmentatus* (Pl. 2, Fig. 25) and *Arctacellularia tetragonala,* which were abundant in Koldaha Shale and Salkhan Limestone, become rare in this assemblage, and ultimately disappear in the upper part of this formation (Figs. 3, 4).

The most conspicuous feature of the Rampur assemblage is the appearance of other acanthomorphic acritarchs, *viz. Vandalosphaeridium reticulatum* and *Trachyhystrichosphaera truncata* (Pl. 2, Figs. 18, 21), coccoid form *Bavlinella faveolata* and ornamented sphaeromorphic acritarchs, *viz. Kildinosphaera verrucata* and *Trachysphaeridium laminaritum* (Pl. 2, Figs. 24, 28). In addition, *a* large number of leiosphaerids with binary fission referable to budding leiosphaerids (Pl. 2, Fig. 19) and vase-shaped microfossils, *Melanocyrillium* (Pl. 2, Fig. 27), also make their appearance in this formation.

Plate 2.

Microfossils from Kheinjua Subgroup (Koldaha Shale, Salkhan Limestone & Rampur Formation).

Koldaha Shale and Salkhan Limestone (magnifications approx. 750x; otherwise mentioned)

1 *Eosynechococcus moorei* Hofmann. S-8/C/1; EFR D49/2. **2** *Tetraphycus diminutivus* Oehler. S-8/C/1; EFR V34/2. **3** *Tappania plana* Yin. DMH-A, 1400-05m/2; EFR F35/4. **4** *Tappania tubata* Yin. DMH-A, 1390-95m/2; EFR G72. **5** *Eoentophysalis* sp. S-9/2; EFR Q44/3. **6** *Satka squamifera* Pyatiletov. S-9/2; EFR P63/1. **7** *Leiosphaeridia tenuissima* Eisenack. S-9/2; EFR H65/3. **8** *Synsphaeridium sorediforme* (Timofeev) Eisenack. S-9/2; EFR O43/2. **9** *Simia annulare* (Timofeev) Mikhaylova. S-9/2; EFR H58/3. **10** *Satka colonialica* Jankauskas. S-9/2; EFR M74/3. **11,14** *Navifusa segmentatus* Prasad and Ashaer. Fig. 11, 930-35, EFR D34/1. Fig. 14, DMH-A, 1720-25m/1; EFR P28. **12** *Siphonophycus robustum* Knoll, Swett and Mark. DMH-A, 930-35m/1; EFR R32/3. **13** *Navifusa majensis* Pyatiletov. DMH-A, 1205-10m/1; EFR N44/1. **15** *Arctacellularia tetragonala* Hofmann and Jackson. DMH-A, 1505-10m/2; EFR N54. **16** *Pterospermopsimorpha insolita* Mikhaylova. DMH-A, 1675-80m/1; EFR F63/1. **17** *Leiosphaeridia kulgunica* Jankauskas. II/55-G10; EFR R32/3.

Rampur Formation (magnifications approx. 750x; otherwise mentioned)

18 *Vandalosphaeridium reticulatum* (Vidal) Vidal. II/168/1; EFR D51. **19** Budding leiosphaerids. II/168/2; EFR L31. **20** *Satka squamifera* Pyatiletov. II/168/2; EFR P70/4. **21** *Trachyhystrichosphaera truncata* German. II/168/2; EFR W57/1; 1000x. **22** *Chlorogloeaopsis contexta* Hofmann and Jackson. DMH-A, 690-95m/1; EFR V54/3; 550x. **23** *Stictosphaeridium implexum* Timofeev. II/168/1; EFR H68/1. **24** *Trachysphaeridium* sp. cf. *T. laufeldi* Vidal. II/168/2; EFR V50; 850x. **25** *Navifusa segmentatus* Prasad and Ashaer. DMH-A, 640-45m/2; EFR U33/3. **26** *Satka colonialica* Jankauskas. II/168/2; EFR B60/2. **27** *Melanocyrillium fimbriatum* Bloeser. II/168/2; EFR K71/4. **28** *Trachysphaeridium laminaritum* (Timofeev) Vidal. II/168/1; EFR G66/3. **29** *Tappania* sp. cf. *T. plana* Yin. II/168/2; EFR J66/4. **30** *Tappania tubata* Yin. II/168/1; EFR J34; 600x.

Plate 2.

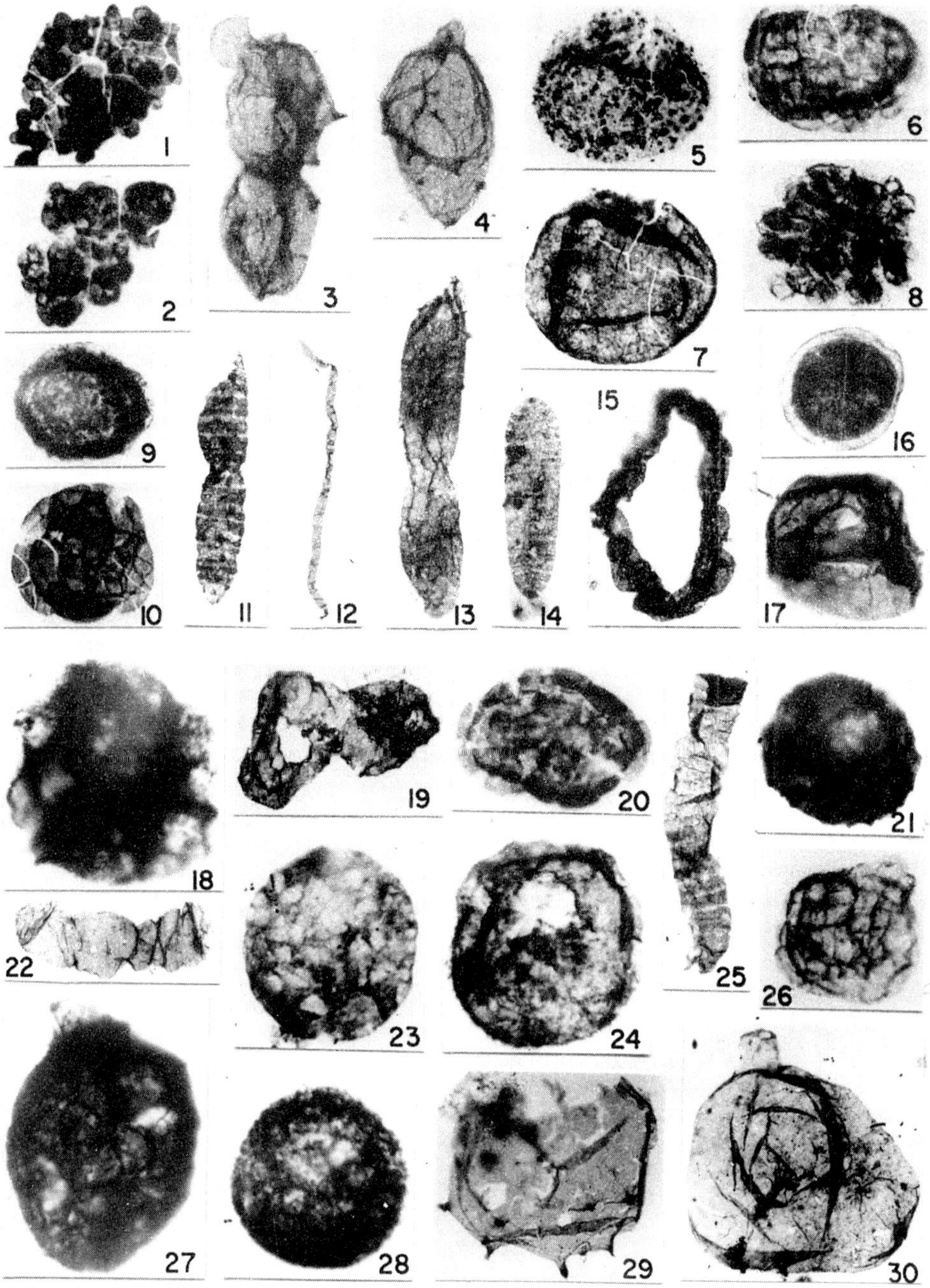
1
2
3
4
5
6
7
8
9
10
11
12
13
14
15
16
17
18
19
20
21
22
23
24
25
26
27
28
29
30

The presence of *Tappania* spp. and *Navifusa segmentatus* largely allow the comparison of Rampur assemblage with Late Mesoproterozoic microfossil assemblages. Nevertheless, this assemblage also includes several taxa, such as *V. reticulatum, T. truncata, B. faveolata,* budding leiosphaerids and vase-shaped microfossils that generally appear in Early Neoproterozoic or near the Meso-Neoproterozoic boundary (Knoll, 1996). In deed, the microfossil assemblage of Rampur Formation shows more akinness to the Early Neoproterozoic assemblages and resembles the Lakhanda (ca. 1000-900Ma) assemblage of Siberia (Jankauskas,1989; German,1990). Forms, such as, *N. majensis, P. insolita, P. saccata, Leiosphaeridia* spp., *A. tetragonala, Satka* spp., *Eomicrocystis* spp. and budding leiosphaerids are common in both the assemblages. The present assemblage also correlates with the Veteranen Group (Spitsbergen) assemblage (Knoll and Swett, 1985) of Early Neoproterozoic (ca. 900-800 Ma) as majority of the forms are common in both the assemblages.

Thus, the appearance of vase-shaped microfossils, budding leiosphaerids and *V. reticulatum*, that generally appear in Early Neoproterozoic, stongly favours for a Tonian (ca.1000-850 Ma) age to this formation. The associated occurrence of Late Mesoproterozoic marker forms suggests that Rampur Formation ranges in age from latest Mesoproterozoic to Early Neoproterozoic (ca. 1050-850 Ma).

Plate 3.

Organic-walled microfossils from Rohtas Subgroup, and Kaimur & Rewa groups.

Rohtas Subgroup (all magnifications approx. 850x; otherwise mentioned)

1, 5 *Pterospermopsimorpha saccata* Yin. II/388/1; Fig. 1, EFR M32/1; Fig. 5, EFR X55/4. **2** *Bavlinella faveolata* (Schepeleva) Vidal. II/388/1; EFR F45. **3** *Kildinosphaera verrucata* Vidal. II/388/2; EFR G48/2. **4** *Symplassosphaeridium tumidulum* Timofeev. II/388/2; EFR L43/4. **6** *Simia annulare* Mikhalova. II/388/2; EFR P70. **7** *Satka squamifera* Pyatiletov. DMH-A, 565-570m/1; EFR P67/3. **8** *Trachysphaeridium laminaritum* (Timofeev) Vidal. II/388/1; EFR S32/1. **9** *Leiosphaeridia asperata* (Naumova) Lindgren. II/388/2; EFR N56. **10** *Karamia segmentata* Jankauskas, Mikhaylova & German. DMH-A, 650-55/2m; EFR W46; 400x. **11** *Kildinosphaera granulata* Vidal. II/388/2; EFR U60/1. **12** *Stictosphaeridium implexum* Timofeev. II/388/1; EFR G49.

Kaimur Group (all magnifications approx. 550x; otherwise mentioned)

13 *Kildinosphaera granulata* Vidal. DMH-A, 500-505m; EFR P44. **14, 15** *Bavlinella faveolata* Vidal. DMH-A, 570-75m/2. Fig.14, EFR M57. Fig. 15., EFR U37. **16** *Symplassosphaeridium tumidulum* Timofeev. DMH-A, 520-25m/2; EFR T63/3. **17, 21** *Leiosphaeridia minutissima* Jank. DMH-A, 520-25m/2. Fig. 17, EFR J26. Fig.21, EFR V44/3. **18, 25** *Synsphaeridium sorediforme* (Timofeev) Eisenack. Fig. 18, DMH-A, 515-20m/2; EFR E58/3. Fig. 25, DMH-A, 520- 25m/1; EFR P52/2. **19** *Leiosphaeridia crassa* (Naumova) Jankauskas. DMH-A, 520-25m/1; EFR F27/2. **20** *Pterospermopsimorpha saccata* Yin. DMH-A, 520-25m/1; EFR G35. **22** *Leiosphaeridia tenuissima* Eisenack. DMH-A, 535-40m; EFR K48. **23** *Trachysphaeridium laufeldi* Vidal. DMH-A, 520-25m/1; EFR P70/4. **24** *Leiosphaeridiia asperata* (Naumova) Lindgren. DMH-A, 530-35m/1; EFR K39.

Rewa group (all magnifications approx. 850x; otherwise mentioned)

26 *Octoedryxium truncatum* Rudavaskaia. RP-1/1; EFR W40. **27** *Bavlinella faveolata* (Schepeleva) Vidal. RP-1/2; EFR X59/2. **28** *Leiosphaeridia jacutica* (Timofeev) Mikhaylova and Jankauskas. RP-1/1; EFR U29. **29** *Synsphaeridium sorediforme* (Timofeev) Eisenack. RP-1/1; EFR S68/1. **30** *Vandalosphaeridium reticulatum* (Vidal) Vidal. RP-1/2; EFR K64/1. **31** *Leiosphaeridia asperata* (Naumova) Lindgren. RP-1/2; EFR Q40. **32** *Symplassosphaeridium tumidulum* Timofeev. RP-1/1; EFR N33. **33** *Pterospermopsimorpha saccata* Yin. DMH-A, 325-330m. EFR O55/2.

Plate 3.

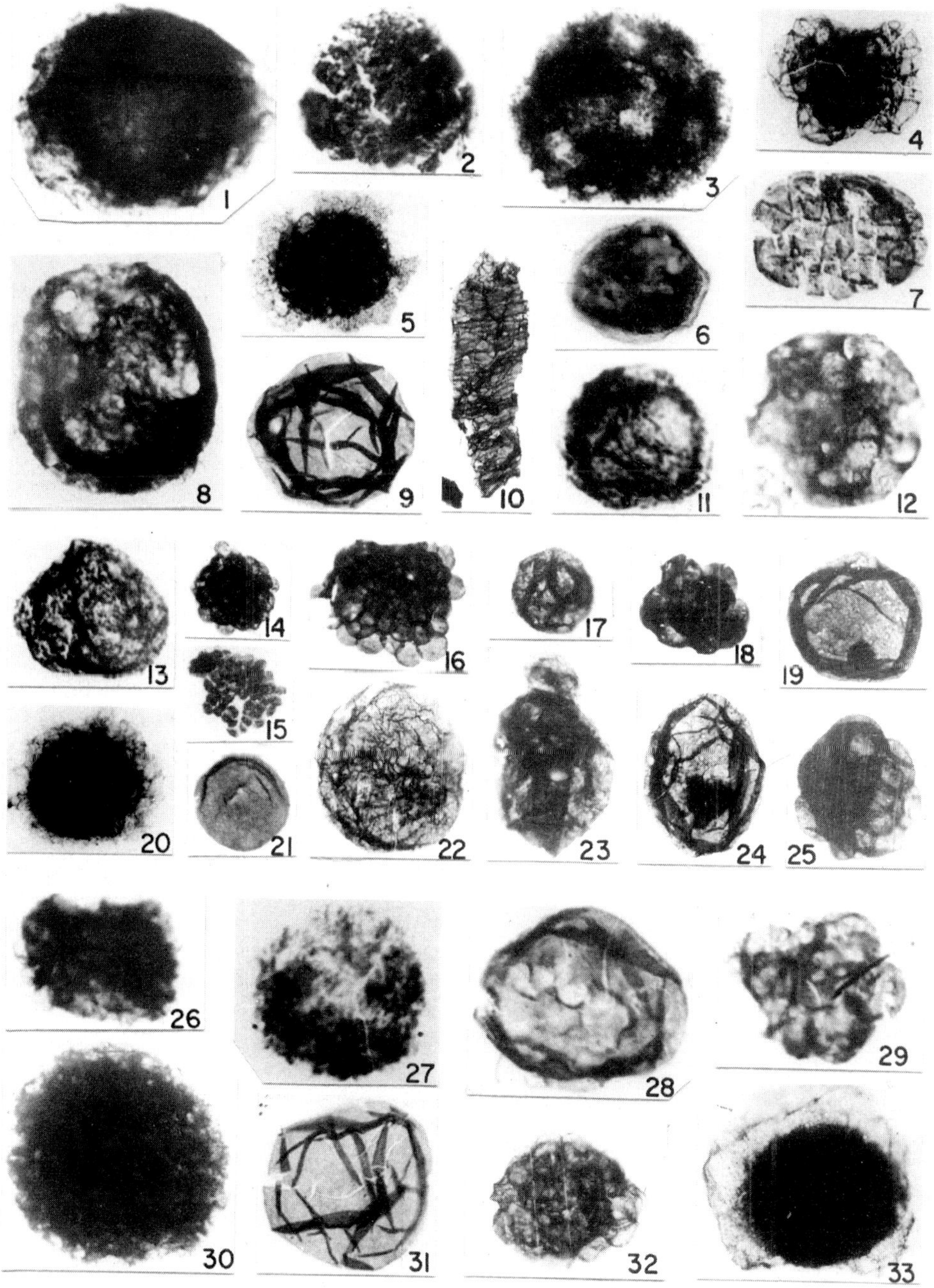
1
2
3
4
5
6
7
8
9
10
11
12
13
14
15
16
17
18
19
20
21
22
23
24
25
26
27
28
29
30
31
32
33

Rohtas Subgroup

Two samples from outcrops (Fig. 2c) and undifferentiated Rohtas sediments in DMH-A well (620-565m) yielded the moderate occurrence of organic-walled microfossils (Pl. 3, Figs. 1-12).

The Rohtas assemblage is mainly represented by abundant sphaeromorphic acritarchs that include *Stictosphaeridium* spp., *Leiosphaeridia* spp., *Kildinopshaera* spp. and *Trachysphaeridium laminaritum.* Important taxa of the underlying Rampur Formation, *viz. Pterospermopsimorpha saccata, P. insolita, Vandalosphaeridium* spp., *Satka* spp., *Simia annulare, Bavlinella faveolata* and budding leiosphaerids persist here. Among these, *Satka* spp., *Eomicrocystis* spp., *Simia annulare, Pterospermopsimorpha insolita* and budding leiosphaerids disappear at different stratigraphic levels within this lithounit. Species of *Tappnia, Navifusa* and *Arctacellularia*, which were the important constituents in the underlying Kheinjua Subgroup, are completely absent in Rohtas Subgroup (Figs. 3 & 4).

The Rohtas assemblage resembles the Middle Neoproterozoic Miroyedikha acritarch assemblage (ca. 850 Ma) of Siberia (Jankauskas,1989; German,1990) since smooth as well as ornamented sphaeromorphic acritarchs are abundant in both the assemblages, and occur in association with coccoid (*Satka* spp., *Synsphaeridium* spp.) and filamentous microfossils. The characteristic budding leiosphaerids are also present in both the assemblages. The Rohtas assemblage also compares with the Middle Neoproterozoic assemblage of Husar-Kanpa formations, recovered from Husar-1 well in central Australia (Hill *et al.*, 2000), as above forms are common in both the assemblages. The species of *Satka, Eomicrocystis, Simia* and *Pterospermopsimorpha insolita* show their disappearance in both the assemblages. It appears that the Rohtas assemblage is little younger in age than the Miroyedikha (ca. 850Ma) assemblage, and resembles Middle Neoproterozoic (ca. 812-777Ma) Husar-Kanpa assemblage of central Australia, and suggests an Early Cryogenian (ca. 850 -750 Ma) age.

Kaimur Group

Though, the outcrop samples are rarely fossiliferous, the undifferentiated Kaimur Group in DMH-A well (565-480m depth-intervals) recorded abundant organic-walled microfossils (Pl. 3, Figs. 13-25).

Leiosphaerids are the dominant constituents of the Kaimur assemblage which are represented by the various species of *Leiosphaeridia,* and followed by the coccoids forms, *viz. Symplassosphaeridium tumidulum, Synsphaeridium sorediforme* and *Bavlinella faveolata* (Pl. 3, Figs. 14-15). Sphaeromorphic acritarchs include *Kildinopshaera* spp., *Pterospermopsimorpha saccata, Trachysphaeridium laminaritum, T. laufeldi, Vandalosphaeridium reticulatum* and *Stictosphaeridium sinapticulum.*

The comparison of Kaimur assemblage is rather difficult, as its is less diversified. However, the assemblage broadly compares with the Mid. to Late Neoproterozoic (ca. 800-700 Ma) Middle Visingso Beds assemblage of Sweden (Vidal, 1976). The age potential taxa, such as *B. faveolata, T. laufeldi, V. reticulatum,* and other forms, *viz. Stictosphaeridium* spp., *T. laminaritum* and *P. saccata* are common in both the assemblages. The occurrence of above taxa in Kaimur Group broadly suggests Middle to Late Neoproterozoic age to it that corresponds to the Middle Cryogenian (ca. 750-700 Ma).

Rewa Group

Three samples from Panna and Jhiri shales in outcrops (Fig. 2b) and undifferentiated Rewa Group from DMH-A well (480-320m), recorded a moderate occurrence of organic-walled microfossils. Microfossil assemblage of Rewa Group is marked by the abundant occurrence of coccoid forms which include *Synsphaeridium* spp., *Symplassosphaeridium* spp. and *Bavlinella faveolata* (Pl. 3, Figs. 27, 29, 32). Different species of *Leiospharidia, Kildinosphaera* and *Trachysphaeridium laminaritum, T.laufeldi* and *Vandalosphaeridium reticulatum* (Pl. 3, Fig. 30) are also common in the assemblage. In addition, polygonomorphic acritarch *Octoedryxium truncatum* (Pl. 3, Fig. 26) appears in this formation.

The microfossil assemblage from Rewa Group is also very poor and less diversified. However, the assemblage is broadly comparable with the Late Neoproterozoic (ca. 700-650 Ma) microfossil assemblage of Upper Visingso Beds from Sweden (Vidal, 1976). Marker taxa, such as *Octoedryxium truncatum, V. reticulatum, T. laufeldi* and *Bavlinella faveolata* are common in both the assemblages. Vidal (1976) and Vidal and Knoll (1983) recorded above taxa in abundance from the Upper Visingso Beds in Sweden, which is dated as ca. 700 - 610 Ma (Vidal, 1976). Thus, a Late Cryogenian (ca.700-650Ma) age is inferred to the Rewa Group. This age inference is corroborated with the facts that the youngest unit of Rewa Group, *viz.* Asan Sandstone, was dated ca. 700 ± 100 Ma (Srivastava and Rajagopalan, 1988), and the Rewa assemblage is succeeded by the *Obruchevella* dominated latest Neoproterozoic-Terminal Proterozoic (Vendian) assemblages of Bhander Group.

Bhander Group

Bhander Group includes four lithounits, *viz.* Ganurgarh Shale, Nagod Limestone, Sirbu Shale and Maihar Sandstone. Samples from Ganurgarh Shale, Nagod Limestone and Sirbu Shale from outcrops (Fig. 2b) and DMH-A well (320-00m) recorded fairly rich and diversified microfossil assemblages with some marker taxa (Pl. 4, Figs. 1-28) that helped in the age inference of Bhander Group.

Ganurgarh Shale

The palynofossil assemblage from Ganurgarh Shale is mainly includes the occurrence of coccoid microfossils, *viz. Synsphaeridium* spp., *Symplassosphaeridium* spp. and *Bavlinella faveolata.* Various species of *Leiosphaeridia, Kildinosphaera, Trachysphaeridium, Chlogloeaopsis* and *Siphonpphycus* are the subdominant ones. However, the most important aspect of Ganurgarh Shale assemblage is the appearance of helically coiled Terminal Proterozoic (Vendian) marker microfossils, referable to *Obruchevella,* in its upper part. It is represented by *O. parva, O. delicata* and *O. valdaica* (Pl. 4, Figs. 4, 5, 8) and rarely occur in the assemblage. Above species of *Obruchevella* occur in associatiation with other Late Neoproterozoic-Terminal Proterozoic marker taxa, *viz. Vandalosphaeridium reticulatum, Trachysphaeridium laufeldi* and *Bavlinella faveolata* which persisted from underlying Rewa Group.

The appearence of above enlisted species of *Obruchevella* in Ganurgarh Shale, though in rarity, is very significant. Particularly, the appearance of *O. parva* and *O. delicata* in this unit is very important as these two species, till now, are recorded from the Terminal Proterozoic (Vendian) and Lower to Middle Cambrian sequences, with their earliest records from the Vendian sediments (Reitlinger, 1948, 1959; Shenfil, 1983; Yakshin and Luchinina, 1981; Kolosov, 1982; Golovenok and Belova, 1983; Cloud *et al.*, 1979; Wang *et al.*, 1983; Song, 1984; Knoll and Ohta, 1988; Knoll, 1992; Mankiewicz, 1992). The oldest record of *Obruchevella valdaica* is known from the Late Mesoproterozoic - Early Neoproterozoic sediments of Greenland (Hofmann and Jackson, 1994) and India (Prasad and Asher, 2001). However, its abundance in Terminal Proterozoic (Vendian) is well established (Knoll, 1996) and observed to disappear in the Late Vendian (Mankiewicz, 1992).

Thus, the appearance of *O. parva* and *O. delicata* in the upper part of this unit suggests an Early Vendian age to the upper part of Ganurgarh Shale. However, the lower part of this unit indicates a latest Cryogenian age of Late Neoproterozoic, as it is characterised by the occurrence of *V. reticulatum, T. laufeldi* and *B. faveolata,* and species of *Obruchevella* are conspicuously absent.

Nagod Limestone

The microfossil assemblage of this formation is quite similar to those from Ganurgarh Shale, but relatively rich. Various species of *Trachysphaeridium, Vandalosphaeridium, Favososphaeridium Leiosphaeridia, Kildinosphaera, Stictosphaeridium* and *Germinosphaera* represent the sphaeromorphic and acanthomorphic acritarchs. Species of *Symplassosphaeridium, Synsphaeridium* and *Bavlinella* include the coccoid forms; whereas, filamentous forms are represented the various species of *Obruchevella*, *Chlorogloeaopsis*, *Oscillatoriopsis* and *Siphonophycus* (Pl. 4, Figs. 9-16).

The most distinguishing aspect of Nagod Limestone assemblage is the abundant occurrence of marker species of Vendian like *Obruchevella, viz. O delicata* and *O. parva* (Pl. 4, Figs. 13, 14). In addition, *Germinosphaera unispinosa*, *Lophosphaeridium* sp. cf. *L. rarum* (Pl. 4, Fig. 12) and *Cristallinium* sp. (Pl. 4, Figs. 11, 15) appear in this formation (Figs. 3, 4).

The oldest abundant occurrence of *O. parva* and *O. delicata* are from the Vendian sediments and range upto Middle Cambrian (Mankiewicz,1992). The presence of *O. parva, O. delicata* and O. *valdaica* in abundance, with associated occurrence of *B. faveolata, T. laufeldi, V. reticulatum* and *F. favosum,* suggests Early Vendian (ca. 600-570 Ma) age to Nagod Limestone. Yet, the occurrence of Late Vendian-Early Cambrian marker taxa, such as *Lophosphaeridium* sp. cf. *L. rarum* and *Cristallinium* sp., extend up to the Late Vendian also. Moreover, the absence of morphologically complex and large acanthomorphic acritarchs of Late Vendian (570-544Ma), that were rcorded from North China (Zhang *et al.*, 1998), Siberian platform (Moczydlowska *et al.*, 1993) and central Australia (Zang and Walter, 1992), suggests that the Nagod Limestone assemblage is little older than the acritarch assemblages of large and complex acanthomorphic acritarchs, and belongs to the Early Vendian (ca. 590-570 Ma). But, it may also extend in the Late Vendian (ca. 560 Ma), as *Lophosphaeridium* and *Cristallinium* appear in this lithounit.

Plate 4.

Organic-walled microfossils from Bhander Group

Ganurgarh Shale (all magnifications approx. 550x; otherwise mentioned)

1 *Oscillatoriopsis psilata* Maithy and Shukla. DMH-A, 215-20m/2; EFR G61/1; 350x. **2** *Chlorogloeaopsis contexta* Hofmann and Jackson. DMH-A, 100-105m/1; EFR W49; 375x. **3** *Trachysphaeridium laufeldi* Vidal. DMH-A, 160-65m/2; EFR S56/4. **4** *Obruchevella delicata* Reitlinger. DMH-A, 165-70m/1; EFR K27. **5** *Obruchevella parva* Song. DMH-A, 150-55m/1; EFR T34/3. **6** *Leiosphaeridia asperata* (Naumova) Lindgren. DMH-A, 215-20m/1; EFR J43. **7** *Octoedryxium truncatum* Vidal. S28/B-1; EFR Q31/3. **8** *Obruchevella valdaica* Jankauskas, Mikhaylova & German. DMH-A, 165-70m/2; EFR O37/3.

Nagod Limestone (all magnifications approx. 750x; otherwise mentioned)

9 *Chlorogloeaopsis kanshiensis* Maithy. DMH-A, 150-55m/2; EFR T33/3; 400x. **10** *Obruchevella valdaica* Jankauskas, Mikhaylova and German.DMH-A, 150-55m/2; EFR T40. **11** *Cristallinium* sp. cf. *C. cambriense* (Slavikova) Vanguestaine. IV/ 613/1; EFR T50/4. **12** *Lophosphaeridium* sp. cf. *L. rarum* Timofeev. DMH-A, 150 -55m/1; EFR R38. **13** *Obruchevella parva* Song. DMH-A, 150-55m/1; EFR K42; 650x. **14** *Obruchevella delicata* Reitlinger. DMH-A, 150-55m/2; EFR M69. **15** *Cristallinium* sp. DMH-A, 150-55m/2; EFR K51 (76μm); 550x. **16** *Favososphaeridium favosum* Timofeev. DMH-A, 115-20m/1; EFR O64; 400x.

Sirbu Shale (all magnifications approx. 700x; otherwise mentioned)

17 *Cristallinium* sp. cf. *C. cambriense* (Slavikova) Vanguestaine. V/4; EFR Q62/2. **18-19, 21-22** *Dictyotidium* sp. V-4/3. Fig.18, EFR V64/3. Fig.19, EFR D34/3. Fig. 21, EFR N72/3. Fig.22, EFR O36/3. **20** *Trachysphaeridium laufeldi* Vidal. DMH-A, 75-80m/1; EFR P70/4. **23** *Cymatiosphaera* sp. DMH-A, 75-80m/1; EFR K31/3 (57μm); 550x. **24** *Vandalosphaeridium reticulatum* (Vidal) Vidal. DMH-A, 15-20m/1; EFR O60; 850x. **25** *Germinosphaera bispinosa* Jankauskas. DMH-A, 15-20m/1; EFR R52/2. **26** *Germinosphaera unispinosa* Jankauskas. DMH-A, 75-80m/1; EFR Q37/2. **27** *Obruchevella parvissima* Song. DMH-A, 15-20m/1; EFR R51/2. **28** *Obruchevella delicata* Reitlinger. DMH-A, 75-80m/2; EFR Q65; 650x.

Plate 4.

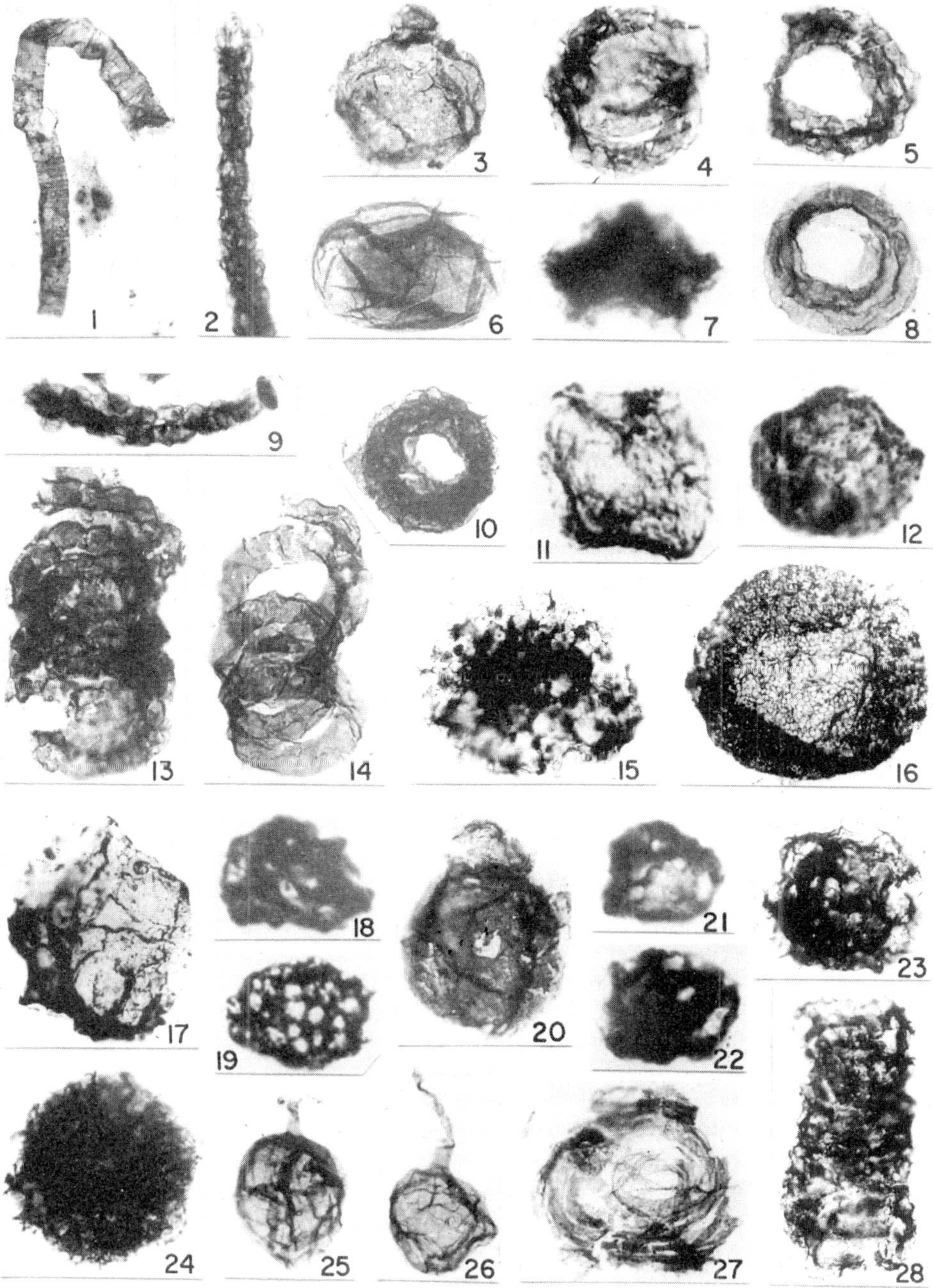

Sirbu Shale

Helically coiled microfossils (*Obruchevella* spp.) are the most common constituents of the Sirbu Shale assemblage, and are represented by *O. parva, O. delicata* and *O. parvissima* (Pl. 4, Figs. 27, 28). Yet, the most distinctive feature of this assemblage is the occurrence of acanthomorphic acritarch *Germinosphaera,* which is represented by *G. unispinosa* and *G. bispinosa* (Pl. 4, Figs. 25, 26). Other age potential taxa, such as *V. reticulatum, B. faveolata* and *Lophosphaeridium* sp. cf. *L. rarum, Cristallinium* sp. and *T. laufeldi* persist here from the Nagod Limestone. In addition, forms referable to *Dictyotidium* sp. (Pl. 4, Figs. 18,19,21,22), *Cymatiosphaera* sp. (Pl. 4, Fig. 23), *Obruchevella parvissima* (Pl. 4, Fig. 27) and *Germinosphaera bispinosa* (Pl. 4, Fig. 25) appear in this formation (Figs. 3, 4).

The Sirbu Shale assemblage resembles the Late Vendian (570-544Ma) microfossil assemblage of Scotia Group, Svalbard (Knoll, 1992), as the species of *Obruchevella, Siphonophycus, Polythricoides* and *Leiosphaeridia* are common to both the assemblages. However, the characteristic large acanthomorphic acritarchs, *viz. Papillomembrana, Ericiasphaera* and *Echinosphaeridium* of Scotia Group are absent here. The occurrence of Vendian marker taxa, *viz. Obruchuvella parva, O. delicata, V. reticulatum, B. faveolata, T.laufeldi, Lophosphaeridium* sp. cf. *rarum* and *Cristallinium* spp. suggests Early to Late Vendian age for the Sirbu Shale assemblage. However, the presence of *Cymatiosphaera* sp., *Dictytiddium* sp. and *Obruchevella parvissima,* which appear in the basal Cambrian, favours an Early Cambrian age to it also. But, the absence of distinctive basal Cambrian marker acritarchs, *viz. Asteridium tornatum, Skiagia ciliosa* and *Comasphaeridium velvetum* in this assemblage, restricts the age of Sirbu Shale to Late Vendian. The absence of marker Lower Cambrian forms of *Obruchevella, viz. O. meishucunensis,* and other species of *Obruchevella* having large filament diameters (> 20 µm) in the assemblage, further suggest that Sirbu Shale assemblage belongs to the Late Vendian (ca. 570 - 544 Ma), and does not encompass into the Early Cambrian.

Maiher Sandstone

Maihar Sandstone represents the youngest lithounit of the Bhander Group and devoid of organic-walled microfossils. It seems that this lithounit was also deposited during Terminal Proterozoic (Late Vendian). Rai (1999) suggested Late Vendian (Ediacaran) age to Maihar Sandstone, based on the record of algal microbial mats.

Discussions

The organic-walled microfossils from Vindhyan sediments of Son valley and DMH-A well from Madhya Pradesh include acritarchs, coccoid and filamentous forms that generally characterise Meso-Neoproterozoic sediments worldwide. Among these, some age marker taxa are also present that helped dating the different lithounits of Vindhyan Supergroup and ascertaining its total age range.

The Basal two lithounits of Semri Group, *viz.* Deoland and Arangi formations (Mirzapur Subgroup) are devoid of organic-walled microfossils. However, the occurrence of abundant filamentous microfossils (*Polythrichoides* and *Karamia*) and unornamented sphaeromorphs (*Leiosphaeridia* spp.) in Kajrahat Limestone, which represents the youngest unit in Mirzapur Subgroup, suggests an Early Mesoproterozoic (Late Calymmian; ca. 1500-1450 Ma) age to it. Earlier record of stromatolites (Misra *et al.*, 1977) from this unit also suggested an Early to Middle Riphean age which are very close to the present suggested age. The older two units, *viz.* Deoland and Arangi formations seem to represent the Early Mesoproterozoic (ca. 1550-1500 Ma) sediments in the Vindhyan Basin, as they immediately underly the Kajrahat Limestone, and the Arangi Formation is characterised by the Early Riphean stromatolites (Kumar, 1976). It appears that the basal Vindhyan sediments are not older than the ca. 1550 Ma, and the sedimention in Vindhyan Basin commenced during Early Mesoproterozoic.

Appearance of *Tappania* (acanthomorphic acritarch), *Satka* and *Eomicrocystis* in Deonar Formation observed to be the most significant biostratigraphic event in Vindhyan stratigraphy. The oldest record of *Tappania* is from the Early Mesoproterozoic (ca.1450Ma) sediments (Jalboi Formation, Roper Group) in northern Australia (Javaux *et al.*, 2001). It was recorded in abundance from Middle to Late Mesoproterozoic sediments of eastern China (Yin,1997) and India (Prasad and Asher, 2001). Its appearance in Deonar Formation indicated an age similar to the Roper Group. However, the record of *Octoedryxium* favours a little younger age of Middle Mesoproterozoic (ca.1450-1350Ma) to the Deonar assemblage than the Roper Group assemblage. The record of *Octoedryxium* in Deonar Formation also suggests that polygonomorphic acritarchs were flourishing even during Middle Mesoproterozoic time. The presence of distinctive Middle Mesoproterozoic microfossil assemblage in Deonar Formation does not support its absolute age of 1630 ± 5 Ma (Ray *et al.,* 2001).

The dominance of acanthomorphic acritarchs (*Tappania* spp.) in Koldaha Shale is also a significant feature. As stated earlier, *Tappania* was recorded in abundance from Middle to Late Mesoproterozoic sediments of China (Ruyang Group, ca. 1200-1100Ma) [Yin, 1997] and India (Bahraich Group, ca.1350-1150Ma) [Prasad and Asher, 2001]. The abundance of *Tappania* in Koldaha Shale and Salkhan Limestone assemblages make their resemblance with Mid. to Late Mesoproterozoic Ruyang and Bahraich microfossil assemblages, suggesting a Late Ectasian-Early Stenian (ca. 1350-1050 Ma) age to the above two units of Kheinjua Subgroup. Though, the presence of *Tappania* in Rampur Formation is indicative of a Late Mesoproterozoic (ca.1100-1000Ma) age, the appearance of *Vandalosphaeridium,* budding leiospharids and *Melanocyrillium* undoubtably suggest an Early Neoproterozoic (ca. 850 Ma) age to this formation. It appears that Rampur Formation ranges in age from latest Mesoproterozoic to Early Neoproterozoic (ca.1050-850Ma). The absolute age datings of Rampur Formation by Kruezer *et al.* (1977) as 1080± 40 Ma corroborates with the present age assignment to it. However, the radiometric dating of this lithounit as 1628±8Ma (Rasmussen *et al.,* 2002) needs reconsideration in view of the record of distinctive latest Mesoproterozoic-Early Neoproterozoic microfossils assemblage.

As discussed earlier, Rohtas assemblage equates with the Middle Neoproterozoic (ca.850 Ma) Miroyedikha assemblage of Siberia (German, 1990), suggesting an Early Cryogenian (Ca. 850-750 Ma) age to it. The absolute dating of Rohtas Limestone as 1599 ± 48Ma (Sarangi *et al.,* 2004) seems incompetable in view of record of distinctive Middle Neoproterozoic microfossil assemblage. Similarly, the absence of Terminal Proterozoic or Early Cambrian organic-walled microfossils does not suggest Late Vendian (Ediacaran) or Early Cambrian age to this lithounit, as suggested by Azami (1998).

The organic-walled microfossil assemblages of Kaimur and Rewa groups are quite different as *Symplassosphaeridium-Synsphaeridium* complex dominates over the *Leiosphaeridia-Kildinosphaera* association of Semri Group. The only innovative feature of Kaimur and Rewa assemblages are the appearance of *Trachysphaeridium laufeldi* and *Octoedryxium truncatum* respectively, which oldest records are from the Late Neoproterozoic (middle Visingso Beds; ca. 700-610Ma) of Sweden (Vidal, 1976a).Thus, the appearance of *T.laufeldi* and *O. truncatum* favours Late Neoproterozoic (ca. 750-650 Ma) age to Kaimur and Rewa groups collectively. The absolute dating of Rewa Group as 700 ± 100 Ma (Srivastava and Rajagopalan, 1988) corroborates the above age inference.

The most important aspect of Bhander Group assemblages is the abundant occurrence of different species of *Obruchevella, viz. O. valdaica, O. parva* and *O. delicata.* They appear in the upper parts of Ganurgarh Shale, and become abundant in the overlying Nagod Limestone and sirbu Shale. Similarly,

Lophosphaeridium spp., *Cristallinium* spp. and *Germinosphaera* spp. appear in Nagod Limestone, and continue to occur in Sirbu Shale. However, some Late Vendian-? Early Cambrian marker taxa, *viz. Obruchevella parvissima, Dictyotidium* spp. and *Cymatiosphaera* sp. appear in the Sirbu Shale.

Though, the occurrence of *V. reticulatum, T. laufeldi* and *B. faveolata* suggests a Late Neoproterozoic (ca. 650-600 Ma) age to the Ganurgarh Shale, the appearence of various species of *Obruchevella* in upper part of Ganurgarh Shale positively extends its age in Early Vendian (ca.590Ma).The abundance of *Obruchevella* spp. in Nagod Limestone, and the appearance of Late Vendian marker taxa, such as *L. rarum* and *Cristallinium* sp. is suggestive of an Early Vendian age to Nagod Limestone, and may extends in Late Vendian also. The persistance of above taxa in the Sirbu Shale, including the various species of *Obruchevella*, and the appearance of other Late Vendian and Early Cambrian marker taxa, *viz. Obruchevella parvissima, Dictyotidium* sp. and *Cymatiosphaera* sp. indicateed Late Vendian (ca. 560-544 Ma) - Early Cambrian age to Sirbu Shale. But, the absence of specific Early Cambrian marker acritarch taxa, *viz. Asteridium tornatum, Skiagia ciliosa* and *Comasphaeridium velvetum*, restricts the age of the Sirbu Shale as Late Vendian.

Conclusions

1. The organic- walled microfossils from Vindhyan sediments of Son Valley and DMH-A suggest that the Vindhyan Supergroup ranges in age from Early Mesoproterozoic (ca. 1550 Ma) to Terminal Proterozoic (ca. 544 Ma), and does not encompass into the Lower Paleozoic (Cambrian).
2. The record of well-developed organic-walled microfossil assemblages of Middle Mesoproterozoic in Deonar Formation, Late Mesoproterozoic-Early Neoproterozoic in Rampur Formation and Middle Neoproterozoic in Rohtas Subgroup does not support older age of Late Paleoproterozoic (ca.1630 Ma) to them as suggested by recent radiometric datings.
3. The recent advocation that Upper Vindhyans are the Lower Paleozoic representatives in the Vindhyan Basin seems untenable in view of record of distinctive Vendian marker species of *Obruchevella* in the Bhander Group, and the absence of Early Cambrian marker microfossils.

Repository

The examined palynological slides are preserved in the Repository Section of Palynology Laboratory (Geology Division) at KDM Institute of Petroleum Exploration, ONGC, Dehradun (India) with registration No. KDMIPE/ PALY/ 2004/ BP/ VINDH. - M.P.

Acknowledgements

Authors are grateful to Shri Y. B. Sinha, Director (Exploration), Oil and Natural Gas Corporation Ltd., for kind permission to publish this paper. Dr. D. Ray, Executive Director-Head, KDMIPE, Mrs. N.J. Thomas, General Manager (GRG) and Shri James Peters, Dy. Gen. Manager (Geology) gave valuable suggestions during the course of this study, for which authors are indebted to them.

References

AHMAD, F. 1962. Paleogeography of Central India in Vindhyan Period. Records of the Geological Survey of India, 87: 513-548.

ANABARASU, K. 2001. Acritarchs from Mesoproterozoic Chitrakoot Formation, Semri Group, Chitrakoot area, Central India. Journal of the Geological Society of India, 57: 179-83.

ASEEVA, E.A. 1974. O spirale I kol'tsevidnykh obrazovaniyakh v verkhnedokembrijskikh otlozheniyakh Podolii [About spiral and ring like structures in the upper Precambrian deposits of Podolina]. Paleontologicheskij Sbornik , 11: 95-98. (in Russian)

AUDEN, J.B. 1933. Vindhyan sedimentation in Son Valley, Mirzapur district. Memoir Geological Survey of India, 62: 141-250.

AZMI, R.J. 1998. Discovery of Lower Cambrian small shelly fossils and brachiopods from the Lower Vindhyan of Son Valley, Central India. Journal of the Geological Society of India, 52: 381-389.

BANERJEE, D.M. and FRANK, W. 1999. Preliminary Ar^{40}/Ar^{39} dates of porcellanite and detrital mica from the Vindhyan of Central India. In: Kumar, S., Jauhri, A.K., Rai, V. and Kulshreshtha, A.K., Eds., Workshop on Vindhyan Stratigraphy and Palaeobiology. Lucknow University, Lucknow, 5. (Abstract)

BISWAS, S.K., BHASIN, A.L. and RAM, J. 1993. Classification of Indian sedimentary basins in the framework of plate tectonics. In: Biswas, S.K., Pandey, J., Dave, A., Maithani, A., Garg, P. and Thomas, N.J., Eds., Petroliferous basins of India. KDM Institute of Petroleum Exploration, Dehradun, India, 1: 1-46.

CHUG, M.L., SINGH, J.N., SHARMA, P.C., SHARMA, N.K., DAS, S.K. and SALUJHA, S.K. 1989. Integrated basin analysis of Foothill Fold Belt and Ganga Foredeep across Indo-Nepal Border. ONGC Report (unpublished), KDMIPE, Dehradun: 1-64.

CLOUD, P., AWRAMIK, S.M., MORRISON, K. and HADLEY, D.G. 1979. Earliest Phanerozoic or latest Proterozoic fossils from the Arabian Shield. Precambrian Research, 10: 73-93.

CRAWFORD, A.R. and COMPSTON, W. 1970. The age of the Vindhyan System of Peninsular India. Geological Society of London, Quarterly Journal, 125: 351-371.

DUTTA, D., DIXIT, G., HUSSIAN, T. and SHUKLA, M. 1998. Geology and stratigraphy of the Vindhyan sediments in Central and Eastern parts of Vindhyan Basin during the field season 1996-97. ONGC Report (unpublished), KDMIPE, Dehradun: 1-53.

FRIEDMAN, G.M., CHAKRABORTY, C. and KOLKAS, M. 1996. $\delta^{13}C$ excursion in the end of Proterozoic strata of the Vindhyan Basin (Central India): its chronostratigraphic significance. Carbonate and Evaporites, 11: 206 - 212.

FRIEDMAN, G. M. and CHAKRABORTY, C. 1997. Stable isotopes in marine carbonates: their implications for the paleoenvironment with special reference to the Proterozoic Vindhyan carbonate (Central India). Journal of the Geological Society of India, 50: 131-159.

GERMAN, T.N., 1985. Nitchatye vodorosli venda [Vendian filamentous algae]. In: Sokolov, B.S. (ed.), Vendskaya sistema 1 [The Vendian System 1] Akademiya Nauk SSSR, Otdelenie geologii, Geofiziki i Geokhimii, Moscow, 146-151.

GERMAN, T.N. (Ed.), 1990. Organic world 1 billion years ago. Nauka, Leningrad: 49. (in Russian)

GOLOVENOK, V.K. and BELOVA, M.I.U. 1984. Riphean microbiota in cherts of the Billyakh Group on the Anabar Uplift. Paleontological Journal, 18 (4): 20-30.

HERON, M., 1932. Vindhyans of Western Rajputana. Records Geological Survey of India, 65 (4): 457-489.

HILL, A.C., COTTER, K.L. and GREY, K. 2000. Mid -Neoproterozoic biostratigraphy and isotope stratigraphy in Australia. Precambrian Research, 100: 281-298.

HOFMANN, H.J. and CHENE, J. 1981. Carbonaceous megafossils from the Precambrian (1800 Ma) near Jixian, northern China. Canadian Journal of Earth Sciences, 18: 443-447.

HOFMANN, H.J. and JACKSON, G.D. 1994. Shale-facies microfossils from the Proterozoic Bylot Supergroup, Baffin Island, Canada. Memoir, Paleontological Society of Canada, 37: 1-35.

HORODYSKI, R.J. 1980. Middle Proterozoic shale-facies microbiota from the lower Belt Supergroup, Little Belt Mountains, Montana. Journal of Paleontology, 54: 649-663.

HORODYSKI, R.J. and DONALDSON, J.A. 1980. Microfossils from the Middle Proterozoic Dismal Lakes Group, Arctic Canada. Precambrian Research, 11: 125-159.

JACKSON, M.J., SWEET, I.P., PAGE, R.W. and BRADSHAW, B.E. 1999. An evaluation of the economic potential of the Northern lawn Hill Platform. In: Bradshaw, B.E. and Scott, D.L., Eds., Integrated basin analysis of the Isa Superbasin using seismic, well-log and geophysical data. Australian Geological Survey, Orgnistion Record, Canberra, Australia (1999) 19.

JANKAUSKAS, T.V. 1989 (Ed.). Mikrofossilii dokembrya SSSR [Precambrian microfossils of the USSR]. Trans. Inst. Geol. Geochronol. Dokembria SSSR Akademia Nauka, Leningrad, 1-188 (in Russian)

JAVAUX, E.J., KNOLL, A. H. and WALTER, M. R. 2001. Morphological and ecological complexity in early eukaryotic ecosystems. Nature, 412: 66-69.

KATHAL, P.K., PATEL, D.R. and ALEXANDER, P.O. 2000. An Ediacaran fossil *Spriggina* (?) from the Semri Group and its implication on the age of the Proterozoic Vindhyan Basin, Central India. Neues Jahrbuch Geologie und Palaeontologie, 6: 321-332.

KNOLL, A.H., 1992. Vendian microfossils in metasedimentary cherts of the Scotia Group, Prins Karls Forland, Svalbard. Paleontology, 35: 751-774.

KNOLL, A.H. 1996. Archean and Proterozoic Paleontology. In: Jansonius, J. and McGregor, D.C., Eds., Palynology: Principles and Applications. AASP Foundation, Tulsa, 1: 51-80.

KNOLL, A.H. and OHTA, Y. 1988. Microfossils in metasediments from Prins Karls Forland, western Svalbard. Polar Research, 6: 59-67.

KNOLL A.H. and SWETT, K. 1985. Micropaleontology of the Late Proterozoic Veteranen Group, Spitsbergen. Palaeontology 28, (3): 451-473 .

KOLOSOV, P.N., 1982. Verkhnedokembrijskie paleoal'gologicheskie ostatki Sibiriskoj platformy [Upper Precambrian Paleo-Algal remains from the Siberian Platform]. Nauka, Moscow, 1-96. (in Russian)

KRALIK, M. 1982. Rb-Sr age determination on Precambrian carbonate rocks of the Carpentarian McArthur Basin, Northern Territory, Australia. Precambrian Research, 18: 157-170.

KREUZER, A. HARRE, W., KURSTEN, M., SCHINTZER, W.A. MURTI., K.S. and SRIVASTAVA, N.K., 1977. K/Ar dates of two glauconites from the Chanderpur Series (Chhatishgarh/India). Geologischen Jahrbuch B, 28: 23-36.

KUMAR, S. 1976. Stromatolites from the Vindhyan rocks of Son Valley - Maihar area, district Mirzapur (U.P.) and Satna (M.P.). Journal of the Geological Society of India, 18: 13-31.

KUMAR, S. 1982. Vindhyan stromatolites and their stratigraphic testimony. In: Valdiya, S.K., Bhatia, S.B. and Gaur, V.K., Eds., Geology of Vindhyanchal. Hindustan Publishing Corporation, New Delhi, 102-112.

KUMAR, S. 1984. Present status of Stromatolite Biostratigraphy: A Review. Geophytology, 14: 96-110.

KUMAR, S. 2001. Mesoproterozoic megafossil *Chuaria - Tawuia* association may represent parts of multicellular plant, Vindhyan Supergroup, Central India. Precambrian Research, 106: 187-211.

KUMAR, S. and SRIVASTAVA, P. 1997. A note on the carbonaceous megafossils from the Neoproterozoic Bhander Group, Maihar area, Madhya Pradesh. Journal of Palaeontological Society of India, 42: 141-146.

MA, G.G., LIU, S.L. and DENG, Z.O. 1980. K-Ar dating of glauconites from Late Proterozoic Ruyang Group in western Henan Proterozoic, and stratigraphical correlation. Bulletin, Yichang Institute of Geology and Mineral Resources, Chinese Academy of Geological Sciences, 1 (2): 103-111. (in Chinese)

MALLET, F.R. 1869. On the Vindhyan series in the north-eastern and central Provinces. Memoir Geological Survey of India, 7 (1): 1-129.

MAITHY, P.K. 1992. Palaeobiology of Vindhyan. Palaeobotanist, 40: 52-72.

MAITHY, P.K. and BABU, R. 1988. The Mid-Proterozoic Vindhyan microbiotas from Chopan, south-east Uttar Pradesh. Journal of the Geological Society of India, 31: 584-590.

MAITHY, P.K. and SHUKLA, M. 1977. Microbiota from the Suket Shales, Rampura, Vindhyan System (Late Precambrian), Madhya Pradesh. Palaeobotanist, 23: 176-128.

MANKIEWICZ, C. 1992. *Obruchevella* and other microfossils in the Burgess Shale: preservation and affinity. Journal of Paleontology, 66 (5): 717-729.

MISRA, R.C., SINGH, S.N. and KUMAR, S. 1977. Two new forms of stromatolites from the Kajrahat Limestone, Dala area, district Mirzapur, U.P. Geophytology, 7: 139-141.

MOCZYDLOWSKA, M., VIDAL, G. and RUDAVASKAIA, V. A. 1993. Neoproterozoic (Vendian) phytoplankton from the Siberian Platform, Yakutia. Paleontology, 36: 495-521.

OEHLER, D. Z. 1978. Microflora from the Middle Proterozoic Balbirini Dolomite (McArthur Group) of Australia. Alcheringa, 2: 269-309.

PEAT, C.R., MUIR, N.D., PLUMB, K.A., MCKIRDY, D.M. and NORVICK, M.S. 1978. Proterozoic microfossils from the Roper Group, Northern Territory, Australia. Bureau of Mineral Resources, Journal of Australian Geology and Geophysics, 3: 1-17.

PRASAD, BALMIKI 1976. Lower Vindhyan formations of Rajasthan. Records Geological Survey of India, 106 (2): 33-53.

PRASAD, BALMIKI 1984. Geology, Sedimentation and Paleogeography of the Vindhyan Supergroup, southeastern Rajathan. Memoir Geological Survey of India, 116 (1):1-107.

PRASAD, BIJAI and ASHER, R. 2001. Acritarch biostratigraphy, and lithostratigraphic classification of Mesoproterozoic and Lower Paleozoic sediments (Pre-Unconformity Sequence) of Ganga Basin, India. Paleotographica Indica (KDMIPE, ONGC, Dehradun), 5: 1-151.

RAI, V. 1999. Discovery of enigmatic microbial mat texures and probable Ediacaran fossils from the Upper Bhander Sandstone Formation, Vindhyan Supergroup, Maihar area, Central India. In: Kumar, S., Jauhri, A.K., Rai V, Kulshreshtha, A.K., Eds., Workshop on the Vindhyan Stratigraphy and Paleobiology, Lucknow University, Lucknow, 44-48.

RAM, J., SHUKLA, S.N., PRAMANIK, A.G., VERMA, B.K, CHANDRA, G. and MURTY, M.S.N. 1996. Recent investigations in the Vindhyan Basin: implications for the basin tectonics. Memoir, Geological Survey of India, 36: 267-286.

RASMUSSEN, B., BOSE, P.K., SARKAR, S., BANERJEE S, FLETCHER, I.R. and MC NAUGHTON, N.J. 2002. 1.6 Ga U-Pb zircon age for the Chorhat Sandstone, Lower Vindhyan, India: Possible implications for early evolution of animals. Geology, 30: 103-106.

RAY, J.S.., MARTIN, M.W., VEIZER, J. and BOWRING, S.A. 2002. U-Pb zircon dating and Sr isotope systematics of the Vindhyan Supergroup, India. Geology, 30: 131-134.

REITLINGER, E.A. 1948. Kembrijskie foraminifery Yakutii [Cambrian foraminifers of Yakutsk]. Byulletin Moscovskogo Obshchestva Ispytatelej Prirody, Otdelenie Geologii, 23: 77-81.

REITLINGER, E.A. 1959. Atlas mikroskopicheskikh ostatkov I problematiki drevnikh tolshch [Atlas of microscopic organic remains and problematica from ancient deposits of Siberia]. Trudy Geologicheskogo Institut. Akademiya Nauka Science, SSSR, 25: 1-63. (in Russian)

SALUJHA, S.K. 1973. Palynological evidence on the age of the Vindhyan Sediments. Proceedings Indian National Science Academy, 39 (1): 62-68.

SALUJHA, S. K., REHMAN, K. and ARORA, C.M. 1971. Plant microfossils from the Vindhyan of Son Valley, India. Journal of the Geological Society of India, 12 (1): 24-33.

SARANGI, S., GOPALAN, K. and KUMAR, S. 2004. Pb-Pb age of earliest megascopic, eukaryotic alga bearing Rohtas Formation, Vindhyan Supergroup, India: implications for Precambrian atmospheric oxygen evolution. Precambrian Research, 132: 107-121.

SASTRI, M.V.A. and MOITRA, A.K., 1984. Vindhyan Stratigraphy - A review. Memoir, Geological Survey of India, 116(2): 109-148.

SEILACHER, A., BOSE, P.K. and PFLUGER, F. 1998. Triploblastic animals more than 1 Billion years ago: trace fossil evidence from India. Science, 282: 80-83.

SHENFIL, V.YU. 1983. Vodorosli v dokembrijskikh otlozheniyakh Vostochnoy Sibiri [Algae in the Precambrian deposits of Eastern Siberia]. Doklady Akademii Nauka SSSR, 269 47: 1-473.

SERGEEV, V.N., KNOLL, A.H. and GROTZINGER, J.P. 1995. Paleobiology of the Mesoproterozoic Billyakh Group, Anabar Uplift, northern Siberia. Paleontological Society Memoir, 39: 1-37.

SHARMA, M., SHUKLA, M. and VENKATACHALA, B.S. 1992. Metaphyte and metazoan fossils from Precambrian sediments of India: a critique. Paleobotanist, 40: 8-51.

SONG, X. 1984. *Obruchevella* from the early Cambrian Meishucun Stage of the Meishucun section, Jinning, Yunnan, China. Geological Magazine, 121 (3): 179-183.

SRIVASTAVA, A.P. and RAJAGOPALAN, G., 1988. F-T age of Rawatbhata area. Journal of the Geological Society of India, 32: 527-529.

VENKATACHALA, B.S., YADAV, V.K. and SHUKLA, M. 1990. Middle Proterozoic microfossils from the Nauhatta Limestone (Lower Vindhyan), Rohtasgarh, India. In: Naqvi, S.M. Ed., Development in Precambrian Geology 8. Elsevier Publication, Amsterdam, 137-165.

VENKATACHALA, B.S., SHARMA, M. and SHUKLA, M. 1996. Age and life of the Vindhyans - Facts and Conjectures. Memoir Geological Society of India, 36: 137-165.

VIDAL, G. 1976. Late Precambrian microfossils from the Visingso Beds in southern Sweden. Fossil and Strata, 9: 1-57.

VIDAL, G. and KNOLL, A.H. 1983. Proterozoic plankton. Geological Society of America Memoir, 161: 265-277.

VINOGRADOV, A.P., TUGARINOV, A.I., ZHYKOV, C.I., STANIKOVA, N., BIBIKOVA, E.V. and KHORRE, K. 1964. Geochronology of the Indian Precambrian. XXIIth International Geological Congress Report, New Delhi, 10: 553-567.

WANG, Y., LU, S., GAO, Z., LIN, W. and MA, G. 1981. Sinian tillites of China. In: Hambrey, M.J. and Harland, W.B., Eds., Earth's Pre-Pleistocene Glacial record. Cambridge University Press, Cambridge, 386-401.

XIAO, S., KNOLL, A.H., KAUFMAN, A.J., YIN, L. and ZHANG, Y. 1997. Neoproterozoic fossils in Mesoproterozoic rocks? Chemostratigraphic resolution of a biostratigraphic conundrum from the North China Platform. Precambrian Research 84: 197-220.

YAKSHIN, M.S. and LUCHININA, V.A. Elenkin [New data on fossil algae of the family Oscillatoriaceae (Kirchn.) Elenkin]. Trudy Instituta Geologii i Geofiziki, Sibirskoe Otdelenie, Akademiya Nauka SSSR, 475: 28-34. (in Russian)

YIN, L. 1997. Acanthomorphic acritarchs from Meso-Neoproterozoic shales of the Ruyang Group, Shanxi, China. Review Palaeobotany Palynology, 98: 15-25.

YIN, L. and GUAN, B. 1999. Organic-walled microfossils of Neoproterozoic Dongjia Formation, Lushan County, Henan Province, North China. Precambrian Research, 94: 121-137.

YIN, L. and YUAN, X. 2002. Review of the microfossil assemblage from the Late Mesoproterozoic Ruyang Group in Shanxi, China. Acta Micropaleontologica Sinica, 20 (1): 39-46.

ZANG, W. and WALTER, M.R. 1992. Late Proterozoic and Cambrian microfossils and biostratigraphy, Amadeus Basin, central Australia. Mem. Assoc. Australas. Paleontology, 12: 1-132.

ZHANG, Z. 1986. Clastic facies microfossils from the Chuanliggou Formation (1800 Ma) near Jixian, North China. Journal of Micropalaeontology, 5 (2): 9-16.

ZHANG, Y., YIN, L., XIAO, S. and KNOLL, A.H. 1998. Permineralised fossils from the terminal Proterozoic Doushantuo Formation, South China. Paeontological Society Memoir, 50: 1-52.

Age of the Vindhyan Supergroup of Central India: An exposition of biochronology vs radiochronology

R.J. Azmi[1], Deepak Joshi[1], B.N. Tiwari[1], M.N. Joshi[2], Kshitij Mohan[3] and S.S. Srivastava[4]

[1]Wadia Institute of Himalayan Geology, 33 GMS Road, Dehra Dun 24800, India
[2]Department of Geology, D.B.S. (P.G.) College, Dehra Dun 248 001, India
[3]Department of Geography, Banaras Hindu University, Varanasi 221 005, India
[4]Department of Earth Sciences, IIT Roorkee, Roorkee 247 667, India

Introduction

The Vindhyan Basin in the central Indian Peninsula is the largest sedimentary basin that covers an area of about 1, 04, 000 sq km with its maximum thickness of about 4400m and comprises a succession of sandstone, shale and limestone/dolomite with a prominent horizon of volcaniclastic sediments (porcellanite) particularly in the lower part. It spans from Agra in the northwest through Chittaurgarh in the west to Sasaram in the east in a sickle-shaped form, girdling the basement of the Bundelkhand Granite-Gneiss Complex (BGC) along with the Bijawar Group lowgrade metamorphics in the western, southern and eastern parts. In the northern and northeastern part a large portion of the Vindhyan sediments is concealed below the Indo-Gangetic alluvium and the Sub-Himalayan Siwalik Range and in the southeastern part by the Deccan Trap (Fig. 1).

Although the Vindhyan sediments drew the attention of geologists around the mid of the 19th century, salient pioneering geological studies were carried out by Auden (1933) in the Son Valley and Chitrakoot area, and by Prasad (1984) in the eastern Rajasthan, which shed light on broad litho-tectonic frame-work of the basin. The lithostratigraphic formalization to the Vindhyan succession as Vindhyan Supergroup (VSG) was done by Sastry and Moitra (1984), Soni *et al.* (1987) and Mathur (1987). Detailed lithostratigraphic divisions for the Son Valley and the Chitrakoot area are shown in Figs. 2 and 4.

Enigma of age of the VSG

Age of the VSG in Indian stratigraphy remained enigmatic for nearly one and half century for the want of robust fossil control, despite the fact that sediments are undeformed and unmetamorphosed. Different ages varying from Paleoproterozoic to Devonian have been assigned to the VSG so far on the basis of scanty body fossils, palynofossils, trace fossils, stromatolites and the geochronology, as well as due to speculations based on lithological similarities from the neighboring sedimentary successions. However, particularly during the early to middle of the 20th century, owing to rare fossil records of Cambrian affinity (Beer, 1919; Rode, 1946; Ghosh and Bose, 1950; Jacob *et al.*, 1953; Sitholey *et al.*,

1953) the initial impression developed that the Vindhyans possibly ranged between the late Precambrian and early Paleozoic age. But, during 1960s and 70s due to extensive radiometric dating of the Vindhyan sediments and the intrusive igneous bodies (Vinogradov *et al.*, 1964; Tugarinov *et al.*, 1965; Crawford and Compston, 1970; Paul *et al.*, 1975; Paul, 1979) and also with the recognition of stromatolites of Riphean age (Valdiya, 1969; Kumar, 1976) the general consensus evolved that the Vindhyan sediments are of Mesoproterozoic-Neproterozoic (~1400 – 600 Ma, Riphean) age and this became so much entrenched in the geological literature that the Vindhyan Basin became a sort of reference Proterozoic basin of the Indian Peninsula. Nevertheless, still time to time, doubts were expressed mostly by biostratigraphers/paleobiologists (eg. Salujha *et al.*, 1971; McMenamin *et al.*, 1983) and also some times by other workers (Haldar and Ghosh, 1981) for such an old age for the VSG. Added to this apprehension, some glaring fossil records of paleobiological and evolutionary significance were mismatching. For example, fossils such as jellyfish (Sisodiya 1982; Singh and Chandra, 1987; Shukla and Sharma, 1990), trace fossils (Sisodiya and Jain, 1984; Chakrabarti, 1990; Shukla and Sharma, 1990; Kulkarni and Borkar, 1996; Sarkar *et al.*, 1996) and vendotaenid ribbon *Tyrasotaenia* (Shukla and Sharma, 1990) though clearly indicative of younger age (Vendian/ Ediacarian - Cambrian) for the VSG, were conceived to be of pre-Ediacaran/Proterozoic age for want of the firm evidence of 'Cambrian explosion' related fossils (e.g., soft-bodied Ediacaran fossils), metazoan traces, metaphytes, skeletalized fossils remains - SSFs, calcareous skeletal algae, etc.), which globally appear in the Vendian/Ediacarian Period and at the Precambrian-Cambrian (PЄ-Є) boundary (544Ma), besides the evidence of the phosphogenesis and Varanger age glaciation (Venkatachala *et al.*, 1996).

The major turning point on the age of the VSG, however, came in October 1998 when incidentally two simultaneous fossil reports (Azmi, 1998a; Seilacher *et al.*, 1998) brought the problem of age of the Vindhyans to the international focus (Brasier, 1998; Azmi, 1998b; Kerr, 1998a, b, 1999; Conway Morris *et al.*, 1998; Bagla, 2000). While Seilacher *et al.* (October, 1998) based on their report of triploblastic animal trace fossils from the Chorhat Sandstone (Kheinjua Subgroup, Lower Vindhyan) claimed to have found the evidence of exceptionally deep origin of multicellular animals (>1100 Ma, ~ 500 million years older than previous world records); Azmi (October, 1998a) discovered Lower Cambrian small shelly fossils (SSFs) from the Rohtasgarh Limestone and suggested a major upward (Vendian – Cambrian) age revision for the VSG. Subsequently, he (Azmi, 1998b) took the evidence of Seilacher *et al*'s reported multicellular organism traces as supporting evidence for the younger age of the VSG. Ironically, both claims could not get instant support in the background of deep rooted notion of Mesoproterozoic age for the Semri Group in the pretext of doubtful biogenicity and/or reproducibility (Editor *JGSI*, 2000 and Bagla, 2000). However, the major gain was that these reports did generate greater interest on the age of the VSG, resulting in number of contemporary researches in radiochronology, C and Sr isotopic signatures and also in paleobiology/biostratigraphy (Banerjee and Frank, 1999; Azmi, 1999b; Kumar, 1999; Kumar *et al.*, 2001; Kumar *et al.*, 2002; Ray *et al.*, 2002; Rasmussen *et al.*, 2002; Ray *et al.*, 2003; Sarangi *et al.*, 2004; Kathal *et al.*, 2000; Anbarasu, 2001; Rai and Singh, 2004; Joshi, 2004). But the strange point that emerged was that while the latest biostratigraphic evidences categorically pointed towards Vendian to early Cambrian age (Azmi, 1999b; Kathal *et al.*, 2000; Anbarasu, 2001a; Azmi *et al.*, 2003; Joshi, 2004), the geochronological and isotopic results (Kumar *et al.*, 2001; Kumar *et al.*, 2002; Ray *et al.*, 2002; Rasmussen *et al.*, 2002; Ray *et al.*, 2003; Sarangi *et al.*, 2004), barring a single age of 617±3. 5 Ma (Ar-Ar) (Banerjee and Frank, 1999), pushed the entire Lower Vindhyan (Semri Group) succession further old into the late Paleoproterozoic age bracket (1600–1700 Ma).

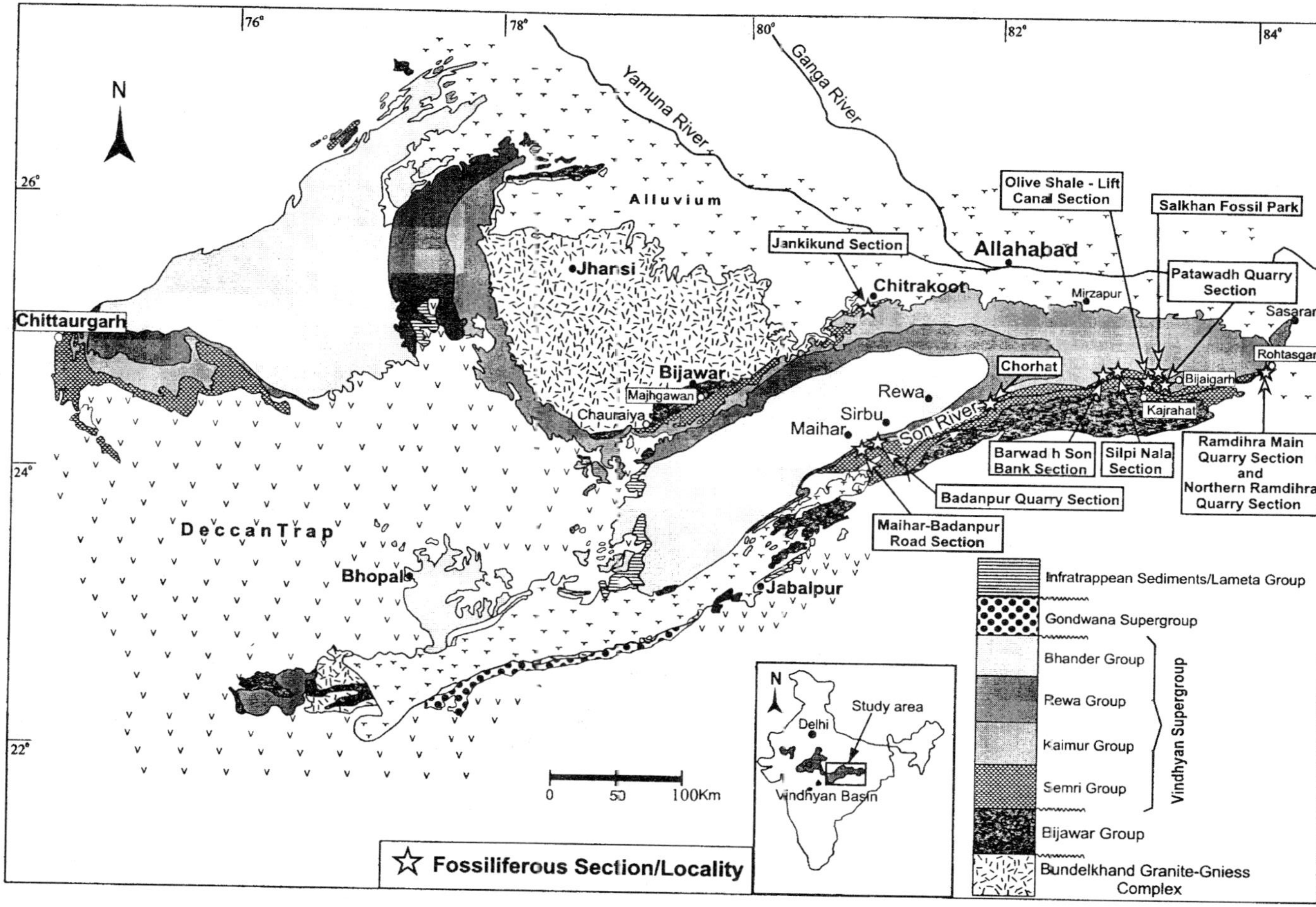

Fig. 1. Geological map of the Vindhyan Basin (after Krishnan and Swaminath, 1959 and Soni *et al.*, 1983) showing locations of the studied fossiliferous sections.

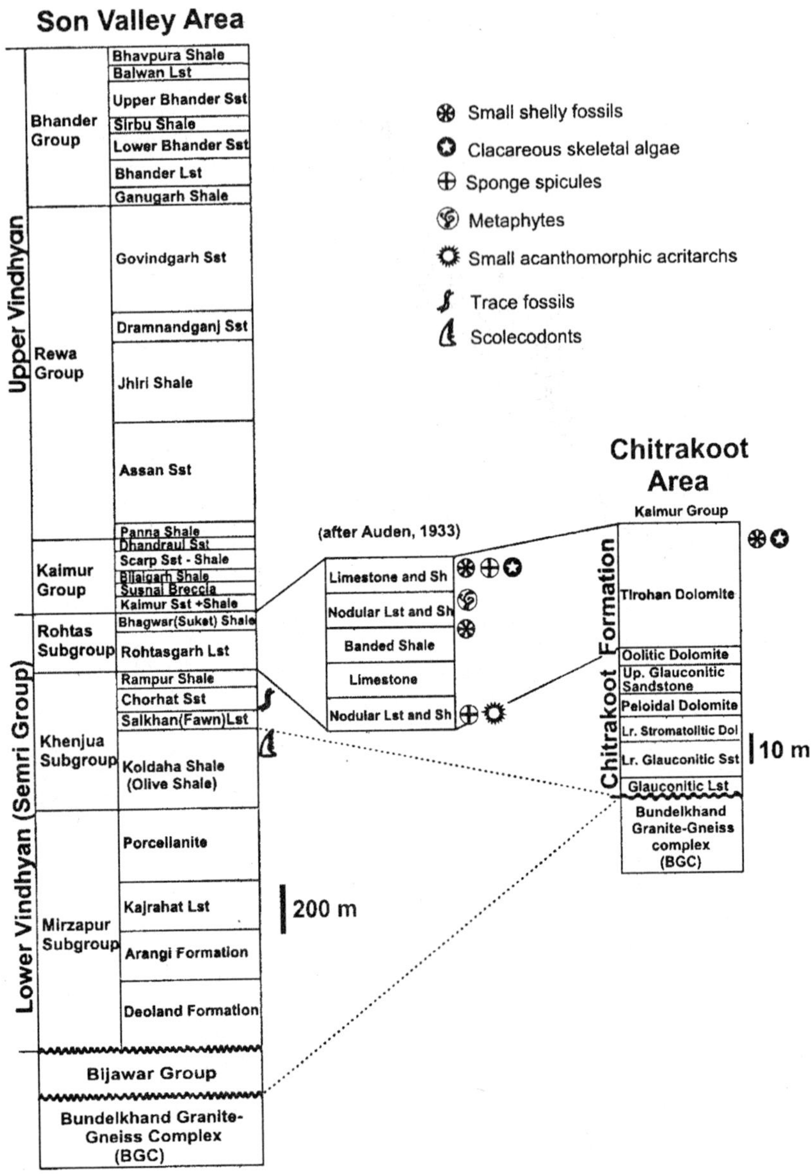

Fig. 2. Generalized lithostratigraphy of the Vindhyan Supergroup in Son Valley (after Sastry and Moitra, 1984) and lithostratigraphic subdivisions of the Rohtas Subgroup (after Auden, 1933). Lithostratigraphic details of the Chitrakoot Formation in Chitrakoot area (after Singh and Pal, 1970; Anbarasu, 2001b) and its correlation with the Lower Vindhyan (Semri Group). Positions of various fossils groups recovered in the present work are marked.

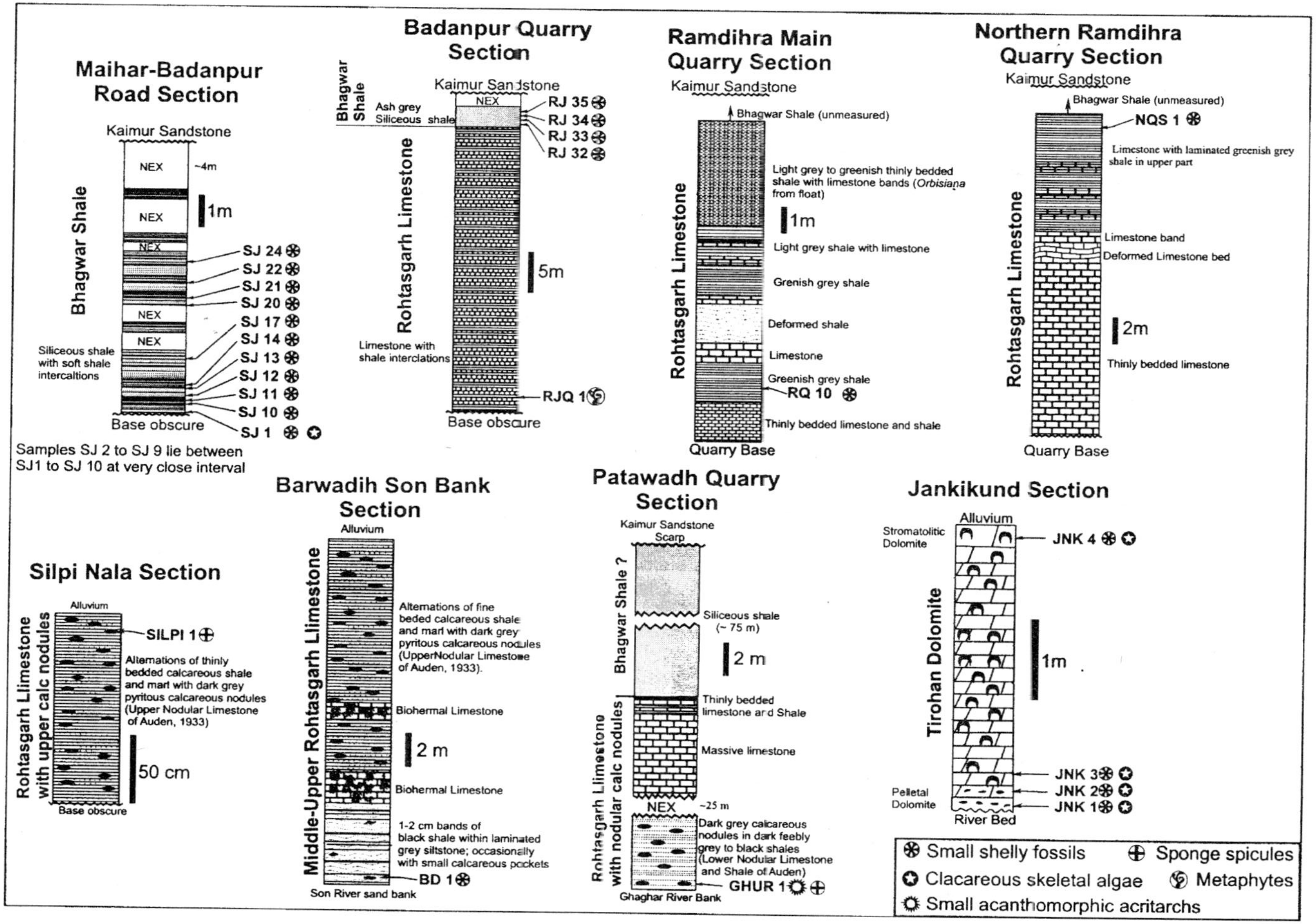

Fig. 3. Lithologs of the studied sections showing positions of the fossil yielding samples.

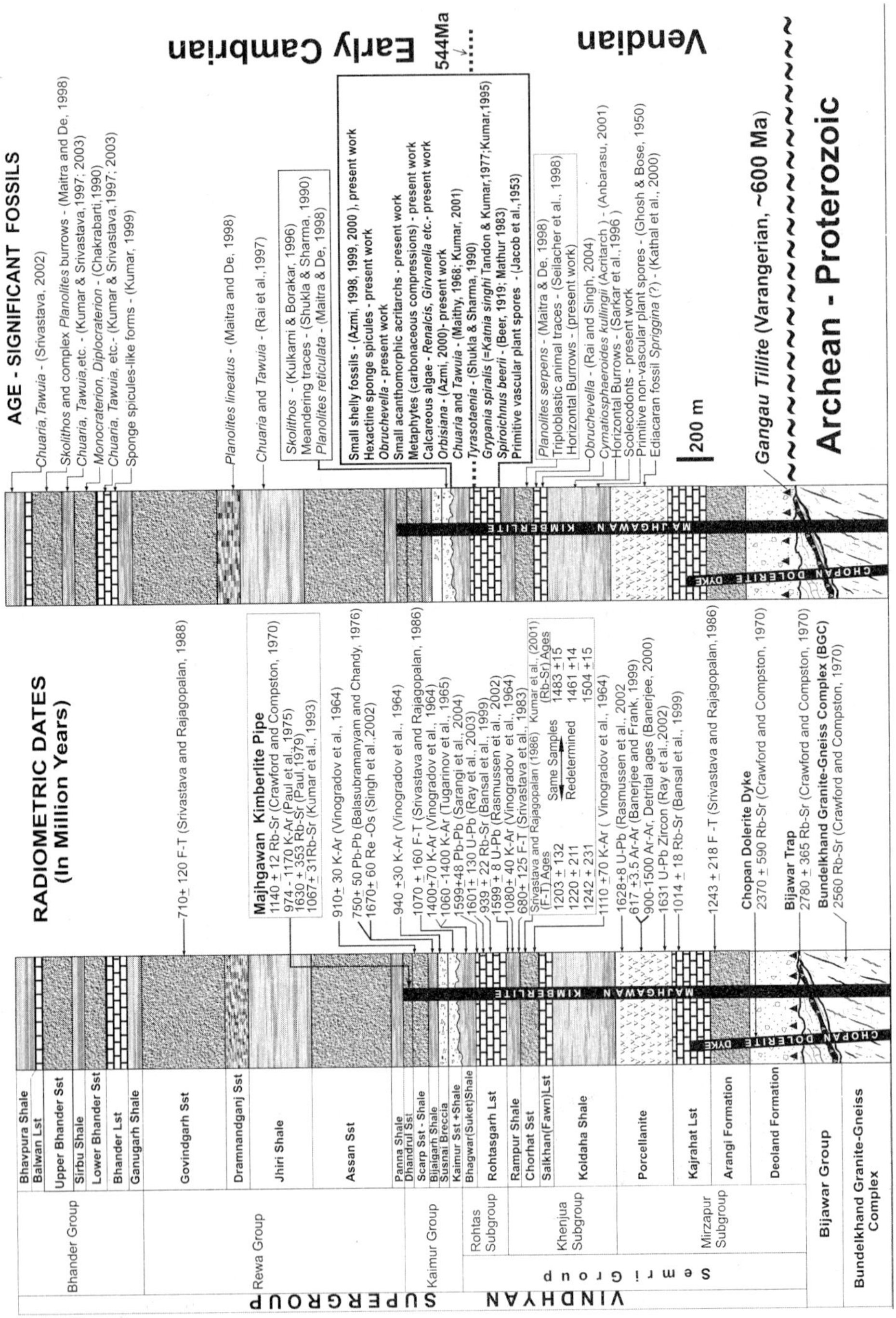

Fig. 4. Inconsistent radiometric dates vs fossil records consistent in indicating Vendian - Early Cambrian age for the Vindhyan Supergroup. Stromatolites being not good age-indicators have not been considered.

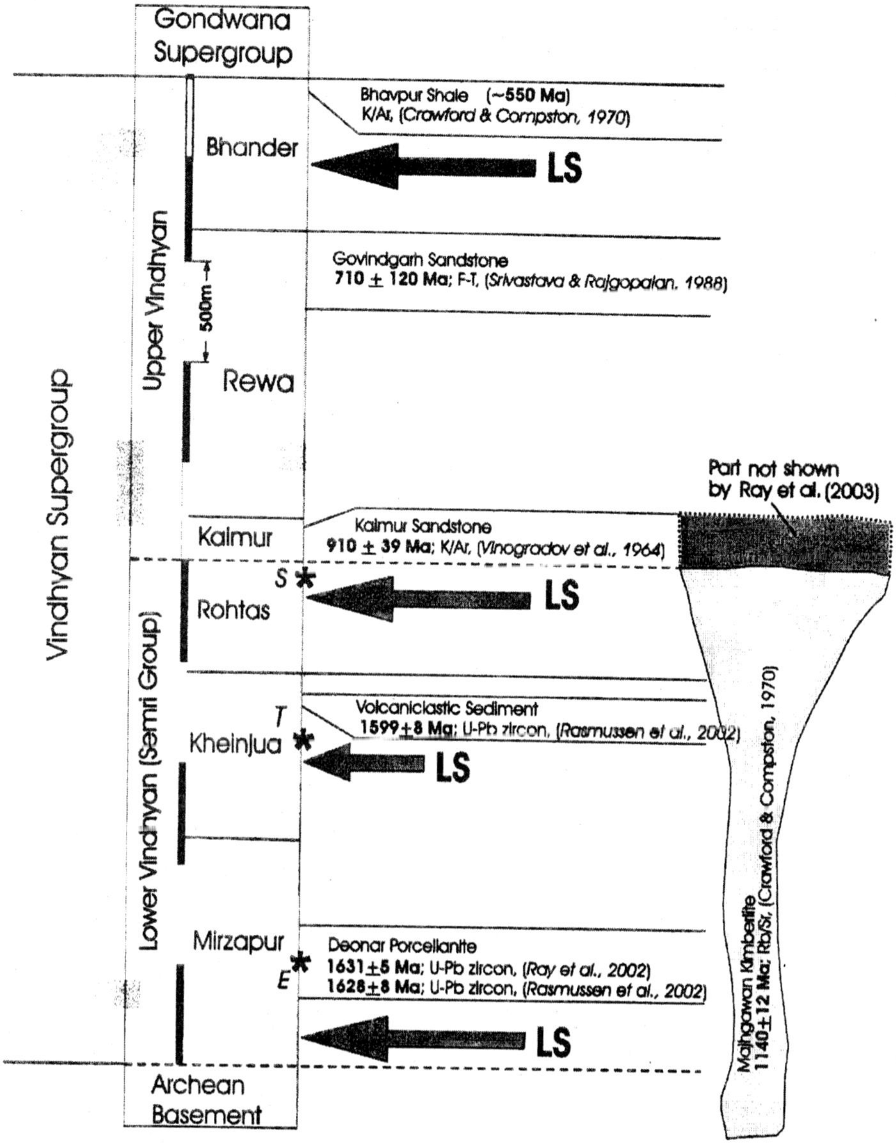

Fig. 5. Corrected version of Fig. 2 in Ray *et al.*, (2003) showing composite stratigraphic column of the Vindhyan Supergroup, where the authors restricted intrusion of the Majhgawan Kimberlite Pipe (MKP) up to the top of the Lower Vindhyan. The actual intrusion level is up to the top of the Kaimur Group, Upper Vindhyan, as we have shown. Stars with *E*, *T*, and *S* represent Ediacaran, trace fossils, and small shelly fossil levels respectively; arrows with LS point positions of major limestone formations. Purported 'reliable' dates of the Ray *et al.*, (2003) in toto are also given.

Latest biostratigraphic materials and results

Here we present our latest biostratigraphic results from the Lower Vindhyan of eastern Vindhyan Basin covering several stratigraphic sections/localities, out of which eight are measured stratigraphic sections that cover the uppermost part of the Semri Group (Rohtas Subgroup) and comprising mainly the Rohtasgarh Limestone and the Bhagwar Shale in the Son Valley area of M. P., U. P. and Bihar, and also the Tirohan Dolomite of the Lower Vindhyan of the Chitrakoot area, M. P. Remaining three sections cover two localities of the Chorhat Sandstone and one locality of the Koldaha Shale (Olive Shale) in the Son Valley. Interestingly, seven sections of Rohtasgarh Limestone and Bhagwar Shale, including those two original SSF yielding sections of Azmi (1998a), are consistent with in yielding earliest Cambrian SSFs (Joshi, 2004). But the phosphatic Tirohan Dolomite of the Chitrakoot area on the northern margin of the Vindhyan Basin is of special significance as it has yielded prolific PЄ-Є boundary SSFs along with rich content of the phosphatized metazoan embryos and earliest Cambrian calcareous skeletal algae. Besides, a number of carbonaceous metaphyte compressions have also been recorded from the shale layers in the lower part of the Rohtasgarh Limestone from the Badanpur Quarry Section, Maihar (M.P.) and also from upper part of the Rohtasgarh Limestone of the Ramdihra Quarry Section, Rohtas (Bihar). In addition, acanthomorphic acritarchs and hexactine sponge spicules have been located in the petrological thin sections of the Lower Nodular Limestone in the Patawadh Quarry Section (Sonbhadra, U.P.) in Son Valley, which belongs to the basal part of the Rohtas Subgroup. One spot sample of the Olive Shale collected during the *"International Field Workshop on the Vindhyan Basin, Central India"* (3-11 December 2002) from the Lift Canal Section near Ghurma, Son Valley, has yielded in thin sections a rich scolecodont bearing lamina. Besides, the horizontal metazoan burrow trace fossils have also been recorded from the Chorhat Sandstone at two locations that also includes the Seilacher *et al*'s (1998) locality, which had exhibited triploblastic animal trace fossils.

Fossil yielding localities and the measured stratigraphic sections indicating positions of fossiliferous samples are shown in Figs. 1 and 3. About different types of fossil assemblages with their chronostratigraphic significance in the context of age of the Lower Vindhyan are given below. Distribution of various fossil groups recorded in different sections of eastern Vindhyan Basin is shown in Table 1. However, only selected fossils are illustrated in Pl. 1-5, which are catalogued in the Repository of Wadia Institute of Himalayan Geology, Dehra Dun bearing WIMF/A numbers.

Small shelly fossils (Pl. 1, 2 and 4, Figs. 26-27)

A very rich assemblage of typical PЄ-Є boundary SSFs has been recorded from at least seven out of eight measured stratigraphic sections in eastern Vindhyan Basin, exposing excellent sections of the Rohtasgarh Limestone, Bhagwar Shale and the Tirohan Dolomite which represent the uppermost Semri Group of the Lower Vindhyan (Figs. 2 and 3), which have yielded SSFs through acid dissolution technique. These are: 1) Jankikund Section in Chitrakoot, M. P., 2) Maihar-Badanpur Road Section near Kaimur Sandstone Hump, Maihar, M. P., 3) Badanpur Quarry Section, Maihar Cement Factory, M. P., 4) Barwadih Son Bank Section and 5) Silpi Nala Section on the left bank of Son river in Chopan area, U. P., 6) Ramdihra Main Quarry Section and 7) North Ramdihra Quarry Section, Rohtas, Bihar.

Although sporadic occurrences of SSFs are recorded from several Son Valley sections, exceptionally high yield has been registered in the Tirohan Dolomite of the Jankikund Section of Chitrakoot, on the northern margin of the Vindhyan Basin, where they are mostly phosphatized owing to the phosphate stringers in the dolomite beds. Abundant preservation of phosphatized animal embryos is the outstanding feature of SSF assemblage of the Tirohan Dolomite (Pl. 2, Figs. 11-14).

Table 1. Distribution of various fossils in the studied sections of eastern Vindhyan Basin (after Joshi, 2004).

Lower Vindhyan sections / Lower Vindhyan fossils groups	Jankikund Section	Maihar-Badanpur Road Section	Badanpur Quarry Section	Silpi Nala Section	Barwadih Son Bank Section	Patawadh Quarry Section	Ramdihra main Quarry Section	Northern Ramdihra Quarry Section	Chorhat Section	Salkhan Fossil Park	Olive Shale-Lift Canal Section
Small Shelly Fossils											
Actinotheca mira (He)	●										
Anabarites trisulcatus Missarzhevsky	●										
Anabarites rotundus Qian	●										
Bathysiphon sp.	●										
Cambrotubulus decurvatus Missarzhevsky	●	●	●								
Coleoloides trigeminatus Missarzhevsky	●	●									
Coleoloides typicalis Walcott	●	●	●								
Halkieria sp.	●						●				
Hyolithellus vladimirovae Missarzhevsky	●										
Mongolodus platybasalis (Yang & He)	●										
Mongolodus rostiformis Missarzhevsky	●										
Platysolenites antiquissimus Eichwald	●	●					●				
Protohertzina anabarica Missarzhevsky	●										
Protohertzina siciformis Missarzhevsky	●										
Rugatotheca sp.	●										
Sabellidites sp.	●										
Siphogonuchites sp.	●										
Spirellus columnaris Jiang	●	●									
Spirellus shankari (Singh & Shukla)	●	●	●								
Spirellus tarsus Jiang	●										
Vindhyanitubulus semriensis n.gen.et n.sp.					●						
Animal Embryos											
Archaeooides granulatus Qian		●	●								
Megasphaera inornata Chen & Liu	●	●	●								
Megasphaera ornata Shuhai & Knoll	●		●								
Megaclonophycus onustus Xue et al.	●										
Olivooides multisulcatus Qian	●							●			
Parapandorina raphospissa Xue et al.		●									
Calcareous Skeletal Algae											
Endoconchia angusta Runnegar		●									
Girvanella problematica Nichlson & Etheridge	●										
Korilophyton inopinatum Voronova	●										
Obruchevella meishucunensis Song	●										
Protoortonella flabellata (Voronova)	●										
Renalcis polymorphus (Maslov)	●										
Small Acanthomorphic Acritarchs											
Bavlinella sp.						●					
Micrhystridium minutum Downie						●					
Micrhystridium dissimilare Volkova						●					
Micrhystridium tornatum Volkova						●					
Micrhystridium cf. *Parvum* Volkova						●					
Micrhystridium obscurum Volkova						●					
Micrhystridium sp.						●					
Veryhachium sp.						●					
Sponge Spicules											
Hexactines				●		●					
Tetracts						●					
Diaxon triact						●					
Metaphytes											
Flabellophyton strigata Yuan Xunlai et al.			●								
Flabellophyton sp.			●								
Konglingiphyton erecta Chen & Xiao			●								
Orbisiana sp.							●				
cf. *Lanceoforma striata* Walter et al.			●								
			●								
Trace Fossils											
Horizontal megascopic burrows									●	●	
Scolecodonts											●

The numerous SSF taxa recorded from these sections are: *Actinotheca* sp., *Anabarites trisulcatus* Missarzhevsky, *A. rotundus* Qian, *Cambrotubulus decurvatus* Missarzhevsky, *Hyolithellus vladimirovae* Missarzhevsky, *Hyolithellus* sp., *Coleoloides trigeminatus* Missarzhevsky, *C. typicalis* Walcott, cf. *Microlites shotanensis* He and Yang, *Rugatotheca* sp., *Sabellidites* sp., *Protohertzina anabarica* Missarzhevsky, *P. siciformis* Missarzhevsky, *Mongolodus platybasalis* (Yang and He), *M. rostriformis* Missarzhevsky, *Halkieria* sp., *Siphogonuchites* sp., *Bathysiphon* sp., *Platysolenites antiquissimus* Eichwald, *Spirellus columnaris* Jiang, *S. shankari* (Singh & Shukla), *S. tarsus* Jiang. Animal embryos: *Megasphaera inornata* Chen and Liu, *M. ornata* Shuhai and Knoll, *Olivooides multisulcatus* Qian, *Parapandorina raphospissa* Xue *et al.*, *Megaclonophycus onustus* Xue *et al.*, *Archaeooides granulatus* Qian.

These SSFs are typical of the basal Cambrian *Anabarites-Hyolithellus-Protohertzina* Assemblage Zone of Nemakit-Daldynian and Meishucunian Stages (Xing *et al.*, 1984). Most of the elements of this SSF assemblage are well known to occur in the pre-trilobite Lower Cambrian strata of Siberia, Mongolia, Kazakhstan, China, Indian Lesser Himalaya, Mackenzie Mountains (Canada), Nevada-California (USA) and in GSSP for P€-€ Boundary, Newfoundland (Matthews and Missarzhevsky, 1975; Azmi *et al.*, 1981; Azmi, 1983, 1996; Brasier and Singh, 1987; Cowie and Brasier, 1989; Qian and Bengston, 1989).

Calcareous skeletal algae (Pl. 3)

Abundant phosphatized calcareous algae have been obtained, for the first time, both in the macerated material as well as in the petrological thin sections of the Tirohan Dolomite of the Chitrakoot area of M. P., along with diverse earliest Cambrian SSFs. Among the calcareous algae, *Girvanella problematica* is the most abundant taxon followed by *Renalcis polymorphus*. These are also the oldest rock-forming calcareous algae. There are other rare calcareous algae, namely *Korilophyton inopinatum*, *Protoortonella flabellata* and *Obruchevella meishucunensis* in the assemblage. *Endoconchia angusta* has also been recorded encrusting on *Parapandorina raphospissa* metazoan embryo from the Bhagwar Shale of Maihar-Badanpur Road Section.

Plate 1.

All illustrated specimens are from Tirohan Dolomite, Jankikund Section, Chitrakoot, except of Fig. 10, which is from Rohtasgarh Limestone, Ramdihra Main Quarry Section, Rohtasgarh. Scale bars represent 100μm unless otherwise mentioned.

1a. *Cambrotubulus decarvatus* Missarzhevsky (JNK-4, WIMF/A 261), scale bar = 50μm. 1b, close view of aperture, scale bar= 10μm. **2.** *Hyolithellus vladimirovae* Missarzhevsky (JNK 3, WIMF/A 262). **3a.** *Hyolithellus vladimirovae* Missarzhevsky (JNK 2, WIMF/A 263). **3b**, closeup view of circular cross section. **3c**, oblique front view of lid (operculum). **4a.** *Hyolithellus vladimirovae* Missarzhevsky (JNK 2, WIMF/A 264). **4b**, showing circular aperture. **5.** *Hyolithellus vladimirovae* Missarzhevsky (JNK 4, WIMF/A 265). **6a.** *Anabarites trisulcatus* Missarzhevsky (JNK 4, WIMF/A 270). **6b**, oblique view showing proximal end. **7a.** *Anabarites trisulcatus* Missarzhevsky (JNK 4, WIMF/A 272). **7b**, showing triradial symmetry at proximal end. **8a.** *Protohertzina siciformis* Missarzhevsky (JNK 2, WIMF/A 293). **8b**, showing inner cavity in apical part. **9a.** *Mongolodus rostriformis* Missarzhevsky (JNK 2, WIMF/A 294). **9b**, subrounded cross section at apical part. **10.** *Halkieria* sp., showing cross section (RQ 10, WIMF/A 296). **11.** *Halkieria* sp., showing flaring base (JNK 3, WIMF/A 295). **12a.** *Protohertzina anabarica* Missarzhevsky (JNK 1, WIMF/A 292). **12b**, showing inner cavity. **13a.** *Mongolodus platybasalis* (Yang and He), (JNK 4, WIMF/A 291). **13b**, top view showing concavity in inner lateral side. **13c**, outer lateral view. **13d**, oblique top view showing the convex lateral side.

Plate 1.

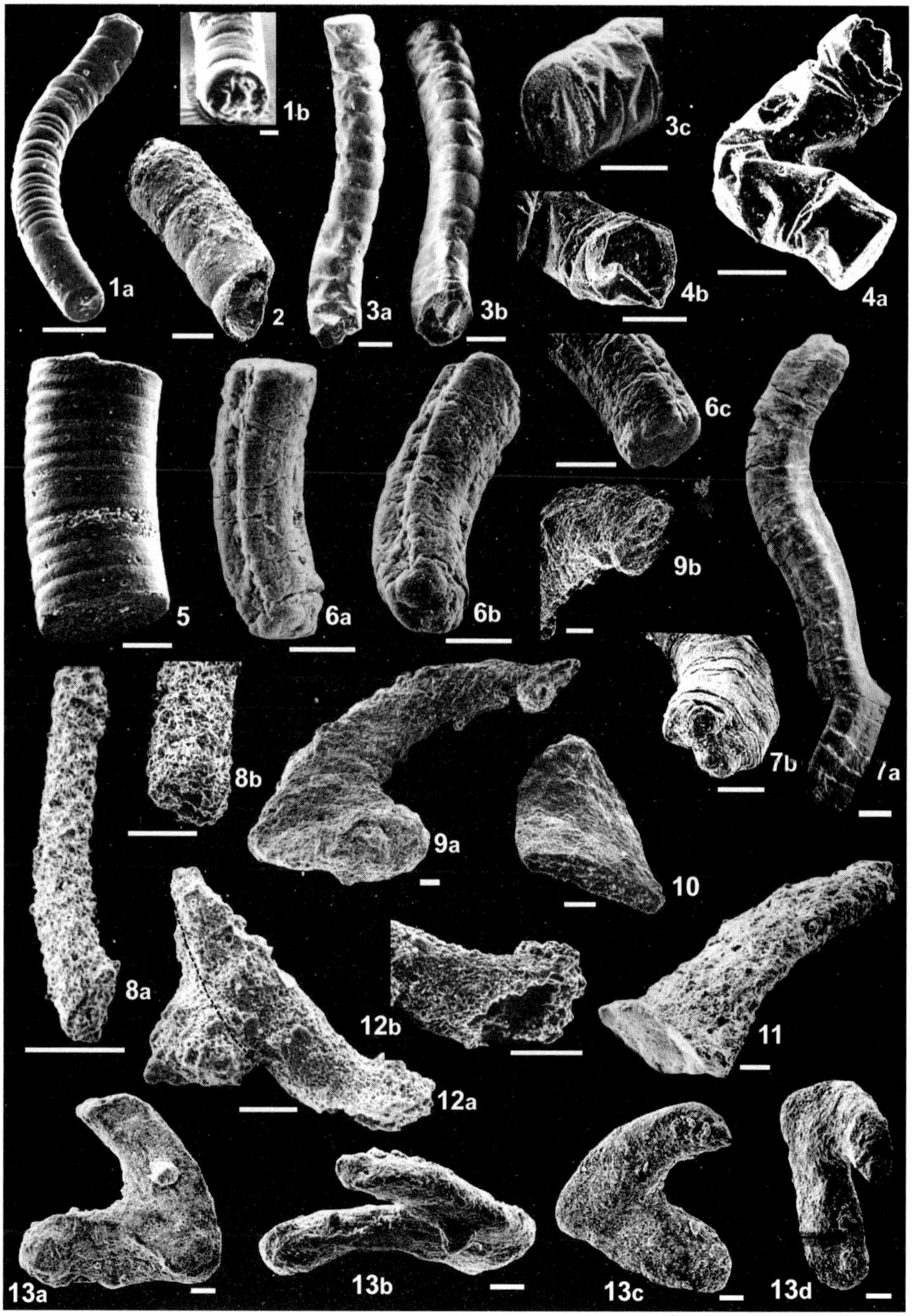
1a
1b
2
3a
3b
3c
4a
4b
5
6a
6b
6c
7a
7b
8a
8b
9a
9b
10
11
12a
12b
13a
13b
13c
13d

Girvanella and *Renalcis* have globally appeared in the latest Vendian and earliest Cambrian respectively, corresponding to the global skeletalization event near the PЄ-Є boundary (Riding and Voronova, 1984). *Girvanella* has maximum recorded stratigraphic range from the latest Vendian to Cretaceous, whereas *Renalcis* ranges from the earliest Cambrian to Devonian (Mankiewicz, 1992a, b). The other two genera *Korilophyton* and *Protoortonella* are restricted to Lower Cambrian only (Riding and Voronova, 1984; Mankiewicz, 1992a, b). The type species *Girvanella problematica,* was described from the Mid-Ordovician of Scotland (Nicholson and Etheridge, 1878), but its initial appearance in the basal part of the Lower Cambrian (Nemakit-Daldynian) of the Siberian Platform (Voronova and Missarzhevsky, 1969; Raaben, 1981) is well established. The species is also known from the Lower Cambrian of Newfoundland (Edhorn, 1979), Antarctica (Wrona, 2004) and the Mid-Ordovician of Tarim Basin, China (Riding and Fan, 2001). *Renalcis polymorphus* also quite widely occurs in the Lower Cambrian strata, such as in Siberia (Luchinina, 1989), Mongolia (Kruse *et al.*, 1996), and in the Upper Krol of the Garhwal-Kumaun Lesser Himalaya (Shukla, 1984; Kalia, 1989). The stratigraphic occurrence of *Korilophyton inopinatum* and *Protoortonella flabellata* is more restricted to the basal part of the Lower Cambrian horizons (mostly the Nemakit-Daldynian and basal Tommotian Stages) of the Siberian Platform (Voronova and Missarzhevsky, 1969; Voronova and Luchinina, 1990; Mankiewicz, 1992a, b); the former species also occurs at this level in Inner Mongolia (Kruse *et al.*, 1996).

Plate 2.

Scale bars represent 100μm unless otherwise mentioned.

1a. *Vindhyanitubulus semriensis* n. gen. et n. sp. (holotype), (BD 1, WIMF/A 288). **1b**, closeup view showing closely packed elongate hexagons, Rohatsgarh Limestone, Barwadih Section, Son Valley **2a.** *Vindhyanitubulus semriensis* n. gen. et n. sp. (paratype), (BD 1, WIMF/A 289). **2b**, top view showing circular cross section with inner and outer wall. 2c, details of elongate hexagons on the outer, Rohatsgarh Limestone, Barwadih Section, Son Valley **3.** *Vindhyanitubulus semriensis* n. gen. et n. sp. (paratype), (BD 1, WIMF/A 290), Rohatsgarh Limestone, Barwadih Section, Son Valley **4.** *Rugatotheca* sp., broken specimen showing flexible wall and hollow interior, (JNK 4, WIMF/A 277), Tirohan Dolomite, Jankikund Section, Chitrakoot. **5.** *Platysolenites antiquissimus* Eichwald (SJ 1, WIMF/A 304), Bhagwar Shale, Maihar-Badanpur Road Section, Maihar. **6.** *Cambrotubulus decarvatus* Missarzhevsky (SJ 6, WIMF/A 285), Bhagwar Shale, Maihar-Badanpur Road Section, Maihar. **7.** *Bathysiphon* sp. (JNK 4, WIMF/A 305), Tirohan Dolomite, Jankikund Section, Chitrakoot. **8.** *Bathysiphon* sp. (SJ 8, WIMF/A 306), Bhagwan Shale, Maihar-Badanpur Road Section, Maihar. **9.** *Platysolenites antiquissimus* Eichwald (JNK 4, WIMF/A 301), Tirohan Dolomite, Jankikund Section, Chitrakoot. **10.** *Platysolenites antiquissimus* Eichwald (JNK 2, WIMF/A 302), Tirohan Dolomite, Jankikund Section, Chitrakoot. **11.** *Megasphaera inornata* Chen & Liu (JNK 4, WIMF/A 319), Tirohan Dolomite, Jankikund Section, Chitrakoot. **12a, b.** *Olivooides multisulcatus* Qian (JNK 4, WIMF/A 333). a and b represent part and counter part of the specimen, Tirohan Dolomite, Jankikund Section, Chitrakoot. **13.** *Archaeooides granulatus* Qian (SJ 7, WIMF/A 325), Bhagwar Shale, Maihar-Badanpur Road Section, Maihar. **14a, b.** *Olivooides multisulcatus* Qian (NQS 1, WIMF/A 332), Rohatsgarh Limestone, Northern Ramdihra Qaurry Section, Rohtasgarh. **15.** *Spirellus shankari* Singh & Shukla (SJ 11, WIMF/A 348), scale bar=50μm, Bhagwar Shale, Maihar-Badanpur Road Section, Maihar. **16.** *Spirellus columnaris* Jiang (SJ 6, WIMF/A 343), Bhagwar Shale, Maihar-Badanpur Road Section, Maihar. **17.** *Spirellus shankari* (Singh & Shukla) (SJ 6, WIMF/A 346), Bhagwar Shale, Maihar-Badanpur Road Section, Maihar. **18.** *Spirellus shankari* (Singh & Shukla) (SJ 20, WIMF/A 349), Bhagwar Shale, Maihar-Badanpur Road Section, Maihar. **19.** *Spirellus shankari* (Singh & Shukla) (RJ 33, WIMF/A 352), Bhagwar Shale, Badanpur Quarry Section, Maihar. **20a, b.** *Spirellus tarsus* Jiang (JNK 4, WIMF/A 354), scale bar=200μm, Tirohan Dolomite, Jankikund Section, Chitrakoot.

Plate 2.

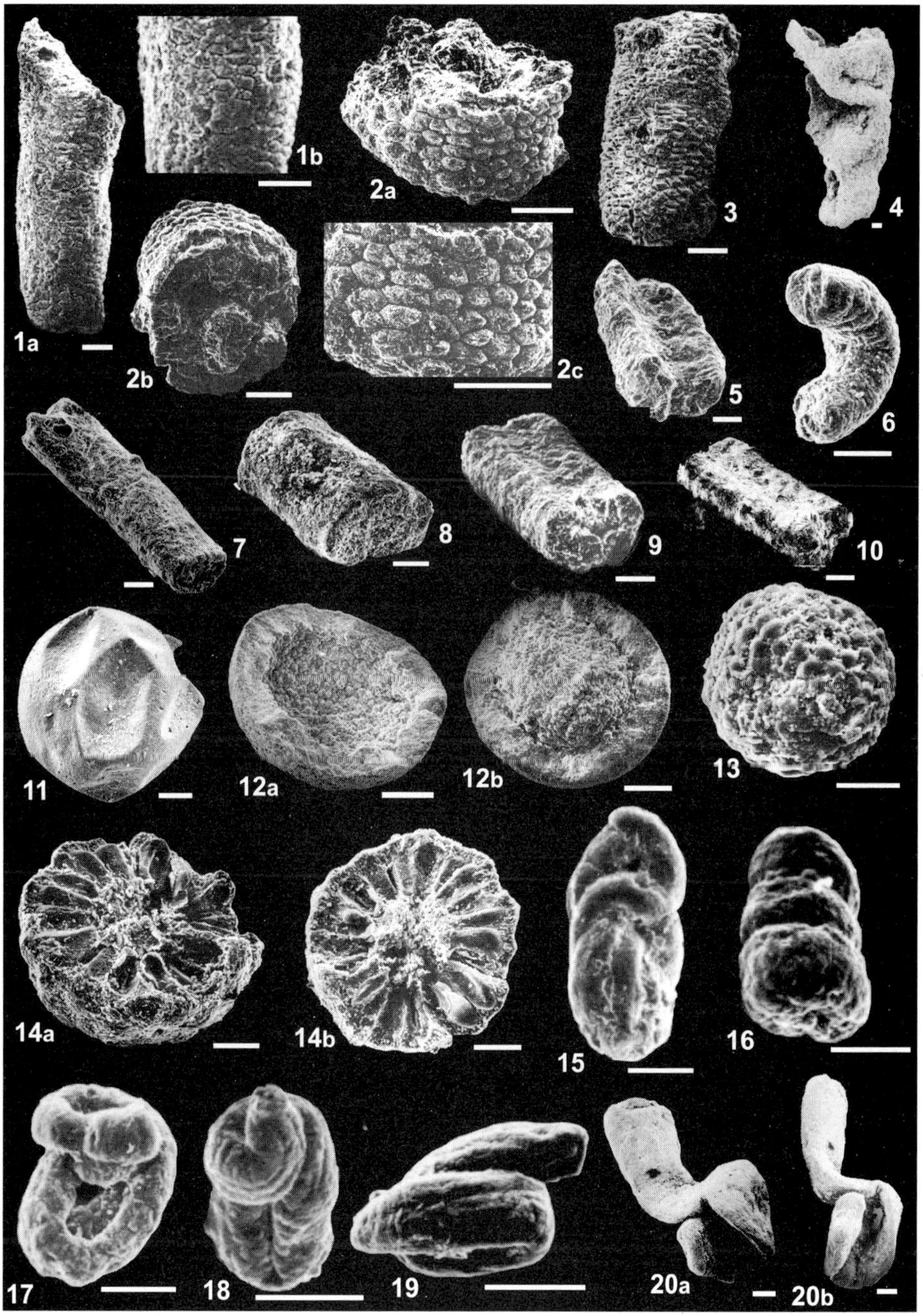
1a
1b
2a
2b
2c
3
4
5
6
7
8
9
10
11
12a
12b
13
14a
14b
15
16
17
18
19
20a
20b

Sponge spicules (Pl. 4, Figs. 9-17)

Among the sponge spicules only hexactines are represented in the present material. These occur in the Lower Calcareous Nodule as well as the Upper Calcareous Nodule horizons of the Rohtasgarh Limestone. The only difference is that the spicules in the Lower Calcareous Nodules (Pl. 4, Figs. 9-14) are much smaller (~10 μm) than in the Upper Calcareous Nodules (Pl. 4, Figs. 15-17). In the Patawadh Quarry Section along the bank of Ghaghar River in Son Valley, numerous small size hexactines are seen in the petrological thin sections of the Lower Calcareous Nodules in association with the small acanthomorphic acritarchs. The hexactine sponge spicules from the Upper Calcareous Nodules of Silpi Nala Section at the bank of Son River have been recovered through dissolution maceration technique. These spicules are comparatively much larger (~ 1mm) and heavily pyritized.

Although monoaxons and hexactinellid type sponge spicules have been seldom recorded from the early Vendian-Ediacarian strata of China, Mongolia, Australia and India, the earliest non-controversial hexactinellids of latest Ediacarian age (ca. 543-549 Ma - very close to the P€-€ boundary) are only known to date from the upper Tsagaan Oloom Formation of the southwestern Mongolia (see Brasier *et al.*, 1997 for detail). The Mongolian specimens are preserved in iron oxide within early diagenetic chert concretions. Our finding of small size hexactine specimens from the Lower Calcareous Nodule horizon of Patawadh Quarry Section is similar to that of Mongolian type preservation as these are also diagenetically replaced by iron oxide. Silpi hexactinellid spicules, however, are diagenetiacally replaced by pyrite. The Early Cambrian age for the Silpi yielding horizon is further constrained because immediately below this in the grey shale horizon a calcareous lenticular band has yielded mongolitubulid tubes, *Vindhyanitubulus semriensis* gen. et sp. nov. (described at the end). Mongolitubulid type sculptured tubes are so far known only from the Early Cambrian strata (see in Bengtson *et al.*, 1990 and Wrona, 2004).

Small acanthomorphic acritarchs (Pl. 4, Figs. 1-8)

Acritarchs are organic-walled microfossils produced mostly by phytoplanktonic protists (Knoll and Walter, 1992). In preservation, acritarchs are usually found as flattened and folded vesicle compressions. They show morphological divergence in their chronological records. The earliest forms recorded from the late Paleoproterozoic (~1800Ma) rocks are large spheroids (Zhang, 1986). Although rare, the earliest acanthomorphic acritarchs (~100 μm) are known from ~1500Ma (Early Mesoproterozoic) (Javaux *et al.*, 2001). Significant increase in their diversity has been noted during the Late Riphean – Early Vendian (pre-Ediacaran) times (Knoll and Walter, 1992). The Early Vendian acanthomorphic acritarchs were distinctly large (Vidal, 1990), which almost disappeared during the Ediacaran faunal radiation, leaving simple leiospherids and small acanthomorphic acritarchs of low diversity in the Late Vendian (Knoll and Walter, 1992). A major radiation of acritarchs marks the beginning of the Cambrian, wherein acanthomorphic acritarchs are further reduced to much smaller size (<25μm, mostly *Micrhystridium*) and are quite distinct from the Neoproterozoic species (Volkova, 1969; Tappan, 1980; Downie, 1982; Knoll and Swett, 1987; Vidal, 1990; Moczydlowska, 1991; Knoll and Walter, 1992; Tiwari and Pant, 2004).

A rich assemblage of small acanthomorphic acritarchs has been recovered in petrological thin sections from the Lower Nodular Limestone unit of the Rohtasgarh Limestone of Patawadh area, exposed along the right bank of the Ghaghar River (Fig. 3). *Micrhystridium* spp. are dominant in the assemblage. *Veryhachium*-like acritarchs are also seldom present (Pl. 4, Fig. 5). The acritarch assemblage comprises *Micrhystridium dissimilare, M. minutum, M. tornatum, M.* cf. *parvum, M. obscurum* (?) and several other specimens of *Micrhystridium* sp.

Micrhystridium (<20μm) rarely occur in the Neoproterozoic rocks and flourish in the Early Cambrian. *M. dissimilare* is characteristic species of the Early Cambrian sequences. (Volkova 1969; Vidal and Knoll 1983; Vidal *et al.*, 1994). *M. minutum, M. tornatum, M.* cf. *parvum, M.* cf. *obscurum* and several other specimens of *Micrhystridium* sp. are also well recorded in the Lower Cambrian of Poland, Ieberia, Siberia and India (Volkova 1969; Downie, 1982; Knoll and Swett, 1987; Tiwari, 1999), which are also present in the Rohtasgarh Lower Nodular Limestone acritarchs assemblage.

Veryhachium is a distinctive Phanerozoic acritarch. The lower stratigraphic range of the genus is from the Cambrian (Sarjent and Stancliffe, 1994). *Veryhachium* is also reported from the Chert-Phosphorite Member of Lower Tal Formation (earliest Cambrian) and the Vaishnodevi Limestone (Jammu Limestne) of the Lesser Himalaya (Tiwari, 1996; Venkatachala and Kumar, 1998). However, some reports of *Veryhachium* from older than the Cambrian are considered speculative (Venkatachala and Kumar, 1998 and references therein).

Scolecodonts (Pl. 4, Figs. 18-22)

Polychaeta is the oldest known class of annelids (Sokolov, 1990) Scolecodonts- the fossilized elements of polychaete jaws- mainly occur in the Early Ordovician and diversify from the Upper Ordovician (Szaniawski, 1996). Although annelid body impressions are known in the Ediacaran (Middle to Upper Vendian) as well as in the Middle Cambrian Burgess Shale faunas, so far there has been only a single record of scolecodont, *Redkinia spinosa* Sokolov, from the Middle Vendian Redkino Series of Moscow Syncline (Sokolov, 1990). However, the affinity of these structures has been doubted (Burzin, 1995).

In the present material, profuse occurrence of scolecodonts has been observed (Pl. 4, Figs. 18-22) in the thin sections of Olive Shale (=Koldaha Shale) of the Lower Vindhyan of Son Valley (spot samples of Olive Shale collected from Stop 11 of the Olive Shale Lift Canal Section during the *International Field Workshop on the Vindhyan Basin, Central India,* December, 2003). Because this scolecodont find is from stratigraphically lower horizon than the Chorhat trace fossils (Seilacher *et al.*, 1998), and the Rohtasgarh Limestone and Bhagwar Shale SSFs of earliest Cambrian age (Azmi, 1998a, 1999b; Joshi, 2004 and the present work), the Olive Shale scolecodont yielding horizon would comfortably fall in Vendian age-range. The presence of Ediacaran fossil *Spriggina* (?) (Kathal *et al.*, 2000) from still below Palkawan Shale (equivalent to Porcellanite Formation) further strengthens the case of Vendian age for the Olive Shale. Accordingly, the present find of scolecodonts in the Lower Vindhyan would correspond to that of the Middle Vendian scolecodont record from the Moscow Basin.

Plate 3.

All illustrated specimens are from Tirohan Dolomite, Jankikund Section, Chitrakoot. 1-3 are SEM photomicrographs of isolated specimens, rest are in thin sections.

1a. *Girvanella problematica* Nicholson and Etheridge, (JNK 4, WIMF/A 356), scale bar=100μm. **1b**, close view of tubules, scale bar=10μm. **1c**, another view, scale bar=100μm. **2.** *Girvanella problematica* Nicholson & Etheridge (JNK 4, WIMF/A 359), scale bar=100μm. **3a.** *Girvanella problematica* Nicholson & Etheridge (JNK 4, WIMF/A 357), scale bar=1mm. **3b**, detail of nodules, scale bar=100μm. **4a.** *Girvanella problematica* Nicholson & Etheridge (JNK 2, WIMF/A 361), scale bar=200μm. **4b**, reoriented close up of tubules of the rectangle area in 4a, scale bar=15μm. **5.** *Girvanella problematica* Nicholson & Etheridge (JNK 2, WIMF/A 365), scale bar=20μm. **6.** *Renalcis polymorphus* (Maslov), (JNK 2, WIMF/A 366), scale bar=100μm. **7.** *Renalcis polymorphus* (Maslov), (JNK 2, WIMF/A 367), scale bar=100μm. **8.** *Renalcis polymorphus* (Maslov), (JNK 4, WIMF/A 369), scale bar=50μm. **9.** *Korilophyton inopinatum* Voronova (JNK 4, WIMF/A 371), scale bar=100μm. **10.** *Korilophyton inopinatum* Voronova (JNK 4, WIMF/A 373), scale bar=100μm.

Plate 3.

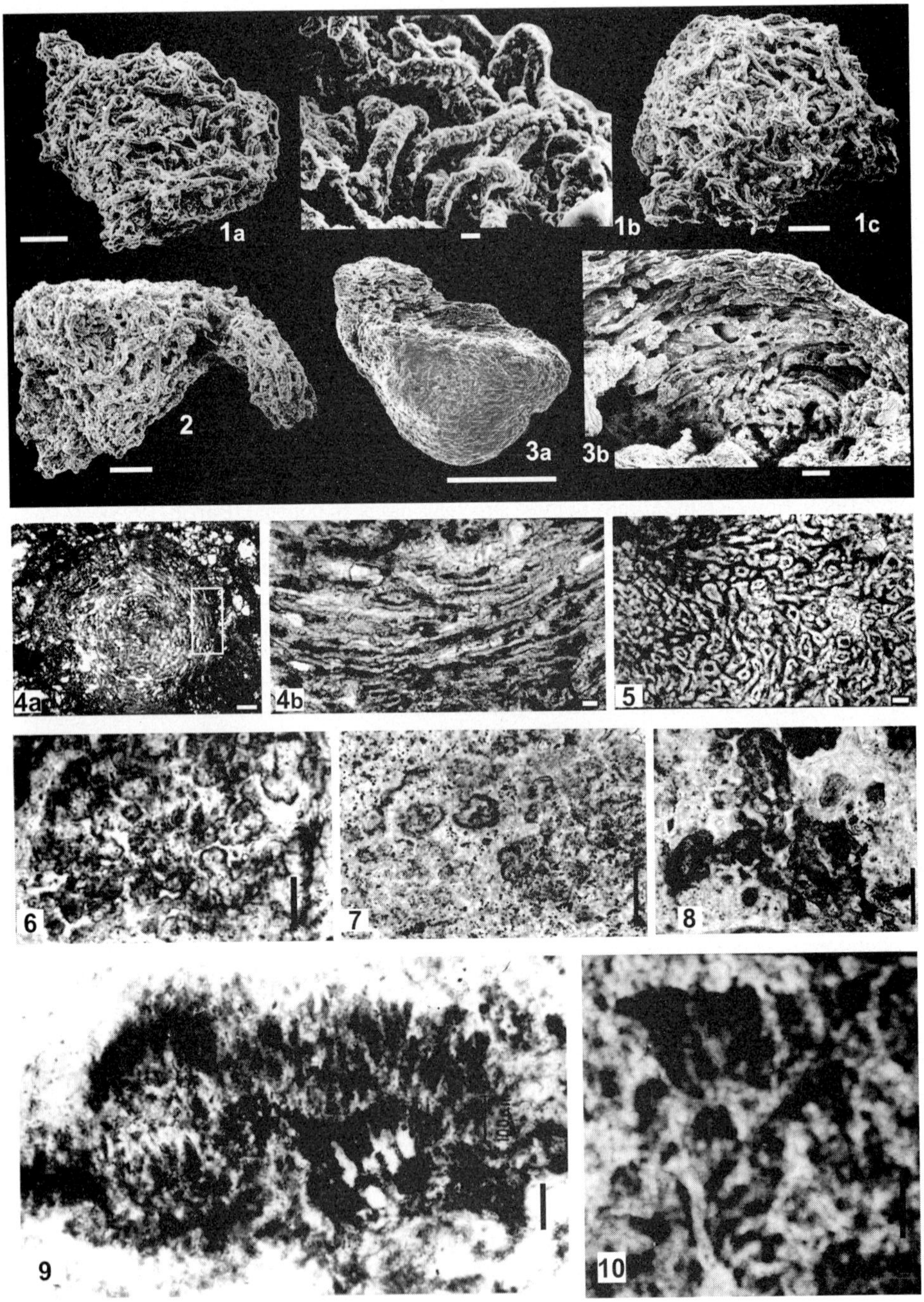

Plate 4.

Illustrated specimens of Figs. 1-14 and 18-22 are in thin sections, 15-17 SEM photomicrographs, 23-27 hand specimens. Scale bar=10μm for Figs. 1-14 and 19-22; 100μm for Figs. 15-18; 1mm for Fig. 23; 5mm for Figs. 24-25; 150μm for Figs. 26-27;

1. *Micrhystridium tornatum* Volkova (GHUR 1, WIMF/A 375), Lower Calcareous Nodule horizon of Rohtasgarh Limestone, Patawadh Quarry Section, Son Valley. **2.** *Micrhystridium* cf. *parvum* Volkova (GHUR 1, WIMF/A 376), Lower Calcareous Nodule horizon of Rohtasgarh Limestone, Patawadh Quarry Section, Son Valley. **3.** *Micrhystridium* cf. *parvum* Volkova (GHUR 1, WIMF/A 377), Lower Calcareous Nodule horizon of Rohtasgarh Limestone, Patawadh Quarry Section, Son Valley **4.** *Micrhystridium* cf. *obscurum* Volkova (GHUR 1, WIMF/A 379), Lower Calcareous Nodule horizon of Rohtasgarh Limestone, Patawadh Quarry Section, Son Valley **5.** *Veryhachium* sp. (GHUR 1, WIMF/A 381), Lower Calcareous Nodule horizon of Rohtasgarh Limestone, Patawadh Quarry Section, Son Valley **6.** *Micrhystridium tornatum* Volkova (GHUR 1, WIMF/A 380), Lower Calcareous Nodule horizon of Rohtasgarh Limestone, Patawadh Quarry Section, Son Valley **7.** *Micrhystridium dissimilare* Volkova (GHUR 1, WIMF/A 378), Lower Calcareous Nodule horizon of Rohtasgarh Limestone, Patawadh Quarry Section, Son Valley **8.** *Micrhystridium* cf. *minutum* Downie (GHUR 1, WIMF/A 386), Lower Calcareous Nodule horizon of Rohtasgarh Limestone, Patawadh Quarry Section, Son Valley **9.** Hexactinellid sponge spicule (GHUR 1, WIMF/A 389), Lower Calcareous Nodule horizon of Rohtasgarh Limestone, Patawadh Quarry Section, Son Valley. **10.** Hexactinellid sponge spicule (GHUR 1, WIMF/A 395), Lower Calcareous Nodule horizon of Rohtasgarh Limestone, Patawadh Quarry Section, Son Valley. **11.** Tetract (GHUR 1, WIMF/A 387), Lower Calcareous Nodule horizon of Rohtasgarh Limestone, Patawadh Quarry Section, Son Valley. **12.** Hexactinellid sponge spicule (GHUR 1, WIMF/A 397), Lower Calcareous Nodule horizon of Rohtasgarh Limestone, Patawadh Quarry Section, Son Valley. **13.** Hexactinellid sponge spicule (GHUR 1, WIMF/A 390), Lower Calcareous Nodule horizon of Rohtasgarh Limestone, Patawadh Quarry Section, Son Valley. **14.** Hexactinellid sponge spicule (GHUR 1, WIMF/A 396), Lower Calcareous Nodule horizon of Rohtasgarh Limestone, Patawadh Quarry Section, Son Valley. **15.** Hexactinellid sponge spicule (SILPI 1, WIMF/A 400), Upper Calcareous Nodule horizon of Rohtasgarh Limestone, Silpi Nala Section, Son Valley. **16.** Hexactinellid sponge spicule (SILPI 1, WIMF/A 401), Upper Calcareous Nodule horizon of Rohtasgarh Limestone, Silpi Nala Section, Son Valley. **17a, b.** Hexactinellid sponge spicule (SILPI 1, WIMF/A 399), Upper Calcareous Nodule horizon of Rohtasgarh Limestone, Silpi Nala Section, Son Valley. **18.** Concentration of scolecodonts and other allied forms in a thin section of Koldaha (Olive) Shale in plain polarized light, Olive Shale-Lift Canal Section, Son Valley (see enlargement in Figs. 19-22). **19.** Cluster of scolecodonts (Olive 1, WIMF/A 402), Koldaha (Olive) Shale (Olive 1), Olive Shale-Lift Canal Section, Son Valley. **20.** cf. *Redkinia spinosa* Sokolov, (Olive 1, WIMF/A 414), Koldaha (Olive) Shale, Olive Shale-Lift Canal Section, Son Valley. **21.** Scolecodont, at different focal depths (Olive 1, WIMF/A 417), Koldaha (Olive) Shale, Olive Shale-Lift Canal Section, Son Valley. **22.** Scolecodont, at different focal depths (Olive 1, WIMF/A 420), Koldaha (Olive) Shale, Olive Shale-Lift Canal Section, Son Valley. **23.** *Orbisiana* Sokolov (metaphyte ribbon), (RMD-F1, WIMF/A 433), from float of the upper part of the Rohtasgarh Limestone, Ramdihra Main Quarry Section, Rohtasgarh. **24.** *Konglingiphyton* erecta Chen and Xiao (RJQ-1, WIMF/A 421), Rohtasgarh Limestone, Badanpur Quarry Section, Maihar. **25.** *Flabellophyton strigata* Yuan Xunlai *et al.*, (RJQ 1, WIMF/A422), Rohtasgarh Limestone, Badanpur Quarry Section, Maihar. **26.** Embedded tube (*Hyolithellus* sp.) and embryo (Olivooides) (JNK 4, WIMF/A 430), Tirohan Dolomite, Jankikund Section, Chitrakoot. **27.** Embedded embryo (*Olivooides*) (JNK 4, WIMF/A 429), Tirohan Dolomite, Jankikund Section, Chitrakoot.

Plate 4.

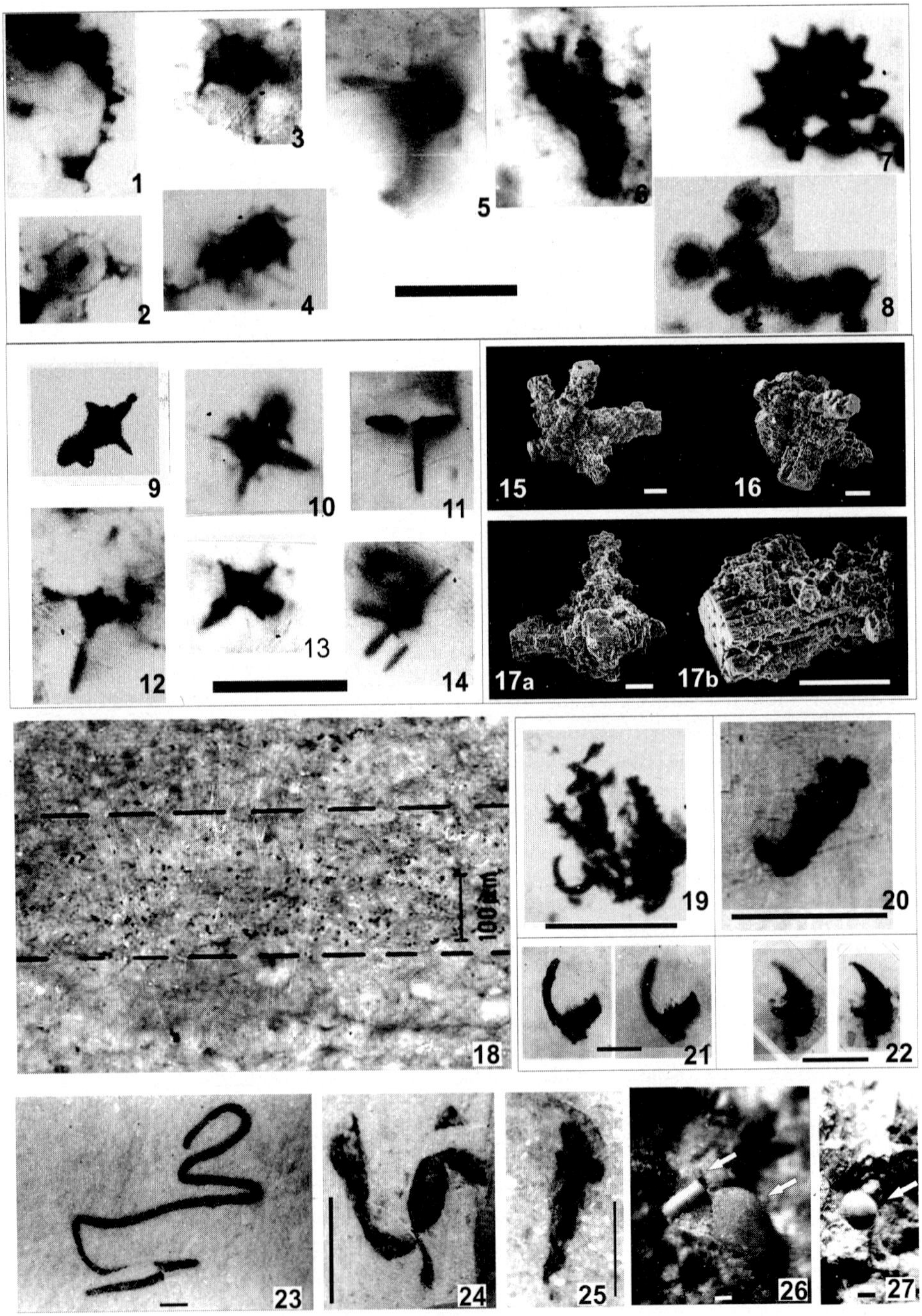
1
2
3
4
5
6
7
8
9
10
11
12
13
14
15
16
17a
17b
18
100 μm
19
20
21
22
23
24
25
26
27

Megascopic carbonaceous metaphyte compressions (Pl. 4, Figs. 23-25)

Megascopic carbonaceous compressions of algal metaphytes (multicellular eukaryotic algae) are well known for their profuse occurrence in the Rohtas Subgroup of the uppermost Lower Vindhyan sediments. Among these, circular and oblong compressions are common, which have recently been identified as *Chuaria circularis* and *Tawuia dalensis,* respectively by several workers (Maithy 1968; Rai *et al.*, 1997; Kumar and Srivastava, 1997, 2003; Kumar, 2001; Srivastava, 2002), though great uncertainties existed since these were first recorded in the Suket Shale by H. C. Jones in 1909, and were later described by Chapman (1935) as inarticulate primitive brachiopods, *Fermoria* and *Protobolella.* Recently, *Chuaria* and *Tawuia* have also been recorded from the Upper Vindhyans as well (Rai *et al.*, 1997; Kumar and Srivastava 1997, 2003). The other distinctly spiriliform ribbon compressions in the Rohtas metaphyte assemblage recorded earlier as *Katnia singhi* by Tandon and Kumar (1977) and recently by Kumar (1995) are the *Grypania spiralis* (Walcott). Besides, vendotaenid ribbon *Tyrasotaenia* sp. has also been identified in Suket Shale by Gnilovskaya – the author of the genus (*In* Shukla and Sharma, 1990). To this assemblage of Rohtas metaphytes here we add some well preserved compressions of *Flabellophyton strigata, Konglingiphyton erecta,* cf. *Lanceoforma striata* and *Orbisiana* sp., which, except the last one, occur mostly in the light grey thin shale intercalations of the Rohtasgarh Limestone, about ~35m below the SSF bearing levels of Bhagwar Shale in the Badanpur Quarry Section of the Maihar area. A single *Orbisiana* ribbon (Pl. 4, Fig. 23) is well preserved on a small slab of light grey siliceous shale that was collected from the float of the Ramdihra Main Quarry Section of eastern Son Valley in Rohtas area (Bihar). This specimen (earlier appeared as carbonaceous algal ribbon fossil in *Science*, 25 August 2000, p. 1273) was examined by Prof. Hans Hofmann (*pers. comm.*, December 2002, *International Field Workshop on the Vindhyan Basin, Central India*) who identified it to be *Orbisiana,* which is known from the Redkino and Valdai Series of Middle Vendian and Upper Vendian to Lower Cambrian of East European Platform, respectively. The Badanpur Quarry metaphyte assemblage comprising *F. strigata*, *K. erecta* and cf. *L. striata* is comparable to that of the upper Sinian beds of the Lantian and Doushantuo Formations of South China, where, excepting *L. striata*, these occur above the Nantuo Tillite and some 150 to 450m below of the PЄ-Є boundary (Xunlai, *et al.*, 1999). In the Rohtasgarh Limestone unit of the Badanpur Quarry, however, these metapyhtes occur close to the PЄ-Є boundary, about 35m below the occurrence of the Lower Cambrian SSFs in the Bhagwar Shale. In a personal communication (email dated 29 August, 2000) from Prof. Xunlai Yuan (Nanjing Institute of Geology and Paleontology, China) to one of the authors (DJ) on the Badanpur Quarry metaphyte assemblage, he opined, "I think that your findings are very important. I guess that the megafossils bearing rock could be around 600Ma in age, just slightly after the Neoproterozoic Marinoan glaciation".

Megascopic burrow trace fossils (Pl. 5)

The locality of Chorhat Sandstone of the Lower Vindhyan from where Seilacher *et al.* (1998) and earlier also Sarkar *et al.* (1996) had described trace fossils was visited in order to scan through the exposed surfaces. One of the samples (Pl. 5) collected from the same unit and the same locality near Chorhat Village (note that the same outcrop is not claimed here) during the Vindhyan Field Workshop (see in Srikantia 1999), organized by the Wadia Institute of Himalayan Geology, Dehra Dun in collaboration with the Geological Survey of India and the *Paleo*ntological Society of India, Lucknow, showed a clear cut evidence of horizontal burrow impressions on the ripple bearing upper surface of the buff-brown siltstone. This sample has been carefully examined by Prof. Franz Fürsich during his visit to Dehra Dun. He is of the opinion that the specimen shows clear evidence of trace fossils. According to him, more or less oval cross sections are seen in structures mostly as negative epireliefs, but occasionally as full relief. Burrow-fill (Diameter ~3mm) of the same material as of the surrounding is evident. Sinuous course structures, often crossing each other, seem to be unbranched. They often cut across the ripple crests. He further felt that the burrows were lying in the softer material and that is why the burrow-fill has a sharp boundary, bordered by depressions on both the sides. He finally concludes that 'the burrow is most likely the product of a deposit feeding worm-like organism'.

Biochronological assessment of the VSG

The small shelly fossils, calcareous skeletal algae, sponge spicules, small acanthomorphic acritarchs, scolecodonts, megascopic carbonaceous metaphyte compressions and the megascopic burrow trace fossils leave no doubt that the Rohtasgarh Limestone and the Bhagwar Shale (Rohtas Subgroup) and their equivalent Tirohan Dolomite in Chitrakoot area representing the uppermost part of the Lower Vindhyan succession (Semri Group) belong to the PЄ-Є boundary (544 Ma) age transition interval. Unprecedentedly, the Rohtas Subgroup has been biochronologically so well constrained that it could be used as a datum in the Vindhyan stratigraphy and can be of immense help in ascertaining the age of the VSG. As in Fig. 4, it is evident that the diverse fossils have been recorded from the Rohtas Subgroup or from slightly lower horizons, which we believe as manifestation of the 'Cambrian explosion' event at this level. Our data also suggests that the occurrence of the metazoan traces in the Chorhat Sandstone (Seilacher *et al.*, 1998; present work, Pl. 5) is clearly within the Vendian time-frame as scolecodonts of Redkinian (Middle Vendian) affinity (Pl. 4 Figs. 18-20) occur immediately below in the Koldaha (Olive) Shale horizon. Therefore, the appearance time of megascopic multicellular animals in the Lower Vindhyan succession is as per the global evolutionary trend. Further, the additional fossil occurrences like Ediacaran fossil *Spriggina* (?) from the Palkawan Shale (=Porcellanite Formation) in Chauraia area, M. P. (Kathal *et al.*, 2000), *Cymatiosphaeroides kullingii* from the basal part of the Chitrakoot Formation (Anbarasu, 2001a), *Obruchevella* from the Salkhan Limestone (Rai and Singh, 2004), *Tyrasotaenia, Chuaria* and *Tawuia* from the Rohtas Subgroup (Maithy, 1968; Shukla and Sharma, 1990; Kumar, 2001) also well support for Vendian to earliest Cambrian age for the Lower Vindhyan succession. It is now well established that the *Chuaria* and *Tawuia* also widely occur in the Lower Cambrian (Brasier *et al.*, 1979; Hamdi, 1989; Zang and Walter, 1992; see also in Brasier, 1999).

However, the occurrence of megascopic spiriliform carbonaceous ribbons of *Grypania spiralis* (Walcott) in the Rohtas Subgroup needs elaboration as such forms occur in a wide age-range and only at a few localities of the world. *Grypania* was initially recorded from 1300 Ma old Belt Supergroup of Montana (Walter *et al.*, 1976). In China, this form occurs in ~1400 Ma old Gaoyuzhuang Formation of Jixian (Hofmann, 1992). However, *Grypania* has recently been recorded from 2.1 b. a. old Negaunee Iron-Formation of Michigan (Tsu-Ming and Runnegar (1992). In India, *Grypania* is exclusively known from the Rohtasgarh Limestone of Katni area of eastern Vindhyan Basin (Kumar, 1995). Earlier, *Grypania* from the Vindhyan was described as *Katnia singhi* Tandon and Kumar (1977) as annelid remain (see also in Conway Morris, 1989 and Tsu-Ming and Runnegar, 1992), which bear striking similarity with the Jixian specimen (earlier identified as trace fossil, see in Du Rulin and Tian Lifu, 1985, Pl. 1, Fig. 28). Since we have now biochronologically constrained the Rohtasgarh Limestone to be of PЄ-Є boundary transition age, it is therefore certain that the upper age limit of *Grypania* has to be extended to latest Vendian/earliest Cambrian. However, we observe that the *Grypania* ribbons of Rohtasgarh Limestone are highly variable in shape and there is a remarkable development of transversely densed annulations, which remind their resemblance with annulations on *Sabellidites*. Further, the Vindhyan *Grypania* specimens (Kumar, 1995) have striking similarity to the forms described as *Calyptrina striata* Sokolov, *Paleolina evenkiana* Sokolov and *Sabellidites cambriensis* Yanichevsky, from the Jinshanzhai Formation of very late Vendian age of North Anhui, China (Xing *et al.*, 1985).

Plate 5.

1a. Horizontal burrow trace fossil on rippled surface of Chorhat Sandstone (CH-F1, WIMF/A 432), Chorhat (Seilacher *et al*'s, 1998 locality). Note the burrow fill. **1b.** Close up of the burrow fill.

Similarly, recent record of large acanthomorphic acritarchs *Tappania* from Deonar Formation (Porcellanite) and from Kheinjua Subgroup (Prasad *et al.*, 2003) too needs a closer scrutiny for its significance as Mesoproterozoic age indicator. *Tappania* was originally described as a new genus by Yin Lei-ming (1997) from the Late Mesoproterozoic Ruyang Group (~ 1000 Ma) of Shanxi, China. This genus has subsequently been recorded from Early Mesoproterozoic Roper Group (~ 1500 Ma) of northern Australia (Javaux *et al.*, 2001). In India, the genus has been recorded by Prasad and Asher (2001) from the lowgrade metasediments of Sarda Formation in the lower part of the Bahraich Group (=lowgrade metasediments of Bijawar and Jaunsar Groups of Vindhyan Basin and Lesser Himalaya, respectively), which overlies the Bundelkand Granite-Gneiss Complex in the subsurface succession of the Ganga Basin. In VSG, since *Tappania* occurs in the Lower Vindhyan which unconformably overlies the Bijawar Group rocks, its age has to be younger in any case. As we have already demonstrated that the *Tappania* bearing strata (Porcellanite and Koldaha Shale) are of Middle Vendian age, it is therefore rational to extend the upper age-range of the taxon up to this age.

The paleontological records from the Upper Vindhyans, in comparison to the Lower Vindhyans, are sparse. However, the most commonly referred fossils comprise, as given in Figure 4, vertical and horizontal trace fossils (namely *Planolites* and *Skolithos*, Kulkarni and Borkar, 1996; Maitra and De, 1998), burrowing structures (like *Monocraterion* and *Diplocraterion*; Chakrabarti, 1990), distinct tightly meandering and overlapping circular trails from the Kaimur Sandstone (Shukla and Sharma, 1990); *Chuaria* and *Tawuia* (Rai *et al.* 1997; Kumar and Srivastava, 1997, 2003) and just a single report of sponge spicule-like forms (Kumar, 1999). Although it may be difficult to ascertain a definite age for the Upper Vindhyans based on these fossils alone, it can be certainly inferred a post-Nemakit-Daldynian age, because the Upper Vindhyans conformably overlie the Rohtas Subgroup rocks at several places. Thus at the face value of these fossils, a pre-trilobite Early Cambrian may be the upper age limit for the Upper Vindhyans. And further, if we consider that the age assessment by Salujha *et al.* (1971) based on palynofossils is correct, the upper age limit of the Upper Vindhyans could be even Late Cambrian to Ordovician.

The biochronological assessment presented above thus suggests that age-range for the VSG would now be from the Vendian (Varangerian) glaciation event (~600 Ma) to maximum up to either the top of pre-trilobite stage of Tommotian (early part of Early Cambrian) or younger up to the Ordovician.

It is important to mention here that so far the stromatolites have been extensively used as a major evidence for assigning the Meso-Neoproterozoic age for the Vindhyan succession and the seemingly equivalent carbonate deposits in the Indian Peninsula and the Lesser Himalaya (Valdiya, 1989). But recent micropaleontological discoveries of protoconodonts and microbiota from the Lesser Himalayan succession have convincingly proved that the stromatolite bearing Gangolihat Dolomite (Deoban Formation), which used to be held as of Early – Middle Riphean age and invariably correlated with the Semri Group (Lower Vindhyan), has now turned up as Vendian–Early Cambrian in age (Azmi and Paul, 2004; Tiwari and Pant, 2004). Therefore, due to proven lost of credibility the stromatolites from the Vindhyan succession have not been discussed in the present biochronological assessment. Undeniably, they are of immense help for interpreting the paleoenvironment and intrabasinal correlation.

Discussion and conclusions

Till recently the Vindhyan Supergroup was widely held to be of Mesoproterozoic to Neoproterozoic age (Riphean, ~1400–600 Ma) mainly based on stromatolites and radiometric dates (see review in Venkatachala *et al.*, 1996). However, Azmi's (1998a) discovery of Lower Cambrian small shelly fossils (PЄ-Є boundary markers) from the Rohtasgarh Limestone called for a major upward revision in the age of the VSG. Though this report attracted international attention (in the background of Seilacher *et al*'s, 1998 startling claim of >1. 1 b. a. old triploblastic animal traces), it did not get instant acceptance on the grounds of doubts expressed about the biogenicity of some of the fossils and/or on the excuse of non-reproducibility (Conway Morris *et al.*, 1998; Bhatt, 1999, 2003; Bhatt *et al.*, 1999; Srikantia, 1999; Editor *JGSI*, 2000). Nevertheless, Azmi had stoutly defended his findings in a series of discussions and reports (Azmi, 1998b, 1999a, b, c; *in* Kerr, 1998a, b, 1999; *in* Editor *JGSI*, 2000; *in* Bagla, 2000; see also discussion on Bhatt *et al.*, 1999 by Kathal and Trivedi, 1999). Now in the present work we have presented amassing evidence of rich fossil assemblages (small shelly fossils, calcareous skeletal algae, sponge spicules, small acanthomorphic acritarchs and metaphyte carbonaceous compressions) from the Rohtas Subgroup that lends credibility to the definite presence of the P€ - € boundary (~544 Ma) strata in the Rohtasgarh Limestone unit of the uppermost Lower Vindhyan (Semri Group) as was earlier proposed by Azmi (1998a). Besides, the occurrence of unquestionable megascopic metazoan burrows in the underlying strata of Chorhat Sandstone and scolecodonts of Redkinian affinity (Sokolov, 1990) in the Koldaha (Olive) Shale, and also the evidence of glaciation (*Gangau Tillite*) at the base of the Vindhyan succession correspond well with their common global presence in the Vendian period (as mentioned before, scolecodonts commonly appear in the Late Cambrian/Ordovician but one record in the Middle Vendian of Moscow Syncline is known). Contrary to this biochronological scenario, the most intriguing are the latest high precision radiochronological dates of ~1600 Ma (Fig. 4; barring one date of 617 Ma by Banerjee and Frank, 1999) in the Lower Vindhyan succession (Rasmussen *et al.*, 2000; Ray *et al.*, 2002, 2003; Sarangi *et al.*, 2004), which not only contradict all previous geochronological dates from the Lower Vindhyan, but make it impossible to reconcile with the biochronologically constrained age estimates for the VSG. These dates therefore need due attention as several inferences drawn based on them are inappropriate, which neither go along the established observed facts of the Vindhyan stratigraphy nor with the geology of the region. These we would like to discuss below:

1. Radiomatric dates of Majhgawan Kimberlite Pipe (MKP) were rightly given due importance in analyzing age of the intruded horizons but were found conflicting with the available dates of intruded sedimentary horizons (Fig. 4). For arriving at a plausible account of the age of the VSG, Ray *et al.* (2003, Fig. 2) showed intrusion of diamondiferous Majhgawan Kimberlite Pipe (MKP) only up to the top of the Rohtas Subgroup of Lower Vindhyan, which is contrary to the fact as it intrudes up to the topmost unit (Baghain Sandstone, = Dhandraul Sandstone) of the Kaimur Group of Upper Vindhyan (Fig. 5; see detail in Mathur, 1962; Haldar and Ghosh, 1978; Babu, 1998; Bose *et al.*, 2001; Kumar *et al.*, 2002; Sarangi *et al.*, 2004). This way they circumvented a well known geochronological bizarre (earlier unsatisfactorily explained by Crawford and Compston, 1970, p. 358) where *younger* Kaimurs [910 ± 39 Ma (K/Ar), Vinogradov *et al.*, 1964] are being intruded by *older* MKP [1140 ± 12 Ma (Rb/Sr), Crawford and Compston, 1970]!

Because of our robust biochronological control on Rohtas Subgroup (Rohtasgarh Limestone and Bhagwar Shale), intrusion of MKP has to be post - Nemakit-Daldynian in age, i. e., <530 Ma. Pertinent observations of Haldar and Ghosh (1981) regarding time lag of the order of 500 million years in phlogopite crystallization and intrusion are apparently valid in this context.

2. We do not subscribe to Ray *et al*'s proposition of a "long lasting hiatus" (~700 million years as per numerical ages in their Fig. 2) separating the Lower and Upper Vindhyans since it occurred to them as a sequel to showing of intrusion level of MKP and by taking into account only those seven dates which were supportive to latest Paleoproterozoic age (~1600 – 1631 Ma) for the Lower Vindhyan and Neoproterozoic (~910 – 550 Ma; here Crawford and Compston, 1970 is incorrectly quoted for ~550 Ma based on 'K/Ar' systematics) age for the Upper Vindhyan (compare large number of available radiometric dates, Fig. 4). Arguably, since Lower to Upper Vindhyan succession in

western Vindhyan Basin is transitional (e. g., Chittaurgarh and Rampura sections of southeastern Rajasthan; see in Prasad, 1984, p. 38-39; Kumar, 2003, p. 286), therefore, Ray *et al*'s proposition of a long hiatus throughout the Vindhyan Basin is untenable. Sarangi *et al.*, (2004), however, interpreted Kaimur Group of >1100 Ma taking into account of age of the intruding MKP. These authors thus also do not apparently go by the idea of a long lasting hiatus between the Lower and Upper Vindhyans.

3. In support of the Late Paleoproterozoic to Neoproterozoic age Ray *et al.* (2003) present proxy signature of $^{87}Sr/^{86}Sr$ ratios obtained from the Vindhyan carbonates. But so far available $^{87}Sr/^{86}Sr$ data is as inconsistent as that of the geochronological data. For example, while Ray *et al.* (2003) obtained 0. 70599 as $^{87}Sr/^{86}Sr$ ratio from the Bhander Limestone (Upper Vindhyan) and inferred Mid-Neoproterozoic age, Kumar *et al.* (2002) got almost the same value (0. 7059 ± 6) from the *Lower Vindhyan* carbonates to which they had assigned Mesoproterozoic age. Although Ray *et al.* (2003) have attempted to explain such contradictions due to methodology; it reveals the vulnerability of $^{87}Sr/^{86}Sr$ ratio based age estimates in the present context.

4. Since the Upper Vindhyans are predominantly coarse siliciclastics and the Lower Vindhyans with almost the same thickness comprise much larger intervals of carbonates and fine siliciclastics including porcellanites (compare lithologies in Fig. 4), our biostratigraphic perception suggests that the Lower Vindhyan took short span of Vendian to earliest Cambrian (~60 million years) with *Gangau Tillite* (or the 'Basal Conglomerate' of the Deoland Formation) representing Early Vendian (Varangerian, ~600 Ma) glaciation in Central India (regarding the glaciation see Bose *et al.*, 2001 and references therein) and the Upper Vindhyans were deposited in a still shorter span of Early Cambrian, between <544 to ~530 Ma (preceding the appearance of trilobites). Interestingly, W. D. West (1981) too had estimated about 60 million years duration for the deposition of the entire VSG, which is very close to our estimate of about 70 million years (Vendian to pre-trilobite Early Cambrian). This view point is at variance with that of the Ray *et al*'s (2003) wherein though Lower Vindhyan was deposited within a short duration ('in a few tens of millions of years', ~ 100 million years), the Upper Vindhyan spans for most of the Neoproterozoic (~350 million years). And if we believe in the lower age limit of the Kaimur Group to be >1100Ma as suggested by Sarangi *et al.* (2004), it would imply that the Upper Vindhyan will range still longer, *i.e.*, from late Mesoproterozoic to Neoproterozoic (~1000Ma). However, both the propositions of Ray *et al.* and Sarangi *et al.* are not acceptable.

5. Our age estimates for the VSG forge a relationship with that of the Lesser Himalayan Blaini – Krol – Tal succession (Mussoorie Group) in which the *Blaini Tillite* at the base marks the beginning of the Vendian and Krol – Tal transition is around the P€ / € boundary as per the SSFs (Azmi, 1983, 1996; Brasier and Singh, 1987) and microbiota (Tiwari, 1999); however, the upper limit of Tal is more extended in Early Cambrian since the Lower Cambrian trilobites are already there (see in Tandon *et al.*, 1988 and Jiang *et al.*, 2002).

The Paleoproterozoic age (~1700Ma) proposed by Rasmussen *et al.* (2002), Ray *et al.* (2002 and 2003) and Sarangi *et al.* (2004) for the base of Vindhyan Supergroup does not even fit in the context of regional geology and stratigraphy, because the opening of the Vindhyan Basin is subsequent to the *Delhi Orogenic Cycle* closing at ~1450 Ma (Roy and Jakhar, 2002). Believing in Late Paleoproterozoic age for the Vindhyans means we are heading to yet another enigma in Indian Stratigraphy that *the Vindhyans are older than the Delhi Orogeny!*

In view of the above, we are of considered opinion that the Late Paleoproterozoic to Neoproterozoic age for the VSG as per even most of the latest (1999-2004) radiochronological dates and the Sr isotope systematics is erroneous. On the basis of consistent fossil records which are in conformity with the evolutionary biotic events around the P€ – € boundary successions world over, we emphasize that the age of the VSG would range from the Terminal Proterozoic (Vendian) to minimum pre-trilobite Early Cambrian (Fig. 4). Apparently the triploblastic animal traces (Seilacher *et al.*, 1998) and the Ediacaran fossil *Spriggina* (?) (Kathal *et al.*, 2000) appeared at appropriate levels in the Vindhyan stratigraphy and hence need not be construed as evidence for the deep origin (Paleoproterozoic/Mesoproterozoic) of multicellularity among animals. And it would be more appropriate to consider ~1600 Ma dates as age of the provenance rather than the age of the Lower Vindhyan (Semri Group). We now strongly recommend a thorough review of the Vindhyan geochronological dates in the background of our robust fossil data and the regional geological set up.

Systematic Description of New Taxon

Phylum, Class, Order uncertain

Family **Mongolitubulidae** Rozonov, 1986 (*nomen nudum*)

Vindhyanitubulus Azmi, Joshi, D. and Srivastava, n. gen.

Etymology: After Vindhyan Mountains of central India.

Type and only species: *Vindhyanitubulus semriensis* Azmi, Joshi, D. and Srivastava, n. sp.

Diagnosis: Small, millimeter size, isolated, cylindrical to slightly tapering, calcareous (?) bilamellar tubes with characteristic sculpture of transversely elongate hexagons all over the external surface. Inner wall comparatively much thinner than the outer sculptured wall.

Remarks: *Vindhyanitubulus* can be easily differentiated from the other Lower Cambrian ornamented mongolitubulid tubes such as *Mongolitubulus* Missarzhevsky (1977), *Tommotitubulus* Fedorov (1986) and *Rushtonites* Hinz (1987) by its characteristic transversely elongate hexagonal surface sculpture.

Vindhyanitubulus semriensis Azmi, Joshi, D. and Srivastava, n. sp.

(Pl. 2, Figs. 1-3)

Etymology: After Semri Group (Lower Vindhyan) of central India.

Material: Holotype WIMF/A 288 (Pl. 2, Fig. 1) from BD-1 of Rohtasgarh Limestone; two figured paratypes WIMF/A 289, 290 (Pl. 2, Figs. 2 and 3) and 3 other fragmentary specimens.

Occurrence: Barwadih Son Bank Section (sample BD-1), middle part of Rohtasgarh Limestone of Lower Vindhyan (Semri Group), central India.

Type locality: Barwadih Son Bank Section, near Silpi village, Son Valley.

Type horizon and age: Banded Shale, middle part of Rohtasgarh Limestone, Lower Vindhyan (Semri Group), central India; Nemakit-Daldynian (Early Cambrian).

Diagnosis: As for the genus.

Description: Cylindrical to slightly tapering calcareous (?) tubes of yellowish-brown in colour with characteristic ornamentation of closely packed, transversally elongate, regular to irregular hexagons. Margin profile undulatory . Cross section circular but variable in case of compressed specimens (Pl. 2, Fig. 3). Undulatory nature of ornamented surface indicates weakly mineralized wall. Wall appears to be bilamellar with inner relatively much thinner (~ 5 μm) than the much thicker outer wall (~20 μm). Inner wall surface is apparently granular. Tube diameter in the widest part ranges between 300 to 350 μm.

Acknowledgements

We thank Prof. B. R. Arora, Director, Wadia Institute of Himalayan Geology, Dehra Dun, for providing necessary facilities for this work and permission for its publication. DJ is indebted to the Head, Department of Earth Sciences, IIT Roorkee for extending all necessary facilities during the progress of a part of this work. Discussions and correspondence with Professors Franz Fürsich, Ole Hints, Hans Hofmann, J. Keith Rigby, Hubert Szaniwski and Yuan Xunlai have been very useful. We are grateful to Professor M. S. Srinivasan for his consistent encouragement in this endeavor.

References

ANBARASU, K. 2001a. Acritarchs from Mesoproterozoic Chitrakoot Formation, Semri Group, Chitrakoot area, Central India. Journal Geological Society of India, 57: 179-183.

ANBARASU. K. 2001b. Facies variation and depositional environment of the Mesoproterozoic Vindhyan sediments. Journal Geological Society of India, 58: 341-350.

AUDEN, J. B. 1933. Vindhyan sedimentation in the Son Valley, Mirzapur District. Memoir Geological Survey of India, 62: 141-250.

AZMI, R. J. 1983. Microfauna and age of the Lower Tal Phosphorite of Mussoorie Syncline, Garhwal Lesser Himalaya. Himalayan Geology, 11: 373-409.

AZMI, R. J. 1996. Evidence for soft tissue basal support in earliest Cambrian protoconodonts from the Lesser Himalaya: conodont function and affinity. XV Indian Colloquium on Micropalaeontolgy Stratigraphy, Dehra Dun: 457-463.

AZMI, R. J. 1998a. Discovery of Lower Cambrian small shelly fossils and brachiopods from the Lower Vindhyan of Son Valley, Central India. Journal Geological Society of India., 52: 381-389.

AZMI, R. J.1998b. Fossil discoveries in India. Science, 282: 6

AZMI, R. J. 1999a. Discussion. Discovery of Lower Cambrian small shelly fossils and brachiopods from the Lower Vindhyan of the Son Valley, Central India. Journal Geological Society of India, 53: 125-130.

AZMI, R. J. 1999b. Discussion. Discovery of Lower Cambrian small shelly fossils and brachiopods from the Lower Vindhyan of the Son Valley, Central India. Journal Geological Society of India, 53: 488-500.

AZMI, R. J. 1999c. Discussion. Discovery of Lower Cambrian small shelly fossils and brachiopods from the Lower Vindhyan of the Son Valley, Central India. Journal Geological Society of India, 53: 730.

AZMI, R. J. and PAUL, S. K. 2004. Discovery of Precambrian-Cambrian boundary protoconodonts from the Gangolihat Dolomite of Inner Kumaun Lesser Himalaya: Implication on age and correlation. Current Science, 86: 1653-1660.

AZMI, R. J., JOSHI, M. N. and JUYAL, K. P. 1981. Discovery of the Cambro-Ordovician Conodonts from the Mussoorie Tal Phosphorite: its significance in correlation of the Lesser Himalaya. In: Sinha, A. K., Singh, B. and Singh, M. Eds., Contemporary Geoscientific Researches in Himalayas. Dehra Dun, India, 1: 245-250.

AZMI, R. J., JOSHI, D., TIWARI, B. N., JOSHI, M. N. and SRIVASTAVA, S. S. 2003. Age of the Vindhyan Supergroup of Central India: An exposition of biochronology vs. geochronology. In: Abstract Volume, XIX Indian Colloquium on Micropaleontology and Stratigraphy and Symposium on Recent Development in Indian Ocean Paleoceanography and Paleoclimate, BHU, Varanasi, : 141-143.

BABU, T. M. 1998. Diamonds in India. Geological Society India, 331p.

BAGLA, P. 2000. Team rejects claim of early Indian fossils. Science, 289: 1273.

BALASUBRAMANYAN, M. N. and CHANDY, K. C. 1976. Lead isotope studies of galena from some occurrences in India. Record Geological Survey of India, 107: 141-148.

BANERJEE, D. M. and FRANK, W. 1999. Preliminary $^{40}Ar/^{39}Ar$ dates of porcellanite and detrital mica from the Vindhyan of central India. In: Volume of Abstracts: Workshop on Vindhyan Stratigraphy and Palaeobiology, Lucknow, India, 5.

BANERJEE, D. M. 2000. How old is the Vindhyan Basin? IGC, Rio, Brazil (Abs.)

BANSAL, M., VIJAN, A. R., MAHAR, Y. S., GHOSH, N., KRISHNA, PRASAD, S. and PRABHY, B. N. 1999. Rb-Sr dating of Lower Vindhyan sediments from Son Valley, Madhya Pradesh, India. Indian Mineralogist, 33: 1-9.

BEER, E. J. 1919. Note on a spiral impression on Lower Vindhyan Limestone. Record Geological Survey of India, 50 (2): 139p.

BHATT, D. K.1999. Discussion. Journal Geological Society of India, 54: 434-435.

BHATT, D. K., 2003. On the report of the small shelly fossils and Brachiopoda form the Vindhyan strata.

BHATT, D. K., GOPAL, S., GUPTA, S., SONI, M. K., MOITRA, A. K., DAS, D. P. and DE, C. 1999. Fossil report from the Semri Group, Lower Vindhyan. Journal Geological Society of India, 53: 717-723.

BENGTSON, S. 1992. Proterozoic and earliest Cambrian skeletal metazons. In: Schopf, W. J. and Cornelius, K., Eds. The Proterozoic Biosphere, Cambridge University Press: 397-411.

BENGTSON, S., CONWAY MORRIS, S., COOPER, B. J., JELL, P. A. and RUNNEGAR, B. N. 1990. Early Cambrian fossils from South Australia. Memoir Association Australian Palaeontologists, 9: 364p.

BOSE, P. K., SARKAR, S. CHAKRABARTY, S. and BANERJEE, S. 2001. Overview of the Meso to Neoproterozoic evolution of the Vindhyan basin central India. Seimentary Geology, 141: 395-419.

BRASIER, M. D. 1998. From deep time to late arrivals. Nature, 395: 547-548.

BRASIER, M. D. 1999. "Deep questions about Vindhyan fossils", Discussion on Discovery of small shelly fossils from the Lower Vindhyan by R. J. Azmi. Journal Geological Society of India, 53:727-730.

BRASIER, M. D. and SINGH, P. 1987. Microfossils and Precambrian- Cambrian boundary stratigraphy at Maldeota, Lesser Himalaya. Geological Magazine, 124 (4): 323-345.

BRASIER, M. D., GREEN, O. and SHIELDS, G. 1997. Ediacarian sponge spicule clusters from Southwestern Mongolia and origins of the Cambrian fauna. Geology, 25 (4): 303-306.

BRASIER, M. D., PEREJON, M. A. and SAN JOSE, M. A. 1979. Discovery of an important Precambrian-Cambrian sequences in Spain. Estudios Geologicas, 35: 323-345.

BURZIN, M. B. 1995. Redescription of the enigmatic microfossils Cochleatina from the Upper Vendian of the East European Platform. Paleontological Journal, 29 (2A): 50-80.

CHAKRABARTI, A. 1990. Traces and dubiotraces: examples from the so-called Late Proterozoic siliciclastic rocks of the Vindhyan Supergroup around Maihar, India. Precambrian Research, 47:141-153.

CHAPMAN, F.1935. Primitive fossils, possible atrematous and neotrematous brachiopod, from the Vindhyan of India. Record Geological Survey of India, 69: 109-120.

CONWAY MORRIS, S.1989. Early metazoans. Sci. Prog., Oxford, 73: 81-99.

CONWAY MORRIS, S., JENSON, S. and BUTTERFIELD, N.1998. Fossil discoveries in India: continued. Science, 282: 1265.

COWIE, J. W. and BRASIER, M. D. (Eds.)1989. The Precambrian-Cambrian boundary. Oxford monographs on Geology and Geophysics, 12: 213p.

CRAWFORD, A. R. and COMPSTON, W.1970. The age of the Vindhyan System of Peninsular India. Quaternary Journal Geological Society of London, 125: 351-371.

DOWNIE, C.1982. Lower Cambrian acritarchs from Scotland, Norway, Greenland and Canada. Transactions, Royal Society. Edinburgh, Earth Sciences, 72: 257-285.

DU RULIN and TIAN LIFU1985. Algal macrofossils from the Qingbaikou System in the Yanshan Range of North China. Precambrian Research, 29: 5-14.

EDHORN, A. S.1979. Girvanella in the "button algae" horizon of the Fourteau Formation (Lower Cambrian), Western Newfoundland. Bulletin des Centers de Recherches Exploration-Production Elf-Aquitanine, 3: 557-567.

EDITOR JGSI, 2000. Vindhyan fossil controversy. Journal Geological Society of India, 55: 675- 680.

FEDOROV, A. B. 1986. Novye trubchatye poblematiki iz stratojpa Tommotskogo yarusa. (New problematical tubes form the type Tommotian section). Paleont. Zh. 1986 (2)110-111. (Russian).

GHOSH, A. and BOSE, A. 1950. Microfossils from the Vindhyans. Science and Culture 15: 330-331.

HALDAR, D. and GHOSH, D. B.1978. Tectonics of the Kimberlites around Majhgawan, Madhya Pradesh, India. Miscellaneous Publication Geological Survey of India, 34: 1-13.

HALDAR, D. and GHOSH, D. B.1981. A discussion on the uncertainties in the dating of the Vindhyan and Bijawar rocks. Miscellaneous Publication Geological Survey of India, 50: 223-228.

HAMDI, B.1989. Startigraphy and palaeontology of the late Precambrian to early Cambrian in the Alborz Mountains, northern Iran. Geological Survey of Iran, Report No. 59.

HINZ, I. 1987. The Lower Cambrian microfauna of Comely and Rushton, Shorpshire, England. Paleontographica A 198: 41-100.

HOFMANN, H. J. 1992. Proterozoic carbonaceous films. In: Schopf, W. J. and Cornelius, K., Eds., The Proterozoic Biosphere, Cambridge University Press:349-358.

HOFMANN, H. J., and BENGTSON, S. 1992. Stratigraphic distribution of megafossils. In: Schopf, W. J. and Cornelius, K., Eds., The Proterozoic Biosphere, Cambridge University Press: 501-506.

JACOB, K., JACOB, C. and SRIVASTAVA, R. N. 1953. Spores and tracheids of vascular plants from the Vindhyan System, India: the advent of vascular plants. Nature, 172: 167-168.

JAVAUX, E. J., KNOLL, A. H., and WALTER, M. R. 2001. Morphological and ecological complexity in early eukaryotic ecosystems. Nature, 412: 66-69.

JIANG, G., BLICK, N. C., KAUFMAN, A. J., BANERJEE, D. M. and RAI, V. 2002. Sequence stratigraphy of the Neoproterozoic Infra Krol Formation and Krol Group, Lesser Himalaya, India. Journal of Sedimentary Research, 72 (4): 524-542.

JOSHI, D. 2004. Biostratigraphy and age of the Lower Vindhyan (Semri Group) of the eastern Vindhyan Basin, India. Unpublished Ph. D. Thesis, IIT Roorkee, India: 165p.

JONES, H. C.1909. In general report, Record Geological Survey of India, 38: 66.

KALIA, P. 1989. Cambrian calcareous algae from the Krol Formation, Nainital in Kumaun Himalaya. Himalayan Geology, 13, 109-118.

KATHAL, P. K. and Trivedi, R. K. 1999. Discussion. Fossil report from Semri Group, Lower Vindhyan. Journal Geological Society of India, 54: 434.

KATHAL, P. K., PATEL, D. R. and ALEXANDER, P. O. 2000. An Ediacaran fossil Spriggina (?) from the Semri Group and its implication on the age of the Proterozoic Vindhyan Basin, Central India. N. Jb. Geol. Palaeont. Mh., 6: 321-332.

KERR, R. A.1998a. Fossils challenge age of billion- years-old animals. Science, 282: 601-602.

KERR, R. A. 1998b. Earliest animals old ones more? Science, 282: 1120.

KERR, R. A. 1999. Earliest animals growing younger? Science, 284: 412-413.

KRISHNAN, M. S. and SWAMINATH, J. 1959. The great Vindhyan Basin of northern India. Journal Geologiacl Society of India. 1: 10-30.

KNOLL, A. H., and SWETT, K. 1987. Micropaleontlogy across the Precambrian-Cambrian boundary in Spitsbergan. Journal of Paleontology, 61: 898-926.

KNOLL, A. H. and WALTER, M. R.1992. Latest Proterozoic stratigraphy and earth history. Nature, 356: 673-678.

KULKARNI, K. G. and BORKAR. V. D.1996. Comments on the range of Skolithos from the Kaimur Group. In: Pandey, J., Azmi, R. J., Bhandari, A. Dave, A., Eds., XV Indian Colloquium Micropaleontology Stratigraphy. Dehra Dun: 519-523.

KUMAR, A., KUMARI, P., DAYAL, A. M., MURTHY, D. S. N. and GOPALAN K. 1993. Rb-Sr ages of Proterozoic kimberlites of India: evidence for contemporaneous emplacements. Precambrian Research, 62: 227-237.

KUMAR, A., GOPALAN, K. and RAJAGOPALAN, G. 2001. Age of the Lower Vindhyan sediments, Central India. Current Science, 81: 806-809.

KUMAR B., DAS SHARMA S., SREENIVAS, B., DAYAL, A. M., RAO, M. N., DUBEY, N. and CHAWLA, B. R. 2002. Carbon oxygen and strontium isotope geochemistry of Proterozoic carbonate rocks of the Vindhyan Basin, central India. Precambrian Research, 113: 43-63.

KUMAR, S. 1976. Significance of stromatolites in the correlation of Semri Series (Lower Vindhyans) of Son Valley and Chitrakut area, U. P. Journal Palaeontological Society of India, 19: 24-27.

KUMAR, S. 1995. Megafossils from the Mesoproterozoic Rohtas Formation (the Vindhyan Supergroup), Katni area central India. Precambrian Research, 72: 171-184.

KUMAR, S. 1999. Siliceous sponge spicule-like forms from the Neoproterozoic Bhander Limestone, Maihar area, Madhya Pradesh. Journal Palaeontological Society of India, 44: 141-148.

KUMAR, S. 2001. Mesoprterozoic megafossils Chuaria-Tawuia association may represent part of a multicellular plant, Vindhyan Supergroup, Central India. Precambrian Research, 106: 187-211.

KUMAR, S. 2003. Discussion. Carbon oxygen and strontium geochemistry of Prterozoic carbonate rocks of the Vindhyan Basin, Central India-comment. Precambrian Research, 121: 285-288.

KUMAR, S. and SRIVASTAVA, P. 1997. A note on the carbonaceous megafossils from the Neoproterozoic Bhander Group, Maihar area, Madhya Pradesh. Journal Palaeontological Society of India, 42: 141-146.

KUMAR, S. and SRIVASTAVA, P. 2003. Carbonaceous megafoosils from the Neoproterozoic Bhander Group, Central India. Journal Palaeontological Society of India, 48: 139-154.

KRUSE, P. D., GANDIN, A., DEBRENNE, F. and WOOD R. 1996. Early Cambrian bioconstruction in the Zavkhan Bason of wetern Mongolia. Geological Magazine, 133 (4): 429-444.

LUCHININA, V. A. 1989. Calcareous algae in Vendian and Lower Palaeozoic stromatolites of the Siberian Paltform. Himalayan Geology, 13: 257-267.

MAITHY, P. K. 1968. The occurrence of micro remains from the Vindhyan Formation of India. Palaeobotanist 17: 48-51.

MAITRA, A. K. and DE, C. 1998. Primitive trace fossils from Vindhyans. Geological Survey of India News, 29: 47-48.

MANKIEWICZ, C. 1992a. Proterozoic and Early Cambrian calcareous algae. In: Schopf, W. J. and Cornelius, K., Eds. The Proterozoic Biosphere, Cambridge University Press, .359-367.

MANKIEWICZ, C. 1992b. Proterozoic and Early Cambrian calcareous algae. In: Schopf, W. J. and Cornelius, K., Eds. The Proterozoic Biosphere, Cambridge University Press, 981-998.

MATHUR, S. M. 1962. Geology of the Panna diamond deposits. Record Geological Survey India, 87: 787-818.

MATHUR, S. M. 1983. A reappraisal of trace fossil described by Vredenburg (1908) and Beer (1919) in rocks of the Vindhyan Supergroup. Record Geological Society of India, 113: 111-113.

MATHUR, S. M. 1987. Geochronology and biostratigraphy of the Vidhyan Supergroup. Special Publciation Geological Survey of India, 11: 23-44.

MATTHEWS, S. C. and MISSARZHEVSKY, V. V. 1975. Small shelly fossils of late Precambrian and early Cambrian age: a review of recent work. Journal Geological Society London, 131: 289-304.

McMENAMIN, D. S., KUMAR, S. and AWRAMIK, S. M. 1983. Microbial fossils from the Kheinjua Formation, Middle Proterozoic Semri Group (Lower Vindhyan) Son Valley Area, Central India. Precambrian Research, 21: 247-271.

MISSARZHVSKY, V. V. 1977. Konodonty (?) I fosfatnye problematiki kembriya Mongolii I Sibiri. (Conodonts (?) and phosphatic problematica from the Cambrian of Mongolia and Siberia) In: Bespozovonochnye Paleozoa Mongolii, (eds. L. P. Tatarinov et al.,) 10-19. Russian.

MOCZYDLOWSKA, M. 1991. Acritarch biostratigraphy of the Lower Cambrian and the Precambrian-Cambrian boundary in southern Poland. Fossils and Starta, 29: 1-127.

MOCZYDLOWSKA, M. 2001. Early Cambrian phytoplankton radiation and appearance of metazoans. Palaeoworld, 13: 293-296.

NICHOLSON, H. A. and ETHERIDGE, R. 1878. A monograph of the Silurian fossils of the Girvan District in Ayrshire. W. Blackwood and Sons, Edinburgh,, 1, 135

NOWLAN, S., NARBONNE, GUY M. and FRITZ, W. H. 1985. Small shelly fossils and trace fossils near the Precambrian-Cambrian boundary in the Yukon Territory, Canada. Lethaia, 18: 233-256.

PAUL, D. K.1979. Isotopic composition of strontium in Indian kimberlites. Cosmochemica et Geochemica Acta, 43: 389-394.

PAUL, D. K., REX, D. C. and HARRIS, P. G. 1975. Chemical characteristics and K-Ar ages of Indian kimberlites, Bulletin Geological Society of America, 56: 364-366.

PRASAD, B. 1984. Geology sedimentation and paleogeography of the Vindhyan Supergroup, southeastern Rajasthan. Memoir Geological Survey of India, 116: 1-107.

PRASAD, B. and ASHER, R. 2001. Acritarch biostratigraphy and lithostratigraphic classification of Proterozoic and Lower Paleozoic sediments (Pre-Unconformity sequence) of Ganga Basin. Palaeontographica Indica, No 5 KDMIPE, ONGC Ltd. Dehradun, 151p.

PRASAD, B., UNIYAL, S. N. and ASHER, R. 2003. Meso-Neoproterozoic organic walled microfossils from Vindhyan sediments of Sone Valley, Madhya Pradesh. Abstract Volume, XIX Indian Colloquium on Micropaleontology and Startigraphy, BHU, Varanasi. 157-158.

QIAN YI and BENGTSON, S. 1989. Palaeontology and biostratigraphy of the early Cambrian Meishucunian stage of Yunnan Province, south China. Fossils & Strata No. 24. Oslo : Universitetsforlaget.

RAABEN. M. E., (Ed.). 1981. The Tommotian Stage and the Cambrian lower boundary problem. Amerind Publication, New Delhi, 359p.

RAI, V. and SINGH, V. K. 2004. Discovery of Obruchevella Reitlinger1948 from the Late Palaeoproterozoic Lower Vindhyan sucession and its significance. Journal Palaeontologicla Society of India, 49: 189-196.

RAI, V., SHUKLA, M. and GAUTAM, R. 1997. Discovery of carbonaceous megafossils (Chuaria-Tawuia assemblage) from the Neoproterozoic Vindhyan succession (Rewa Group), Allahabad – Rewa area, India. Current Science, 73 (a): 783-788.

RASMUSSEN, B., BOSE, P. K., SARKAR, S., BANERJEE, S., FLETCHER, I. R. and MCNAUGHTON, N. J. 2002. 1. 6 Ga U-Pb zircon age for the Chorhat Sandstone, Lower Vindhyan, India: possible implications for early evolution of animals. Geology, 30: 103-106.

RAY, J. S., MARTIN, M. W., VEIZER, J. and BOWRING, S. A. 2002. U-Pb zircon dating and Sr isotope systematics of the Vindhyan Supergroup, India. Geology, 30: 131-134.

RAY, J. S., VEIZER, J. and DAVIS, W. J. 2003. C, O, Sr and Pb isotope systematics of carbonate sequences of the Vindhyan Supergroup, India: age, diagenesis, correlations and implications for global events. Precambrian Research, 121: 103-140.

RIDING, R. and FAN, J. 2001. Ordovician calcified algae and cyanobacteria, northern Tarim Basin subsurface, China. Palaeontology, 44: 783-810.

RIDING, R. and VORONOVA, L. 1984. Assemblages of calcareous algae near the Precambrian/Cambrian boundary in Siberia and Mongolia. Geological Magazine, 121: 205-210.

RODE, K. P. 1946. A new find of fossils in Vindhyan rocks of Rohtas Hills in Bihar. Current Science, 15: 247-248.

ROY, A. B. and JAKHAR, S. R. 2002. Geology of Rajasthan (Northwest India) Precambrian to recent. Scientific Publishers (India), Jodhpur.

ROZANOV, A. Y. 1986. Problematica of the early Cambrian. In: Hoffmann, A. and Nitecki, M. H., Eds., Problematic Fossil Taxa. Oxford University Press, New York: 87-96.

SALUJHA, S. K., REHMAN, K. and RAWAT, M. S. 1971. Fossil palynomorphs from the Vindhyans of Rajasthan (India). Review Palaeoboany Palynology, 11: 65-68.

SARANGI, S., GOPALAN, K. and KUMAR, S. 2004. Pb-Pb age of earliest megascopic eukaryotic alga bearing Rohtas Formation, Vindhyan Supergroup, India: Implications for Precambrian atmospheric oxygen evolution. Precambrian Research, 132: 107-121.

SARJEANT, W. A. S. and STANCLIFFE, R. P. W. 1994. The Micrhystridium and Veryhachium complexes (Acritarcha: Acanthomorphitae and Polygonomorphitae): a taxonomic reconsideration. Micropaleontology, 40: 1-77.

SARKAR, S., BANERJEE, S. and BOSE, P. K. 1996. Trace fossils in the Mesoproterozoic Koldaha Shale, Central India and their implications. N. Jb. Geol. Palaeont. Mh., 7: 425-438

SASTRY, M. V. A. and MOITRA, A. K. 1984. Vindhyan stratigraphy – A review. Memoir Geological Survey of India, 116: 109-148.

SEILACHER, A., BOSE, P. K. and PFLÜGER, F. 1998. Triploblastic animals more than 1 billion years ago: trace fossil evidence from India. Science, 282: 80-83.

SITHOLEY, R. V., SRIVASTAVA, R. N. and VERMA, C. P. 1953. Microfossils from the Vindhyans. Proc. Natl. Inst. Sci. India, 19:195-202.

SHUKLA, M. 1984. Microstromatolites from the "Calc-Zone of Pithoragarh", Kumaun Himalaya. Geophytology, 14: 240-241.

SHUKLA, M. and SHARMA, M. 1990. Palaeobiology of Suket Shale, Vindhyan Supergroup age implications. Special Publication Geological Survey of India, 28: 411-434.

SISODIYA, D. S. 1982. Fossil impression of Jelly fish in the Nimbahera Limestone, Semri Group of Vindhyan Supergroup of rock. Current Science, 51 (22): 1070-1071.

SISODIYA, D. S. and JAIN, V. K. 1984. A note on trace fossil in Kaimur Group of rocks Mandssaur distt. M. P. Record Geological Survey of India 133 (6): 110-112.

SINGH, S. K. and CHANDRA. G. 1987. Fossil jelly fish from Lower Vindhyan Rocks of Rohtas, India. Special Publication Geological Survey of India 11: 114-117.

SINGH, S. N. and PAL, O. P. 1970. Geology around Chitrakut area, Distt. Banda, U. P. Journal Palaeontological Society of India, 14: 77-85.

SINGH, S. K., RAI SANTOSH, K. and TRIVEDI, J. R. 2002. 187 Re-187Os systematics of black shales from the Vindhyans: Implications to their geochronology. 10th ISMAS-WS Puri, 242-244.

SOKOLOV, B. S. 1990. Vendian Polychaeta. In: Sokolov, B. S. and Iwanowski, A. B., Eds., The Vendian System, 1. Springer-Verlag Berlin Heidelberg, 244-246.

SONI, M. K., CHAKARBARTY, S. and JAIN, V. K. 1983. Vindhyan Supergroup – A review. In: Purana Basins of Penninsular India. Memoir Geological Society of India, 6: 87-138.

SRIKANTIA, S. V. 1999. Workshop on the Vindhyan stratigraphy and paleobiology, March 19-20, 1999, Lucknow. Journal Geological Society of India, 53 (6): 724-726.

SRIVASTAVA, A. P., SAINI, H. S. and RAJAGOPALAN, G. 1983. Suitability of glauconite for fission track dating. Current Science, 52: 866-869.

SRIVASTAVA, A. P. and RAJAGOPALAN, G. 1988. F-T ages of Vindhyan glauconitic sandstone beds exposed around Rawatbhata area, Rajasthan. Journal Geological Society of India, 32: 527-529.

SRIVASTAVA, A. P. and RAJGOPALAN, G. 1986. F-T Dating of Precambrian deposits of Vindhyan Group at Chitrakut, Banda Distt. (U. P. -M. P.). In: K. K. Sharma, Ed., Nuclear Tracks: Application to Earth Sciences, Space Physics and Nuclear Physics, Dehra Dun, 41-52.

SRIVASTAVA, P. 2002. Carbonaceous megafossils from the Dholpura Shale, Uppermost Vindhyan Supergroup, Rajasthan: An age implication. Journal Palaeontologicla Society of India, 47: 17-105.

SZANIAWSKI, H. 1996. Scolecodonts. In: Jansonius, J. and McGregor, D. C., Eds., Palynology: principles and applications. American Association of Stratigraphic Palynologists Foundation, 1, 337-354.

TANDON, K. K. and KUMAR, S. 1977. Discovery of annelid and arthropod remains from Lower Vindhyan rocks (Precambrian of Central India). Geophytology, 7: 126-128.

TANDON. S. K., THAKUR V. C., NANDA, A. C., AZMI, R. J., BAGATI, T. N., TEWARI, V. C. and KUMAR, R. 1988. Excursion Guide: Himalayan sequence of Dehra Dun-Mussoories Sector. Geologicla Society of India. 90p.

TAPPAN, H. 1980. The Paleobiology of the Plant Protists. 1028, W. H. Freeman and Company. San Francisco.

TIWARI, M.1996. Precambrian-Cambrian boundary microbiota from the Chert Phosphorite Member of Tal Formation in the Korgai Syncline, Lesser Himalaya, India. Current Science, 71 (9): 718-719.

TIWARI, M. 1999. Organic-walled microfossils from the Chert-Phosphorite Member, Tal Formation, Precambrian-Cambrian boundary, India. Precambrian Research, 97: 99-113.

TIWARI, M. and PANT, C. C. 2004. Neoprterozoic silicified microfossils in Infra Krol formation, Lesser Himalaya, India. Himalayan Geology, 25 (1): 1-21.

TIWARI, M. and PANT, C. C. 2004. Organic-walled microfossils from the Neoproterozoic black phosphatic stringers in the Gangolihat Dolomite, Lesser Himalaya, India. Current Science, 87: 1733-1738.

TUGARINOV, A. I., SHANIN, L. L., KAZAKOV, G. A. and ARAKELYNIAN, M. M. 1965. On the glauconite ages of the Vindhyan System (India) (In Russian) Geokhimiya, 6, 652-660 (English abstract: Geochem. Intl., 504).

TSU-MING HAN and RUNNEGAR, B. 1992. Megascopic eukaryotic algae from the 2. 1 billion-year-old Negaunee Iron Formation, Michigan. Science, 257: 232-235.

VALDIYA, K. S. 1969. Stromatolites of the Lesser Himalayan carbonate formations and the Vindhyan. Journal Geoogical Society of India, 10: 1-25.

VALDIYA, K. S. 1989. Precambrian stromatolite biostratigraphy of India - A review. Himalayan Geology, 13: 181-213.

VENKATACHALA, B. S., and KUMAR, A. 1998. Occurrence of Veryhachium and its implication on the age of the Vaishnodevi Limestone, Jammu. Current Science, 75 (5): 431-432.

VENKATACHALA, B. S., SHARMA, M. and SHUKLA, M. 1996. Age and life of the Vindhyans – facts and conjectures. In: Bhattacharyya, A. Ed., Recent Advances in Vindhyan Geology. Memoir Geological Society of India, 36: 137-165.

VIDAL, G. 1990. Giant acanthomorphic acritarchs from the Upper Proterozoic in southern Norway. Paleontology, 33: 291-292.

VIDAL, G. and KNOLL, A. H. 1983. Proterozoic plankton. Geological Society of America Memoir, 161: 265-277.

VIDAL, G., PALACIOS, T., GAMETZ-VINTANED, J. A., BALDA, M. A. D., and GRANTS, S. W. F. 1994. Neoproterozoic-early Cambrian geology and palaeontology of Iberia. Geological Magazine, 131 (6): 729-765.

VINOGRADOV, A. P., TUGARINOV, A. I., ZIKHOV, C. I., STANIKOVA, N. I BIBIKOVA, E. V. and KHORRE, K. 1964. Geochronology of Indian Precambrian. Report 22nd IGC New Delhi, 10, 553-567.

VOLKOVA, N. A. 1969. Acritarchs of the Russian Platfrom. In: Raaben, M. E., Ed., The Tommotian Stage and the Cambrian Lower Boundary Problem. Amerind, New Delhi, 259-273. English translation, 1981.

VORONOVA, L. G. and MISSARZHEVSKY, V. V. 1969. Finds of algae and worm tubes in Precambrian – Cambrian boundary beds in the north of the Siberian Platform. Doklady Akad Nauk SSSR, 184: 206-209.

VORONOVA, L. G. and LUCHININA, V. A. 1990. In: Sokolov, B. S. and Iwanowski, A. B., Eds., The Vendian System, Springer-Verlag Berlin Heidelberg,, 1: 193-203.

WALTER, M. R., OEHLER, J. H. and OEHLER, D. Z. 1976. Megascopic algae 1300 million years old from the Belt Supergroup. Montana a reinterpretation of Walcott's Helminthoidichnites. Journal of Plaeontology, 50: 877-881.

WEST, W. D. 1981. Vindhyans of Central India. Miscallaneous Publication Geological Survey of India, 50: 1-4.

WRONA, R. 2004. Cambrian microfossils from glacial erratics of King George Island, Antarctica. Acta Palaeontologica Polonica, 49: 13-56.

XING, Y., DING, Q., LUO, H., HE, T., and WANG, Y. 1984. The Sinian- Cambrain boundary of China and its related problems. Geological Magazine, 121 (3): 155-170.

XING, Y., DING, Q., LIN, W., YAN, Y. and ZHANG, L. 1985. Metazoans and trace fossils. In: Xing, Y., Duan, C., Liang, Y., and Cao, R., Eds., Late Precambrian Palaeontology of China. Geological Memoris, Series 2, (2), Geological Publishing House Beijing (in Chinese with English summary): 182-192, pl. 38-40.

XUNLAI, YUAN, JUN, LI AND RUIJI, CAO. 1999. A diverse metaphyte assemblage from the Neoproterozoic black shales of South China. Lethaia, 32: 143-155.

YIN LEI-MING. 1997. Acanthomorphic acritarchs form Meso-Neoproterozoic shales of the Ruyang Group, Shanxi, China. Review of Palaeontology and Palynology, 98: 15-25.

ZANG WEN-LONG and WALTER, M. R. 1992. Late Proterozoic and Early Cambrian microfossils and biostratigraphy, northern Anhui and Jiangsu, central eastern China. Precambrain Research, 57: 243-323.

ZHANG. Z. 1986. Clastic facies microfossils from the Chaunlingguo Formation (1800 Ma) near Jixian, North China. Journal of Micropalaeontology, 5: 9-16

ZHU, S. and CHEN, H. 1995. Megascopic multicellular organisms from the 1700-million- years-old Tuanshanzi Formation in the Jixian Area, North china. Science, 270: 620-6232.

Acritarchs, their occurrence and significance in the Yong Limestone Formation of the Tethyan Garhwal Himalaya, Uttaranchal, India

H.N. Sinha[1], S.S. Srivastava[2] and Bijai Prasad[3]

[1]Department of Geology, Vinoba Bhave University, Hazaribagh 825301 (Jharkhand), India
[2]Department of Earth Sciences, IIT Roorkee, Roorkee (U.A.) 248001, India
[3]KDM Institute of Petroleum Exploration, Dehradun (U.A.) 247667, India

Introduction

The Yong Limestone Formation of the Tethyan Garhwal Himalaya is well exposed in and around Sumna (30°40′ N, 80°50′ E) of the Chainoli district of Uttaranchal. The formation conformably lies above the Shiala Formation. The Yong Limestone was referred to as "Nodular Limestone" and included in the pre-Silurian Shiala "Series" (Heitn and Gansser, 1939) or Somna "Series" (Dave and Rawat, 1968) (Fig. 1). Shah and Sinha (1974) recognized its biohermal nature and considered it to be a separate lithostratigraphic unit and named it as "Yong Limestone Formation" named after deserted village of Yong in the Yong valley, where, one of the reefs is exposed near Sumna village. The reef is about 750 m across and has an exposed thickness of 45 m. This formation is rich in fossils with poor to fair preservation of fauna and flora respectively containing largely massive stromatoporids, corals, bryozoans, crinoids, branched algae, chitinozoa, acritarchs, melanosclerites, and scolecodonts (Shah and Sinha, 1974; Kumar *et al.*, 1977; Khanna *et al.*, 1985, 1996a, b; Sinha *et al.*, 1998).

. This formation has been assigned an Late Ordovician to Early Silurian age by Shah and Sinha (1974). However, Sinha *et al* (1996a) have considered the Yong Limestone Formation to be of Ludlovian (Late Silurian) age based on index forms of acritarchs. This formation is characterized by imperceptible bedding features and lack of sedimentary structures possibly due to current remobilization. Shah and Sinha (1974) and Khanna *et al* (1985) suggested domal disposition of this unit causing lack of sedimentary structure. However, Kumu *et al* (1977) suggested that the formation represents deposits of a zero-energy carbonate flat, where, corals and bryozoa flourished.

The method of preparation of strato-litho-petrographic column and the traverse direction for sample collection in and around Sumna village has been described earlia in Sinha *et al.* (1996a) (Fig. 2). The Yong Limestone is mainly characterized by a grayish-green biohermal limestone (wackestone-bioclast) with thin shale bands imperceptibly merging into a biostromal limestone.

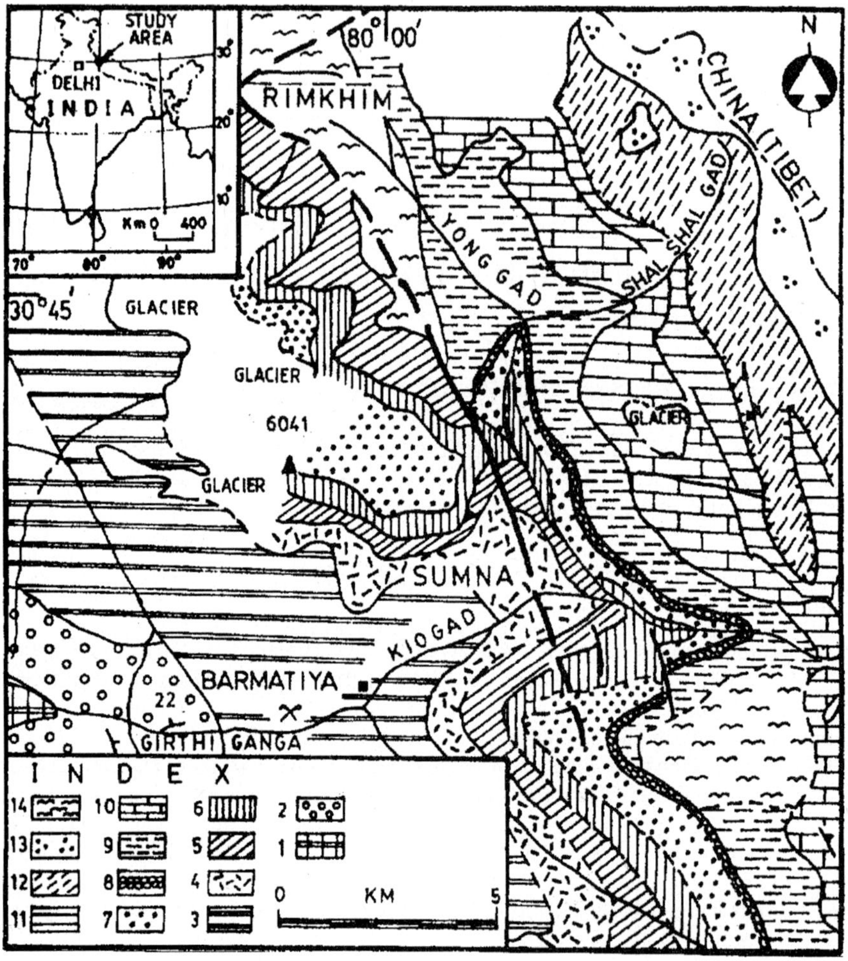

Fig. 1. Geological map of the Sumno-Rimkhim area of the Tethyan Zone of the higher Garhwal Himalaya, India (Sinha, 1989). 1 = Martoli Group; 2 = Ralam Formation; 3 = Garbyang Formation; 4 = Shiala Formation; 5 = Yong Limestone Formation; 6 = Variegated Formation; 7 = Muth Quartzite Formation; 8 = Kuling Shale-Fenestella Shale; 9 = Kalapani Limestone Member/Kuti Formation; 10 = Kioto Formation; 11 = Lapthal Formation; 12 = Spiti Shale; 13 = Giumal Sandstone; 14 = Alluvial.

Acritarchs and their significance

Acritarchs were first discovered by White (1862) in cherts of the Silurain-Carboniferous limestone from the New York State (Servais, 1996). Extensive studies on the significance of acritarcha began in the early 1930s by Eisenack. Eisenack (1931) initially understood that these microfossils were the eggs of some planktic organisms which were deposited on the sea bottom.

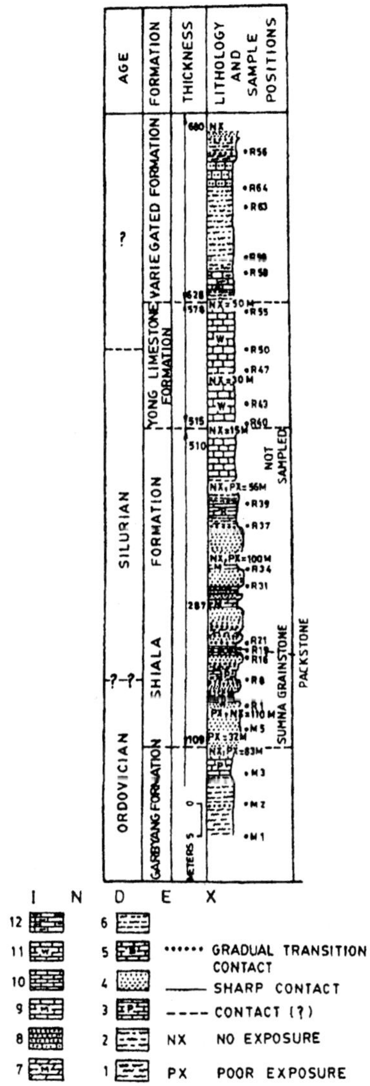

Fig. 2. Integrated strato-litho-petrographic column of the Sumna-Rimkhim section along the right bank of Yong gad valley showing sample positions. 1 = siltstone; 2 = alteration of siltstone and siltyshale; 3 = limestone (packstone); 4 = sandstone (calcareous quartz arenite); 5 = limestone (boundstone, algal); 6 = siltyshale; 7 = dolostone (mudstone, dolomite); 8 = limestone, sandy (mudstone, arenaceous); 9 = limestone (mudstone); 10 = limestone; 11 = limestone (wackstone, bioclast); 12 = limestone (mudstone, argillaceous). Petrographic units described under bracket (after Sinha *et al.* 1998).

Evitt (1963) erected the informal category of the 'group' Acritarcha which takes its name from Greek word akriotos meaning uncertain, doubtful and arche means origin, *i.e.*, acritarchs are those organic-walled microfossils which have unknown biological affinities and cannot be placed into any existing classification or organisms confidently. "Even today, acritarch classification has not yet passed a phenetic stage and remains arbitrary and artificial" (Servais, 1996).

The first definition of acritarchs was given by Evitt (1963), and more recently, by Fensome *et al.*. (1990) which is as follows: "unicellular, or apparently unicellular, non-colonial organic-walled microfossils of unknown or highly questionable affinity, occurring in palynological assemblages". Recently, Servais *et al.* (1977) expressed their views that acritarchs represent a varied selection of biological groups and the extent of their variation is not known at present.

Most acritarchs are found in marine or brackish water sediments. However, there is a report of non-marine forms of acritarchs from Permain to the Holocene (Riding and Duxbury, 1993). They range in size from 5 to 500 μm. Acritarchs are the dominant group of palynomorphs and are found in marine Lower Paleozoic rocks of many continents. It has been found that the Cambrain to Silurian was the favorable time for the evolution of acritarchs (Servais, 1996) and they reached acme of diversity in sculptural and structural complexity during Early Ordovician (Traverse, 1988).

The significance of acritarcha is governed by their micro size range, biodiversity, morphological complexity, abundance (10 g of rock sample yields 1000 of individuals), fast evolution leading to availability of many stratigraphic significant forms and wide geographic distribution. All these criteria help in resolving and understanding the problems of paleoceanography and oil exploration through high resolution biostratigraphy, correlation of strata (local and regional), refined dating of sedimentary sequences, paleoceology, paleogeography and paleoenvironment (Jain, 1991).

Limitation of acritarchs

From the above discussion, it can be said that the acritarchs are useful in biostratigraphy and refined dating mainly of Lower Paleozoic rocks, particularly, where, no other tool is available. Besides, they are very helpful in the interpretation of paleoceology, *i.e.* indicating relative distances from paleoshorelines and in some cases, the paleosalinity. There is a small proportion of acritarch that are sensitive to temperature and, thus, provide good indicators of paleoclimate (Cocks and Verniers 1998). Apart from its immense utility, there are many acritarchs which have very wide range of distributions turning to cosmopolitan in nature and, thus, not useful in solving paleobiogeographical problems. Also, many acritarchs are long ranging in a age and, thus, not helpful in correlation (cocks and Verniers, 1998). Many acritarchs are very small in size (5μ m) and chances of reworking of such taxa from older formation are very high. Lastly, there is a large number of unresolved taxonomic problems with acritarchs which may allow further use of acritarchs in understanding ecological and paleobiogeographical problems (Cocks and Verniers, 1998).

Palynofacies

The published literature shows that there is a scarcity of data on environmental influence of acritarchs. However, it has been found that the acritarch abundance and diversity both increase with distance from shore (Molyneux and Al-Hajri, 2000) and maximum abundance of acritarch lies in open-marine facies, whereas, marginal-marine facies contains rare to no acritarchs (Smith and Saunders, 1970). Diversity is also found maximum in outer neritic condition, *i.e.* in offshore marine shelf facies (Dorning, 1981, Al-Ameri 1983). A few workers on acritarchs have found that sphaeoromorphs dominated in near-shore shallow water conditions (Gray and Boucot, 1972; Jacobson, 1979; Dorning, 1981, Al-Ameri, 1983). Dorning (1981) reported sphaeoromorphs dominant in deep-water basinal assemblages also. Thus, sphaeromorph acritachs have wide range of distribution from inshore to offshore regions and more diverse assemblages are reported from intervening environments (Molyncus and Al-Hajri, 2000).

Sinha *et al.* (1998) reported a rich assemblage of acritarchs from the Shiala and Yong formations. The assemblages from the Shiala Formation include sphaeromorphs, acanthomorphs, netromorphs, polygonomorphs, diacromorphs and herkomorphs, whereas, the overlying Yong Limestone Formation yielded sphaeromorphs, acanthomorphs, polygonomorphs, and diacromorphs. Thus, there is a striking contrast in abundance and species diversity in Shiala Formation when compared with Yong Limestone Formation (Sinha *et al.*, 1998). The above comparison reveals that the distribution of rich and poor microphytoplankton assemblages may co-exist in neighbouring areas. Riegel (1974) suggested that this type of microphytoplankton distribution is controlled by broad difference in facies mainly and not by phylogenctic or climatic changes. In the present case, Shiala is mainly composed of clastics with minor non-clastics, whereas, Yong Limestone is mainly non-clastics with thin bands of clastics. Thus, we find facies differ remarkably in their phytoplankton content. However, when we consider the facies factor for the control and distribution of acritarch, then sedimentological and biological factors play an important role in dispersal and fossilization of cysts (Riegel, 1974). However, these two factors may not be the major cause of the distribution of acritarchs in Shiala and Yong formations because of the fact that the contrast in phytoplankton content depends upon the original living communities too. The diversity and abundance of original living communities indicate an ecologic maturity of the phytoplankton community, caused perhaps by a slight increase in nutrient levels and a better adaptation to the nutrient levels than existed (Wicander, 1975) during Shiala as compared to the Yong Limestone Formation.

Reigel (1974) illustrates that very little is known about the actual relationship between the living communities and fossilized cyst in modern seas. Here, if we assume that a casual relationship existed between the phytoplankton content and ecologic conditions, which may happen (Riegel, 1974), we may be able to conclude that the development of acritarchs was inhabited under adverse conditions during the deposition of the Yong Limestone Formation as compared to the underlying Shiala Formation. These adverse conditions are current remobilization (as indicated by lack of sedimentary structures) as well as irregular salinities and nutrient supply in a restricted basin during the deposition of the Yong Limestone Formation.

Plate 1.

1,2 *Polygonium gracilis,* Slide No. R-47 (1), coord. 37.2 x 91.5; R-47 (3), coord 44.4 x 102.2. Polygonal areas 14.4 µm and 12 µm, process length 5-6 µm. **3,6** *Mycrhystridium* sp. Slide No. R-47 (3), coord. 56.3 x 97.5. R-47 (3), coord. 34.6 x 103.8. Central body diameter 12 µm and 12 µm, process length 6 to 9.6 µm. **4,10** cf. *Unellium* sp. Slide No. R-47 (1), coord. 40 x 104; R-47 (1), coord 42.2 x 105. Central body diameter 13 µm and 12 µm, process length 6-7 µm. **5,12,14** *Multiplicisphaeridium* sp. Slide No. R-47 (3), coord. 64.5 x 94.3; R-47 (3), coord. 42.5 x 102.1; R-47(3) coord. 60.5 x 109.8. Central body diameter 16 µm, 12 µm and 16 µm. Process length 5,12,10 µm. **7** cf. *Baltisphaeridium* sp. Slide No. R-47 (1), coord. 34 x 97.5; Central body diameter 12 µm process 6 µm. **8,11** *Diacromorphitae* sp. Slide No. R-47 (3), coord 66.1 x 98.6; Central body diameter 15 µm, average length of process 7 µm, R-47 (1), coord. 52.1 x 95; Central body diameter 12 µm, process 6 µm. **9** *Lophosphaeridium* sp. Slide No. R-50 (1), coord. 62.5 x 97; Total diameter 16 µm. **13** *Acanthodiacrodian*, Slide No. R-43 (1), coord. 48.5 x 98; Central body diameter 14 µm, process (largest 16 µm). **15** *Micrhystridium stellatum* Slide No. R-47 (1), coord. 37 x 96; Central body diameter 8 µm, process length largest 6 µm. **16** Pores in an organic plate, Slide No. R-47 (1).

Plate 1.

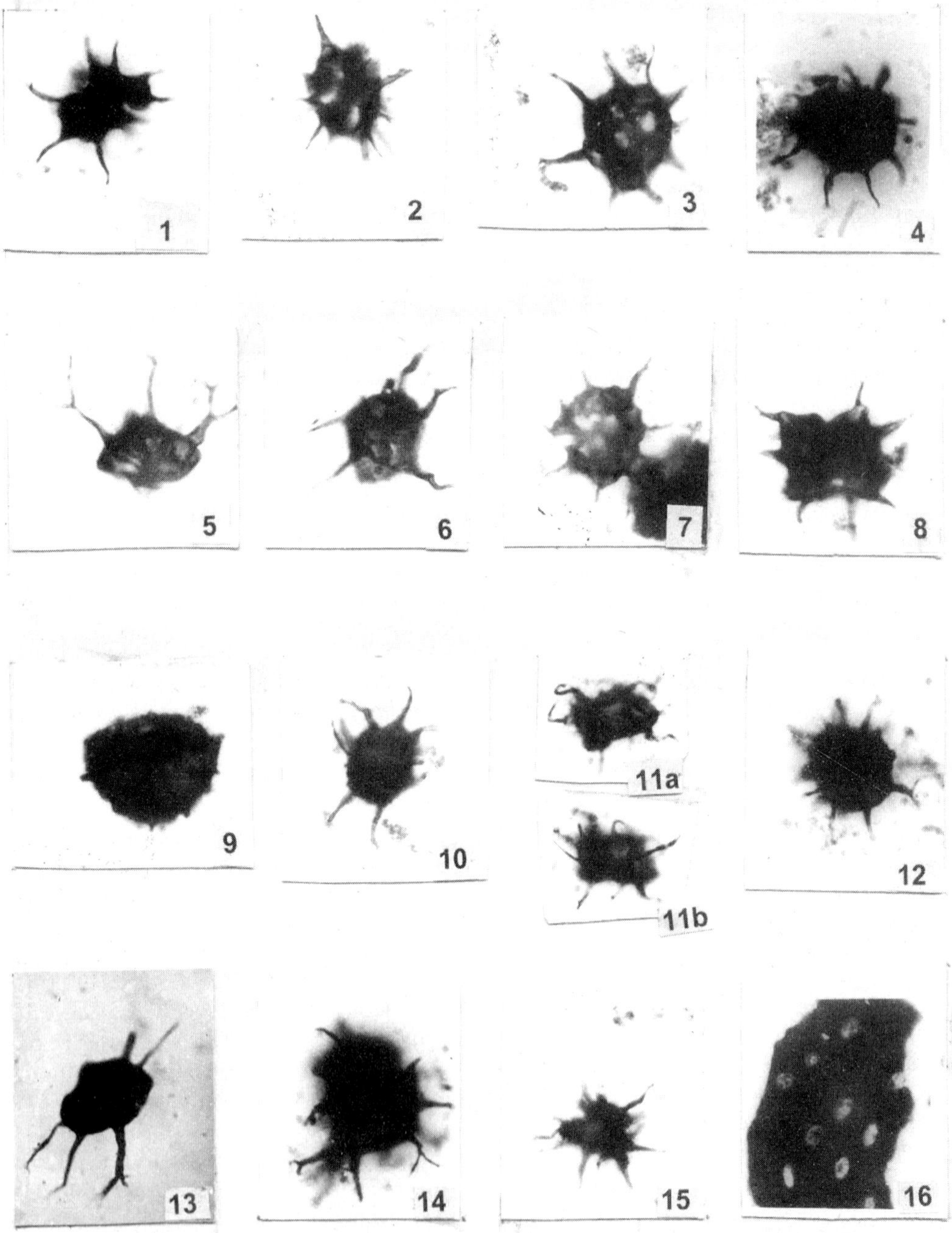

However, the acritarchs recovered from the Yong Limestone are dark (Pl. 1). The darkening of the vesicle is seen in all the co-existing playnomorphs (Sinha *et al.*, 1996 b) recorded from this formation. The darkening of palynomorphs could take place by:

(a) reworking from older rocks,
(b) deposition of these specimens in shallow water conditions and, thus, oxidized more prior to the final burial, and
(c) Thermal maturity due to progressive increase of the carbon ratio.

Out of these three reasons, described above, the second and third reasons are the main causes of darkening of the cyst wall. The chances of reworking is minimum because fossils are well preserved and not abraded, thus, the question of reworking does not arise (Pl. 1).

Conclusions

A rich assemblage of acritarchs has been recovered from Yong Limestone formation and Shaila Formation from the Tethyan Garhwal Himalaya. Based on the assemblage and stratigraphically important taxa, a Middle to Late Silurian age has been assigned to Yong Limestone formation. The Ordovician/Silurian boundary lies within Shaila Formation and thus the earlier view that the boundary lies at the base of Yong Limestone formation needs rethinking.

Acknowledgements

The first author is thankful to Dr. Laxmi Shiveshwari for excusing him from domestic responsibility during writing of the manuscript. The neat typing of the manuscript by Mr. V.V.R. Rao is thankfully acknowledged.

References

A1-AMERI, T.K. 1983. Acid-resistant microfossils used in the determination of Palaeozoic environments in Libya. PPP, V41, 103-116.

COCKS, L.R.M. and VERMIERS, J. 1998. Applicability of planktonic and nektic fossils to paleographic reconstructions. Acta Universitatis Carolinae-Geologica, 42 (3/4), 97-108.

DAVE, V.K.S. and RAWAT, B.S. 1968. A preliminary account of the geological traverses in Bhapkund-Malari-Niti area, Chamoli Distt., U.P., Res. J. Roorkee Univ., 10(3-4) 97-108.

DORNING, K.J. 1981. Silurian acritarch distribution in the Ludolvianshelf sea of South Wales and the Welsh Borderland. In: J.W. Neale and Braiser (Eds.), Microfossils from Recent and fossil shelf seas. Wiley, Chichester, England, 31-36.

EISENACK, A. 1931. Neue Mikrofossilien des baltischen Silurs. I. Paleontol., 13: 74-118.

EVITT, W.R. 1963. A discussion and proposals concerning fossil dinoflagelattes hystrichospers, and acritarchs, II. Proc. Natl. Acad. Sci., 49: 298-302.

FEMSOME. R.A., WILLIMS, G.L. BARSS, M.S., FREEMAN, J.M. and HILL, J.M. 1990. Acritarchs and fossil parsinophytes: an index to genera, species and intraspecific taxa. A.A.S.P. Contrib. Ser., 25: 1-77.

GRAY, J. and BOUCOT, A.J. 1972. Palynological evidence bearing on the Ordovician-Silurian paraconformity in Ohio. Bulletin of the Geological Society of America, 83, 1299-1313.

HEIM, A. and GANSSER, A. 1939. Central Himalaya, geological observations of the Swiss expedition 1936. Mem. Soc. Helv. Sci. Nat. 73(1), p. 245.

JACOBSON, S.R. 1979. Acritarchs and paleoenvironmental indicators in Middle and Upper Ordovician rocks from Kentucky, Ohio and New York. J1. Paleontol., 53, 1197-1212.

JAIN, K.P. 1991. Fossil phytoplankton: Importance in Paleocenography. Current Science, 61 (9-10), 697-700.

KHANNA, A.K., SINHA, A.K. and SAH, S.C.D. 1985. Yong Limestone of Tethys Himalaya – its stratigraphic status and palynological fossils. Geol. Soc. Ind., 26(3): 191-198.

KUMAR, S., SINGH, I.B. and SINGH, S.K. 1977. Lithostratigraphy, structure, depositional environment, paleocurrent and trace fossils of the Tethyan sediments of Malla Johar area, Pithoragarh-Chamoli district U.P. India. J. Palaeont. Soc Ind., 20: 396-435.

MOLYNEUX, S.G. and A1-HAJRI, S. 2000. Palynology of problematic Lower Palaeozoic lithofacies in Central Saudi Arabia. Said A1-Hajri and Bernad Owens (Eds), Gulf Patro Link CIMP, 18-41.

RIDING, J.B. and DUXBURY, S. 1993. A new non marine acritarch from the Middle Jurassic of Britain. Spec. Pap. Paleontol., 48: 67-78.

RIEGEL, W. 1974. Phytoplankton from the Upper Emsian and Eifelian of the Rhineland Germany–A preliminary report. Review of Palaeob. Palynol., 18, 29-39.

SERVAIS, T. 1996. Some considerations on acritarch classification. Review of Palaeobotany and palynology, 93, 9-22.

SHAH, S.K. and SINHA, A.K. 1974. Stratigraphy and tectonics of the 'Tethyan' zone in a part of western Kumaun Himalaya: Himalayan Geology, 14: 1-27.

SINHA, H.N., SRIVASTAVA, S.S. and PRASAD, B. 1996a. Record of prolific and indubitable acritarchs from the Lower Palaeozoic strata of the Tethyan Garhwal Himalaya and age implication. Current Science, Vol. 71(1), 24-31.

SINHA, H.N., SRIVASTAVA, S.S., KAPOOR, P.N. and CHAUHAN, R. 1996b. Chitinoza and mealnosclerite from the Lower Paleozoic sequence of the Tethyan Garhwal Himalaya- A note on their identification and distinction. Current Science, 71(5), 412-416.

SINHA, H.N., PRASAD, B and SRIVASTAVA, S.S. 1998. Ordovician-Silurian acritarch bistratigraphy of the Tethyan Garhwal Himalaya, India, Review of Paleobotony and Palynology, 103, 167-199.

SMITH, N.D. and SAUNDERS, R.S. 1970. Paleoenvironments and their control of acritarch distribution: Silurian of east-central Pennsylvania. J. Sediment. Petrol., 40(1): 324-333.

TRAVERSE, A. 1988. Paleopalynogy. Unwin Hyman, London, p. 600.

WICANDER, E.R., 1975. Fluctuations in a Late Devonian–Early Mississippian phytoplankton flora of Ohio, USA, PPP, 17, 89-108.

WHITE, M.C. 1962. Discovery of microscopic organisms in the siliceous nodules of the Paleozoic rocks of New York Am. J. Sci. Arts: 333.

Palynostratigraphy and depositional environment of Lower Gondwana sediments in Raigarh basin, Chhattisgarh, India

Ram Awatar

Birbal Sahni Institute of Paleobotany, 53, University Road, Lucknow 226 007, India

Introduction

The Raigarh Coalfield lies at the middle part of the Upper Mahanadi Gondwana Master basin with Ib-River Coalfield in the east, Mand and Hasdo-Arand coalfields in the west and north-west part of Raigarh Gondwana Basin. The total area of this coalfield is ~900 Sq km, bounded by latitude: 21°. 45 00' and 22°. 42′ 00' N and longitude 83°. 01 00' and 83°. 45 00' E in Raigarh district, Chhattisgarh. Geological Survey of India (GSI) has drilled a number of boreholes in search of coal to assess the total coal reservoir in of the area.

The Paleobotanical and paleontological data from Raigarh Coalfield is meagerly known. Pal (1988) reported different species of *Glossopteris, viz., G. indica, G. raniganjensis, G. major, G.gondwanensis, G.angustiofolia, Vertebraria indica* and *Schizoneura sp.* from the basal red bed of the Kamthi Formation. Subsequently, Bandhopadhyay (1990), recorded the *Glossopteris* and *Sameropsis* seeds in association with fossil fauna - Fenestillid bryozoa, ovoid, non-calcareous Ostracod and fresh to brackish water Estherids- *Paleolimniadiopsis* sp. of Late Permian age. In addition to these, Meena (1999), Jana *et al.* (2002) and Chakraborty (2003) have recorded the Early and Late Permian palynofossils from this coalfield. Recently, Chakraborti and Chakraborty (2001) recorded the Triassic plant megafossil- *Baiera, Pterophyllum* and microfossils *viz., Alisporites, Falcisporites Klausipollenites, Brachysaccus, Chordasporites,Verucosisporites* and *Weylandites* etc from Baronakund area, Raigarh Coalfield. In the present study Early Permian palynoflora is being reported from the out crop section, for filling the gap of information related to the Karharbari Formation.

Geology

The sedimentary sequence of Raigarh Coalfield was classified into Talchir, Barakar and Kamthi formations (Raja Rao, 1983). Later on, Chakraborti (2001), classified the lithosequence of Mand-Raigarh Basin into Talchir, Barakar, Barren Measures, Raniganj and Kamthi formations. The lithological details are given in Table 1.

Material and Methods

Seven samples were collected from a 3.40 m thick sedimentary sequence exposed, in a nala cutting (locally known as Jhugna Bhwana Nala), located about 1.5 km S-W of Sohanpur Village, between latitude 22° 25' 09" and 83° 20' 00"; (topo sheet No 64 N/7), Fig. 1. For recovery of palynofossils, samples were crushed into small pieces (1-2 mm) and kept into Hydrofluoric acid

(40%), to dissolve the siliceous component of rock samples; followed by HNO_3 to dissolved the carbonaceous matter and finally 5% alkali was used for cleaning the debris from the surface of spore/pollen grains. The macerated residues were smeared on cover slip with polyvinyl alcohol and mounted on the glass slides with Canada balsam. Details of the samples are given below.

Table 1. Generalised geological succession of the Mand-Raigarh Basin (after Chakraborti 2001)

Formation/Member	Thickness (m)	Lithology
		Pebbly to gravely bed, lateritic soil.
Deccan Trap		Doleritic to basaltic flows, dykes and sills.
Kamthi ?	200 ± 50	Unclassified younger sediments.
Kamthi (Sensu stricto)	300	Mega cross-bedded, profuse clay clasts bearing pebbly sandstone with ferruginous sandstone band, red clay, siltstone, claystone and conglomerate
		---Unconformity------
Raniganj (Lr. Kamthi)	200 ± 30	Interbanded fine- grained sandstone, claystone shale carbonaceous shale to the east and fine to coarse grained sandstone, silty shale, purple ferruginous sandstone to the west. Coal Seam (2.41- 3.68 m) trick with two bands.
Barren Measures	150 ± 30	Claystone , shale, siltstone, fine-grained sandstone, ironstone and carbonaceous shale band. Arenites increases to due west.
B Upper Member **A**	200 ± 30	Interbanded/interbedded sequence of fine to medium grained grey shale, claystone, siltstone, carbonaceous shale coal seams with major ironstones. (Seam No. V to VI)
R Middle Member **A**	150 ± 50	Arkose, coarse to granular to pebbly sandstone sandstone with minor shale/ carbonaceous shale.
K Lower Member **A** **R**	100 ± 30	Fine to medium grained pink garnet or/ quartz bearing sandstone with gray shale and siltstone with one local seam..
		---Unconformity------
Karharbari		Unmapable
Talchir	100	Matrix- based variegated green conglomerate, splintery shale, siltstone, thickly bedded fine-grained yellow to khaki sandstone with a basal boulder bed.
		---Unconformity------
		Basement rocks

Table-2 Details of samples and their lithology.

Sr. No.	Sample No.	Lithology	Thickness (m)
1.	JBNS-1	Claystone (gray)	0.50 (bottom sample)
2.	JBNS-1a	Claystone (gray)	0.50
3.	JBNS-2	Carbonaceous shale	0.25
4.	JBNS-3	Siltstone with carb. Shale	0.75
5.	JBNS- 4	Claystone (yellow)	0.25
6.	JBNS-5	Coaly shale	0.40
7.	JBNS-6	Siltstone	0.25 (top sample)

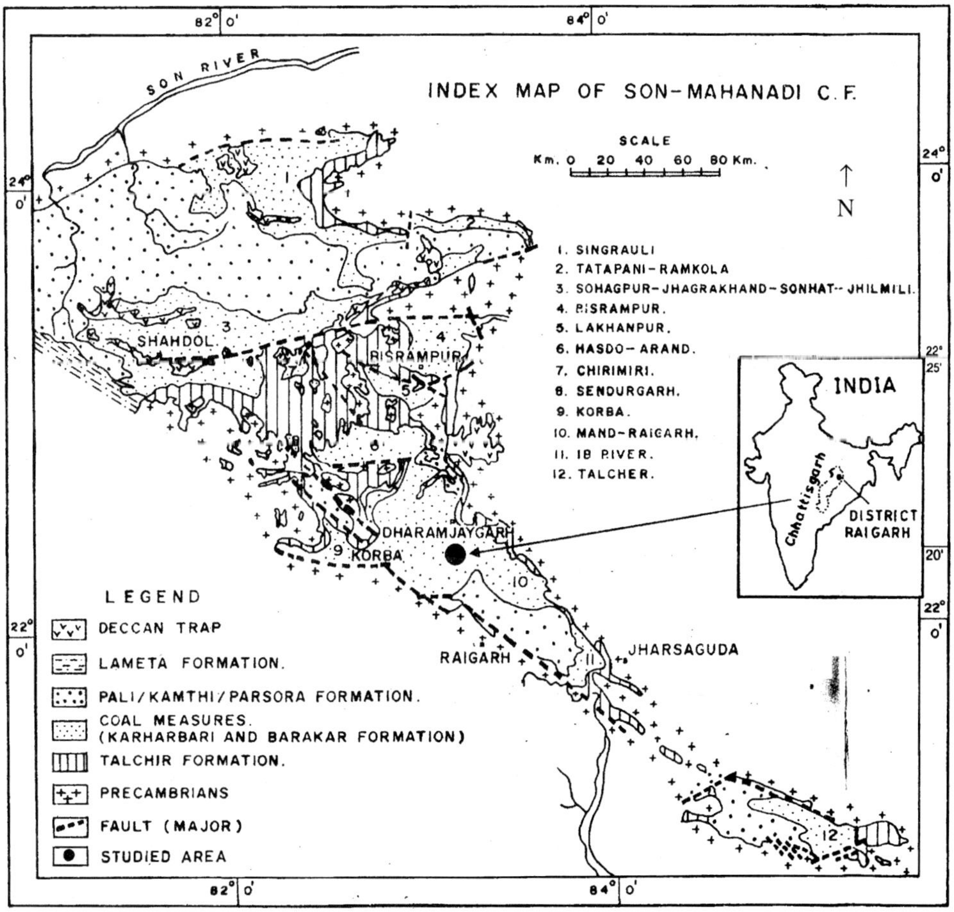

Fig. 1. Geological map of Son- Mahanadi Basin, showing location of studied samples.

Palynostratigraphy

Total 24 genera and 30 species were identified from the productive samples. A chick list of the genera and species is given below:

Balmeella tetragona, Brevitriletes communis*, Callumispora gretensis*, C. fungosa*, Crescentipollenite fuscus*, Circumstriatites talchirensis*, C. ovatus, Dentatispora sp., Faunipollenites various*, F. singraulenisis*, Ginkgocycadophytus novus, Indotriradites korbaensis, Jayantisporites pseudozonatus*, Lacinitriletes sp, Leiospharidia talchirensis*, Microbaculispora sp, Microfoveolatispora foveolata*, Parasaccites korbanesis*, P. obsicurus*, Plicatipollenites indicus*, P. trigonalis*, Potonieisporites magenus*, Quadrisporites horridus*, Sahnites gondwanensis,* Striatopodocarpites decorus*, S. circularis, Tetraporina sp., Tiwariasporis flavatus*, Vestigisporites sp., Virkkipollenites indicus.* Palynomorphs marked with asterisk (*) are illustrated in Plate-1.

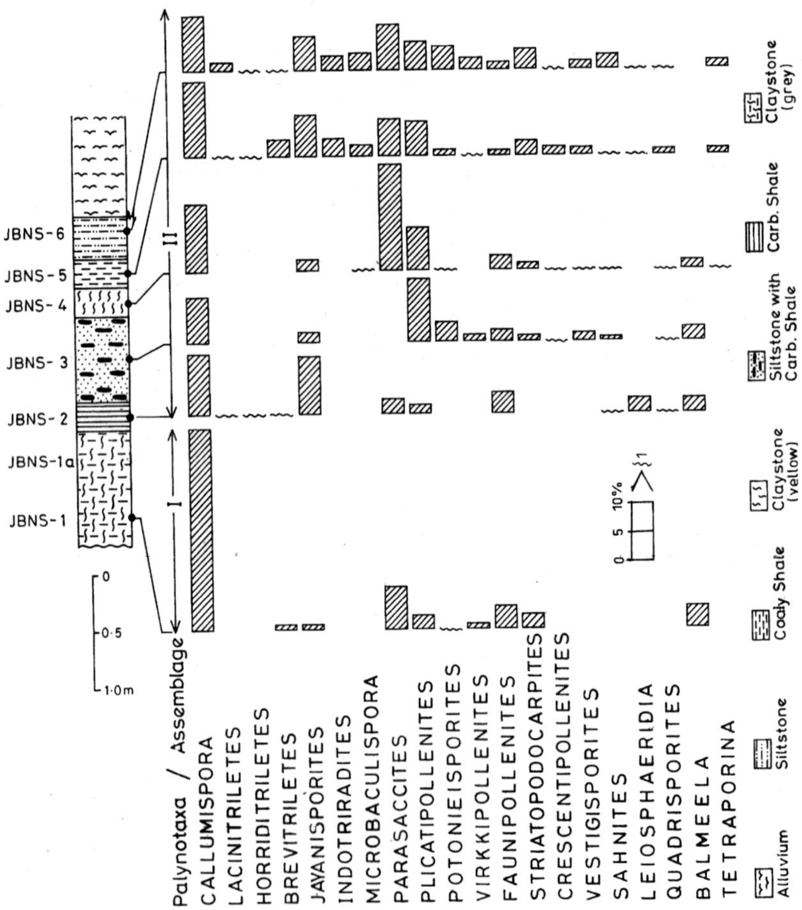

Fig. 2. Percentage frequency of important palynofossils recorded in JBNS, Mand-Raigarh Coalfield.

On the basis of the quantitative analyses of the palynological composition, two distinct palynozones have been identified, depicted in (Fig. 2). These assemblages are described as:

Assemblage-I: *Callumispora gretensis* palynozone (Sample No. JBNS-1). This assemblage is characterized by the dominance of the genus *Callumispora* (60%), in association with *Parasaccites, Plicatipollenite,* zonate trilet and striate disaccate pollen. The index species identified at this label are – *Callumispora gretensis* and *Circumstriatites talchirensis*. The *Callumispora* dominating palynoflora in association with monosaccates is characteristic of Lower Karharbari palynoflora.

Assemblage-II: *Parasaccites- Callumispora* Zone (Sample nos. JBNS1a-JBNS- 6). In this zone *Callumispora* declined and the monosaccates (*Parasaccites, Plicatipollenites, Virkkipollenites* and *Potonieisporites*) increases considerably in association with striate disaccate pollen. The marker palynofossils recorded in the assemblage are- *J. pseudozonatus, Indotriradites korbanesis,* and *Microbaculispora sp.*in association *Faunipollenites various, Striatopodocarpites decorus, and Crescentipollenites fuscus*. The assemblage is correlates with Upper Karharbari palynoflora.

Discussion

As such, this area of Mand-Raigarh Coalfield, has similar trends of palynological history in the Karharbari Formation, when compared with other basins of Indian Gondwana. The miofloral Assemblage-I, described from the sediments of the JBNS is dominated by *Callumispora* in association with *Parasaccites, Plicatipollenites,* and zonate trilete spores. Similar palynofloral assemblages are known from the Lower Karharbari seam of the Giridih Coalfield (Srivastava, 1973); Jayanti Coalfield (Lele & Makada,1972); Chirimiri Coalfield, (Srivastava, 1980b, Paradol-Chirimiri railway cutting); North Karanpura Coalfield (Srivastava, 1980a, Honhe area); Umaria Coalfield (Srivastava and Anand-Prakash, 1984, Umrar River), and Johilla, Coalfield, (Anand-Prakash and Srivastava, 1984, Pali Coal Mine,) and Johilla Coalfield (Tiwari and Ram-Awatar, 1989, Johilla River Section). The above comparison shows that the sediments, exposed along Jhugna Bhawana Nala Section is closely comparable with Lower Karharbari playnofloral, however, in the former assemblages percentage of *Jayantisporitesis* is dominate and in the present assemblage this taxa is poorly recorded.

Plate -1

All the figures are enlarged ca x 500, unless otherwise mentioned.

Fig. 1. *Callumispora fungsoa* Bharadwaj and Tiwari, 1977 emend. Tiwari *et al.*, 1989. **Fig. 2.** *Microfoveolatispor foveolata* Tiwari emend. Tiwari and Singh, 1981. **Fig. 3.** *Jayantisporites pseudozonatus* Lele and Makada, 1972. **Fig. 4.** *Brevitriletes communis (*Bhardwaj and Srivastava) Tiwari and Singh, 1981. **Fig. 5.** *Callumispora grettiensis* (Bharadwaj and Srivastava) emend. Tiwari *et al.*, 1989. **Fig. 6.** *Leiosphaeridia crescentica* Sinha, 1969. **Fig. 7.** *Quadrisporites horridus* Hennely emend. Potonie and Lele, 1961. **Fig. 8.** *Sahnites gondwanensis* (Mehta) Pant 1955 emend. Tiwari and Vijaya, 1984. **Fig. 9.** *Tiwariasporis flavatus* Maheshwari and Kar, 1967. **Fig.10.** *Faunipollenites varius* Sinha, *1969 emend.* Tiwari *et al.*, 1989. **Fig.11.** B*almeella tetragona* Pant and Mehta, 1963. **Fig.12.** *Parasaccites obscurus* Tiwari 1965 emend. Tiwari *et al.*, 1989. **Fig.13.** *Plicatipollenites indicus* Lele, 1964. **Fig. 14.** *Plicatipollenites trigonalis* Lele, 1964. **Fig. 15.** *Crescentipollenites fuscus* Bharadwaj, Tiwari and Kar, 1971. **Fig. 16.** *Striatopodocarpites decorus* Bharadwaj and Salujha, 1964. **Fig. 17.** *Circumstriatites talchirensis* Lele and Makada, 1972. **Fig. 18.** *Faunipollenites singrauliensis* Sinha, 1972 emend. Tiwari *et al.*, 1989. **Fig. 19.** *Potonieisporites neglectus* Potonie and Lele, 1961. **Fig. 20.** *Parasaccites korbensis* Bose and Maheshwari, 1968 emend. Tiwari *et al.*, 1989.

Plate-1

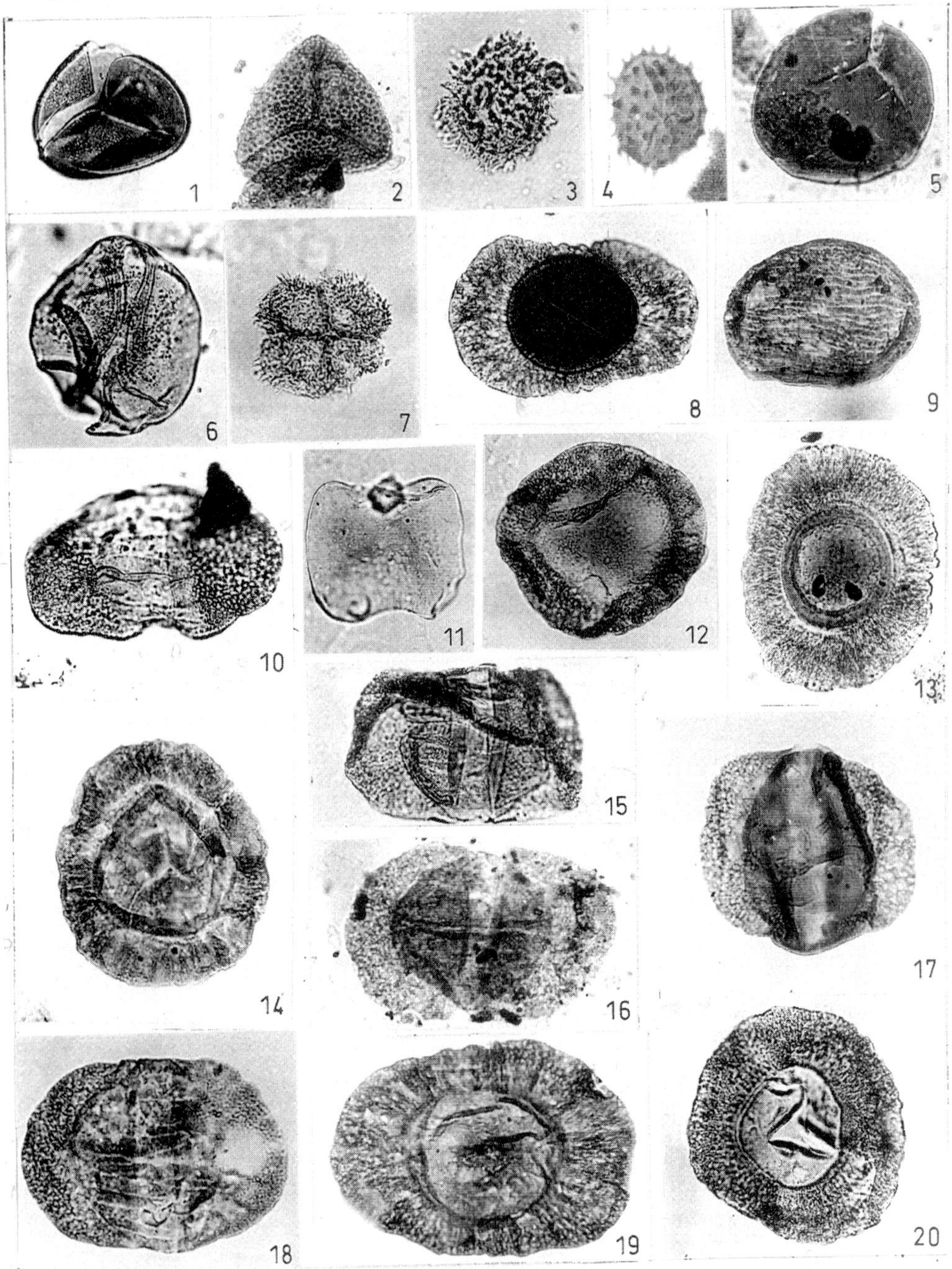

The JBNS (Sample no. 2-10), shows maximum development of redial monosaccates (*Parasaccites* and *Plicatipollenites*) which reappear in Upper Karharbari member along with striate disaccates, smooth and zonate trilete groups (Figure-3). The palynoflora *Callumispora* declines in abundance in the younger sequence of the palynoassemblage and monosaccates increases considerably. Similar succession of the palynofloral assemblages have also been recorded in Raniganj Coalfield (Tiwari, 1973, Sonbada Nala Section, zone-2); Korba Coalfield (Bharadwaj & Srivastava, 1973, B.H. NCKB 19, zone-2); Chirimiri Coalfield (Srivastava, 1980b). The coeval palynoassemblages have also been recorded from Karanpura Coalfield (Kar, 1973); Manuguru Coalfield (Srivastava & Jha, 1992, palynozone-2); Budharam Coalfield (Srivastava & Jha, 1995, Palynozone-3) from Godavari Coalfield. The *Parasaccites*+ *Callumiapora* dominating palynoflora in association with striate- disaccates palynofloral assemblage also recorded from Mand-Raigarh, Coalfield (Jana *et al.*, 2002, zone-2), fairly resemble with present assemblage, except that *Paravesicaspora* is fairly rich in former assemblage, while absent in the JBNS assemblage. The similar palynoassemblage have also been recorded from Umaria Coalfield (Anand-Prakash & Srivastava, 1984, zone-2), Johilla Coalfield, (Tiwari and Ram-Awatar, 1989), shows a fair degree of resemblance with present assemblage. Thus the two palynoassemblages (I&II) together compare well with *Crucisaccites monoletus* assemblages-zone of (Tiwari & Tripathi 1992) of Indian Gondwana.

The panorama of the climate of a particular area in a given time is recorded mainly on the basis of fauna and flora of that region. Besides, lithology, sedimontology, paleogeographical and paleolatitudinal position are the other parameters, which enable to better understand the paleoclimate of a region.

It is well established that the paleogeographical position of India was mainly between 20°-45° S latitude, during Early Permian time. The climate during Early Permian period was extremely cold, which has already proved by the sedimentlogical study (Singh, 1976), mega- and micro floral studies (Lele, 1976, Tiwari & Tripathi, 1988) and oxygen isotope (Dutta & Suttner, 1986). During the Early Karharbari time the climate was slightly ameliorated with high humidity which was favourable for the luxuriant growth of vegetation, therefore, thick coal seams have been recorded in Karharbari Formation (Tiwari & Tripathi, 1988). Kemp (1973), established four stages describing the distribution of microfloras within the glaciated Gondwana area. These are- Pre-stage1, Stage 1 microflora, Stage 2 microflora, and Stage3 microflora. Among these, only Stage 2 microflora, are the most widespread in their geographical distribution, occurring on all Gondwana continents, (Yang,1993). In stage 2, the common palynoflora identified as Index fossils for the Early Permian palynofloral events are- *Microbaculispora tentula, Protohaploxypinus sp (=Faunipollenites* sp.), *Brevitriletes* sp., in association with *Horriditriletes* ramosus, *Jayantisporites pseudozonatus.* These miospores have also been recorded in the deglaciation sequence of Early Permian sediments of Tengchang Group, south-west China and also in other continents of Gondwana realm. (Wopfner & Jin, 1993).

Some characteristic palynoflora Viz., *Microbaculispora* sp., *Protohaploxypinus sp (=Faunipollenites* sp.), *Brevitriletes* sp., *Horriditriletes* sp. and *Jayantisporites peudozonatus* have already been recorded from the other Indian basins, and presently from the sediments of Jhugna-Bhwana Nala Section, clearly evidence the deglaciation palynofloral event proposed by Wopfner & Jin, 1993. JBNS, could be one of the example related to the above hypothesis as many palynoflora are common in the present assemblages. More data is required on the biostratigraphy and paleoclimatology to confirm the above facts from India Gondwana.

Acknowledgements

The author expresses his gratitude to the Director, Birbal Sahni Institute of Paleobotany, Lucknow, for permission to publish the present work. Thanks are due to Mr. Y.P Singh for help to preparing the MS on computer.

References

AAND-PRAKASH and SRIVASTAVA, SURESH, C. 1984. Miofloral studies of the Lower Gondwana sediments in Johilla Coalfield, Madhya Praadesh, India. Palaeobotanist, 32 : 243-252.

BANDHOPADHYAY, R.K. Fossil fauna from Raigarh Gondwana Caolfield, Madhya Pradesh. Coal News, Geological Survey of India.

BHARADWAJ, D.C. and SRIVASTAVA, SURESH C. 1973. Subsurface palynological succession in Korba Coalfield, M.P., India. Palaeobotanist, 20: 137-151.

CHAKRABORTI, B. 2001. New palaeobotonical evidences for the Early Triassic to Middle Triassic age for the lowermost red bed of the Kamthi Formation, Raigarh Gondwana Basin, M.P. Geological Survey of India Special Publication Number, 54: 113-122.

CHAKRABORTI, B. and CHAKRABORTY, SHAMPA. 2001.Triassic floral assemblage from Baronakund, raigarh Gondwana Basin, Madhya Pradesh. Geological Survey of India Special Publication Number, 54:103-122.

CHAKRABORTY, SHAMPA. 2003. Miofloral assemblages of the subsurface Lower Gondwana rocks of Permian in parts of Mahanadi Basin, India. Acta Palaeontologica Sinica, 42: 13-21.

DUTTA, P.K. and SUTTNER, L.J. 1986. Alluvial sandstone composition and palaeoclimate II. Authentic mineralogy. Journal of Sedimentlogy Petrology, 56: 346-358.

JANA, B.N. BHATTACHARYYA, A.P. and CHAKRABORTI, B. 2002. Permian palynological succession from Mand-Raigarh Coalfield, Chhattisgarh. Journal of the Geological Society of India, 59: 537-546.

KAR, R.K. 1973. Palynological delimitation of Lower Gondwanas in the North Karanpura sedimentary basin, India. Palaeobotanist, 20:300-317.

KEMP, E.M.1973. The palynology of Late Paleozoic glacial and deposits of Gondwanaland, Gondwana, Papers from the third Gondwana Symposium, Canberra, Australia.

LELE, K.M. and MAKADA, R. 1972. Studies in Talchir flora of India–7, palynology in Talchir Formation in the Jayanti Coalfield, Bihar. Geophytology, 2: 41-73.

MEENA, K.L. 1999. Palynostratigraphic studies of subsurface Raniganj and Barrenmeasures sediments from Mand-Raigarh Coalfield, Son-Mahanadi, Madhya Pradesh, India. Palaeobotanist, 48: 217-224.

PAL, A.K. 1988. Report on the palaeontological study in the Mand-Raigarh Coalfield, Madhya Pradesh. Geological Survey of India, unpublished report.

RAJA RAO, C.S.1983.Coal resources of Madhya Pradesh, Jammu and Kashmir: Mand-Raigarh Coalfield: Bulletin Geological Survey of India, Volume III, Series A. Number, 45:12-20.

SINGH, I.B. 1976. Mineralogical evidences for the climatic vicissitudes in India during Gondwana times. Geophytology, 6:174-185.

SRIVASTAVA, SURESH C. 1973. Palynostratigraphy of Giridih Coalfield. Geophytology, 3: 166-183.

SRIVASTAVA, SURESH C. 1980a. Miofloral succession of the Lower Gondwana in the North Karanpura Coalfield. Geophytology, 10: 29-33.

SRIVASTAVA, SURESH C. 1980 b. Palynostratigraphy of Lower Gondwana sediments in Chirimiri Coalfield, (M.P.). Geophytology, 10: 62-71.

SRIVASTAVA, SURESH C. and JHA, NEERJA. 1992. Palynostratigraphy of Permian sediments in Manuguru area, Godavari Graben, Andhra Pradesh, India. Geophytology, 22:103-110.

SRIVASTAVA, SURESH C. and JHA, NEERJA 1995. Palynostratigraphy and correlation of Permian-Triassic sediments in Budharam Area, Godawari Graben, India. Journal Geological Society India, 46:647-653.

SRIVASTAVA, S.C. and ANAND-PRAKASH 1984. Palynological succession of the Gondwana sediments in Umaria Coalfield, Madhya Pradesh. Palaeobotanist, 32: 26-34.

TIWARI, R.S. 1973. Palynological succession in Barakar Type area. Geophytology, 3: 166-183.

TIWARI, R.S. and RAM-AWATAR 1989. *Sporae-dispersae* and correlation of Gondwana sediments of Johilla Coalfield, Son Valley Graben, Madhya Pradesh. Palaeobotanist, 37: 94 -114.

TIWARI, R.S and TRIPATHI, A. 1988. Palynological zones and their climatic inference in the coal-bearing Gondwana of peninsular India. Palaeobotanist, 36: 87-101.

TIWARI, R.S. and TRIPATHI, A. 1992. Marker Assemblage-Zones of spore and pollen species through Gondwana Palaeozoic and Mesozoic sequence in India. Palaeobotanist 40: 194-236.

WOPFNER, H. and JIN, X.C. 1993. Baoshan and Tengchong blocks of western Yunnan (China) in the Late Palazoic moasic of the eastern Tethys. IGCP, 321- Gondwana depression and Asian accretation, Third International Symposium and Field Excursion: 47-49.

YANG, W.P. 1993. The Earliest Permian palynological events in the Gondwana and its palaeoclimatic implication. Permophiles, 33: 15 16.

Biostratigraphic significance of megaspores from the Triassic sequence of Peninsular India

Amit K. Ghosh and Jayasri Banerji

Birbal Sahni Institute of Palaeobotany, 53 University Road, Lucknow 226 007, U.P., India

Introduction

Pteridophytes can be categorized into two groups *viz*., homosporous and heterosporous on the basis of their spore production. Some of the pteridophytes produce two different types of spores *i.e.*, micro- and megaspores; the former types give rise to male gametophytes and the latter types give rise to female gametophytes. These are called homosporous and heterosporous respectively. Heterospory is an effective means of reproduction, which was evolved during Late Early Devonian (Scott and Hemsley, 1996). Amongst the present day Pteridophytes it is evidenced in Lycopsida (*Selaginella* and *Isoetes*) and Pteropsida (*Marselia*, *Salvinia* and *Azolla*). According to Chaloner (1967) there is evidence of progressive increase in maximum spore size through the Early and Middle Devonian and then a gradual segregation into small spores (micro/miospores) and large spores (megaspores) *i.e.*, heterospory. Most of the Paleozoic megaspores were produced by heterosporous lycopsids (Bateman and Dimichele, 1994). However, heterospory is not an exclusive phenomenon of lycopsides, during the Carboniferous and Permian it has also been noticed in the Sphenopsids, Noeggerathiopsids and Filicopsids (Pant and Mishra, 1986). Dispersed isospores with a diameter greater than 200 µm are usually considered as megaspores (Zerndt, 1934). According to Pant and Mishra (1986) this lower size limit is not absolute. In accordance wih Bateman and Dimichele (1994) a spore size of 200 µm is sufficient enough to designate megaspores as probably containing archegoniate-bearing gametophytes. Dispersed megaspores are well represented in the Indian Gondwana sediments ranging from early Permian to Early Cretaceous. Significant contributions towards the study of Meoszoic megaspores of Indian Gondwana sediments have been made by Sukh-Dev, 1961; Singh *et al.*, 1964; Maheshwari and Banerji, 1975; Banerji *et al.*, 1978; Pant and Basu, 1979; Banerji *et al.*, 1984; Pal, 1991; Pal *et al.*, 1997; Jana and Ghosh, 1997; Sannigrahi and Pal, 1997; Pal and Sannigrahi, 2002 and Jana, 2004). In the last couple of decades improved techniques *e.g.*, SEM studies of exosporium, nature of inner body etc. have received much attention to the megaspore workers dealing with morphotaxonomy. However, implications of megaspores in biostratigraphic zonation have received very little attention. A number of species of megaspores have proved useful for stratigraphic zonation and correlation, particularly in Europe and at generic level, some of the megaspore taxa are obviously more natural and stratigraphically useful (Kovach and Batten, 1989). It may be mentioned here that a number of megaspore species have proved useful for biostratigraphic zonation and correlation of the Triassic succession in Europe. Qualitatively five different well diversified megaspore assemblages have been reported from the Triassic succession of Damodar Valley Basin and Son valley Basin of Peninsular

India. Significance of megaspores in Triassic biostratigraphy is yet to be established from India and as a matter of fact an attempt has been made to conform the biostratigraphic zonation vis-à-vis correlation of the five megaspore assemblages recorded from Peninsular India.

Stratigraphy of the Triassic succession in peninsular India

In the Peninsular India, Triassic succession is known to occur in Damodar Valley, Koel Valley, Rajmahal, South Rewa, Satpura, Pranhita-Godavari and Mahanadi Valley basins (Fig.1). However, in none of the areas the sequence is complete. Therefore, various lithounits from different areas have been correlated on the basis of paleontological and lithological attributes. A generalized classification of Triassic Gondwana in India is represented in Table 1 (based on Lele, 1969; Bose, 1974; Roy-Choudhury *et al.*, 1975; Sastry *et al.*, 1977; Maheshwari *et al.*, 1978; Pal, 1984a, 1984b; Ram-Awatar, 1988; Kutty *et al.*, 1988; Tiwari and Ram-Awatar, 1990; Tripathi, Tiwari and Kumar, 1990; Srivastava and Jha, 1990; Pal *et al.*, 1991; Pal and Ghosh, 1997).

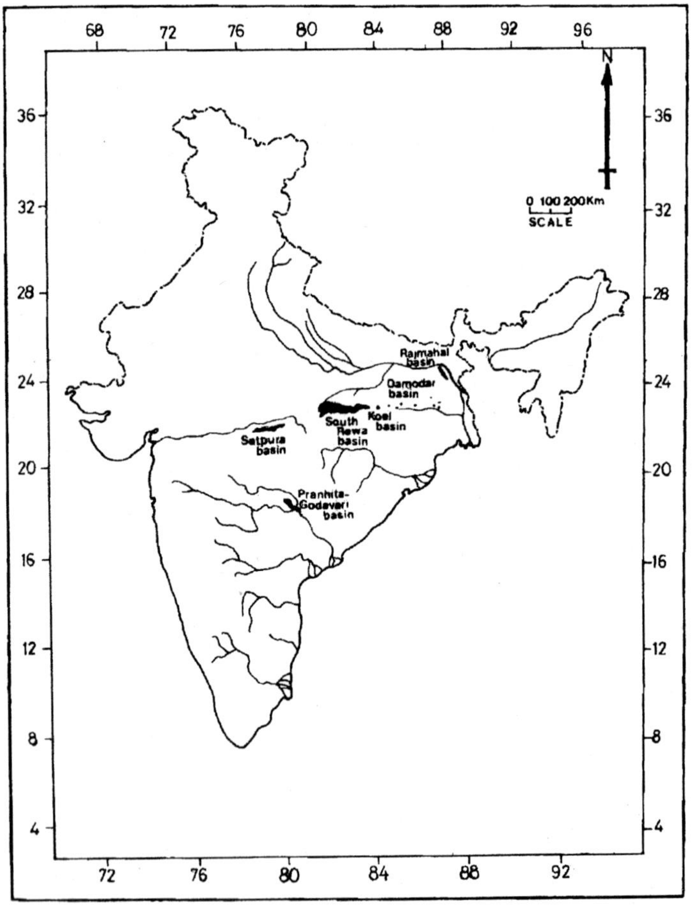

Fig. 1. Map of India showing Triassic occurrences in Peninsular region

Megaspore assemblages from the Triassic succession of peninsular India

Assemblage I. Megaspores from the Maitur (Lower) Member of Panchet Formation (Damodar Valley Basin)

From the Lower Member of Panchet sequence *i.e.*, from the Maitur member of Panchet Formation exposed in the Nonia Nala Section (near Asansol, District Burdwan, West Bengal) megaspores were first described by Maheshwari and Banerji (1975). Later on, Sannigrahi and Pal (1997) further analyzed the megaspore assemblage and studied the frequency distribution of the taxa. The megaspore assemblage recovered from the Maitur Member of Panchet Formation is represented by *Banksisporites gondwanensis* Maheshwari & Banerji, *B. minuticorpus* Maheshwari & Banerji, *B. panchetensis* (Maheshwari & Banerji) Banerji, Kumaran & Maheshwari, *Biharisporites maiturensis* Maheshwari & Banerji, *B. luguensis* Pal, Ghosh & Sannigrahi, *Verrutrilets obscurus* (Maheshwari & Banerji) Banerji, Kumaran & Maheshwari, *V. distinctus* (Maheshwari & Banerji) Banerji, Kumaran & Maheshwari, *Talchirella sinuata* Maheshwari & Banerji, *T. dubia* Maheshwari & Banerji, *T. flavata* (Kar) Bharadwaj & Tiwari, *Pantiella bosei* Maheshwari & Banerji, *P. bharadwajii* Maheshwari & Banerji, *P. maheshwarii* Sannigrahi & Pal, *Maiturisporites indicus* Maheshwari & Banerji, *M. distinctus* Maheshwari & Banerji, *M. spinotriletus* Maheshwari & Banerji, *M. jayasrii* Sannigrahi & Pal ?*Nathorstisporites* and a tetralete form.

Table-1 A generalized classification of Triassic Gondwana in peninsular India

Standard scale		Gondwana basins	DAMODAR VALLEY BASIN	KOEL VALLEY BASIN	RAJMAHAL BASIN	SOUTH REWA BASIN	SATPURA BASIN	PRANHITA-GODAVARI VALLEY BASIN	MAHANADI VALLEY BASIN
LOWER JURASSIC						Hartala Hill Beds		Dharmaram Formation	
TRIASSIC	UPPER	RHEATIAN				Tiki Formation			Pathargarh Beds
		NORIAN			Dubrajpur Formation			Maleri Formation	
		CARNIAN	Supra-Panchet Formation	Mahadeva Formation		Parsora Formation	Bagra Beds	Bhimaram Formation	
	MIDDLE	LADINIAN					Denwa Formation		
		ANISIAN				Nidpur Beds	Pachmari Formation	Yerrapalli Formation	
	LOWER	SYTHIAN	Panchet Formation	Panchet Formation		Panchet Formation /	Almod Formation	Mangli Beds /	Sarimunda Hill Beds
UPPER PERMIAN			Raniganj Formation	Raniganj Formation		Pali Formation	Bijori Formation	Kamthi Formation	Hinjrida Ghati/ Kamthi Fm.

Assemblage II. Megaspores from the Deoli (Upper) Member of Panchet Formation (Koel Valley Basin)

Pal *et al.*, (1997) described megaspores from the Deoli Member, which is considered as an younger member of Panchet Formation. The Deoli (Hirapur) Member (Gee, 1932; Maheshwari and Banerji, 1975) of Panchet Formation is exposed on the southern bank of Dhardharia Nala at the base of Jurwa Village, East Bokaro Coalfield (Presently Jharkhand State). The megaspore assemblage of the Deoli Member of Panchet Formation from the Lugu Hill Section in East Bokaro Coalfield is represented by *Banksisporites tenuis* (Dikstra) Dettmann, *B. dettmannae* Banerji, Kumaran & Maheshwari, *B. gondwanensis* Maheshwari & Banerji, *B. panchetensis* (Maheshwari & Banerji) Banerji, Kumaran & Maheshwari, *Biharisporites luguensis* Pal, Ghosh & Sannigrahi, *Verrutriletes jurwaensis* Pal, Ghosh & Sannigrahi, *Talchirella pantii* Pal, Ghosh & Sannigrahi, *Pantiella bharadwajii* Maheshwari & Banerji, *Maiturisporites indicus* Maheshwari & Banerji, *M. endosporiferous* Pal, Ghosh & Sannigrahi, *M. bokaroensis* Pal, Ghosh & Sannigrahi, *Umiaspora granulosa* Pal, Ghosh & Sannigrahi.

Assemblage III. Megaspores from the Nidpur beds of South Rewa Gondwana Basin

Early Middle Triassic rocks of Nidpur beds are exposed on the western bank of Gopad River near village Nidpur. The megaspore assemblage described by Pant and Basu (1979) from the Nidpur beds of South Rewa Gondwana Basin is characterized by *Grambastisporites nidhpurensis* Pant & Basu, *Srivastavaesporites* (=*Banksisporites*) *triassicus* Pant & Basu, *Trikonia emarginata* Pant & Basu, *Mamillaespora sidhiensis* Pant & Basu, *Nidhitriletes spinosa* Pant & Basu.

Assemblage IV. Megaspores from the Lower Horizon (Harai locality) of Tiki Formation, South Rewa Gondwana Basin

Tiki Formation is named after the village Tiki in Shahdol District, Madhya Pradesh. Banerji *et al.*, (1978) described a megaspore assemblage from the Janar Nala Section (Lower Horizon, Harai locality) of Tiki Formation. The megaspore assemblage comprises *Triletes* sp., *Bokarosporites janarensis* Banerji, Kumaran & Maheshwari, *Banksisporites pinguis* (Harris) Dettmann, *B. dettmannae* Banerji, Kumaran & Maheshwari, *B. panchetensis* (Maheshwari & Banerji) Banerji, Kumaran & Maheshwari, *B. sinuosus* Dettmann, *B. tenuis* (Dijkstra) Dettmann, *Banksisporites* sp. cf. *B. gondwanensis* Maheshwari & Banerji, *Banksisporites sp., Biharisporites sprsus* Banerji, Kumaran & Maheshwari, *Biharisporites* sp., *?Biharisporites* sp., *Verrutriletes distinctus* (Maheshwari & Banerji) Banerji, Kumaran & Maheshwari, *V. minuticorpus* Banerji, Kumaran & Maheshwari, *V. obscurus* (Maheshwari & Banerji) Banerji, Kumaran & Maheshwari, *Bacutriletes* sp., *Horstisporites areolatus* (Harris) Potonié, *Erlansonisporites triassicus* Banerji, Kumaran & Maheshwari, *E. singhii* Banerji, Kumaran & Maheshwari, *?Erlansonisporites* sp., *Hughesisporites variabilis* Dettmann, *Nathorstisporites hopliticus* Jung, *Nathorstisporites* sp.

Assemblage V. Megaspores from the Upper Horizon (Giar locality) of Tiki Formation, South Rewa Gondwana Basin

From the Upper horizon *i.e.*, from Giar locality of Tiki Formation occurrence of megaspores belonging to the genera *Banksisporites, Verrutriletes, Nathorstisprites, Hughesisporites, Bacutriletes, Erlansonisporites, Minerisporites, Maexisporites* and *Aneuletes* has been reported by Pal, 1991; Pal & Sannigrahi, 2002.

Table – 2. Distribution pattern of megaspore genera in the Triassic sequence of Peninsular India.

Age, Horizon, Formation	Lower Triassic Panchet Formation		E. Mid. Triassic Nidpur Beds	Upper Triassic Tiki Fmormation	
Name of Genera	Maitur Member Assem. - I	Deoli Member Assem. – II	Assem. – III	Harai locality Assem.–IV	Giar locality Assem.–V
Jhariatriletes	----------				
Tetralete Form	----------				
Banksisporites	----------	----------	----------------------	----------	----------
Talchirella	----------	----------			
Biharisporites	----------	----------		----------	
Maiturisporites	----------	----------		----------	
Pantiella	----------	----------			
Verrutriletes	----------	----------		----------	----------
Nathorstisporites	-?--?--?-			----------	----------
Umiaspora		----------			
Grambastisporites			----------------------		
Trikonia			----------------------		
Mamillaespora			----------------------		
Nidhitriletes			----------------------		
Triletes				----------	
Bokarosporites				----------	
Horstisporites				----------	
Hughesisporites				----------	----------
Bacutriletes				----------	----------
Erlansonisporites				----------	----------
Minerisporites					----------
Maexisporites					----------
Aneuletes					----------

Correlation

The distribution pattern of megaspore genera recorded from the Triassic sequence of Peninsular India is shown in Table 2. Among the Triassic megaspore assemblages from Peninsular India, the two assemblages from the Lower Triassic of Panchet Formation *i.e.*, Maitur (Lower Member) and Deoli (Upper Member) designated as Assemblage I and Assemblage II are characterized by many common genera *viz*., *Banksisporites, Biharisporites, Talchirella, Maiturisporites, Pantiella* and *Verrutriletes.* Even at least three species, *Banksisporites panchetensis, Pantiella bharadwajii* and *Maiturisporites distinctus* are also common. Unlike Maitur assemblage (Assemblage I), *Banksisporites dettmannae* and *B. tenuis* are present in the Deoli assemblage (Assemblage II). These species are so far known from the

Upper Triassic rocks only. Moreover, occurrence of the genus *Umiaspora* in the Assemblage II is also significant.

The megaspore Assemblage III from the Nidpur beds (Early Middle Triassic) is characterised by *Banksisporites, Grambastisporites, Trikonia, Mamillaespora* and *Nidhitriletes*. Except *Banksisporites* and *Mamillaespora* rest of the genera is unique to Nidpur beds and is not known from elsewhere. Though the megaspore genus *Srivastavaesporites* (=*Banksisporites*) is common to Maitur, Deoli, Nidpur and Tiki assemblages, its two species in the Nidpur assemblage (Assemblage III) are different.

The two assemblages from the Upper Triassic of Tiki Formation *i.e.*, Harai and Giar localities designated as Assemblage IV and Assemblage V respectively are characterized by many common genera *viz*., *Banksisporites, Verrutriletes, Bacutriletes, Erlansonisporites, Hughesisporites* and *Nathorstisporites*. Occurrence of the characteristic Rhaetic – Liassic elements like *Banksisporites pinguis, Horstisporites areolatus, Erlansonisporites triassicus* and *Nathorstisporites hopliticus* in Assemblage IV indicates that the Harai assemblage is much older than Aseemblages I, II and III. The megaspore assemblage (Assemblage V) from the Giar locality of Tiki Formation is slightly older as evidenced by the appearance of genera *Minerisporites, Maexisporites* and *Aneuletes*.

Comparison with the Lower Triassic megaspores of Europe

The Permian megaspore assemblages of India, Brazil and Central Africa are characterized by presence of cavate trilete forms in which the mesosporium has variously aligned cushions. The megaspores having these characteristic features are represented by the genera *Duosporites, Talchirella, Dijkstraea, Gulatriletes* and *Surangeaesporites*. Out of these forms only the genus *Talchirella* extends in the Lower Triassic (Maitur Member and Deoli Member of Panchet Formation; Assemblage I and II). The genus is also reported from the Lower Triassic of Romania (Antonescu and Lantz, 1973) and Poland (Marcinkiewicz, 1976). 'Cushioned' megaspores also have been reported from the Nidpur megaspore assemblage (Assemblage III), however, none of these 'cushioned' forms are present in the Upper Triassic megaspore assemblages of South Rewa Gondwana Basin (Assemblage IV and V).

Comparison with the Upper Triassic megaspores of Australia and Europe

A few Rhaetic – Liassic megaspores are known from Australia (Dettmann, 1961). The following Australian species are common to the Upper Triassic megaspore assemblages of Tiki Formation (Assemblages IV and V): *Banksisporites pinguis, B. sinuosus, Hughesisporites variabilis* and *Nathorstisporites hopliticus*. Besides, the Australian assemblage is also comparable to the Upper Triassic assemblages of India (both the assemblages of Tiki Formation) in having the presence of the genus *Bacutriletes*. However, both the Australian assemblage and the Assemblage IV of Tiki Formation is/are evidenced by the presence of genus *Horstisporites*. Thus, it is evident that the Upper Triassic megaspore assemblage from Tiki Formation is more akin to Rhaetic – Liassic.

The common elements between the Indian Upper Triassic and Greenland megaspore (Harris, 1935) assemblages are *Banksisporites pinguis, Bacutriletes* sp. *Horstisporites areolatus* and *Nathorstisporites hopliticus*. The same species are also known from Germany (Jung, 1960). Moreover, from the Keuper of Germany the megaspore genera *Banksisporites*, *Biharisporites*, *Verrutriletes Horstisporites* and *Nathorstisporites hopliticus* have been reported (Reinhardt and Fricke, 1969).

The Rhaetic – Liassic of Poland (Marcinkiewicz, 1960, 1961, 1962, 1971) has the following genera common to the Upper Triassic megaspore assemblage of Tiki Formation (Assemblage IV): *Banksisporites, Biharisporites, Verrutriletes, Bacutriletes, Horstisporites* and *Nathorstisporites*. The Polish assemblage also is represented by the genera *Minerisporites, Maexisporites, Thomsonia, Aneuletes* and *Dictyothylakos*. Amongst these the genera *Minerisporites, Maexisporites* and *Aneuletes* are known from the Giar megaspore assemblage (Assemblages V) of Tiki Formation.

Conclusion

From the distribution pattern (Table 2) it is evident that the megaspore taxa are endemic up to the Early Middle Triassic (Nidpur beds) starting from the Lower Triassic (Maitur Member and Deoli Member of Panchet Formation). Right from the Upper Triassic (Tiki Formation: Megaspore Assemblage IV and V) cosmopolitan genera *viz*., *Erlansonisporites*, *Horstisporites* etc. start appearing. Megaspores are also significant from biostratigraphic point of view as evidenced in the present case study on Triassic sequence of Peninsular India. Further study, generation of more data is required for finer biostratigraphic zonation. However, emphasis should be given both on qualitative and quantitative analysis and correlation with global data on megaspore biostratigraphy.

References

ANTONESCU, E. and LANTZ, TAUGOURDEAU. 1973. Considérations sur des megaspores et micro spores du Trias inférieur et Moyen de Roumanie. Palaeontographica B, 144: 1-10.

BANERJI, J., JANA, B. N. and MAHESHWARI, H. K. 1984. The fossil flora of Kachchh – II, Mesozoic megaspores. Palaeobotanist, 33: 190-227.

BANERJI, J., KUMARAN, K. P. N. and MAHESHWARI, H. K. 1978. Upper Triassic sporae dispersae from the Tiki Formation: Megaspores from the Janar Nala Section, South Rewa Gondwana Basin. Palaeobotanist, 25: 1-26.

BATEMAN, R.M. and DIMICHELE, W. A.1994. Heterospory: The most iteractive key innovation in the evolutionary history of plant kingdom. Biological Review, 69: 345-417.

BOSE, M. N. 1974. Triassic floras. In: Surange *et al.* (Eds.). Aspects and Appraisal of Indian Palaeobotany, Birbal Sahni Institute of Palaeobotany, Lucknow, pp. 285-293.

CHALONER, W. G. 1967. Spores and land-plant evolution. Review of Palaeobotany & Palynology, 1: 83-93.

DETTMANN, M. E. 1961. Lower Mesozoic megaspores from Tasmania and South Australia. Micropalaeoeontology, 7: 71-86.

GEE, E. R. 1932. The geology and coal resources of Raniganj Coalfield. Mem. Geol. Surv. India, 61: 1-343.

HARRIS, T. M. 1935. The flora of Scresby Sound, East Greenland. Meddr. Greenland, 112(1): 154-167.

JANA, B. N. 2004. Palaeopalynology of Kachchh – I: Mesozoic Megaspores. In: Srivastava P. C. (Eds.) Vistas in Palaeobotany and Plant Morphology: Evolutionary and Environmental Perspectives, Professor D. D. Pant Memorial Volume, pp. 221-227.

JANA, B. N. and GHOSH, A. K. 1997. A megaspore assemblage from the Athgarh Formation and its bearing on the age of the formation. Paleobotanist, 46(1,2): 149-155.

KAVACH, W. L. and BATTEN, D. J. 1989. Worldwide stratigraphic occurrences of Mesozoic and Tertiary Megaspores. Palynology, 13: 247-277.

KUTTY, T. S., JAIN, S. L. and ROY-CHOWDHURY, T. 1988. Gondwana Sequence of northern Pranhita-Godavari Valley: its stratigraphy and vertebrate faunas. Palaeobotanist, 36: 214-229.

LELE, K. M. 1969. The problem of Middle Gondwana in India. Proc. 22nd Int. Geol. Congress, New Delhi, 9: 181-202.

MAHESHWARI, H. K. and BANERJI, J. 1975. Lower Triassic palynomorphs from the Maitur Formation, West Bengal, India. Palaeoentographica B, 152: 149-190.

MAHESHWARI, H. K., KUMARAN, K. P. N. and BOSE, M. N. 1978. The age of the Tiki Formation: with remarks on the miofloral succession in the Triassic Gondwana of India. Palaeobotanist, 25: 254-265.

MARCINKIEWICZ, TERESA. 1960. Analiza megasporowa osadów jurajskich okolic Gorzowa Slaskiego-Praszki. Kwart. Geol. Warszawa, 4(3).

__________,1961. Wyniki badán megasporowych liasu wotworze wiertnicym Gorzów Wielkopolski. Kwart. Geol. Warszawa, 5(4).

__________, 1962. Megaspory retyku I liasu z wiercenia Mechow K/Kamienia Pomorskiego I ich warto é stratyoraficzna. Pr. Inst. geol. Warszawa, 30(III).

__________, 1971.The stratigraphy of the Rhaetian and Lias in Poland based on megaspore investigations. Pr. Inst. geol. Warszawa, 65: 1-57.

__________, 1976. Distribution of megaspore assemblage in Middle Bundsandstein of Poland. Acta Palaeont. Pol., 21(2): 191-200.

PAL, P. K. 1984a. Some fragmentary plant remains from the Hartala Hill, South Rewa Gondwana Basin, India. Palaeobotanist, 32(2): 126-129.

__________,1984b. Triassic plant remains from the Tiki Formation, South Rewa Gondwana Basin, India. Palaeobotanist, 32(3): 253-309.

__________, 1991. Megaspores from the Upper Triassic beds of Giar, Tiki Formation, India. Global Environment and Diversification of plants through Geological time. Professor Birbal Sahni Centenary International Conference, Allahabad (Late Abstract).

__________, 1997. Megafloral zonation of the Permian-Triassic sequence in the Kamthi Formation Talcher Coalfield, Orissa. Palaeobotanist, 46(1,2): 81-87.

PAL, P. K. and SANNIGRAHI, A. 2002. Megaspores from the Upper Triassic sequence of Son Valley Basin and their stratigraphic significance. National Conference on Biodiversity–Past and Present, Palaeobotanical Society, Lucknow (Abstract), p. 49.

PAL, P. K., GHOSH A. K. and SANNIGRAHI, A. 1997. Megaspores from the Panchet Formation of East Bokaro Coalfield, India. Journal of the Palaeontological Society of India, 42: 61-69.

PAL, P.K., CHAKRABORTY, U., GHOSH, A. K. and GHOSH, A. 1991. Triassic plant megafossils from the Kamthi Formation of Talcher Coalfield, India – a new report. Indian Journal of Geology, 63(2): 119-125.

PANT, D. D. and MISHRA, S. N. 1986. On lower Gondwana megaspores from India. Palaeontographica B, 198: 13-73.

PANT, D. D. and BASU, N., 1979. On some megaspores from the Triassic of Nidpuri, India. Review of Palaeobotany & Palynology, 28: 203-221.

RAM-AWATAR. 1988. Palynological dating of Supra-Barakar formations in Son Valley Graben. Palaeobotanist, 36: 133-137.

REINHARDT, P. and FRICKE, D. 1969. Megasporen aus dem Unteren Keuper Mecklenburgs. Mber. Dt. Akad. Wiss. Berl., 11: 399-411.

ROY-CHOWDHURY, M. K., SASTRY, M. V. A., SHAH, S. C., SINGH, G., and GHOSH, S. C. 1975. Triassic floral succession in the Gondwana of Peninsular India. Gondwana Geology, 3rd Int. Gond. Symp. Canberra: 149-157.

SANNIGRAHI, A. and PAL, P. K. 1997. Megaspore succession through the Raniganj-Panchet transition in the Raniganj Coalfield, India. Bull. Bot. Surv. India. 39(1-4): 157-178.

SASTRY, M. V. A., ACHARYA, S. K., SHAH, S. C., SATSANGI, P. P., GHOSH, S. C., RAHA, P. K., SINGH, O. and GHOSH, R. N. 1977. Stratigraphic Lexicon of Gondwana Formations of India. Geol. Surv. India, Miscl. Publ., 36: 1-170.

SCOTT, A.C. and HEMSLEY, A. R. 1996. Palaeozoic megaspores. In: Jansonius, J. Mc Gregor, D.C., (Eds.). Palynology, Principles and Applications. American Association of Stratigraphic Palynologists Foundation, pp. 629-639.

SINGH, H. P., SRIVASTAVA, S. K. and ROY, S. K. 1964. Studies on the Upper Gondwana of Cutch – I. Mio- and macrospores. Palaeobotanist, 12(3): 182-306.

SRIVASTAVA, S. C. and JHA, N. 1990. Permian-Triassic palynofloral transition in Godavari Graben, Andhra Pradesh. Palaeobotanist, 38: 92-97.

SUKH-DEV. 1961. The fossil flora of the Jabalpur Series–3. Spores and pollen grains. Palaeobotanist, 8: 43-56.

TIWARI, R. S. and RAM-AWATAR 1990. Palynodating of Nidpur beds, Son Graben, Madhya Pradesh. Palaeobotanist, 38: 105-121.

TRIPATHI, A., TIWARI, R. S. and KUMAR, P. 1990. Palynology of the subsurface Mesozoic sediments in Rajmahal Basin, Bihar. Palaeobotanist, 37(3): 367-388.

TRIPATHI, C. and SATSANGI, P. P. 1963. Lystrosaurus fauna of the Panchet Series of the Raniganj Coalfield. Pal. Ind. Geol. Surv. India, 37(3): 1-53.

ZERNDT, J. 1934. Les megaspores du basin houller Polonais. Partie1. Bull. Int. Acad. Pol. Sci. Lett. Ser., B1: 1-56.

Santonian in India

K. Ayyasami[1], S. Anantharaman[1] and Ryoji Wani[2]

[1]Palaeontology Division, Geological Survey of India, Bandlaguda, Hyderabad 500068, India
[2]Department of Geology, National Science Museum, 3-23-1, Hyakunincho, Shinjuku-ku, Tokyo 169-0073, Japan

Introduction

The Cretaceous rocks of southern India are represented by both marine and non-marine rocks and occur as detached patches along the southeastern coast of Peninsular India. Among them, the outcrop in the Ariyalur area is the most important one as it is the largest and exposes the rocks ranging in age from Aptian to Maastrichtian. Here, the lower Cretaceous rocks are non-marine in nature and occupy very small patches between the Archaean metamorphic rocks and the marine sedimentaries, recognised as Upper Gondwana beds (Sivaganga/Terani Formation) yielding plant fossils. The upper Cretaceous rocks are chiefly marine and yield numerous invertebrate, vertebrate and microfossils. The marine sediments are rich in foraminifera and ammonites. When correlated with the European Cretaceous, a few faunal breaks in biotic sequence based on fossil ammonites have been recognised (Ayyasami and Rao, 1984; Ayyasami and Banerji, 1984.. Among the Upper Cretaceous stages, the Santonian Stage ranging about 2.3 million years, is important as there was a substantial sea level change which was perceived all over the world (Hancock and Kauffman, 1977) within a time span of about 3 million years. The type area of the Santonian Stage is in Aquintaine region of France. It is not fully developed there and ammonites are rare. For this reason, the Subcommission on Cretaceous stratigraphy, a part of the IUGS, looked elsewhere for the Santonian-Campanian boundary stratotype (Birkelund *et al.*, 1984). The Santonian rocks are found in the United Kingdom, Germany, France, Italy, Gosau, the United States of America, Madagascar, Western Australia and Japan suggesting their widespread distribution on the globe. In India, the rocks of Santonian age are reported from the Cretaceous outcrops of the Cauveri Basin in the areas of Ariyalur (Sastry *et al.*, 1968); Vriddachalam (Banerji, 1966); Pondicherry (Banerji, 1968) and Thanjavur (Banerji, 1973) (Table 1). The Oil and Natural Gas Corporation seemed to have detected the rocks of this age in some of the borewells drilled in the offshore area in Bombay High Oilfield (ONGC, *personal communication*).

Cretaceous Lithostratigraphy of the Ariyalur Area

The lithostratigraphy of the Ariyalur area is well documented by Blanford (1862); Sundaram and Rao (1986) and Sundaram *et al.* (2001). The last mentioned authors divided the rocks into three groups and eight formations with some members and their classification is followed here (Table 2).

Table 1. Santonian in the Ariyalur Cretaceous

AUTHOR	GROUP/AREA	SANTONIAN FOSSIL/ZONE
Stoliczka, 1865-71	Ariyalur Group	*Marsupites* spp., *Inoceramus cripsianus*
Banerji, R. K., 1966	Pondicherry Vriddachalam	*Globotruncana concavata* Zone
Sastry *et al.*, 1968	Trichinopoly Group	*Placenticeras tamulicum* Zone
Ayyasami & Rao, 1979	Ariyalur	Absence of Santonian fossils in the Trichinopoly Group
Chidambaram, L., 2001	Garudamangalam Formation (= Trichinopoly Group)	*Globotruncana coronata* Range Zone
Ayyasami *et al.* (present work)	Ariyalur Group	*Texanites roemeri, Marsupites milleri, Inoceramus balticus*
Ayyasami *et al.* (present work)	Vriddachalam	*Inoceramus balticus*
Ayyasami *et al.* (present work)	Thanjavur	*Micraster coranguinum, Sphenoceramus* sp.

Table 2. Lithostratigraphic classification of the Cretaceous rocks of southern India proposed by various authors

Group (Blanford, 1862)	Ariyalur (modified after Sundaram et al. 2001)	Vriddhachalam (Sundaram&Rao, 1987)	Pondicherry (Rajagopalan, 1965)	Thanjavur	Sivaganga (Banerji, 1982)	Age
ARIYALUR GROUP	Niniyur Formation	No equivalent found in outcrop area	Pondicherry Formation	--	--	Palaeocene
	Kallamedu Fm. Kallankurichchi Fm.	Aladi Fm. Mattur Fm.	Mettuveli Fm. Valudavur Fm.	Ariyalur Group (No formal formational name)	--	Maastrichtian
	Sillkkudi Fm. – Kilpalavur Member	Patti Fm. Parur Fm.	--		--	Campanian to Late Santonian
TRICHINOPOLY GROUP	Anaipadi Fm.		--	--	--	Coniacian
	Kulakkalnattam Fm.		--	--	--	
UTTATTUR GROUP	Karai Fm. – Kunnam Member	--	--	--	--	Turonian
	Karai Fm. – Odiyam Member		--	--	--	Cenomanian
	Dalmiapuram Fm.		--	--	--	Albian
	Aroygyapuram Fm.		--	--	--	
--	Terani Fm.	(Present in subsurface only)	--	--	Sivaganga Fm.	Aptian
ARCHAEAN GRANITIC/GNEISSIC BASEMENT						

Position of Santonian in Ariyalur Area

Stoliczka (1865-71) described *Marsupites* cf. *M. ornatus*, *M. milleri* and *Inoceramus crispianus* from the Ariyalur Group, now considered belonging to Santonian (Plate – 1). Sastry *et al.* (1968)

recognised the occurrence of Santonian rocks in the upper part of the Trichinopoly Group represented by the *Placenticeras tamulicum* Zone. However, Sinha and Das sarma (1970) changed the generic position of *Placenticeras* to *Proplacenticeras* and thus, the range of this Zone has come to be restricted to Coniacian. Following this observation on placenticeratids and on an independent collection of invertebrates, Ayyasami and Rao (1978) argued the absence of Santonian rocks in the Trichinopoly Group. Further, the biostratigraphic zonation based on heteromorph ammonite group by Ayyasami (1990) and on inoceramid bivalves by Ayyasami and Rao (1996) also indicated the absence of Santonian in the Trichinopoly Group. However, Chidambaram (1985; 2000) suggested the presence of Santonian in the Garudamangalam Group based on borehole samples. He identified foraminifera of *Globotruncana coronata* Range Zone covering the Early and Late Santonian. Banerji (1966) classified the rocks of Vriddachalam and Pondicherry areas and assigned *Globotruncana concavata* Zone to represent Santonian in the two subbasins of Cauveri Basin. But, later workers (Govindan, 1970; Ayyasami and Anantharaman, 2001) disputed the presence of *Globotruncana concavata* Zone in these two subbasins and rejected the idea.

Santonian and Campanian Stages in the Ariyalur Area

The presence of Santonian in the Trichinopoly Group was disputed by Ayyasami and Rao (1978) who demonstrated the presence of *Proplacenticeras tamulicum, Puzosia indopacifica, Peroniceras dravidicum, Scaphites yokoyamai* and other fossils of Coniacian age in the upper beds of the Trichinopoly Group. However, Chidambaram (1985, 2000) argued in favour of the presence of Santonian in the Garudamangalam Formation (= Trichinopoly Formation of ONGC, 1976), equivalent to the upper part of the Trichinopoly Group of Blanford (1862). He recognized Early to Late Santonian horizon in a borehole (Kunnam well no. 1 at coordinates: 11° 12' 22": 79° 02' 15") drilled at a location south of Kunnam village, considered by him to fall within the Garudamangalam Formation. An enlarged Fig. 1 of Chidambaram (1985) shows the location of the borehole in the Melmattur Sandstone of the Ariyalur Group itself (Fig. 1, herein) Further, the plotting of the coordinates of the location in a map by Sastry *et al.* (1964) also falls in the area where Ariyalur Group of rocks occurs (Fig. 2). The borehole, now with a defunct hand pump, is actually situated on the western side of the road, instead of on the eastern side as shown in the map by Chidambaram (1985), thus, may require correction. Hence, the absence of Santonian in the Trichinopoly Group as advocated by Ayyasami and Rao (1978) is confirmed and the observations by Chidambaram (2000) need re-evaluation.

The absence of the Santonian in Trichinopoly Group naturally yields scope to find it in the younger Ariyalur Group. As mentioned earlier, the fossils described by Stoliczka (1865-71 from the Ariyalur Group, namely, *Marsupites* cf. *M. ornatus, Marsupites milleri, Inoceramus cripsianus* are referable to Santonian. *Texanites roemeri, Inoceramus balticus* (= *I. cripsianus* Stoliczka) and *Anapachydiscus arrialoorensis* were collected in the basal beds of the Sillakkudi Formation of the Ariyalur Group near Orattur and *Marsupites milleri* at Gudalur by the authors (*see also* Ayyasami, 2001) (Plate – 2). In Thanjavur area, *Inoceramus crispianus* was reported by Vredenburg (1910) from Budalur locality where only the rocks of the Ariyalur Group occur. This fossil is now assigned to *Inoceramus balticus* (Plate – 3). The authors now report *Kossmaticeras theoboldianum, Micraster coranguinum, Damesites sugata* and *Sphenoceramus* sp. from a well cutting near Tatchankurichchi in Thanjavur area. In the Vriddachalam area, the authors report *Inoceramus balticus* from Sendamangalam. By comparison and correlation, the basal Ariyalur bed may be assigned to Late Santonian age in all the three subbasins, namely, Ariyalur, Thanjavur and Vriddachalam subbasins of the Cauveri basin. The Cretaceous rocks of Pondicherry are assigned to only Maastrichtian by Kennedy and Henderson (1992) and Anantharaman and Ayyasami (2001).

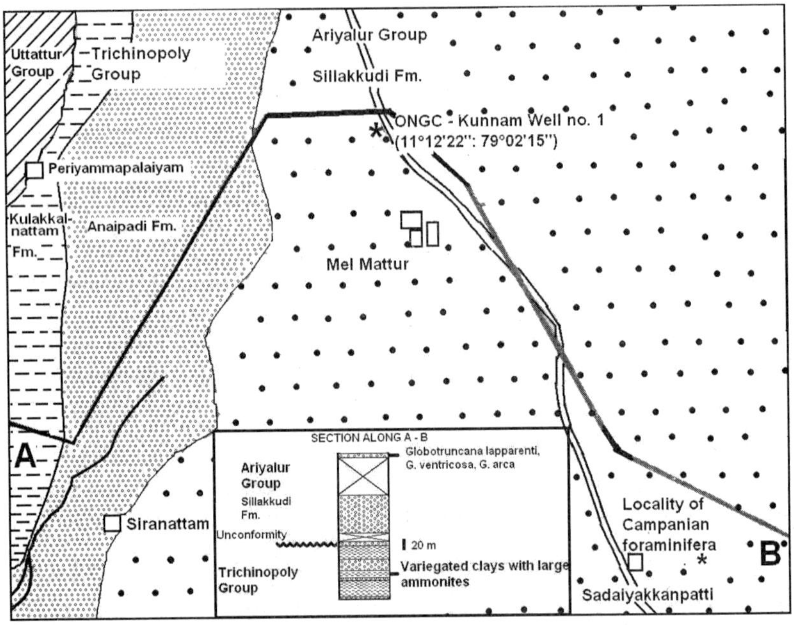

Fig. 1. Enlarged part of a section of the geological map of Chidambaram (1985)

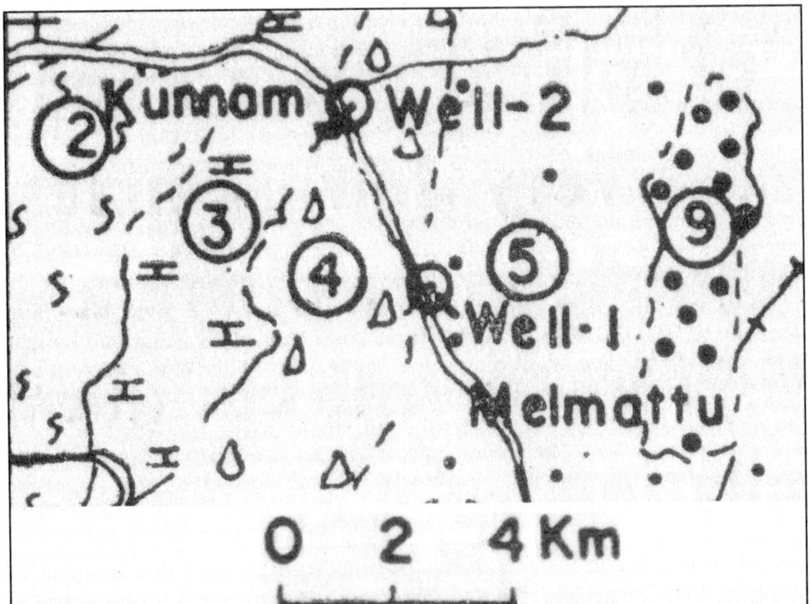

Fig. 2. Geological map and geological section along A–B (Modified after Sastry *et al.*, 1964)

Plate 1.

Plate 1.

A. *Marsupites milleri* Loc: Olaipadi. GSI, Kolkata. Type collection No. 1725
B. & C. *Marsupites* conf. *ornatus* Loc: Olaipadi. GSI, Kolkata Type collection No. 1726 & 1727
D. *Inoceramus balticus* (=*I. cripsianus* Stoliczka) Loc: Olaipadi. GSI, Kolkata. Type collection 1269

Plate 2.

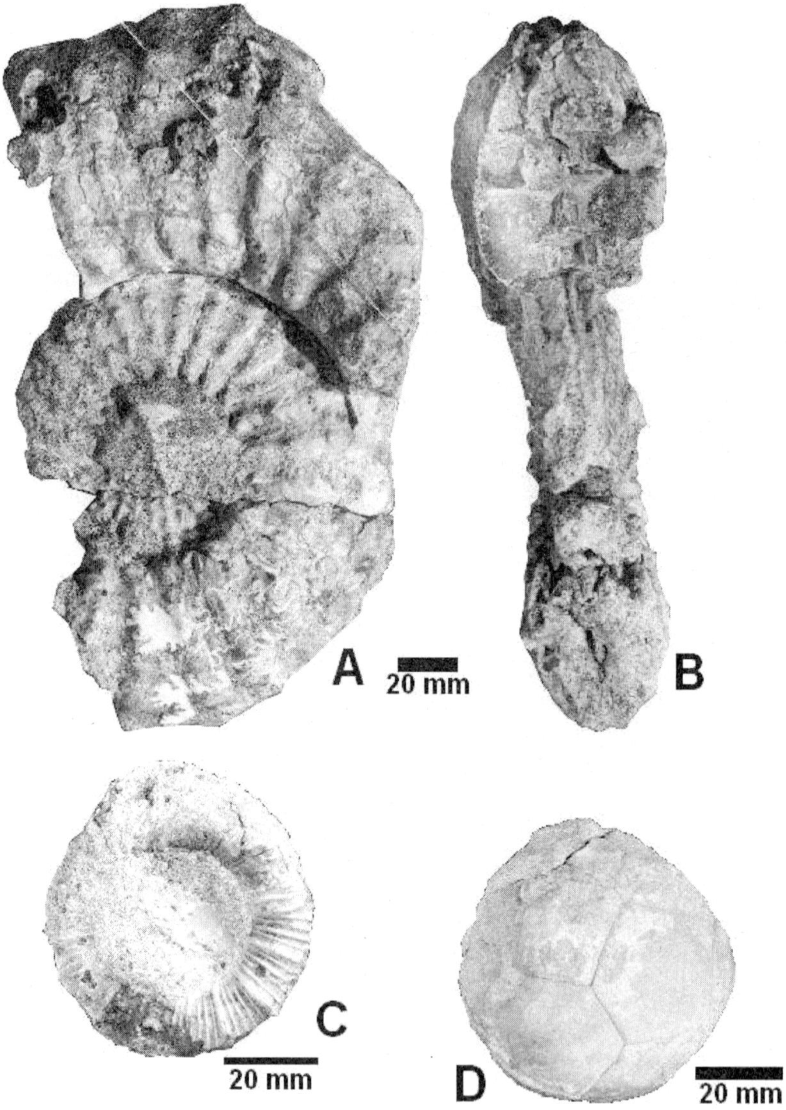

Plate 2. Santonian fossils from Ariyalur area

A. & B. *Texanites roemeri*. Loc: Orattur. GSI, Hyderabad collection. Side and cross sectional views.
C. Gyptoxoceras sp. Loc: Karambium. GSI, Hyderabad collection.
D. *Marsupites milleri* Loc: Gudalur. GSI, Hyderabad collection.

Plate 3.

Plate 3. Santonian fossils from Thanjavur area

A. & B. *Kossmaticeras theobaldianum.* Loc: Tatchankurichchi, Thanjavur. National Science Museum, Tokyo collection. Venter and side views.

C. *Inoceramus cripsianus.* Loc: Budalur, Thanjavur. GSI, Kolkata. Duplicate collection. K12/921

Undoubted rocks of Campanian age are present in the Ariyalur Group as documented by Sastry *et al.* (1968) and many others. The *Karapadites karapadense, Pachydiscus pre-egertoni, Inoceramus simplex* and *Kossmaticeras madrasinus* (Plate – 4) are the Campanian fossils collected to the east of Karapadi very near the type section of Sillakkudi Formation of Sastry *et al.* (1972). From the foregoing observations, a possible Santonian – Campanian boundary within Sillakkudi Formation of the Ariyalur Group is expected in the Ariyalur area.

Plate 4.

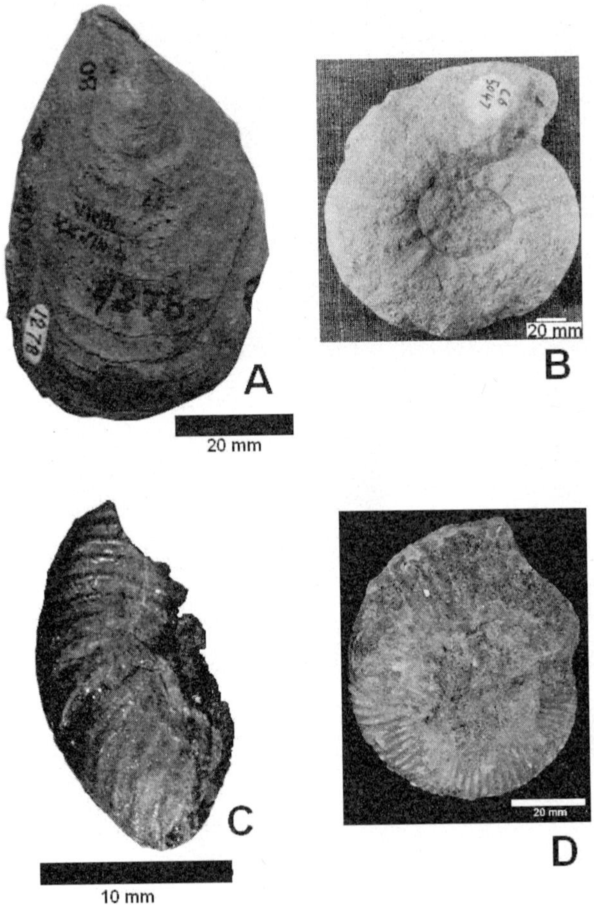

Plate 4. Campanian fossils from Ariyalur area

A. *Inoceramus simplex.* Loc: SW of Mullur. GSI, Kolkata. Type collection no. 1278
B. *Pachydiscus pre-egertonianus.* Loc: Orattur. GSI, Hyderabad collection.
C. *Karapadites karapadense.* Loc: Karapadi. GSI, Hyderabad collection.
D. *Kossmaticeras madrasinus.* Loc: Karapadi. GSI, Hyderabad collection.

A Possible Boundary Stratotype at the Santonian – Campanian Transition in the Ariyalur Area

Gale *et al.* (1987) made a study on the Santonian – Campanian boundary problems in many sections over many continents and also listed sections that stood as candidates for the Santonian – Campanian boundary. The list included Isle of Wight and Dorset sections in England where *Marsupites testudinarius* of the Santonian was followed by *Offaster pilula* of Campanian. In Germany, the Lagerdorf section yielded *Gonioteuthis granulata* and *Marsupites testudinarius* marked the end of Santonian and no characteristic fossil had been identified for the Campanian. In Italy, in the Gubbio section, *Globotruncana elevata* of the Santonian was followed by *Broinsonia parka parca* of the Campanian. The McIntyre gully in the famous Gingin locality in Western Australia yielded *Marsupites testudinarius* of Santonian age, but no Campanian was known from there. In all these sections, no consensus was arrived to select a global stratotype and the meeting in 1995 (Lamolda and Hancock, 1996) ended with the search for global stratotype in Navarra, Spain; Seaford Head, England and Ten Mile Creek, Texas. Yasukova (2002) argued that none of the criteria recommended at the Second International Symposium on Cretaceous Stage Boundaries, Brussels in 1995 – namely, the extinction of the genus *Marsupites*, first appearance of the ammonite *Placenticeras bidorsatum* and the evolutionary first appearance of the belemnite *Gonioteuthis granulataquadrata* could be applied in northeast Russia, Far East Russia or Japan because the taxa in question had not been recorded in these areas. He further stated that the base of Campanian should be linked directly or indirectly to the 33R/34N palaeomagnetic boundary.

The Cretaceous rocks of Ariyalur area offer a scope for the selection of a global stratotype for the Santonian – Campanian boundary. The basal bed of the Ariyalur Group yielded *Inoceramus balticus* and *Anapachydiscus ariyalurensis* at Kannanur, Karambium and Varagur. *Texanites roemeri* was reported from Orattur (Ayyasami, 2001) and *Marsupites milleri* from Gudalur. Therefore, the basal bed of the Sillakkudi Formation can be safely dated as Santonian. Having dated the basal bed of the Sillakkudi Formation, the continuity of section across the sandstone unit is searched for tracing the boundary. There is a nala section to the north and east of Mel Mattur village. It satisfies the criteria demanded by the Subcommission on Cretaceous Stratigraphy (Table 3). A brief account of the section is given below:

Table 3. Comparison at Santonian – Campanian Boundary Fossils In Other Parts of The World

Country	Locality	Santonian	Campanian
England	Isle of Wight and Dorset	*Marsupites testudinarius*	*Offaster pilula*
Germany	Lagerdorf	*Gonioteuthis granulata, Marsupites testudinarius*	-
Italy	Gubbio	*Globotruncana elevata*	*Broinsonia parka parca*
Western Australia	Gingin	*Marsupites testudinarius*	-
U. S. A.	Austin, Texas	*Submortoniceras tequesquitense* range	
U. S. A.	Western Interior	*Desmoscaphites bassleri*	*Scaphites leei III*
India	Ariyalur, Vriddachalam and Thanjavur	*Texanites roemeri, Inoceramus balticus, Marsupites milleri, Marsupites* conf. *ornatus, Micraster coranguinum*	*Karapadites karapadense, Hemiaster cristatus, Inoceramus simplex*

International Commission on Cretaceous Stratigraphy at its Brussels meeting (1995) gave these recommendations:

1. Extinction of *Marsupites*
2. Appearance of *Placenticeras bidorsatum*
3. Appearance of *Gonioteuthis granulataquadrata*

In comparison to the Brussels' recommendations, the Cretaceous rocks of southern India in Ariyalur, Thanjavur and Vriddachalam show at the Santonian – Campanian transition, the following events:

1. Extinction of *Marsupites* (top Santonian)
2. Extinction of *Micraster* (top Santonian)
3. Extinction of *Kossmaticeras* (top Santonian)
4. Extinction of *Damesites* (top Santonian)
5. Extinction of *Inoceramus balticus* (top Santonian)
6. Appearance of *Karapadites karapadense* (Beginning Campanian)
7. Appearance of *Hemiaster cristatus* (Beginning Campanian)
8. Appearance of *Inoceramus simplex* (Beginning Campanian)

The Sillakkudi Formation of the Ariyalur Group ranges from the upper Santonian to Campanian as seen above. A nala bed begins at about 2 km north of Mel Mattur village and traverses to south. Chidambaram (1985) gave the borehole locality of Kunnam well no.1, very near to the road and also to this nala, from where he collected Santonian microfossils (Fig. 1). The borehole data showing the presence of *Globotruncana bulloides*, *Globigerinelloides asperses*, *Heterohelix reussi*, *Bolivinoides strigillatus*, *Ellipsonodasaria alexanderi*, *Gavelinella clementiana convexa*, *G. lorneiana*, *G. monterelensis*, *G. pseudoexcoleta*, *G. stelligera*, *G. tourainensis*, *Globorotalites michelinianus* of *the Globotruncana coronata* Range Zone are of Santonian age for the upper part of the borehole samples from 66 m to the top. Hence, the location of the borehole is still in Santonian. The location of the borehole is on sandstone. Along the nala adjacent to it, the gritty sandstone and clay alternate while sandstone is the predominant lithology. The nala is traced southwards to Sadaiyakkanpatti. Sastry *et al.* (1964) reported *Globotruncana ventricosa* from a sample collected from a locality to the east of Sadaiyakkanpatti. These observations show the boundary of these two Upper Cretaceous stages may be traced in the nala section running almost parallel to the road from Mel Mattur to Sadaiyakkanpatti and offers scope for the Santonian – Campanian boundary in Ariyalur outcrop area.

Acknowledgement

The authors are thankful to the Director General of the Geological Survey of India, Kolkata for permitting the senior author to attend and present the paper at the XIX colloquium on Micropalaeontology and Stratigraphy at the Banaras Hindu University, Varanasi. We thank Prof. M. S. Srinivasan, Professor Emeritus, BHU for fruitful discussions after the presentation of the paper at the colloquium. The views expressed are that of the authors and not necessarily of the departments to which they belong.

References

ANANTHARAMAN, S. and AYYASAMI, K. 2001. A revised age for the Cretaceous rocks of Pondicherry. Indian Minerals, 55, 85-90.

AYYASAMI, K. 1990. Cretaceous heteromorph ammonoid biostratigraphy of southern India. Newsletters on Stratigraphy, Stuttgart. 33 , no.2/3, 111-118.

AYYASAMI, K. and BANERJI, R.K. 1984. Cenomanian-Turonian transition in the Cretaceous of southern India. Bulletin of the Geological Society of Denmark, Copenhagen, .33, 21-30.

AYYASAMI, K. and RAO, B.R.J. 1980. On the possible faunal break at the contact of the Trichinopoly and Ariyalur Groups, Tiruchirappalli district, Tamil Nadu. Proceedings of the VII Indian Colloquium on Micropalaeontology and Stratigraphy, Madras, 1-5.

AYYASAMI, K. and RAO, B.R.J. 1984. Observation on the biostratigraphic zones of the Trichinopoly Group (Upper Cretaceous), Tamil Nadu. Proceedings of the X IndianColloquium on Micropalaeontology and Stratigraphy, Pune, 223-230.

AYYASAMI, K. AND RAO, B. R. J. 1996. Dispersal of Cretaceous inoceramids in Gondwanaland. Proceedings of the 9th International Gondwana Symposium, Hyderabad, 1, 339-354.

BANERJI, R. K. 1966. The genus Globotruncana and biostratigraphy of the Lower Ariyalur Stage (Upper Cretaceous) of Vriddachalam, South India. Journal of the Geological Society of India, 7, 51-57.

BANERJI, R. K. 1968. Late Cretaceous foraminiferal biostratigraphy of Pondicherry area, South India. Memoirs of the Geological Society of India, 2, 30-49.

BANERJI, R. K. 1973. Foraminiferal biostratigraphy and geological evolution of the Thanjavur subbasin, South India. Journal of the Geological Society of India, 14, no.3, 257-274.

BANERJI, R. K. 1982. Sivaganga Formation: its sedimentology, micropalaeontology and sedimentation history. Journal of the Geological Society of India, 23, 450-457.

BIRKELUND, T., HANCOCK, J. M., HART, M. B., RAWSON, P. F., REMANE, J., ROBASZYNSKI, F., SCHMID, F. and SURLYK, F. 1984. Cretaceous Stage boundaries– Proposals. Bulletin geological Society of Denmark, 33, 3-20.

BLANFORD, H. F. 1862. On the Cretaceous and other rocks of South Arcot and Trichinopoly districts. Memoirs of the Geological Survey of India, 4, 217p.

CHIDAMBARAM, L. 1985. Foraminiferal assemblages from upper part of Uttatur Group (Paravay Formation) and Trichinopoly Group (Garudamangalam Formation), Cauvery Basin, South India. Bulletin of the Geological, Mining and Metallurgical society of India, 52, pt.1, 12-19.

CHIDAMBARAM, L. 2000. Middle Turonian – Santonian foraminiferal biostratigraphy of Trichinopoly Group, South India. Memoirs of the Geological Society of India, 46, 173-204.

GALE, A. S., MONTGOMERY, P., KENNEDY, W. J., HANCOCK, J. M., BURNETT, J. A. and Mc ARTHUR, J. M. 1995. Definition and global correlation of the Santonian – Campanian boundary. Terra Nova, 7, 611-622.

HANCOCK, J. M. and KAUFFMAN, E. G. 1977. The great transgressions of the Late Cretaceous. Journal of the Geological Society of London, 136, 175-186.

KENNEDY, W. J. and HENDERSON, R. A. 1992. Non-heteromorph ammonites from the Upper Maastrichtian of Pondicherry, South India. Palaeontology, 36, pt. 2, 381-442.

LAMOLDA, M. A. and HANCOCK, J. M. 1996. The Santonian Stage and substages. Bulletin de l'institut royal des sciences naturelles de Belgique, Sciences de la Terre, Supplement 66, 95-102.

MOHABEY, D. M. 2001. Indian dinosaur eggs: a review. Journal of the Geological Society of India, 58, 479-508.

ONGC. 1976. Rock Stratigraphy of Cauvery basin. Mass Media and Publications Ltd., Chennai. 11p.

RAJAGOPALAN, N. 1965. Late Cretaceous and Early Tertiary stratigraphy of Pondicherry, S. India. Journal of the Geological Society of India, 6, 108 - 121.

SASTRY, M. V. A., MAMGAIN, V. D. and RAO, B. R. J 1964. Progress report of the Geological Survey in the Tiruchirapalli district, Madras. GSI unpublished report for the field season 1962-63, 31p.

SASTRY, M. V. A., MAMGAIN, V. D. and RAO, B. R. J. 1972. Ostracod fauna of the Ariyalur Group (Upper Cretaceous), Tiruchirapalli district, Tamil Nadu. Memoirs of the Geological Survey of India, Palaeontologia Indica, 40, 59p.

SASTRY, M. V. A., RAO, B. R. J. and MAMGAIN, V. D. 1968. Biostratigraphic zonation of the Upper Cretaceous formations of Trichinopoly district, South India. Memoirs of the Geological Society of India, no.2, 10-17.

SINHA, N. K. and DAS SARMA, D. C. 1967. Placenticeras tamulicum (Kossmat) from the Cretaceous of South India. Indian Minerals, 21, no.4, p 369.

STOLICZKA, F. 1865-71. Cretaceous fauna of South India. Memoirs of the Geological Survey of India, Palaeontologia indica, Ser. 4 & 6.

SUNDARAM, R., HENDERSON, R. A., AYYASAMI, K. and STILWELL, J. D. 2001. Lithostratigraphic revision and palaeoenvironmental assessment of the Cretaceous System exposed in the onshore Cauvery Basin, southern India. Cretaceous Research, 22, no.6, . 743-762.

SUNDARAM, R. and RAO, P. S. 1987. Stratigraphy of the Upper Cretaceous rocks of Vriddhachalam area, South Arcot district, Tamil Nadu. Geological Survey of India, Special Publication no. 11, 515-522.

VREDENBURG, E. 1910. Cretaceous rocks of Tanjore. Records of the Geological Survey of India, 40, 336-339.

YAZUKOVA, E. 2002. Ammonite and inoceramid radiations after the Santonian – Campanian bioevent in Sakhalin, Far East Russia. Lethaia, 35, 51-60.

Microfloral remains from Deccan Intertrappeans: Implications on Cretaceous-Tertiary (K-T) transition and paleoecology

Ratan Kar and R.S. Singh

Birbal Sahni Institute of Paleobotany, Lucknow 226007, India

Introduction

The Deccan Traps are one of the largest continental basalt provinces of the world and cover an area of approximately 5,0000 sq, km. in peninsular India. Considering the bulk of lava that erupted and the huge area it covered, the Deccan volcanism was a relatively shorter episode. Past geological studies (K/Ar and Ar/Ar dating) from the trap provinces provided majority of ages in time span of 69-65 Ma. However, recent studies from the west coast basaltic province at Goa (Widdowson *et al.*, 2000) and from the columnar joints at Andheri, Mumbai (Sheth, *et al.* 2001) have provided much younger ages of 62 Ma and 60 Ma years respectively. Considering the post K/T boundary (65 Ma) dates, the Deccan volcanism continued, at least in some parts of peninsular India, during Early Paleocene and is thus a major Cretaceous –Tertiary boundary (KTB) event (Courtillot *et al.*, 1986, 1988, 1990; Duncan and Pyle, 1988; Sahni and Bajpai, 1988). The Deccan volcanic activity has also been linked to the mass extinction at the terminal Cretaceous (McLean, 1985; Officer and Drake, 1985; Courtillot *et al.*, 1988, 1990; Sutherland, 1994). However, one school of thought (Alvarez, *et al.* 1980) proposes the 'impact theory' as the cause of mass extinction at the KTB. Whatever be the reasons of mass extinction, the KTB represents a very crucial phase of earth history. It marked the extinction of many organisms including the dinosaurs, the then dominant species on the earth, and the consequent evolution and establishment of other forms specially the angiosperms (flowering plants).

Deccan Intertrappean Beds

During the quiescent phases of volcanic activity, thin sedimentary beds were deposited in the shallow lakes and water bodies formed by the obstruction of the drainage system by the lava flows. Such intertrappean beds, sandwiched between successive lava flows are exposed in isolated patches mostly on the peripheral areas of the Deccan volcanic province. Good sections of these beds are present at Ranipur and Padwar near Jabalpur, Mahgaon-Kalan in Madhya Pradesh, Nagpur, Anjar, Dayapar and adjacent areas in Kutch, Naskal in Andhra Pradesh, Gurmatkal in Karnataka and Lalitpur in Uttar Pradesh (Fig. 1).

The flora and fauna of these volcano-sedimentary deposits have evoked considerable interest since several decades. Varied flora and fauna seems to have flourished on the cooled basaltic terrain, during the lull in volcanic activity. Ostracodes, molluscs, charophytes, blue- green algae, aquatic ferns, fishes, dinosaurs, other vertebrates and varied plant groups thrived in a period close to the Cretaceous–Tertiary volcanic episode. The above mentioned fossil findings are known from the studies of Lydekker (1890), Sahni (1984a, 1984b), Jain and Sahni (1983, 1985), Srivastava *et al.* (1986), Sahni and Bajpai (1988), Prasad and Sahni (1988), Ghevaria (1988), Prasad (1989), Rana (1990), Prasad and Khajuria (1996), Bhatia *et al.* (1990, 1996), Khajuria and Prasad, (1998), Bajpai and Prasad (2000), Mohabey (1996, 2000, 2001), Mohabey and Udhoji (1996), Khosla and Sahni, (2000, 2003). Palynology of the intertrappean beds has been worked out by Chitaley (1951), Venkatachala and Sharma (1974a,b, 1984), Baksi and Deb (1981), Mathur and Sharma (1990), Prakash *et al.* (1990), Bonde and Kumaran, (1993), Dogra *et al.* (1994, 2004), Sahni *et al.* (1996), Kar and Srinivasan (1998), Kar *et al.* (1998) , Singh and Kar (2002, 2003), Kar and Singh (2003) , Samanta and Mohabey (2003). All the fossil findings reflect a Maastrichtian age for the intertrappean beds, with the exception of Lalitpur (Singh and Kar, 2002, 2003), from where Palaocene marker fossils were recorded.

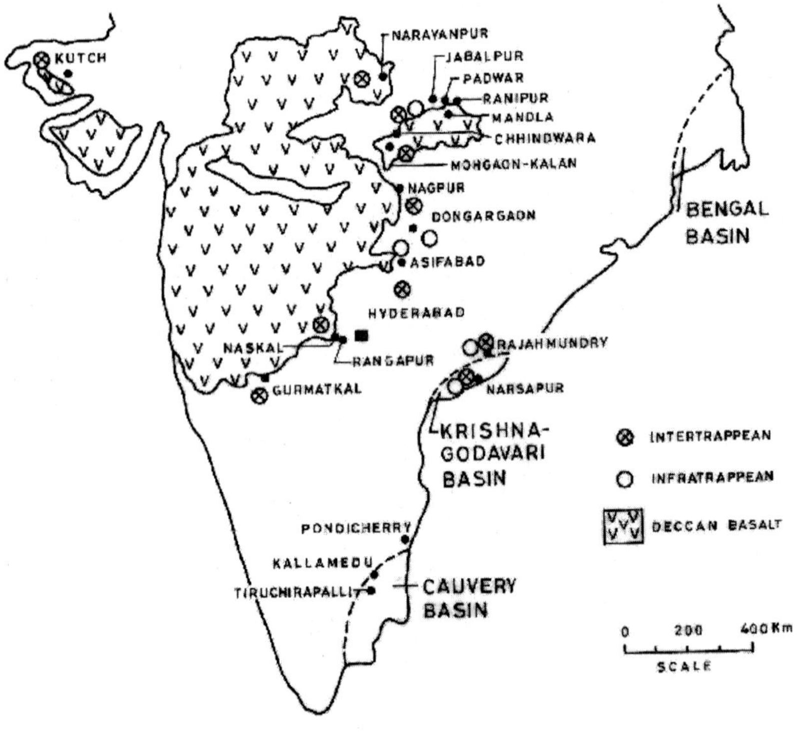

Fig. 1. Map showing the extent of Deccan volcanic province and important infra and intertrappean localities (after Sahni *et al.,* 1996).

Importance of palynological studies with reference to the Intertrappean Beds

Initially, the bulk of paleobotanical work was limited to mega fossil studies where in many fossil woods were reported from the so called 'Deccan intertrappean' (?) beds and were assigned a Paleocene-Eocene age (Prakash, 1960; Bande and Prakash, 1982). However, all the mega floral remains were collected in scattered condition and not from measured sections and hence do not fulfill the demand of precision required for finer stratigraphic correlation. Moreover, the mega floral remains seem to have more affinity to the younger elements and were therefore assigned a Tertiary (mostly Eocene) age. This is of course most unlikely and is not compatible with other fossil records, which advocate a Maastrichtian age.

On the other hand, microfossil remains like pollen and spores are more effective for finer stratigraphic resolutions. Pollen-spores are produced in huge quantities and due to their resistant nature are abundantly preserved in almost all types of sediments and represent the principal groups useful for biostratigraphic correlation. The population of fossil pollen and spores in a sedimentary rock sequence, is thus, the optimum representation of contemporaneous flora. The application of palynology is even more effective for the correlation of intertrappean beds due to dominantly non-marine nature of the deposits.

Materials and methods

Mostly argillaceous sediments having carbonaceous content were taken for palynological analysis. Standard chemical techniques (maceration) were employed for the release of palynomorphs from the sediments. 20-50 gram of samples were crushed to 2mm size, washed with water and kept in plastic jars. Hydrofluoric acid (40%) was then added and kept for 4-5 days. Occasional stirring and adding of fresh Hydrofluoric acid was done to get rid of the silica content. Though, shale and carbonaceous shales reacted to Hydrofluoric acid treatment normally, chert samples took much longer time (2-3 weeks). Thereafter, the material was thoroughly washed with water and Nitric acid (commercial 40%) was added. The time for Nitric acid treatment varied (3-7 days) depending upon the nature of carbonaceous matter – the material was observed at regular intervals under the microscope in watch glass until the palynomorphs became clear. If required, the acid free residue was washed with 5% KOH solution. The residue was then passed through a 400 mesh sieve and washed thoroughly with water. The macerates were mixed in polyvinyl alcohol and strewn on the cover slips and mounted in Canada balsam.

Area of study

Of the typical infra/intertrappean localities, palynological studies have been undertaken from Jabalpur, Ranipur, Padwar, Mohgaon-Kalan, Anjar, Lalitpur and Naskal. It has been observed that at many other localities, the interprappean beds have not yielded pollen-spores. Also, from many localities mentioned above, the sediments have otherwise yielded an abundance of animal fossils like molluscs, ostracodes, fish and dinosaur remains but are generally poor in palynomorphs.. The reason for this is not well understood as pollen-spores are supposedly present in abundant manner especially in the fresh water deposits. It may be due to the alkaline nature of the lakes (Khadkikar *et al.* 1999), as pollen-spores are damaged under such conditions. Also, the lakes and water bodies were mostly ephemeral in nature and the sediments were frequently subjected to sub-aerial exposures. This may have caused the palynomorphs to have got oxidized and are hence poorly recorded in the sediments.

All the known intertrappean localities have yielded a Maastrichtian flora, other than Lalitpur which recorded Paleocene markers. Here, three intertrappean beds at Naskal, Lalitpur and Padwar are looked into, with reference to the utility of palynofossils in the study of K-T transition and its bearing on the ecological conditions prevalent during that point of time.

Naskal Intertrappean

The Intertrappean beds at Naskal have been studied in detail by Prasad (1989), Prasad and Khajuria (1996), Khajuria and Prasad (1998). They recorded a rich fauna comprising of molluscs, ostracodes, mammals, reptiles, amphibians and fishes and advocated a Maastrichtian age for the sediments. Sahni *et al.* (1996) and Kar *et al.* (1998) have commented on some preliminary palynological studies from the area. The intertrappean beds are exposed about 2km northeast of the village of Naskal, Andhra Pradesh and represent the southeastern fringe of the Deccan basalt province. The studied section attains a thickness of about 1m and comprises carbonaceous shale, brown mudstone, chert, massive white mudstone and cross bedded mudstone. The carbonaceous shale has yielded palynofossils, which though quantatively poor, contain certain Maastrichtian marker forms (Fig. 2). The palynoflora is represented by *Mulleripollis bolpurensis, Gabonisporites vigourouxii, Polypodiisporites* sp., *Maculatospora* sp., *Minerisporites* sp., *Cyathidites australis, Cycadopites* sp., *Liliacidites* sp., *? Aquilapollenites indicus, Tricolpites reticulats, Azolla cretaceae, Ariadnaesporites* sp., *Pharagmothyrites* sp. The palynological assemblage closely resembles the Deccan intertrappean assemblage described from Mohgaoan-Kalan, Padwar and Ranipur as all of them contain Maastrichtian marker species *viz. Gabonisporites vigourouxii, Mulleripollis* and *Azolla*. Though, mostly continental forms have been recorded, the presence of dinoflagellate cysts indicate that the depsitional environment was influenced by brackish water conditions. A warm and humid climate is deduced by the presence of fungal spores and fruiting bodies and some pteridophytic spores during the deposition of the sediments.

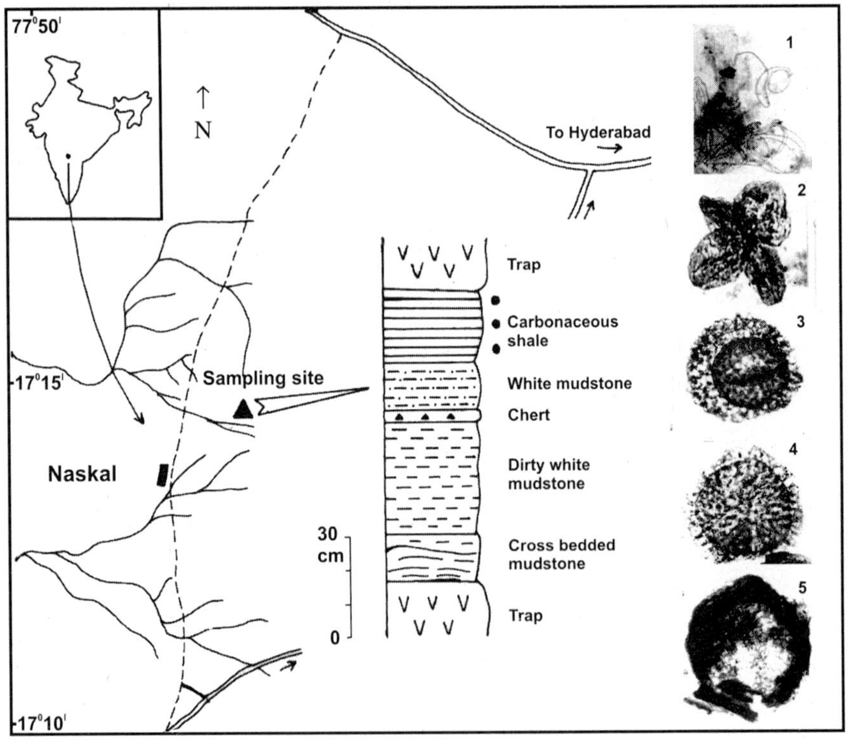

Fig. 2. Location map and litholog of Naskal intertrappean beds (modified after Sahni *et al.* 1996) showing marker Maastrichtian palynofossils (1. *Ariandnaesporites* sp., 2. *Mulleripollis bolpurensis,* 3-4. *Gabonisporites vigourouxii,* 5. *Azolla cretacea).*

Padwar Intertrappean

A good section of intertrappean sediments is present in the subsurface near the village of Padwar, 15 km east of Jabalpur, Madhya Pradesh (Fig. 3). Prakash *et al.* (1990), Sahni *et al.* (1996) and Kar *et al.* (1998) have worked out the dug well samples and have recorded a Maastrichtian palynofloral assemblage. Recently we visited the village and collected samples from the old working and also from a newly dug well in the vicinity (200m east of the old well). The thickness and the lateral extent of the sediments indicate that they were deposited in a relatively large water body. The lake and its surroundings were biologically productive as is evident in the presence of lignitic layer at the base. It seems that thereafter the lake was subjected to shallowing and sub-aerial exposure where in three meters volcanic ash bed was deposited. Biological activity in the water body seems to have diminished as evidenced by the absence of carbonaceous sediments. Favourable conditions again returned and lignitic and carbonaceous shales were deposited in the upper part of the sequence. From the lower part of the sequence, a diverse palynoflora has been recorded which includes *Botryococcus branniini, Nostoc sp., Brazilea sp., Todisporites major, Cyathidites minor, Osmundacidites cephelus, Acrostichumsporites meghalayensis, Gabonisporites vigourouxii, Lycopodiumsporites sp., Lygodiumsporites sp., Ariadnaesporites ariadnae, Laevigatosporites sp., Costatheca sp., Triporoletes reticulatus, Araucariacites australis, Spinizonocolpites echinatus, Mulleripollis bolpurensis and Aquilapollenites bengalensis.* The upper part of the sequence, though having good carbonaceous matter, has so far not yielded marker pollen-spore taxa. Fresh samples have again been collected and further work is in progress for palynostratigraphic resolution in hope that the upper part of the sequence may contain the transitional K-T flora.

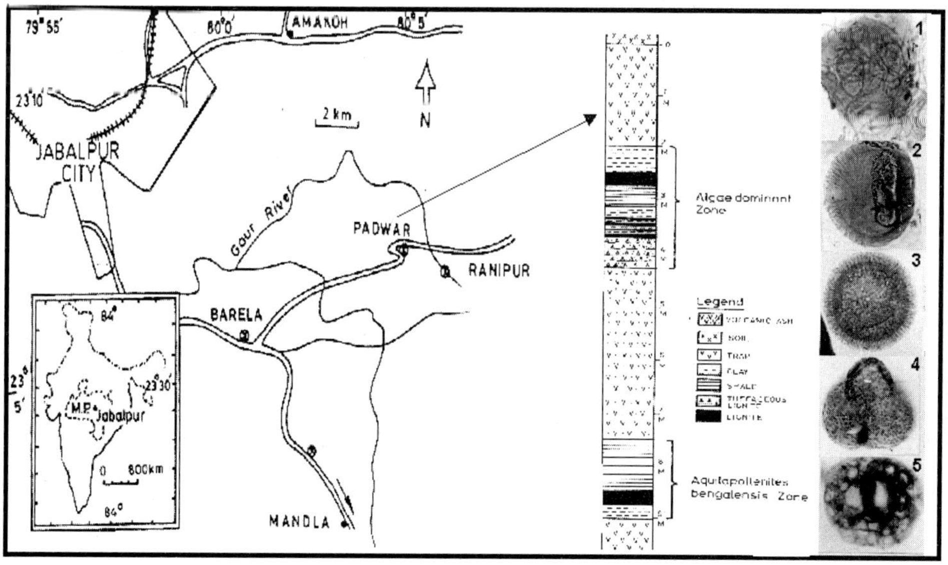

Fig. 3. Location map and litholog of Padwar intertrappean section (modified after Sahni *et al.* 1996 and Kar *et al.* 1998) showing marker Maastrichtian palynofossils (1. *Ariadnaesporites ariadnae*, 2. *Aquilapollenites bengalensis*, 3. *Gabonisporites vigourouxii*, 4. *Acrostichumsporites meghalayensis*, 5. *Azolla cretacea*).

Lalitpur Intertrappean

This represents the northern most extension of Deccan Trap province. The intertrappean beds are exposed in a nala cutting ENE of the village Papro (Fig. 4). As in other intertrappean localities, the basal part of the section here is not formed by the trap basalts, but by the rocks of the Vindhayan Group. The Vindhyan sediments are overlain by a layer of volcanic ash over which the trap basalts lie. The intertrappeans are represented by a chert bed overlying the basalts. The section is capped by a laterite layer, formed by the weathering and leaching of the topmost basaltic rocks. The chert bands have yielded palynofossils which includes *Pharagmothyrites eocaenica, Plochmopeltinites cooksonii, Cyathidites australis, C. minor, Todisporites major, T. kutchensis, Klukisporites* sp., *Foveosporites* sp., *Biretisporites convexus, B. bellus, Lithoparaphyses paproensis, Cingulatisporites* sp., *Dandotiaspora dilata, D. plicata, D. pseudoauriculata, Spinizonocolpites echinatus, Lakiapollis ovatus, Matanomadhiasulcites* sp., Though, the assemblage is quantitavely poor, it is of immense stratigraphic significance as Paleocene marker fossils have been recorded. This is the only record of any fossil, younger than Maastrichtian, from the intertrappean beds. The assemblage varies from other known Deccan intertrappean assemblages of Mohagaon-Kalan , Padwar , Ranipur, Naskal and Anjar from where Maastrichtian marker palynofossils have been recorded. The palynoassemblage of Lalitpur is comparable to the assemblages recorded from that of Madh formation of Kutch (Saxena, 1980; Kar, 1985); Lakadang Sandstone of Meghalaya (Dutta and Sah, 1970; Kar and Kumar, 1986) in having Paleocene marker fossils viz., *Dandotispora dilata, D. plicata, D. pseudoauriculata, Matanomadhiasulcites* and *Lakiapollis ovatus*. The depositional environment was marked by some degree of salanity /marine influence, as evidenced by the presence of halophytic palynomorph remains viz. *Spinizonocolpites echinatus* and *Lithoparaphyses paproensis*, which are related to mangrove elements *Nypa* and *Achrostichum* respectively.

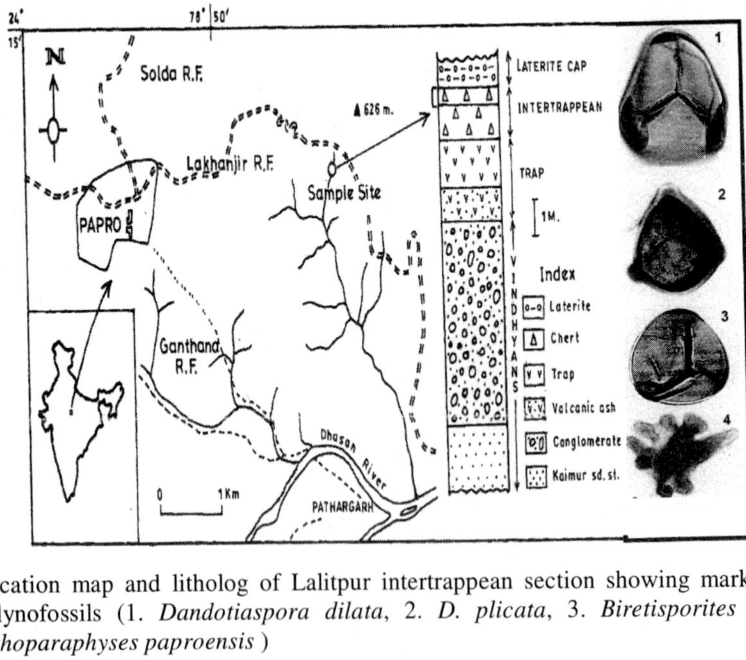

Fig. 4. Location map and litholog of Lalitpur intertrappean section showing marker Paleocene palynofossils (1. *Dandotiaspora dilata*, 2. *D. plicata*, 3. *Biretisporites convexus*, 4. *Lithoparaphyses paproensis*)

Discussion

Though varied paleontological studies have been undertaken for quite sometime now from a number of intertrappean localities and conclusive data thereby generated, the K-T boundary from a section in the Deccan intertrappeans is still lacking. Efforts should be made to record Maastrichtian and Danian (or Paleocene) markers from the same section. Absence of a continuous section is due to the fact that generally intertrappean beds are thin (1-2m), which record only one palynoevent, reflecting one age – either Maastrichtian or Paleocene. Complete continuous sections spanning across the KTB are rare. Efforts have to be made to identify sections which are thick and a number of traps and intertraps are present in succession. As successive intertrappean beds are encountered, which may represent different time zones, it may cut across the KTB and could be stratigraphically important.

The absence of a K-T boundary section is a big lacuna in Indian geology, for which many groups have been trying hard for many years. The difficulty is due to the lack of stratigraphic control above the KTB *i.e.* in Early Tertiary. So far the fossils (dinosaur bones, eggshells, teeth, ostracodes etc.) recorded from various intertrappean sections are confined only upto Maastrichtian, *i.e.* below the KTB. In this context the discovery of Paleocene palynofossils from the intertrappeans of Lalitpur is of paramount significance as it is the only record of Tertiary fossils from the intertrappean beds.

Acknowledgements

R.K. is grateful to DST, New Delhi, for sponsoring a Fast-Track Young Scientist Project. The authors thank Prof. Ashok Sahni for his constant encouragement and to the Director, Birbal Sahni Institute of Paleobotany for permitting R.K. to attend the I.C.M.S at Varanasi and for providing the necessary research facilities.

References

ALVAREZ, L.W., ALVAREZ, W., ASARO, F. and MICHEL, H.V.1980. Extraterrestrial cause for the Cretaceous-Teretiary extinctions. Science 208: 1095-1098.

BAJPAI, S. and PRASAD, G.V.R. 2000. Cretaceous age for Ir-rich Deccan intertrappean deposits: palaeontological evidence from Anjar, western India. Journal of the Geological Society, London 157: 257-260.

BAKSI, S.K. and DEB, U. 1981. Palynology of the Upper Cretaceous of the Bengal Basin, India. Review of Palaeobotany and Palynology 31: 335-365.

BANDE, M.B. and PRAKASH, U. 1982. Palaeoclimate and palaeogeography of central India during the Early Tertiary. Geophytology 12:152-165.

BHATIA, S.B., PRASAD, G.V.R. and RANA, R.S. 1990. Charophytes from the Deccan intertrappean beds near Rangpur, Andhra Pradesh, India. Palaeobotanist 37: 316-323.

BHATIA, S.B., PRASAD, G.V.R. and RANA, R.S. 1996. Maastrichtian non-marine ostracodes from peninsular India: Palaeogeographic and age implications. Memoir Geological Society of India 37: 297-311.

BONDE, S.D. and KUMARAN, K.P.N. 1993. A liliaceous inflorescence from the Deccan intertrappean beds of India. Current Science 65: 776-778.

COURTILLOT, V., BESSE, J., VANDAMME, D., MONTIGNY, R., JAEGER, J.J. and CAPETTA, H. 1986. Deccan flood basalts at the Cretaceous/Tertiary boundary. Earth and Planetary Science Letters 80: 361-374.

COURTILLOT, V., FERAUD, G., MALLUSKI, H., VANDAMME, D., MOREAU, M.G. and BESSE, J. 1988. Deccan flood basalts and the Cretaceous/Tertiary boundary. Nature 333: 843-846.

COURTILLOT, V., VANDAMME, D., BESSE, J., JAEGER, J.J. and JAVOY, M. 1990. Deccan flood basalts at the Cretaceous/Tertiary boundary; data and inferences. Geological Society of America, Special Paper 247: 401-409.

CHITALEY, S.D. 1951. Fossil microflora from the Mohgaon-Kalan beds of the Madhya Pradesh, India. Proceedings of the National Institute of Sciences, India 17: 373-381.

DOGRA, N.N., SINGH, R.Y. and KULSHRESHTHA, S.K. 1994. Palynostratigraphy of infratrappean Jabalpur and Lameta formations (Lower and Upper Cretaceous) in Madhya Pradesh, India. Cretaceous Research 15: 205-215.

DOGRA, N.N., SINGH, R.Y. and SINGH,R.Y. 2004. Palynological assemblage from the Anjar intertrappeans, Kutch district, Gujarat: Age implications. Current Science 86:1596-1597.

DUNCAN, R.A. and PYLE, D.G. 1988. Rapid eruption of the Deccan flood basalts at the Cretaceous/Tertiary boundary. Nature 333: 841-843.

DUTTA, S.K. and SAH, S.C.D. 1970. Palynostratigraphy of the Tertiary sedimentary formations of Assam: 5. Stratigraphy and palynology of the South Shillong Plateau. Palaeontographica 131B: 1-62.

GHEVARIYA, Z.G. 1988. Intertrappean dinosaurian fossils from Anjar area, Kachchha district, Gujarat. Current Science 57: 248-251.

JAIN, S.L. and SAHNI, A. 1983. Upper Cretaceous vertebrates from Central India and their palaeogeographical implications. Proceeding Seminar Cretaceous of India, BSIP, Lucknow: 66-83.

JAIN, S.L. and SAHNI, A. 1985. Dinosaur egg shell fragments from the Lameta Formation of Pisdura, Chandrapur District, Maharashtra. Geoscience Journal 6: 211-220.

KAR, R.K.1985. The fossil flora of Kachchh-IV. Tertiary palynostratigraphy. Palaeobotanist 34:1-279.

KAR, R.K. and KUMAR, M. 1986. Palaeocene palynostratigraphy of Meghalaya, India. Pollen Spore 28: 177-218.

KAR, R.K., SRINIVASAN, S. 1998. Late Cretaceous palynofosils from the Deccan intertrappean beds of Mohgaon – Kalan, Chhindwara District, Madhya Pradesh. Geophytology 27: 17-22.

KAR, R.K., SAHNI, A., AMBWANI, K., SINGH, R.S., 1998. Palynology of Indian onshore–offshore Maastrichtian sequences in India: implication for correlation and palaeobiogeography. Indian Journal of Petroleum Geology 7: 39-49.

KAR, R. and SINGH, R.S. 2003. First Occurrence of fossil paraphyses resembling Acrostichum Linn. from the Lalitpur intertrappean sediments (Palaeocene), Uttar Pradesh, India. Palaeobotanist 52: 81-85.

KHADKIKAR, A.S., SANT, D.A., GOGTE, V. and KARANTH, R.V. 1999. The influence of Deccan volcanism on climate: insights from lacustrine intertrappean deposits, Anjar, western India. Palaeogeography, Palaeoclimatology, Palaeoecology 147: 141-149.

KHAJURIA, C.K., PRASAD, G.V.R. 1998. Taphonomy of a Late Cretaceous mammal – bearing micro vertebrate assemblage from the Deccan intertrappean beds of Naskal, peninsular India. Palaeogeography, Palaeoclimatology and Palaeoecology 137: 153-172.

KHOSLA, A., SAHNI, A. 2000. Late Cretaceous (Maastrichtian) ostracodes from the Lameta Formation, Jabalpur Cantonment area, Madhya Pradesh, India. Journal of the Palaeontological Society of India 45: 57-78.

KHOSLA, A., SAHNI, A. 2003. Biodiversity during the Deccan volcanic eruptive episode. Journal of Asian Earth Sciences 21: 895-908.

LYDEKKER, R. 1890. Notes on certain vertebrate remains from the Nagpur District (with description of a fish-skull by A.S.Woodward). Record Geological Survey of India 23: 20-22.

MATHUR, Y.K., SHARMA, K.D. 1990. Palynofossils and age of the Ranipur intertrappean bed, Gaur River, Jabalpur, M.P. Proceedings of the IGCP 216 & 245 Seminar-cum-Workshop, Chandigarh: 58-59.

McLEAN, D.M. 1985. Deccan traps mantle degassing in the terminal Cretaceous marine extinctions. Cretaceous Research 6: 235-259.

MOHABEY, D.M. 1996. Depositional environment of Lameta Formation (Late Cretaceous) of Nand-Dongargaon inland basin, Maharashtra: the fossil and lithological evidences. Memoir Geological Society of India 37: 363-386.

MOHABEY, D.M. 2000. Indian Upper Cretaceous (Maestrichtian) dinosaur eggs: their parataxonomy and implications in understanding the nesting behaviour. First International Symposium on Dinosaur eggs and babies, Isona, Spain, 1999, 139-154.

MOHABEY, D.M., 2001. Dinosaur eggs and dungs (faecal mass) from Late Cretaceous of Central India: dietary implications. Geological Survey of India Special Publication 64: 605-615.

MOHABEY, D.M. and UDHOJI, S.G. 1996. Fauna and flora from late Cretaceous (Maastrichtian) non-marine Lameta sediments associated with Deccan volcanic episode, Maharashtra: its relevance to the K-T boundary problem, palaeoenvironments and palaeogeoraphy. Gondwana Geology Magazine 2: 349-364.

OFFICER, C.B. and DRAKE, C.L. 1985. Terminal Cretaceous environmental events. Science 227: 1161-1167.

PRAKASH, U. 1960. A survey of Deccan intertrappean flora of India. Journal of Palaeontology 35: 1027-1040.

PRAKASH, T., SINGH, R.Y., SAHNI, A. 1990. Palynofloral assemblage from the Padwar Deccan intertrappean (Jabalpur), M.P. Proceedings of the IGCP 216 & 245 Seminar-cum-Workshop, Chandigarh: 68-69.

PRASAD, G.V.R., 1989. Vertebrate fauna from the infra and intertrappean beds of Andhra Pradesh: Age implications. Journal of the Geological Society of India 34: 161-173.

PRASAD, G.V.R. and KHAJURIA, C.K. 1996. Palaeoenvironment of the Late Cretaceous Mammal-Bearing Intertrappean Beds of Naskal, Andhra Pradesh, India. Memoir Geological Society of India 37: 337-362.

PRASAD, G.V.R. and SAHNI, A. 1988. First cretaceous mammal from India. Nature 332: 638-640.

RANA, R.S. 1990. Alligatorinae teeth from the Deccan intertrappean beds near Rangpur, Hyderabad District, Andhra Pradesh: Further evidence of Laurasiatic elements. Current Science 59: 49-51.

SAHNI, A. 1984a. Upper Cretaceous-Early Palaeogene palaeogeobiography of India based on terrestrial vertebrate faunas. Memoir Geological Society, France 147: 125-137.

SAHNI, A. 1984b. Cretaceous-Paleocene terrestrial faunas of India. Lack of faunal endemism during drifting. Science 226: 441-443.

SAHNI, A and BAJPAI, S. 1988. Cretaceous-Tertiary boundary events: the fossil vertebrate, palaeomagnetic and radiometric evidence from Peninsular India. Journal Geological Society of India 32: 382-396.

SAHNI, A., VENKATACHALA, B.S., KAR, R.K., RAJANIKANTH, A., PRAKASH, T, PRASAD G.V.R. and SINGH, R.Y. 1996. New Palynological Data from the Deccan Intertrappean Beds: Implication for the Latest Record of Dinosaurs and Synchronous Initiation of Volcanic Activity in India. Memoir Geological Society of India 37: 267-283.

SAMANT, B. and MOHABEY, D.M. 2003. Late cretaceous (Maastrichtian) Non-Marine Dinoflagellate (Peridiniates) and Aquilapollenites bearing Polynoassemblage from a new Deccan Inter-Trap near Daiwah River Section, Chandrapur District, Maharashtra. Gond. Geol. Magz. 18(1): 19-26.

SAXENA, R.K. 1980. Palynology of the Matonomadh Formation in the type area, north western Kutch, India (Part-2). Systematic description of gymnosperms and angiospermic pollen grains. Palaeobotanist 26: 130-143.

SHETH, H.C., PANDE, K. and BHUTANI, R. 2001. ^{40}Ar-^{39}Ar age of a national geological monument: The Gilbert Hill basalt, Deccan Traps, Bombay. Current Science 80: 1437-1440.

SINGH, R.S., KAR, R.K. 2002. Palaeocene palynofossils from the Lalitpur intertrappean beds, Uttar Pradesh, India. Journal Geological Society of India 60: 213-216.

SINGH, R.S. and KAR, R. 2003. Palynological Assemblage from Deccan Intertrappean Bed, Lalitpur, Uttar Pradesh, India. Gondwana Geological Magazine Special Volume 6: 217-223.

SRIVASTAVA, S., MOHABEY, D.M., SAHNI, A. and PANT, S.C. 1986. Upper Cretaceous dinosaur egg clutches from Kheda District (Gujarat, India). Palaeontographica A 193: 219-233.

SUTHERLAND, F.L. 1994. Volcanism around K/T boundary time – its role in an impact scenario for the K/T extinction events. Earth Science Review 36: 1-26.

VENKATACHALA, B.S. and SHARMA, K.D. 1974a. Palynology of the Pondicherry area (Cauvery Basin). New botanist 1: 170-200.

VENKATACHALA, B.S. and SHARMA, K.D. 1974b. Palynology of the Cretaceous sediments from the sub-surface of Vridhachalam area (Cauvery Basin). Geophytology 4: 153-183.

VENKATACHALA, B.S. and SHARMA, K.D. 1984. Palynological zonation in sub-surface sediments in Narsapur well no. 1, Godavari-Krishna Basin, India. Proceedings Tenth Indian Colloquium on Micropalaeontology and Stratigraphy, Pune : 445-466.

WIDDOWSON, M., PRINGLE, M.S. and FERNANDEZ, O.A. 2000. A post K-T Boundary (Early Palaeocene) Age for Deccan-type Feeder Dykes, Goa, India. Journal of Petrology 41: 1177-1194.

Paleocene Vertebrate fauna from the Fatehgarh Formation of Barmer District, Rajasthan, Western India

R.S. Rana[1], K. Kumar[2] and H. Singh[1]

[1]Department of Geology, HNB Garhwal University, Srinagar 246 174 UA, India
[2]Wadia Institute of Himalayan Geology, 33 General Mahadeo Singh Road, Dehradun 248 001 UA, India

Introduction

In Rajasthan, vertebrate fossils have been reported from virtually all Late Paleocene-Eocene successions including the Palana and Marh formations of Bikaner (Jolly and Loyal 1985; Paliwal 1999; Kumar *et al.* 2005), the Sanu Formation of Jaisalmer (Das Gupta 1975) and the Akli (Akli Bentonite) and Kapurdi (Fuller's Earth) formations of the Barmer District (Barooah 1950; Tewari 1968; Sahni and Choudhary 1972, 1975; Verma 1982; Rana *et al.* 2005). However, sampling of older Paleocene and Cretaceous beds for systematic vertebrate faunal studies was not carried out even though such beds have long been mapped and occurrence of fish teeth etc. have been noticed before (Deshmukh and Mishra 1971; Verma 1982). The knowledge of the subcontinent's Cretaceous-Paleocene vertebrate faunas, especially the continental mammals is important for paleobiogeographic inferences in the context of India-Asia collision. Some of the Cretaceous and earliest Paleogene successions of Rajasthan and of the neighbouring Gujarat were deposited in complex intermixed environments that allowed coexistence and or mixing of continental and coastal marine faunas. An extensive sampling of these deposits is, therefore, desirable. The Fatehgarh Formation exposed in the Barmer District of Rajasthan was chosen for such a prospecting because: 1) it is stratigraphically older than the Sanu, Akli, and Palana formations of Rajasthan and the Cambay Shale of Gujarat some of which have yielded an admix of continental and marine fauna (Rana *et al.* 2004; Kumar *et al.* 2005), 2) it was assigned a Cretaceous (Aptian-Albian) age (Pareek 1981, 1984; Sinha-Roy *et al.* 1998), and 3) it has been known to contain vertebrate fossils (Deshmukh and Mishra 1971).

The initial collecting was aimed at study of vertebrate fauna of the Fatehgarh Formation for interpreting its age and depositional conditions, and at reviewing the scope of finding mammalian remains. The vertebrate faunal assemblage is completely dominated by fish both in terms of taxonomic diversity and abundance, and also includes very fragmentary material of a crocodile and a mammal. The mammal will be described separately when more complete material is obtained. This contribution is significant as: 1) it is the first systematic record of vertebrate fauna from the Fatehgarh Formation describing the oldest vertebrate fossils of Rajasthan, 2) it provides a basis for more precise dating of the formation than previously available, 3) it provides important paleoecological and paleoenvironmental information, and 4) it shows a good potential for recovery of Paleocene continental vertebrate remains.

Geology And lithostratigraphy

Rajasthan State is diagonally traversed by the northeast-southwest trending Aravalli range that passes through its middle and acts as a barrier for the easterly progressing Thar Desert. The area to the west of this range has Mesozoic and Tertiary succession in three major sedimentary basins, namely, the Jaisalmer Basin in the northwest, the Palana-Ganganagar Basin in the north and the Barmer Basin in the south-central part of the state. Presently, large tracts of these basins are covered by thick desert sands, and exposures are seen only in small patches (Fig. 1). The Fatehgarh Formation with all the vertebrate fossils that form the basis of this paper was deposited in the Barmer Basin, which is an elongated north-south trending graben extending from Mandai in the north to Barmer in the south. The Barmer Basin is bounded by the Barmer-Birmania high on the west, Jodhpur high on the east, Fatehgarh fault on the north and the Malani Igneous suite on the south. It embodies Jurassic-Eocene sediments that were deposited under shallow marine to fluvial conditions (Sinha-Roy *et al.* 1998). Its general geological set up as compiled from Pareek (1981, 1984) and Sinha-Roy *et al.* (1998) is outlined below:

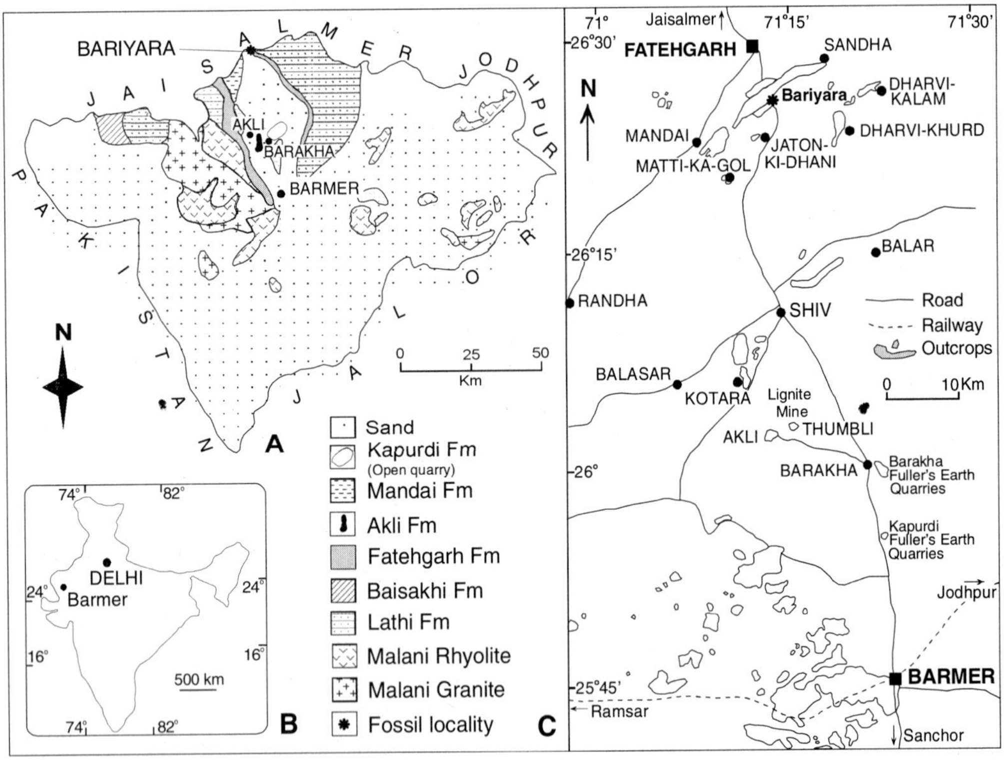

Fig. 1. Geological map of the Barmer District showing the location of Bariyara (A), an inset map of India showing the location of Barmer (B) and a road and rail map of Barmer-Fatehgarh areas showing the location of fossiliferous section at Bariyara (C).

Table 1. Lithostratigraphy of the Rajasthan Basin

Kapurdi Formation (30 m)	Yellowish-white shale like clay with Gypsum bands (Fuller's Earth). *Fossils:* vertebrates, invertebrates and plants	Lower Eocene (Ypresian-Lutetian)
Mandai Formation (100 m)	Conglomerate, dirty white sandstone, ferruginous sandstone and quartzitic sandstone. *Fossils:* Bivalves, gastropods and plants	Paleocene (Thanetian)
Akli Formation (265 m)	Grey and reddish bentonite, greenish grey clay with or without lignite intercalated with siliceous earth and gypsum. *Fossils:* vertebrates, invertebrates and plants	Paleocene (Montian)
	----------Unconformity----------	
Fatehgarh Formation (70 m)	Siltstone, ferruginous siltstone with sub-bentonite, flaggy white kankery sandstone, phosphatic sandstone, pebbly and gritty sandstone and basal conglomerate. *Fossils:* vertebrates and invertebrates	?Cretaceous (Aptian-Albian)
	----------Unconformity----------	
Lathi Formation (360 m)	White and grey sandstone with ferruginous sandstone bands, minor shale, basal conglomerate. *Fossils:* Gastropods, Plants	Jurassic (Lias)

The Lathi Formation, comprising a basal conglomerate and coarse-grained sandstone with hematite nodules, lignite bands and fossilized wood logs, is the oldest litho unit in the Barmer Basin. It is considered as of a Lower Jurassic age based on plant fossils, *Pterophyllum* and *Ptillophyllum* etc. (Verma 1982; Pareek 1984).

The Fatehgarh Formation unconformably overlies the Lathi Formation. It consists of a basal polymictic conglomerate followed by pebbly-gritty, phosphatic and ferruginous sandstones, mudstone, siltstone, and bentonite clay. The pinkish white, parallel laminated sandstone in the lower-middle part of the formation constitutes its major thickness. The yellowish brown to muddy coloured, clayey sandstone in the middle-upper part is rich in gastropods and also contains vertebrates. The top of the formation is marked by a whitish sandstone bed (Fig. 2).

The Fatehgarh Formation is overlain by the Paleocene-earliest Eocene Akli Formation that consists of sandstone and bentonite clays and is rich in diverse invertebrates including foraminifers, vertebrates and plants (Pareek 1981, 1984; Verma 1982). The nature of contact between these two litho units is not well understood but most workers separate these with a hiatus as long as the Late Cretaceous (Pareek 1981, 1984; Sinha-Roy *et al.* 1998). Sisodia and Singh (2000) have shown a 44 m thick Mataji-Ka-Dungar Formation comprising claystone, siltstone and sandstone beds overlying the Fatehgarh Formation, and placed the Akli Formation above that. It is not clear as to whether the Mataji-Ka-Dungar Formation has been carved out of the Akli Formation or it is a separate formation mapped only in some of the sections. Here, we have followed the stratigraphic set up given by Pareek (1984) and Sinha-Roy *et al.* (1998) and these works do not mention the Mataji-Ka-Dungar Formation.

The Fatehgarh Formation crops out on the eastern, northern and western peripheries of the Barmer Basin but the best exposures are in the northwestern part, where they form low lying ridges between Bariyara Village in the east and Jaton Ki Dhani in the west (Fig. 1). Its maximum thickness exposed in the Lordi *Nala* cutting (latitude 26° 25.904' N: longitude 71°12.764' E) near Bariyara Village is around 70 m. Other good sections are exposed at Mandai (latitude 26° 22.554' N: longitude 71° 08.887' E) and Sandha (latitude 25° 29.776' N: longitude 71° 18.662' E) villages (Sisodia and Singh 2000). This study was limited to the Bariyara section (Fig. 2).

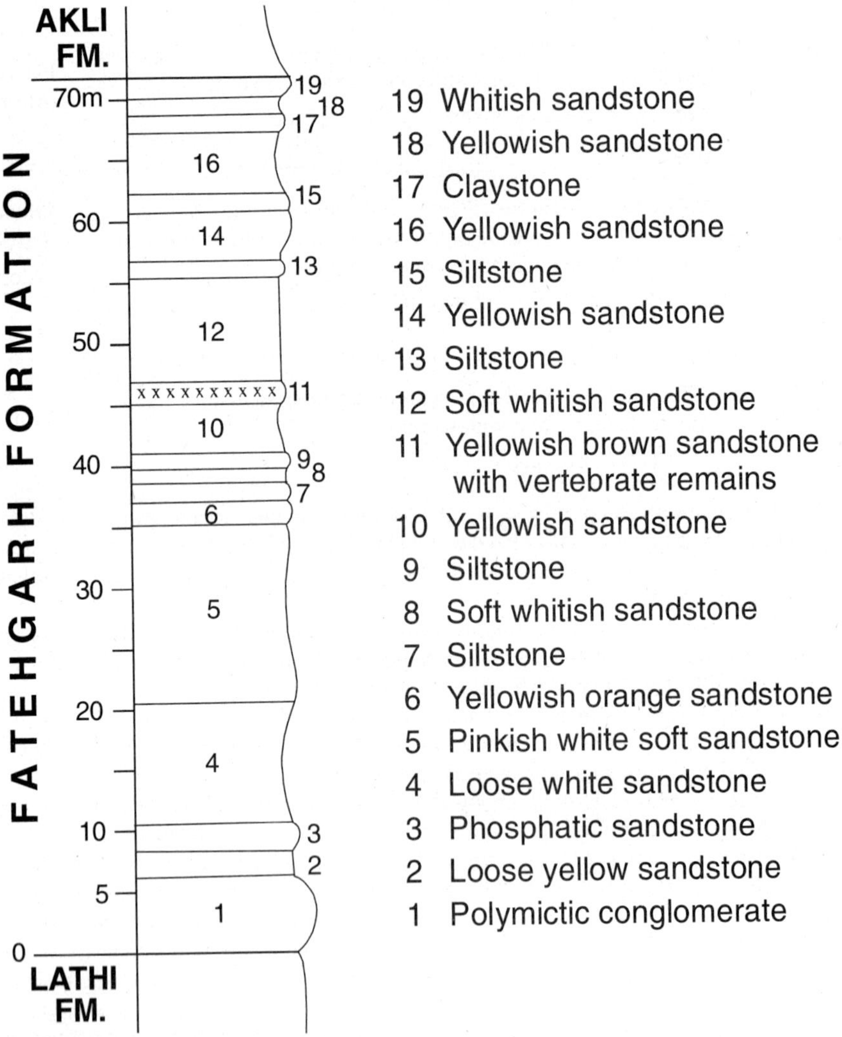

Fig. 2. Measured log of fossiliferous section exposed at Bariyara Village showing the vertebrate fossil yielding level.

Provenance of fossil material

All the fossil specimens described here were recovered from the same stratigraphic level, *i.e.* friable yellowish brown clayey sandstone that is 2 m thick and lies about 46 m above the base and 25 m below the top of the Fatehgarh Formation, and from the same geographic locality, *i.e.*, at Bariyara Village (latitude 26° 26' N: longitude 71° 04' E), about 6 km south of the township of Fatehgarh in the Barmer District of western Rajasthan (Figs. 1, 2). The sight of abundant gastropod shells and of some fish teeth on outcrop surface prompted us to sample this horizon for microvertebrates. None of the other sampled horizons yielded any vertebrate remains.

Material and methods

The fossil material comprises dental plates, isolated teeth, spines, scales, vertebrae, dermal denticles and incisor teeth. Twelve taxa are identified based mainly on their dentitions and diagnostic scales. They are represented by ray fish, *Dasyatis* sp., *Myliobatis* sp.1, *Myliobatis* sp. 2, *Myliobatis* sp. 3; a gar, *Lepisosteus* sp.; a halecostomian, *Pycnodus* sp., teleosteans, Osteoglossidae gen. et. sp. indet., *Apateodus* sp., *Enchodus* sp., *Stephanodus lybicus*, *Eotrigonodon indicus*, and an indeterminable crocodilian. A very fragmentary tooth in the collection has been identified as belonging to a mammal; it will be described later when more complete material is found. In view of the dissociated nature of fossils the specific identification in most cases has not been attempted.

In describing ray fish teeth, terminology of Cappetta (1987) has been used and in classifying the fauna works of Greenwood *et al.* (1966), Cappetta (1987) and Carroll (1988) followed. In measurements of myliobatiform teeth length refers to labio-lingual dimension, and width equals mesio-distal dimension.

All the fossils reported herein were recovered by screen-washing of the bulk samples only with water. By processing 500 kg of rock sample we recovered over a thousand dental and skeletal elements all of which have a distinctive light brownish colour. Isolated teeth of dasyatid ray fish are by far the most common elements in the collection followed by myliobatids, pycnodontids and osteoglossids. Crocodiles and mammals are represented only by a fragmentary tooth each.

Repository: All specimens described and illustrated (FV/R/GU, Fatehgarh Vertebrates/ Rajasthan/ Garhwal University) in this article are housed in the Laboratory of Vertebrate Paleontology, Department of Geology, HNB Garhwal University at Srinagar UA, India.

Systematic Descriptions

Class	CHONDRICHTHYES Huxley 1880
Subclass	ELASMOBRANCHII Bonaparte 1838
Superorder	BATOMORPHII Cappetta 1980
Order	MYLIOBATIFORMES Compagno 1973
Superfamily	DASYATOIDEA Whitley 1940
Family	DASYATIDAE Jordan 1888
Genus	DASYATIS Rafinesque 1810

Dasyatis sp.

Plate 1, figures 1-11

Material: FV/R/GU 6001-6300, isolated median and lateral teeth.

Description and Remarks: Median teeth characterised by prominent transverse crest and short and broad lingual crown-face; lingual visor uniform and not tapering; labial crown-face convex with interconnected ridges and grooves; lateral teeth have longer lingual crown-face with a prominent median ridge flanked by lateral depressions.

Dasyatis is commonly known from most of the Late Cretaceous-Eocene successions in the peninsular India as well as the sub-Himalaya (Loyal 1983; Kumar and Loyal 1987; Kumar 1989; Prasad 1989; Rana 1990; Kumar 1996; Rana *et al.* 2004; Rana *et al.* 2005). *Dasyatis* sp. teeth from the Lower Eocene Kapurdi Formation of Rajasthan (Rana *et al.* 2005, pl. 2, figs. 11-17) have a longer lingual crown-face with a tapering visor and scalloped ornamentation as against nearly uniform visor and interconnected ridge and groove ornamentation in teeth from Fatehgarh. *Dasyatis* sp. 1 teeth from the Lower Eocene Cambay Shale of Vastan, Gujarat differ from Fatehgarh teeth in similar features, whereas *Dasyatis* sp. 2 teeth also from Vastan differ in having a flat labial crown-face with rugged ornamentation besides a tapering lingual visor (Rana *et al.* 2004, fig. 3; 7-12). *Dasyatis,* commonly called whiptail stingrays are essentially bottom dwelling coastal fish often penetrating estuaries and rivers and preferring tropical and subtropical temperatures (Wheeler 1985).

Superfamily MYLIOBATOIDEA Compagno 1973
Family MYLIOBATIDAE Bonaparte 1838
Genus MYLIOBATIS Cuvier 1817

Myliobatis sp. 1
Plate 1, figures 12-14
Material: FV/R/GU 6301-6310, isolated median teeth.
Description: Median teeth four times as broad as long; laterally narrower and lower; labio-lingually arched with convex lingual and concave labial side; labial and lingual crown-faces wrinkled with baso-apical ridges; lingual side has a ridge along the base of crown with a parallel groove above it; occlusal surface laterally tapering and rough with irregular margins; crown-height and root-depth maximum in central part of tooth and decrease laterally; root extends beyond lingual margin of crown; basal surface of root flat with around twenty rectangular root lobes separated by narrower grooves.

Remarks: Myliobatis is the most common eagle ray in the Paleocene-Eocene beds of Rajasthan (Rana *et al.* 2005), Gujarat (Mishra 1980; Bajpai and Thewissen 2002; Rana *et al.* 2004) and the northwest sub-Himalaya (Kumar and Loyal 1987; Kumar 1996). The teeth of *Myliobatis* sp. 1 from Fatehgarh compare well with those of *Myliobatis* sp. 1 from the Cambay Shale, Vastan in labio-lingual arching but differ in having less lingually shifted roots with broader lobes (Rana *et al.* 2004, fig. 3; 19-20). *Myliobatis* sp. from the Kapurdi Formation (Rajasthan) has more lingually shifted root with lingually angular root lobes (Rana *et al.* 2005; pl. 2, fig. 24). *Myliobatis* sp. 1 teeth from Fatehgarh also resemble those of *M. wurnoensis* from the Thanetian of Niger (Cappetta 1972, pl. 8, figs. 10-11) but are up to 75 percent smaller. *Myliobatis tewarii* from the Middle Eocene of Kachchh has flat rectilinear median teeth about 7 times as broad as long with clearly hexagonal crowns (Mishra 1980), whereas *Myliobatis* sp. from the Lower Eocene Panandhro Lignite (Gujarat) has *M. tewarii* like median teeth that are only 5 times or less as broad as long (Bajpai and Thewissen 2002). *Myliobatis* sp. teeth from Eocene of Khasi Hills, Meghalaya (Satsangi and Bora 1980) look identical to those of *Myliobatis tewarii* from Kachchh.

Myliobatis sp. 2
Plate 1, figures 15-16
Material: FV/R/GU 6315-6320, isolated median teeth.
Description: Median teeth up to 50 percent smaller and less labio-lingually arched than those described above as *Myliobatis* sp. 1; occlusal surface hexagonal, more or less flat and unornamented; crown length uniform; root lobes and grooves subequal.

Myliobatis sp. 3
Plate 1, figures 17-24

Material: FV/R/GU 6331-6400, isolated lateral teeth.

Description: Lateral teeth about twice longer than broad; occlusal surface irregularly shaped in some teeth but generally elliptical and distinctly pitted; a small ridge present on lingual side; root as elongated as crown, bilobed or trilobed with numerous fine pores; maximum length of teeth ranges from 1.08 to 3.5 mm; median teeth with matching ornamentation yet to be found.

Remarks: Myliobatis sp. 3 is based only on lateral teeth that cannot be associated with the median teeth of *Myliobatis* sp. 1 and *Myliobatis* sp. 2 described above because their occlusal surface is distinctly pitted as against more or less smooth surface in the two species. Some of the lateral teeth can possibly be associated with either of the two species if it is assumed that all median teeth in the collection may have lost their ornamentation due to heavy wear. This however, seems unlikely because all lateral teeth are distinctly ornamented, justifying their attribution to a separate species.

Similarly shaped teeth with bilobed or trilobed roots have recently been described as *Myliobatis* sp. 2 from the Lower Eocene Cambay Shale of Vastan (Rana *et al.* 2004, fig. 3; 21-24). However, they possess a triangular, flat and rough occlusal surface and wrinkled labial crown edge as against regularly pitted occlusal surface and more or less smooth crown edge in the Fatehgarh species. Similar teeth have been described from the Paleocene Cannonball Formation of South Dakota, USA as *Myliobatis* sp. (Cvancara and Hoganson 1991) but they have nearly smooth and rhomboid occlusal surface as against elliptical and pitted one in the Fatehgarh species.

Myliobatis, commonly called eagle rays, have a circumglobal distribution in tropical seas and more commonly in temperate continental shelves. Like dasyatids they are essentially benthic thriving over muddy and sandy bottoms of coastal waters, bays and estuaries. In India they occur along the coastal areas, the mouth of Ganges and in the Chilka Lake.

Class	OSTEICHTHYES Huxley 1880
Subclass	ACTINOPTERYGII Cope 1887
Infraclass	NEOPTERYGII Regan 1923
Division	GINGLYMODI Cope 1872
Order	LEPISOSTEIFORMES Berg 1940
Family	LEPISOSTEIDAE Cuvier 1825
Genus	LEPISOSTEUS Lacépède 1803

Lepisosteus sp.

Plate 1, figures 25-26

Material: FV/R/GU 6421, an isolated tooth; 6431-6435, isolated ganoid scales.

Description: Tooth conical, slightly recurved with distinct apical and lower parts; apical part smaller and acute but blunt, smooth, and having translucent enameloid; lower part enameloid-less and tubular; it broadens gradually down to the base and possesses parallel longitudinal striations; striations absent near junction of apical and lower parts and more prominent towards base. Ganoid scales typical gar-like, *i.e.*, diamond-shaped; ganoine layer has some surface pores and concentric growth lines.

Remarks: The Fatehgarh gar material is here referred to *Lepisosteus* sp. because this taxon has been reported widely from the Late Cretaceous-?Earliest Paleocene beds of Madhya Pradesh (Hora 1938, pl. 18, figs.2-4), Maharashtra (Jain and Sahni 1983; Gayet *et al.* 1984; Rana 1990), Andhra Pradesh (Prasad and Sahni 1987; Prasad 1989) and Gujarat (Bajpai *et al.* 1990) as also from the Eocene beds of Rajasthan (Kumar *et al.* 2005) and the sub-Himalaya (Rana and Kumar 1990). Most of these records are based on isolated teeth, scales and vertebrae; only *L. indicus* from the Late Cretaceous Lameta Formation of Dongargaon is known by a skull (Woodward 1890, 1908). *Lepisosteus* is widely known from Cretaceous-Eocene sequences of Madagascar, North America, South America, Europe and the Indian subcontinent (Wiley 1976; Astibia *et al.* 1990; Bertini *et al.*

1993; Gottfried and Krause 1998; Weems 1998; Kumar *et al.* 2005), but presently it is restricted to North America (Wheeler 1985).

Division	HALECOSTOMI Regan 1923 *sensu* PATTERSON 1973
Order	PYCNODONTIFORMES Berg 1937
Suborder	PYCNODONTOIDEI Nursall 1996
Superfamily	PYCNODONTOIDEA Poyato-Ariza and Wenz 2002
Family	PYCNODONTIDAE Agassiz 1833
Genus	PYCNODUS Agassiz 1833

Pycnodus sp.

Plate 1, figures 27-30

Material: FV/RGU 6451-6460, isolated vomerine and splenial teeth.

Description: Teeth transversely elongated, elliptical, oval, or triangular with blunt margins; in some elongated teeth occlusal surface bears two transverse ridges that meet each other laterally; in oval and triangular teeth a single circular ridge is present around a depression in the middle of occlusal surface; some teeth lack ridges.

Remarks: Although known from most of the Late Cretaceous-Eocene localities of India pycnodontids were previously not reported from Rajasthan. *Pycnodus* is widely known by several species in the Cretaceous-Early Tertiary deposits of India. *Pycnodus* teeth from Fatehgarh are comparable with those of commonly known species, *P. bicresta* (Plate 1, figs. 27, 29), *P. toliapicus* (Plate 1, fig. 30) and *P. lametae* known from the Late Cretaceous-Eocene successions of the sub-Himalaya (Kumar and Loyal 1987; Kumar 1996) as well as the peninsular India (Woodward 1908; Gayet *et al.* 1984; Prasad and Sahni 1987; Prasad 1989; Rana 1990; Prasad and Khajuria 1990; Prasad and Singh 1991). They seem to belong in more than one species, but their isolated nature does not allow specific assignments.

Plate 1.

Scale bar represents 1 mm in all figures except Figs. 9,22,32 and 43 where it equals 0.1 mm

1-11 *Dasyatis* sp. – (1) FV/R/GU 6002, occlusal/ labial view; (2) 6003, lateral view; (3) 6004, basal view; (4) 6005, occlusal/ lingual view; (5) 6006, labial view; (6) 6007, lingual view; (7) FV/R/GU 6008, lingual view; (8) FV/R/GU 6009, lingual view; (9) 6009, crown ornamentation; (10) 6010, labial view; (11) 6011, lingual view. **12-14** *Myliobatis* sp. 1 – FV/R/GU 6301, median tooth in occlusal (12), lingual (13) and basal views (14). **15-16** *Myliobatis* sp. 2 - FV/R/GU 6315, median tooth in occlusal (15) and basal views (16). **17-24** *Myliobatis* sp. 3 - (17) FV/R/GU 6334, lateral view; (18) 6332, basal view; (19) 6338, lingo-lateral view; (20) 6333, lingual view; (21) 6336, occlusal view; (22) 6336, occlusal surface at high magnification; (23) 6335, lateral view; (24) 6331, lateral view. **25-26** *Lepisosteus* sp. – (25) FV/R/GU 6421, tooth in lateral view; (26) 6431, ganoid scale in surface view. **27-30** *Pycnodus* sp. – (27) FV/R/GU 6451, occlusal view; (28) 6452, basal view; (29) 6455, occlusal view; (30) 6454, occlusal view. **31-38** Osteoglossidae gen. et sp. indet. – (31) FV/R/GU 6543 isolated scale, outer view; (32) 6543, outer view at high magnification; (33) 6541 outer view; (34) 6542, outer view; (35) 6545, inner view; (36) 6501, isolated tooth in lateral view; (37) 6502 lateral view; (38) 6503, lateral view. **39-42** *Apateodus* sp., isolated tooth – (39) FV/R/GU 6602, lateral view; (40) 6603, lateral view; (41) 6604, lateral view; (42) 6605, lateral view. **43** *Enchodus* sp., FV/R/GU 6851, isolated tooth, lateral view. **44-46** *Stephanodus lybicus*, pharyngeal teeth – (44) FV/R/GU 6883, lateral view; (45) 6884, lateral view; (46) 6885, lateral view. **47-49** *Eotrigonodon* sp. pharyngeal and oral teeth in lateral views – (47) FV/R/GU 6951; (48) 6952; (49) 6953. **50** Crocodylidae indet. - FV/R/GU 7001, isolated tooth in lateral view

Plate 1.

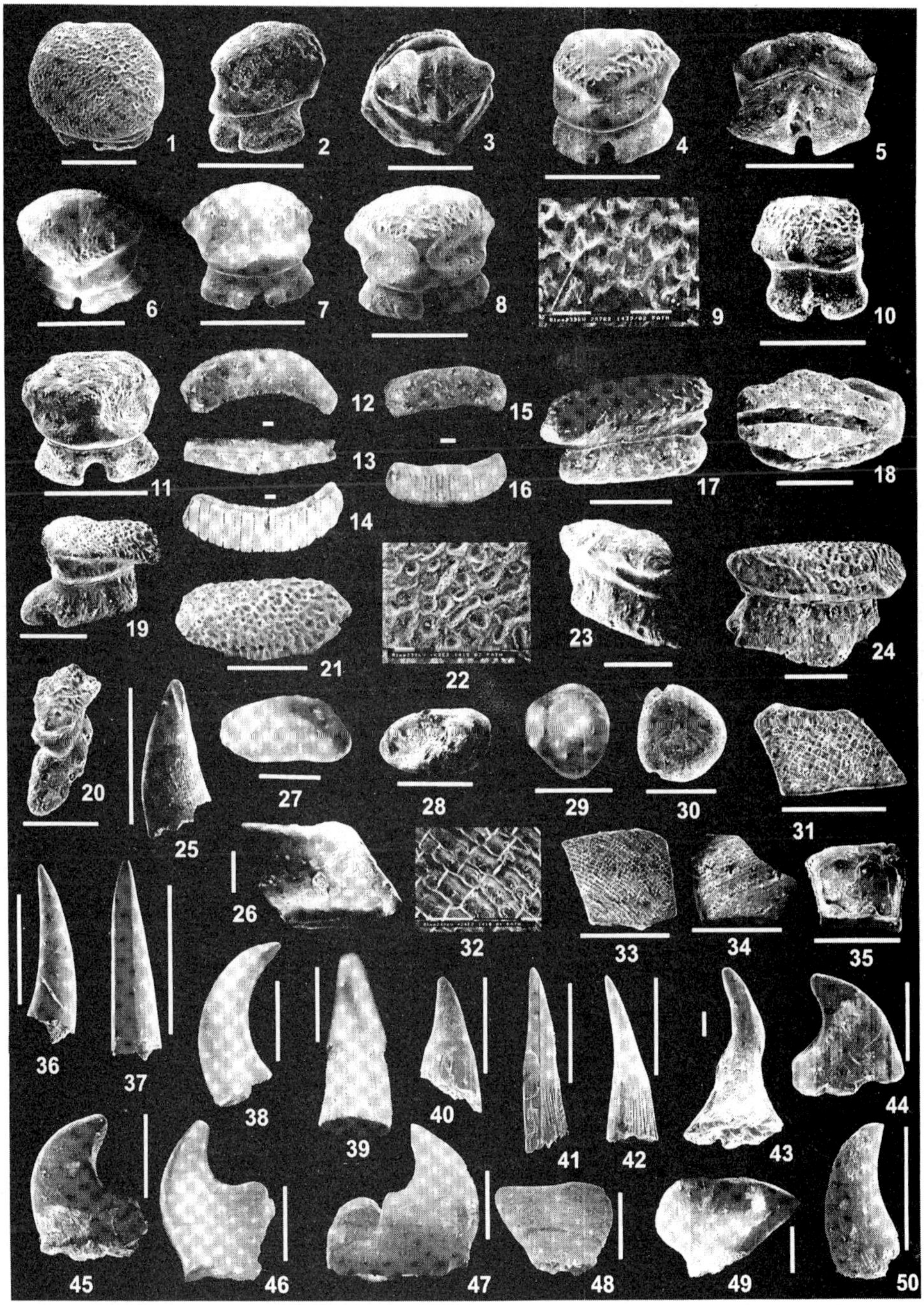
1
2
3
4
5
6
7
8
9
10
11
12
13
14
15
16
17
18
19
20
21
22
23
24
25
26
27
28
29
30
31
32
33
34
35
36
37
38
39
40
41
42
43
44
45
46
47
48
49
50

Pycnodus is typical of pycnodonts, which were characterized by short but deep bodies that looked nearly circular in side view, a downturned beak and small mouth with numerous teeth of varying shapes. Pycndontid fishes thrived in habitats varying from shallow marine, brackish to freshwaters of tropical to temperate regions.

Division	TELEOSTEI Müller 1846
Superorder	OSTEOGLOSSOMORPHA Greenwood, Rosen, Weitzman and Myers 1966
Order	OSTEOGLOSSIFORMES Berg 1940
Suborder	OSTEOGLOSSOIDEI Greenwood, Rosen, Weitzman and Myers 1966
Family	OSTEOGLOSSIDAE Bonaparte 1832

Genus and species indeterminable.

Plate 1, figures 36-38

Material: FV/RGU 6501-6520, isolated teeth; 6541-6600, squamules.

Description: Teeth conical, slightly curved to straight with distinct apical and basal parts; apical part smaller, conical, pointed and covered with transparent enameloid; basal part much longer, sub-cylindrical slightly narrowing upwards, and ornamented by fine longitudinal striations. Squamules rhombic, quadrangular, polygonal or irregular consisting of many thin lamellae seen in lateral view stacked one upon the other; outer side ornamented with linearly arranged rectangular areas with one or two tubercles in each rectangle (Pl. 1, figs. 31-33); inner side concave having a high rim all along its margin and a few tubercles (Pl. 1, fig. 35).

Remarks: In India, osteoglossids were first reported from the Intertrappean beds of Deothan and Kheri based on scales referred to *Musperia* (Hora 1938). Later Gayet *et al.* (1984) and Prasad (1987) assigned similar squamules to *Phareodus*. Jolly and Bajpai (1988) reported three morphotypes of squamules from the Middle Eocene beds of Kalakot, Jammu and Kashmir. The external ornamentation on squamules from Fatehgarh is similar to that on *Phareodus* squamules from Andhra Pradesh (Prasad 1987, fig. 1v) and to Type II scales from Kalakot (Jolly and Bajpai 1988, fig. 3c). Squamules of *Taverneichthys bikanericus*, a new osteoglossid documented recently from the Paleocene Palana Formation of Bikaner are not known (Kumar *et al.* 2005).

Osteoglossid teeth from Bariyara are similar to those described as *Phareodus* sp. from the Intertrappean beds of Andhra Pradesh (Prasad and Singh 1991) and as Osteoglossidae gen. et sp. indet. from the Cambay Shale, Gujarat (Rana *et al.* 2004; fig. 3; 34). Osteoglossidae or the bony tongues are among the largest freshwater fish characterized by bony scales. They are globally known from Cretaceous-Eocene beds but presently restricted to tropical and subtropical regions of South America, central Africa, south-east Asia and northern Australia (Grande 1984; Wheeler 1985).

Order	AULOPIIFORMES Rosen, 1973
Suborder	ENCHODONTOIDEI Berg, 1940
Family	*incertae sedis*
Genus	APATEODUS Woodward 1901

Apateodus sp.

Plate 1, figures 39-42

Material: FV/R/GU 6601-6800, isolated teeth.

Description: Teeth conical and lanceolate with a slight bulge at the junction of apical and basal parts; apical part generally larger, flattened and smooth having sharp lateral edges with shiny enameloid; basal part sub-cylindrical with fine vertical striations; pulp cavity nearly rounded in cross section.

Remarks: In India, *Apateodus* has previously been reported from the Intertrappean beds of Nagpur (Gayet *et al.* 1984), Asifabad (Prasad and Sahni 1987), Naskal (Prasad and Khajuria 1990), and Kachchh (Bajpai *et al.* 1990) and from the Infratrappean beds of Marepalli (Prasad and Singh 1991). The teeth from Fatehgarh are similar to those of *A. striatus* from Asifabad (Prasad and Sahni 1987) and of *Apateodus* sp. from Nagpur (Gayet *et al.* 1984). Similar teeth from the Late Cretaceous Lameta Formation of Nand area (Nagpur District, Maharashtra) have been erroneously referred to *Lepisosteus* sp. by Udhoji *et al.* (1990). Apart from India, *Apateodus* is known from Europe (Woodward 1902; Goody 1969), North America (Wilson and Chalifa 1989) and North Africa (Romer 1966; Carroll 1988).

Family ENCHODONTIDAE Lydekker 1889
Genus ENCHODUS Agassiz 1835

Enchodus sp.
Plate 1, figure 43
Material: FV/R/GU 6851, an isolated tooth.

Description and Remarks: Tooth very small (~0.75 mm in length) and conical with a keel at the junction of apical and basal parts; apex long, pointed and smooth; basal part small and ornamented by fine irregular striations; root broad and expanded. *Enchodus* is widely known in the Late Cretaceous-Early Tertiary successions of peninsular India as well as the sub-Himalaya (Jain and Sahni 1983; Kumar and Loyal 1987; Prasad 1989; Rana 1990; Bajpai *et al.* 1990; Kumar 1996; Rana *et al.* 2004).

Superorder ACANTHOPTERYGII Gouan 1770
Order TETRAODONTIFORMES Berg 1940
Family EOTRIGONODONTIDAE Weiler 1929
Genus STEPHANODUS Zittel 1888

Stephanodus lybicus (Dames) Cappetta 1972
Plate 1, figures 44-46
Material: FV/R/GU 6881-6910, isolated pharyngeal teeth.

Description: Teeth sickle shaped, laterally compressed and cuspidate with well differentiated root and crown; crown covered with translucent enameloid, smooth with a claw-shaped terminal cusp; secondary cusps much smaller, one or two in number; root compressed like crown and quite deep.

Genus EOTRIGONODON Weiler 1929

Eotrigonodon indicus (Lydekker) 1886
Plate 1, figures 47-49
Material: FV/R/GU 6951-6960, isolated pharyngeal and oral teeth.

Description: Pharyngeal teeth differ from those of *S. lybicus* in having broader crown, less curved main cusp and poorly developed or no secondary cusps. Oral teeth laterally expanded and broad with axe like but smooth crowns; outer surface convex, inner surface concave; cutting edge non-serrated and smooth; root half as broad as crown.

Remarks: Isolated teeth of diverse eotrigonodontid fish are common in the Late Cretaceous-Early Tertiary successions of the peninsular India as well as the sub-Himalaya (Lydekker 1886; Jain and Sahni 1983; Kumar and Loyal 1987; Prasad 1989; Rana 1990; Bajpai *et al.* 1990; Kumar 1996; Rana *et al.* 2004). *Stephanodus lybicus* is among the most widely known eotrigonodontid. In India, its remains generally occur associated with those of *E. indicus* and are known from the Kakara and Subathu formations of the northwestern sub-Himalaya (Kumar and Loyal 1987; Kumar 1996), Cambay Shale of Gujarat (Rana *et al.* 2004), Eocene Marh Formation of Rajasthan (Jolly and Loyal 1985) as well as from the Infra and Intertrappean beds of several localities (Prasad 1989; Rana 1990; Bajpai *et al.* 1990). Of the other known genera *Kankatodus* has so far been reported only from the sub-Himalaya (Kumar and Loyal 1987; Kumar 1996), and *Indotrigonodon* and *Pisdurodon* only from the Deccan associated sedimentaries (Jain and Sahni 1983).

Puffer fish, the living kins of eotrigonodontids are shallow water inshore forms that can also live in brackish and freshwaters (Wheeler 1985).

Class REPTILIA Laurenti 1768
Order CROCODYLIA Gmelin 1788
Suborder EUSUCHIA Huxley 1875
Family CROCODYLIDAE Cuvier 1807

Genus and species indeterminate
Plate 1, figure 50
Material: FV/R/GU 7001, an isolated fragmentary tooth
Description and Remarks: Tooth very small, slightly curved with fine vertical striations. No other significant features were observed therefore its generic assignment is withheld. Similar teeth referred to *Crocodylus* sp. have previously been reported from the Deccan trap-associated sedimentary beds of peninsular India (Rana and Sati 2000; pl. II, figs. 14, 15), but they are 2 to 3 times larger.

Discussion

The Fatehgarh Formation of the Barmer Basin has been known as a shallow marine deposit consisting mainly of sandstone and containing abundant gastropod shells and some unidentified fish teeth. In the absence of any faunal studies, there has been little or no discussion on its age and most workers considered it as an Aptian-Albian (Cretaceous) deposit based probably on stratigraphy and or correlation with successions in the contiguous basins (Pareek 1979, 1981, 1984; Sinha-Roy *et al.* 1998). A few works though also mentioned a Cretaceous-Paleocene or Paleocene age for these beds (Das Gupta 1975; Verma 1982). Hence our documentation of diverse fish remains from Fathehgarh beds attains special significance.

Age of Fatehgarh Formation inferred from ichthyofauna

The Fatehgarh vertebrates documented here are dominated by fish both in terms of number of species and number of fossil elements recovered. The crocodilians and mammals are represented by unidentified fragmentary material and therefore are of no biochronologic value. The Pisces, in general, are long ranging and of little help in age interpretations but while comparing with an accurately dated assemblage the presence and or absence of some key forms can often help restrict the fossiliferous horizon within an acceptable time frame. A larger assemblage obviously gives better control.

The Bariyara fish assemblage consisting of eleven taxa has marked similarities with the ichthyofauna known from the Late Cretaceous-?Earliest Paleocene Infra and Intertrappean beds in several localities of the peninsular India (Jain and Sahni 1983; Gayet *et al.* 1984; Prasad and Sahni 1987; Prasad 1989; Rana 1990; Prasad and Khajuria 1990; Prasad and Singh 1991; Rana 1990; Rana and Kumar 1990; Bajpai *et al.* 1990). Presently it looks much less diverse than the assemblages recorded from the Infra and Intertrappean localities, but this could be due to the fact that the new assemblage is based on collections from just 500 kg of sample taken from a single locality. Both these assemblages are identified mostly from well preserved but dissociated dental remains and scales, and both are completely devoid of shark teeth. All fish taxa recognized so far in the Fatehgarh fauna excepting the eagle ray *Myliobatis* are also represented in the Infra and Intertrappean assemblages. The notable Infra and Intertrappean taxa that are absent in Fatehgarh include ray fish *Igdabatis* and

Rhombodus. Both these genera are exclusively Cretaceous, whereas *Myliobatis* ranges from Paleocene to Recent (Carroll 1988). *Igdabatis* a common ray in the Infra and Intertrappean assemblages (Prasad and Cappetta 1993) is known outside India only from the Late Cretaceous beds of Niger (Cappetta 1972). Similarly *Rhombodus* has thus far been reported only from the Cretaceous deposits of North America, Europe and Africa besides India (Cappetta 1987; Prasad and Cappetta 1993).

All Fatehgarh fish taxa except *Myliobatis* range into Cretaceous, but none of them is restricted within the Cretaceous. Similarly all of them except *Apateodus* also range into Eocene and or beyond (Table 1). *Myliobatis* is not known from the pre-Paleocene deposits, and there is no record of *Apateodus* from the post-Paleocene beds. The presence of *Myliobatis*, and the complete absence of marker Cretaceous taxa including the rays *Igdabatis* and *Rhombodus* in the Bariyara assemblage suggest a post-Cretaceous age for the Fatehgarh Formation and the presence of *Apateodus*, a Late Cretaceous-Paleocene fish restricts it within the Paleocene. Rest of the assemblage is consistent with this interpretation. Additional biochronologic studies based on larger assemblages will be necessary for more specific age assignment.

It may be added here that dinosaur remains, so common in the Infra and Intertrappean beds including those in neighbouring Gujarat, have not been noticed in the Fatehgarh Formation, though these two deposits were apparently formed under similar set up. This is another pointer to a post-Cretaceous age for the Fatehgarh Formation.

Pareek (1979, 1984) and Sinha-Roy *et al.* (1998) correlated the Fatehgarh Formation with the Abur Formation of the Jaisalmer Basin, which is rich in Lower Cretaceous-Aptian coquinoids, ammonites and brachiopods The vertebrate fauna documented here does not support this correlation. However, as also pointed out by Das Gupta (1974) the Fatehgarh Formation may in part be correlatable with the subsurface beds of the Sanu Formation, which overlie the Abur Formation and consist mainly of sandstones containing gastropods and fish teeth etc. like the Fatehgarh Formation.

Paleoenvironmental inferences

From the viewpoint of paleoecology, the Bariyara assemblage is characterized by the near absence of obligate pelagic forms. This is particularly depicted by the lack of any sharks, which are fairly common in most shallow marine deposits, including the Cambay Shale, the Akli Formation, the Kapurdi Formation and the Madh Formation etc. of western India as well as the Kakara and Subathu formations of the northwest sub-Himalaya. On the other hand, inshore forms suggesting a lagoonal or estuarine environment constitute the bulk of fauna (Table 1). Also there are a couple of dominantly freshwater taxa in the assemblage, and a land mammal too.

The dasyatid and myliobatid rays prefer shallow coastal waters of tropical to temperate climate feeding mostly on invertebrates and small fishes (dasyatids) and hard-shelled molluscs and crustaceans (myliobatids). The dasyatid and pycnodont fish both of which were earlier considered as exclusively marine are now known to be either secondary or peripheral, *i.e.*, able to move from fresh to saline waters and vice versa (Gayet *et al.* 1993; Poyato-Ariza *et al.* 1998). Pycndontid fish thrived in habitats varying from shallow marine, brackish to freshwaters of tropical to temperate regions. Their crushing type dentition was suited for feeding on clams, snails, coral or other fauna with exoskeleton.

All living osteoglossids are primarily freshwater forms and a vast majority of their fossils have come from freshwater beds, but one Eocene genus, *Brychaetus* is known from marine deposits (Roellig 1974; Taverne 1978). Fossil lepisosteids are known essentially from freshwater deposits (Wiley 1976). Extant gars also prefer freshwater though some species have been seen frequenting coastal marine and brackish waters (Suttkus 1963). Whether osteoglossids and lepisosteids actually lived in the Fatehgarh Sea or were washed from the nearby freshwater habitats is not known.

Table1. Geological age range, paleoenvironment and paleoclimate of the Fatehgarh ichthyofauna.

Genus/ species	Cretaceous	Paleocene	Eocene			Oligocene	Miocene			Pliocene	Pleistocene	Recent	Water Salinity/ Depth					Climate	
			Lower	Middle	Upper		Lower	Middle	Upper				Fresh	Brackish	Estuarine	Shallow Marine	Deep Marine	Tropical	Temperate
Dasyatis	—	—	—	—	—	—	—	—	—	—	—	—	X	X	X	X	--	X	X
Myliobatis		—	—	—	—	—	—	—	—	—	—	—	--	--	X	X	--	X	X
Lepisosteus	—	—	—	—	—	—	—	—	—	—	—	—	X	X	X	--	--	X	X
Pycnodus	—	—	—	—	—								X	--	X	X	--	X	X
Osteoglossidae	—	—	—	—	—	—	—	—	—	—	—	—	X	--	--	--	--	X	X
Apateodus	—	—											--	--	X	X	X	X	X
Enchodus	—	—	—	—	—								--	--	X	X	X	X	X
Stephanodus	—	—	—	—	—								--	--	X	X	--	X	--
Eotrigonodon	—	—	—	—	—								--	--	X	X	--	X	--

Modern relatives (puffer fish) of eotrigonodontid fish are essentially inshore forms occurring around reefs, sea grass beds and estuaries though a few are living a pelagic life while others have invaded freshwaters in many parts of the world (Wheeler 1985).

Fossil enchodontids have been recorded from deep as well as shallow water deposits, but in India they are known mostly from shallow and coastal deposits.

Collectively, the Bariyara assemblage implies a tropical to temperate climate and an essentially shallow coastal basin with freshwater habitats in close vicinity. The lone mammal tooth in the collection, obviously washed into the basin from the land, supports this. This interpretation is also consistent with mixed sand and mud tidal flat environment inferred from the sedimentological studies (Sisodia and Singh 2000).

Conclusions

The first assemblage of microvertebrates comprising *Dasyatis* sp., *Myliobatis* sp. 1, *Myliobatis* sp. 2, *Myliobatis* sp. 3, *Lepisosteus* sp., *Pycnodus* sp., Osteoglossidae gen. et. sp. indet., *Apateodus* sp., *Enchodus* sp., *Stephanodus lybicus*, *Eotrigonodon* sp., and Crocodylidae indet. is documented from the Fatehgarh Formation of the Barmer District, Rajasthan. It compares well with the fish assemblages known from the Late Cretaceous-?Earliest Paleocene Infra and Intertrappean beds of peninsular India but differs in lacking typical Cretaceous elements. It negates the previous assignment of the Fatehgarh Formation to Aptian-Albian and restricts it within the Paleocene though not allowing further refinement at this stage.

The paleoecological information inferred from the occurrence and behaviour of modern relatives of Bariyara biotas indicates a tropical to temperate climate and a shallow coastal water body like an estuary.

The presence of a mammal tooth fragment in the Fatehgarh Formation indicates fair scope of finding more complete remains, which can then be used in its precise dating and understanding the paleobiogeography of that time.

Acknowledgements

Authors are thankful to the Director, Wadia Institute of Himalayan Geology, Dehradun for providing the SEM and other facilities for this work. RSR thanks the Department of Science and Technology for funding this study vide Project no ESS/VES/092/2000.

References

ASTIBIA, H., BUFFETAUT, E., BUSCALIONI, A. D., CAPPETTA, H., CORRAL, C., ESTES, R., GARCIA-GARMILLA, F., JAEGER, J. J., JIMENEZ-FUENTES, E., LE LOEUFF, J., MAZIN, J. M., ORUE-ETXEBARRIA, X., PEREDA-SUBERBIOLA, J., POWELL, J. E., RAGE, J. C., RODRIGUEZ-LAZARO, J., SANZ, J. L. and TONG, H., 1990. The fossil vertebrates from Laño (Basque Country, Spain); new evidence on the composition and affinities of the Late Cretaceous continental faunas of Europe. Terra Nova, 2: 460-466.

BAJPAI, S., SAHNI, A., JOLLY, A. and SRINIVASAN, S. 1990. Kachchh Intertrappean biotas: affinities and correlation. 101-105. In: Sahni, A. and Jolly, A., Eds., Cretaceous event stratigraphy and the correlation of Indian nonmarine strata. Contributions from the Seminar cum Workshop IGCP 216 & 245, Chandigarh, 1-125.

BAJPAI, S. and THEWISSEN, J.G.M. 2002. Vertebrate fauna from Panandhro lignite field (Lower Eocene), District Kachchh, western India. Current Science, 82: 502-509.

BAROOAH, S.K., 1950. Fossil fish and crabs in the Fuller's Earth deposits of Jodhpur State. Current Science, 19: 165.

BERTINI, R. J., MARSHALL, L. G., GAYET, M. and BRITO, P. M. 1993. Vertebrate faunas from the Admantina and Marilia formations (Upper Bauru Group, Late Cretaceous, Brazil) in their stratigraphic and paleobiogeographic context. Neues Jahrbuch für Geologie und Paläontologie Abhandlungen, 188: 71-101.

CAPPETTA, H. 1972. Les poisons Crétacés et Tertiares du basin des Iullemmeden (République du Niger). Palaeovertebrata, 5: 179-251.

__________, 1987. Mesozoic and Cenozoic Elasmobranchii. In: Schultze, H.P., Ed., Hand book of Paleoichthyology, Chondrichthyes II .Gustav Fischer Verlag, Stuttgart, New York, 1-193.

CARROLL, R.L., 1988. Vertebrate Paleontology and evolution. W.H. Freeman and Company, New York, 1-698.

CVANCARA, A.M. and HOGANSON, J.W. 1993. Vertebrates of the Cannonball Formation (Paleocene) in north and South Dakota. Journal of Vertebrate Paleontology, 13 (1): 1-23.

DAS GUPTA, S.K., 1975. A revision of the Mesozoic-Tertiary stratigraphy of the Jaisalmer Basin, Rajasthan. Indian Journal of Earth Science, 2 (1): 77-94.

DESHMUKH, G.P. and MISHRA, S.P., 1971. Geological mapping in parts of Barmer and Jaisalmer districts, Rajasthan (field Season 1969-70). Unpublished Report of the Geological Survey of India.

GAYET, M., RAGE, J.-C. and RANA, R.S., 1984. Nouvelles ichthyofaune et herpétofaune de Gitti Khadan, le plus ancien gisement connu du Deccan (Crétacé/ Paléocène) à microvertébrés. Implications paléogéographiques. Mémoires de la Société géologique de France, 147: 55-65.

GAYET, M., SEMPERE, T., CAPPETTA, H., JAILLARD, E. and LÉVY, A. 1993. La présence de fossiles marins dans le Crétacé terminal des Andes centrales et ses conséquences paléogéographiques. Palaeogeography, Palaeoclimatology, Palaeoecology, 102: 283-319.

GOODY, P.C., 1969. The relationships of certain Upper Cretaceous teleosts with special reference to the myctophoids. Bulletin of the British Museum (Natural History), Section A: Geology, Supplement 7: 1-255.

GOTTFRIED, M.D. and KRAUSE, D.W. 1998. First record of gars (Lepisosteidae, Actinopterygii) on Madagascar: Late Cretaceous remains from Mahajanga Basin. Journal of Vertebrate Paleontology, 18: 275-279.

GRANDE, L., 1984. Palaeontology of the Green River Formation, with a review of the fish fauna. Bulletin of the Geological Survey of Wyoming, 63: 1-333.

GREENWOOD, P.H., ROSEN, D.E., WETTZMAN, S.H. and MYERS, G.S., 1966. Phyletic studies of teleostean fishes with provisional classification of living forms. Bulletin of the American Museum of Natural History, 131: 339-456.

HORA, S.L., 1938. On some fossil fish-scales from the Inter-trappean beds at Deothan and Kheri, Central Provinces. Records of the Geological Survey of India, 73: 267-294.

JAIN, S.L. and SAHNI, A., 1983. Some Upper Cretaceous vertebrates from Central India and their palaeogeographic implications. 66-83. In: Maheshwari, H. K., Ed., Cretaceous of India. Proceedings of the Symposium on Cretaceous of India: palaeoecology, palaeogeography and time boundaries, Indian Association of Palynologists, 1-217.

JOLLY, A. and BAJPAI, S., 1988. Fossil Osteoglossidae from the Kalakot Zone (Middle Eocene): implications for palaeoecology, palaeobiogeography and correlation. Bulletin of the Indian Geologists' Association, 21 (1): 71-79.

JOLLY, A. and LOYAL, R.S., 1985. Record of Microvertebrate from the Middle Eocene Marh Stage of Sri Kolayat-ji, Rajasthan. Bulletin of the Geological Mining and Metallurgical Society of India, 52: 374-384.

KUMAR, K., 1989. A report on the occurrence of microvertebrates in the Subathu Formation (Montian-Early Lutetian) near Nilkanth, Garhwal Himalaya, Uttar Pradesh, India. Current Science, 58: 743-746.

_______, 1996. Microvertebrate assemblage from the Kakara Formation (Paleocene-?Eocene), Himachal Pradesh, northwest Himalaya, India. 493-507. In: Pande, J., Azmi, R.J., Bhandari, A. and Dave, A., Eds., Contributions to XV Indian Colloquium on Micropalaeontology and Stratigraphy, Dehradun, 1-827.

KUMAR, K. and LOYAL, R.S., 1987. Eocene Ichthyofauna from the Subathu Formation, northwestern Himalaya, India. Journal of the Palaeontological Society of India, 32: 60-84.

KUMAR, K., RANA, R.S. and PALIWAL, B.S., 2005. Osteoglossid and lepisosteid fish remains from the Palaeocene Palana Formation, Rajasthan, India. Palaeontology, 48 (6): 1187-1209.

LOYAL, R.S., 1983. On a new species of a stingray fish from Subathu Formation, Subathu, Himachal Pradesh, India. Bulletin of the Indian Geologists' Association, 17 (1): 57-65.

LYDEKKER, R., 1886. The Indian Tertiary and Post-Tertiary Vertebrata, Tertiary fishes. Memoirs of the Geological Survey of India, Palaeontologia Indica, Series 10, 3 (8): 241-264.

MISHRA, V.P., 1980. A new species of Myliobatis and some shark teeth from the Middle Eocene of Kutch, western India. Journal of the Palaeontological Society of India, 23-24: 81-85.

PALIWAL, B.S., 1999. Fish skull from Palana Formation at Hadla-Bhatiyan, District Bikaner, Rajasthan. Current Science, 76: 1536-1538.

PAREEK, H.S., 1979. The Permian-Mesozoic-Palaeogene palaeogeography of the Rajasthan and Gujarat shelf and correlation with that of the Indus Basin. Proceedings of IV International Gondwana Symposium, 1: 23-36.

_______, 1981. Basin configuration and sedimentary stratigraphy of Western Rajasthan. Journal of the Geological Survey of India, 22: 517-527.

_______, 1984. Pre-Quaternary geology and mineral resources of northwestern Rajasthan. Memoirs of the Geological Survey of India, 115: 1-99.

POYATO-ARIZA, F. J., TALBOT, M. R., FREGENAL-MARTÍNEZ, M. A., MELÉNDEZ, N. and WENZ, S. 1998. First isotopic and multidisciplinary evidence from nonmarine coelacanths and pycnodontiform fishes: palaeoenvironmental implications. Palaeogeography, Palaeoclimatology, Palaeoecology, 144: 65-84.

PRASAD, G.V.R., 1987. Squamules of osteoglossid fish from the Intertrappean beds of Pargi, Andhra Pradesh. Current Science, 56 (24): 1270-1272.

_______, 1989. Vertebrate fauna from the Infra- and Inter-trappean beds of Andhra Pradesh: age implications. Journal of the Geological Society of India, 34: 161-173.

PRASAD, G.V.R. and CAPPETTA, H., 1993. Late Cretaceous selachians from India and the age of the Deccan traps. Palaeontology, 36 (1): 231-248.

PRASAD, G.V.R. and KHAJURIA, C.K., 1990. A record of microvertebrate fauna from the Intertrappean beds of Naskal, Andhra Pradesh. Journal of the Palaeontological Society of India, 35: 151-161.

PRASAD, G.V.R. and SAHNI, A., 1987. Coastal plain vertebrate assemblage from the terminal Cretaceous of Asifabad, peninsular India. Journal of the Palaeontological Society of India, 32: 5-19.

PRASAD, G.V.R. and SINGH, V., 1991. Microvertebrates from the Inter-trappean beds of Rangareddi District, Andhra Pradesh and their biostratigraphic significance. Bulletin of the Indian Geologists Association, 24: 1-20.

RANA, R.S., 1990. Palaeontology and palaeoecology of the Intertrappean (Cretaceous-Tertiary transition) beds of peninsular India. Journal of the Palaeontological Society of India, 35: 105-120.

RANA, R.S. and KUMAR, K., 1990. Late Cretaceous-Early Tertiary fish assemblage from peninsular India and Himalayan region: Comments on phylogeny and palaeobiogeography. 47-49. In: Sahni, A. and Jolly, A., Eds., Cretaceous event stratigraphy and the correlation of Indian nonmarine strata. Contributions from the Seminar cum Workshop IGCP 216 & 245, Chandigarh, 1-125.

RANA, R.S. and SATI, K.K., 2000. Late Cretaceous-Palaeocene crocodilians from the Deccan trap-associated sedimentary sequences of peninsular India. Journal of the Palaeontological Society of India, 45: 123-136.

RANA, R.S., KUMAR, K. and SINGH, H., 2004. Vertebrate fauna from the subsurface Cambay Shale (Lower Eocene), Vastan Lignite Mine, Gujarat, India. Current Science, 87 (12): 1726-1733.

RANA, R.S., KUMAR, K., LOYAL, R.S., SAHNI, A., ROSE, K.D., MUSSELL, J., SINGH, H. and KULSHRESHTHA, S.K., 2006. Selachians from the Early Eocene Kapurdi Formation (Fuller's Earth), Barmer District, Rajasthan, India. Journal of the Geological Society of India (in press).

ROELLIG, H. F. 1974. The cranial osteology of Brychaetus muelleri (Pisces Osteoglossidae) Eocene, Isle of Sheppey. Journal of Paleontology, 48: 947-951.

ROMER, A.S., 1966. Vertebrate Paleontology. The University of Chicago Press, Chicago, 1-468.

SAH, S.C.D., 1983, Cretaceous of India-some aspects of time boundaries and paleogeography In: Cretaceous of India: paleoecology, paleogeography and time boundaries, Indian Assoc. of Palynologists (Ed. Maheshwari, H.K.) p. 5-21.

SAHNI, A. and CHOUDHARY, N.K., 1972. Lower Eocene fishes from Barmer, south western Rajasthan, India. Proceedings of the Indian National Science Academy, 38 A (3-4): 97-102.

SAHNI, A. and CHOUDHARY, N.K., 1975. A new Eocene louvar from Barmer, southwestern Rajasthan, India. Journal of the Palaeontological Society of India, 20: 391-395.

SATSANGI, P.P. and BORA, R., 1980. A fossil eagle ray fish from Eocene of Khasi Hills, Meghalaya. Journal of the Geological Society of India, 21: 566-567.

SINHA-ROY, S., MALHOTRA, G. and MOHANTY, M., 1998. Geology of Rajasthan. Geological Society of India, Bangalore, pp.1-278.

SISODIA, M.S. and SINGH, U.K., 2000. Depositional environment and hydrocarbon prospects of the Barmer basin, Rajasthan, India. NAFTA, 51 (9): 309-326.

SUTTKUS, R. D. 1963. Order Lepisostei. Memoir of the Sears Foundation of Marine Research, 1: 61-88.

TAVERNE, L., 1978. Ostéologie, phylogénèse et systématique des téléostéens fossiles et actuels du super-ordre des Ostéoglossomorphes. Part II. Ostéologie des genres Phareodus, Phareoides, Brychaetus, Musperia, Pantodon, Singida, Notopterus, Xenomystus et Papyocranus. Mémoires de la Classe des Sciences, Académie royale de Belgique, Collection in-8°, 2° Série, 42 (6): 1-213.

TEWARI, K.K., 1968. A new fossil percoid fish from the Lower Tertiary Fuller's Earth deposits of Kapurdi, Barmer District, Rajasthan. Journal of the Zoological Society of India, 20 (1-2): 95-103.

UDHOJI, S.G., MOHABEY, D.M. and VERMA, K.K., 1990. Palaeontological studies of Lameta Formation of Nand area and their bearing on K-T boundary problem. 73-74. In: Sahni, A. and Jolly, A., Eds., Cretaceous event stratigraphy and the correlation of Indian nonmarine strata. Contributions from the Seminar cum Workshop IGCP 216 & 245, Chandigarh, 1-125.

VERMA, K.K., 1982. The fossil record and environment of desert covered areas of western India. Proceedings of the Workshop on the problems of the deserts in India held at Jaipur in 1975, Geological Survey of India, Miscellaneous Publication 49: 141-152.

WEEMS, R. E. 1998. 6. Actinopterygian fish remains from the Paleocene of South Carolina. 147-164. In: Sanders, A.E., Ed., Paleobiology of the Williamsburg Formation (Black Mingo Group; Paleocene) of South Carolina (USA). Transactions of the American Philosophical Society, 88 (4): 1-276.

WHEELER, A., 1986. The world encyclopedia of fishes. Mac Donald & Co. Ltd. London, 365 p.

WILEY, E.O., 1976 The phylogeny and biogeography of fossil and recent gars (Actinopterygii: Lepisosteidae). University of Kansas Museum of Natural History Miscellaneous Publication, 64:1-111.

WILSON, M.V.H. and CHALIFA, Y., 1989. Fossil marine actinopterygian fishes from the Kaskapau Formation (Upper Cretaceous: Turonian) near Watino, Alberta. Canadian Journal of Earth Sciences, 26: 2604-2620.

WOODWARD, A. S., 1890. Note on certain vertebrate remains from the Nagpur district with description of a fish skull. Records of the Geological Survey of India, 23: 20-24.

________, 1902-1912. The fossil fishes of the English Chalk. Palaeontographical Society, London, Monographs 56, 57, 61-65.

________, 1908. On some fish remains from the Lameta beds at Dongargaon, Central Provinces. Memoirs of the Geological Survey of India, Palaeontologia Indica, New Series, 3 (3): 1-6.

Palynostratigraphy and paleoenvironment of the Subathu Formation of Dharampur Area, Solan District, Himachal Pradesh

Y. Raghumani Singh[1], N.N. Dogra[2] and R.Y. Singh[3]

[1]P. G. Department of Geology, University of Jammu, Jammu 180006, India
[2]Department of Earth Sciences, Kurukshetra University, Kuruksehtra 136 119, India
[3]Department of Geology, Punjab University, Chandigarh-160 014, India

Introduction

The Subathu Formation is named after the Subathu Cantonment (30^0 59' N: 76^0 01'E) and is exposed along Kalka-Shimla Road from Kalka to Kumarhatti in Himachal Pradesh, North-Western Himalaya, India. The present work is based on the palynological investigations of the Subathu Formation exposed at Koti and Dharampur areas of Solan District in Himachal Pradesh. Five sections namely, Koti, Chakki Ka Mor-Bhojnagar Road along Koshalia river, Dharampur township, Prachin Shiva Temple and Patte Ka Mor (Fig. 1) were measured and systematically sampled for the present study.

The stratigraphic set up of the area is worked out by a number of workers namely Medlicott (1864, 1879), Pilgrim and West (1929), Pascoe (1964), Bhatia and Mathur (1964), Raiverman and Seshvataram (1965), Mathur (1969), Chaudhri (1968, 1975), Singh *et al.* (1978), Karunakaran and Ranga Rao (1979), Raiverman (1979) and Krishnan (1982), Sarkar and Singh (1988), Srikantia and Bhargava (1998) and Singh *et al.* (2003) among others. The Palaeogene rocks in this area are represented by the Subathu, Dagshai and Kasauli Formations, in the ascending order of stratigrphy and are exposed around Dharampur and Koti (Fig. 2). Lithologically the Subathu Formation consists of limestones, well indurated sandstones, olive green splinitary shales, grey marls, black carbonaceous and ash purple shales with occasional white sandstones and coal beds (Chaudhri, 1991). A generalized stratigraphic succession of the Palaeogene rocks in the studied area is given below:

Table 1. Generalized lithostratigraphic succession (after Singh *et al.*, 2003)

Siwalik Supergroup	Characteristic of rocks
Kasauli Formation	Massive sequence of well indurated sandstones with thin intercalations of pale-grey shales.
Dagshai Formation	Alternating sequence of well indurated sandstones and purple clays.
Subathu Formation	Limestones, well indurated sandstones and grey-green shales

Pre-Tertiary Rocks

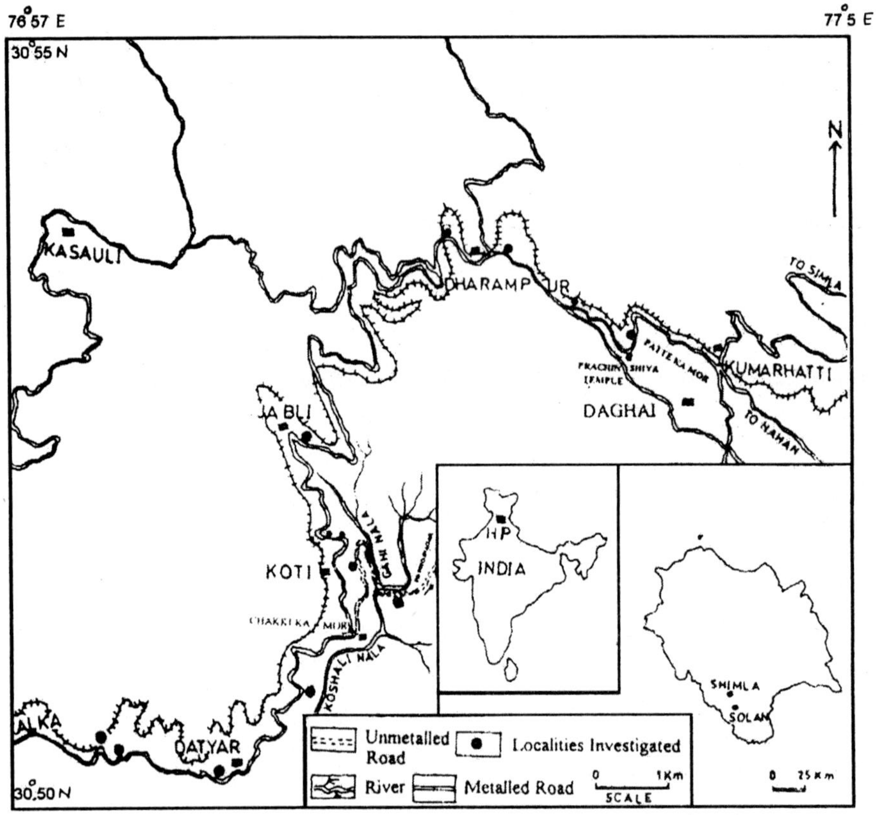

Fig. 1 . Location map of the area showing localities investigated.

Palynology

The palynological assemblage recovered during the present study from the Subathu Formation of this area comprises 158 species referable to 79 form genera. Out of these, 80 species belonging to 33 form genera are algae, 6 genera and 7 species are fungi, 21 species assignable to 7 form genera are pteridophytes while angiosperms constitute 37 species in 24 genera and gymnosperms are represented by 5 species in 2 form genera. Besides above indigenous mioflora, a number of reworked Permocarboniferous palynomorphs and foraminiferal linings have also been found in Koti section and Dharampur township section respectively.

Qualitatively the miofloral assemblage is dominated by algae followed by angiosperms, pteridophytes and gymnosperms. The fungal remains occur in the subordinate amount. But, quantitatively dinoflagellates are associated with gymnosperms, angiosperms, pteridophytes and fungal remains in order of decreasing population. Foraminiferal linings are also present in considerable number.

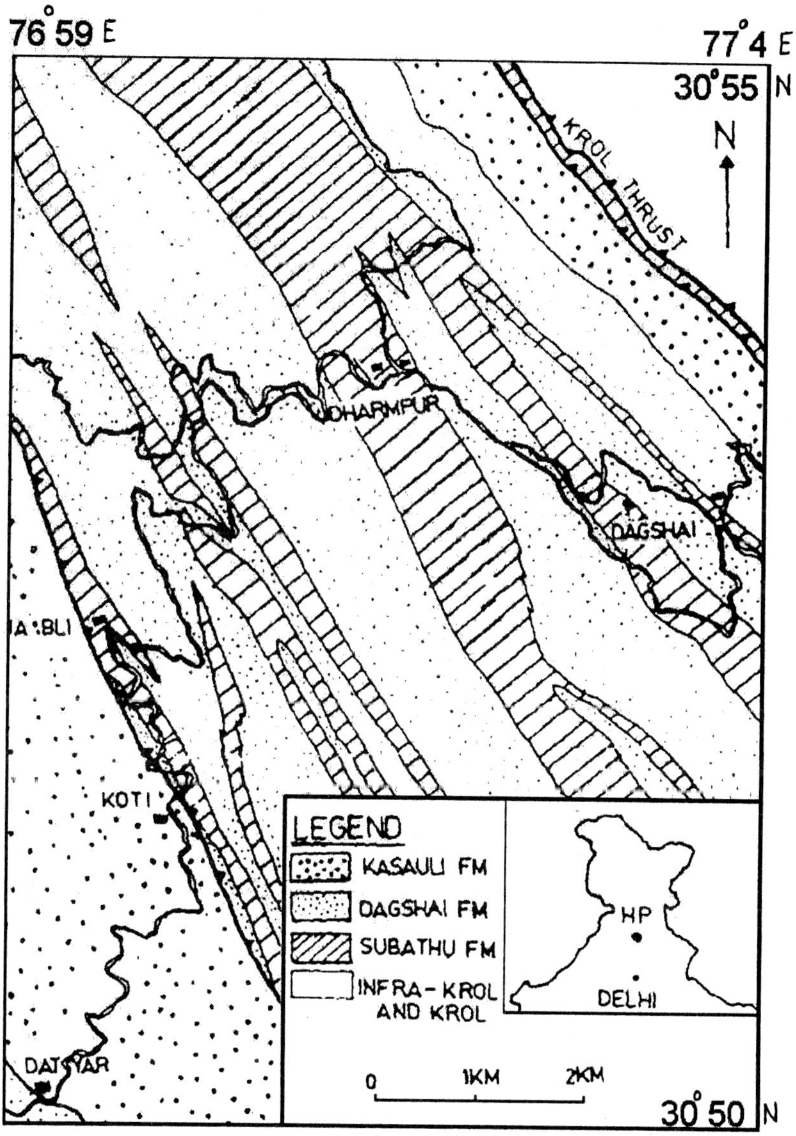

Fig. 2 Geological map of the studied area (after Raviraman, 1979).

Palynostratigraphic Zonation

The palynostratigraphic zonation of the Subathu Formation has been attempted mainly on the basis of the frequency variation pattern observable within the miofloral assemblages recovered from the different litho-horizons of the Subathu Formation exposed around Koti and Dharampur regions, along Kalka-Shimla Road. The FAD, LAD, maximum development and the decline of the miofloral assemblages, appear to serve as reliable basis for delimiting boundaries of palynological zones. The FAD and LAD of certain taxa of palynomorphs proved quite helpful in demarcating the biostratigraphic intervals. In the light of above considerations the Subathu Formation has been differentiated into five distinct assemblage palynozones namely *Homotryblium* spp. Assemblage zone, *Cordosphaeridium* spp. Assemblage zone, *Thalassiphora* spp. Assemblage zone, *Cleistosphaeridium* spp. Assemblage zone and *Pediastrum* spp. Assemblage zon. nov. (Figs. 3 & 4).

A formal description of each zone, together with the characteristic palynotaxa data is attempted here and also compared with biostratigraphic zones established by Singh *et al.* (1978) at Subathu Town in the Shimla hills and Sarkar and Singh (1988) at Baneth-Bagthan areas in the Shimla hills. The details of the palynostratigraphic zones in the Subathu Formation are discussed below in the ascending order.

***Homotryblium* spp. Assemblage zone** Singh, *et al.* (1978)

Type locality. Subathu section, Subathu town area, Shimla Hills.

Present locality. Dharampur township section and Prachin Shiva temple section, Dharampur, Himachal Pradesh.

Lithology. Splintery shale, Shelly limestone and Sandstone

Lower contact. not exposed.

Upper contact. The upper part of *Homotryblium* spp. Assemblage zone is characterized by shelly limestone and is overlain by splintery shale.

Characteristic species of the zone. *Homotryblium abbriviatum, H. pallidium, H. tenuispinosum, H. floripes, Hystrichosphaeridium tubiferum, H. arborispinosum, H. daenie, Cordosphaeridium granulatum, C. multispinosum, C. gracile, Achmosphaeridium multifurcata, Areoligera sentosa, Cleistosphaeridium diversispinosum, C. flexosum, C. fibrospinosum, Trichinidium hirstum, Thalassiphora velata, Th. pelegica, Tennua kutharensis, T. simlanesis, Polysphaeridium subtitle, P. zoharyi, Oligosphaeridium complex, Kopilisphaera subathuenesis* sp. nov., *Cyclonephelium compactum, Glaphyrocysta divaricata, G. exuberans, Gonyaulcysta* sp., *Adnatosphaeridium multispinosum. A. vittatum, Areosphaeridium* sp., *Lingulodinium machaerophorum, Muratodinium fimbriatum, Operculodinium centrocarpum, O.* sp., *Pediastrum magnus, Spiniferites ramous, S. monilis, S. membranaceus, Cycathidites cooksonii, Graminidites media, Lycopodiumsporites pattamorensis, L. palaeocenicus, Lygopodiumsporites barogensis, L. eocenicus, L. enigmaticus, L. problematicus, L.* sp., *Monolites* sp., *Neocouperipollis brevispinosum, Palmidites infrafoveolatus, P. noviculatus, P.* sp., *Schizaeosporites eocenicus, Todisporites minor, T. plicatus, T. rarus* and *T. major.*

Remarks. This zone was established by Singh *et al.* (1978) at Subathu town area in Shimla Hills and was later on delineated in Banethi-Bagthan areas also by Sarkar and Singh (1988). The zone is characterized by the dominance of *Homotryblium pallidium, H. tenuispinosum* and *H. abbreviatum.* It can be differentiated from the overlying *Cordosphaeridium* zone in having very less frequency of *Cordosphaeridium inodes, C. granulatum, C. multispinosum* and dominance of *Trichinidium hirstum* in the later.

***Cordosphaeridium* spp. Assemblage zone** Singh *et al.* (1978)

Type locality. Subathu section, Subathu town area, Shimla Hills

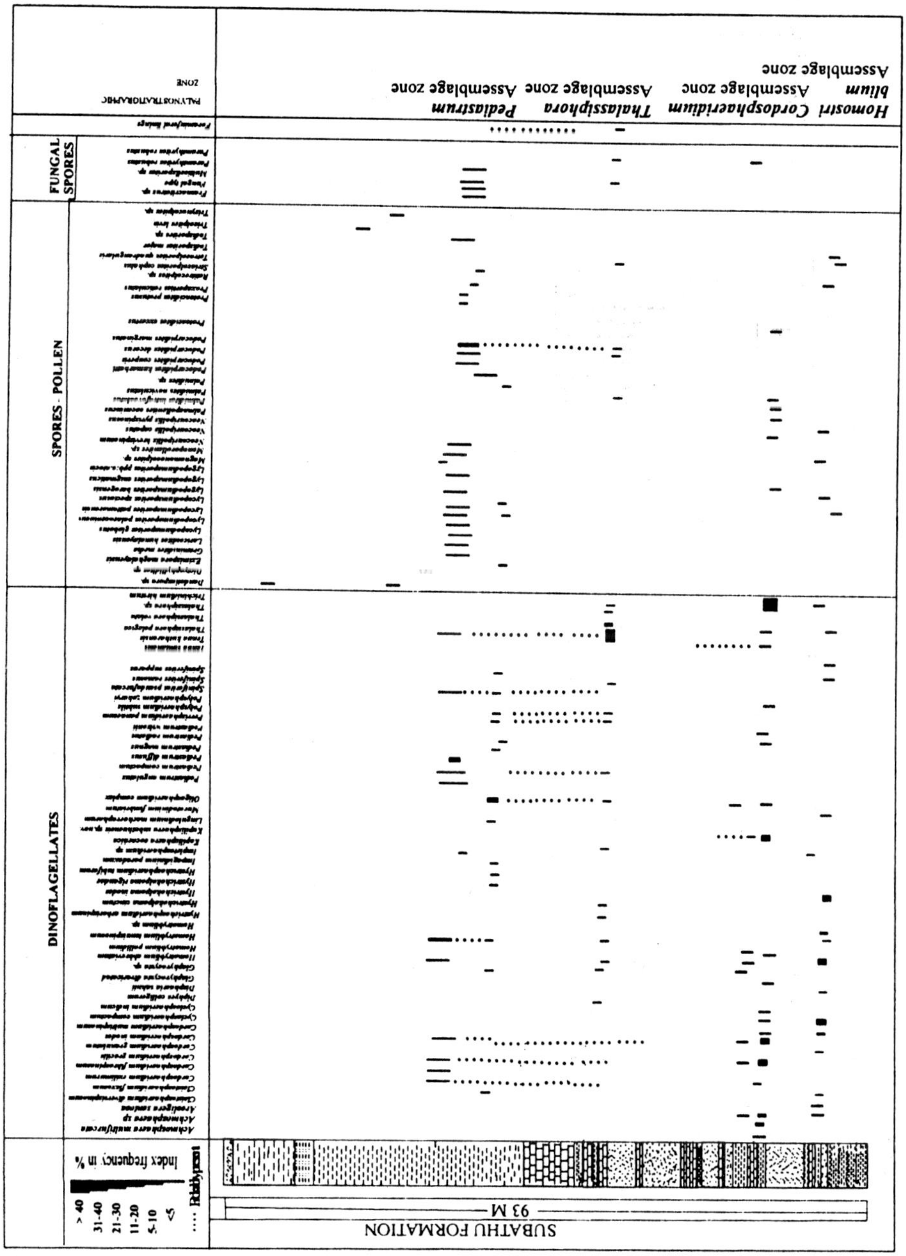

Fig.-3. Stratigraphic distribution of mioflora and palynological zonation of Subathu Formation at Dharampur township section, Dharampur (for litho-index, see fig.4)

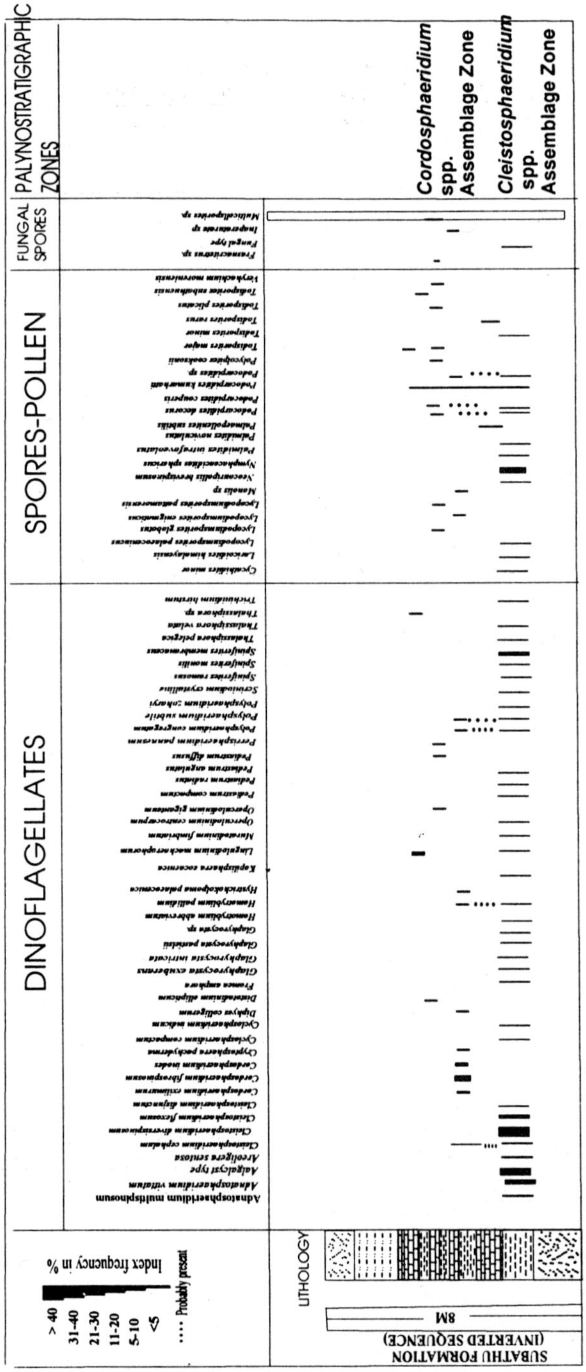

Fig.-4. Stratigraphic distribution of mioflora and palynological zonation of Subathu Formation at Patte Ka Mor Section, near Dharampur.

Litho-Index

Quartzite Shelly Sandstone Limestone/Shelly limestone Sandstone Shale/Splintery Shale

Present locality. Dharampur township section, Prachin shiva temple section and Patte Ka Mor section, Dharampur, Shimla Hills.

Lithology. Splintery shales, limestone and sandstone. Thickness variable from place to place.

Lower contact. Sandstone, shelly limestone and splintery shales.

Upper contact. Shelly limestone and splintery shales.

Characteristic species of the zone. *Cordosphaeridium fibrospinosum, C. granulatum, C. inodes, C. multispinosum, C. gracile, Adnatosphaeridium multispinosum, A. vittatum, Homotryblium abbriviatum, H. pallidium, H. tenuispinosum, Hystrichosphaeridium tubiferum, Achmosphaera multifurcata, Areoligera sentosa, Cleistosphaeridium diversispinosum, C. cephalum, Trichinidium hirstum, Oligosphaeridium complex, Distatodinium ellipticum, Thalassiphora velata, Th. pelegica, Polysphaeridium congregatum, Perrisphaeridium pannesum, Operculodinium centrocarpum, O. giganteum, Kopilisphaera eocaenicus, Cyclonephelium compactum, C. indicum, Glaphyrocysta divaricata, G. exuberans, G. intricate, G. ordinata, Diphyes colligerum, Gonyaulcysta* sp., *Lingulodinium machaerophorum, Pediastrum magnus, P. diffuses, Lycopodiumsporites pattamorensis, L. globutus, L. palaeocenicus, Lygopodiumsporites barogensis, L. engimaticus, Neocouperipollis brevispinosum, N. pyrispinosus, Palmidites noviculatus, Todisporites minor, T. plicatus, T. subathuensis, T. major, Podocarpidites decorus, P. couperi, P. kumarhattiensis* and *Polycopites cooksonii.*

Remarks. This zone was already established by Singh *et al.* (1978) at Subathu township in Shimla Hills and later on was extended by Sarkar and Singh (1988) to Banethi-Bagthan area in Shimla Hills. This zone has here been extended to Prachin Shiva temple and Patte Ka Mor. This zone can be differentiated from the overlying *Thalassiphora* spp. Assemblage zone in having very less frequency of *Cordosphaeridium* spp. in the later and the dominance of *Thalassiphora velata, Thalassiphora pelegica* and *Thalassiphora* sp.

***Thalassiphora* spp. Assemblage zone** (Singh *et al.*, 1978) Sarkar and Singh (1988)

Type locality. Jabli, Kalka-Shimla Highway, Shimla Hills.

Present locality. Dharampur township section, Dharampur and Chakki Ka Mor-Bhojnagar road section, Koti.

Lithology. Splintery shales, limestone and sandstone.

Lower contact. Splintery shale.

Upper contact. Reddish splintery shale, shelly limestone.

Characteristic species of the zone. *Thalassiphora velata, Th. pelegica, Th.* sp. *Muratodinium fimbriatum, Samlandia chlamdophora, Disphaeria sahnii, Distatodinium ellipticum, Glaphyrocysta divaricata, G. exuberans, Homotryblium abbriviatum, H. pallidium, H. tenuispinosum, Hystrichosphaeridium tubiferum, H. arborispinosum, H. daenie, H. granulatum, Hystrichokolpoma riguadae, H. cincular, Adnatosphaeridium multispinosum, A. vittatum, Areosphaeridium* sp., *Cleistosphaeridium complex, Kopilisphaera subathuenesis* sp. nov., *K. eocaenicus, Lingulodinium machaerophorum, Operculodinium centrocarpum, O.* sp., *Spiniferites ramous, S. monilis, S. membranaceus, Trichinidium hirstum, Pediastrum radiatus, P. wilsonii, P. compactum, Neocouperipollis pyrispinosus* and *T. major.*

Remarks. The *Subathu sahnii* spp. Assemblage zone was first established by Singh *et al.* (1978) at Jabli, Kalka-Shimla Highway, Shimla Hills. Later on Sarkar and Singh (1988) renamed this zone as *Thalassiphora* spp. Assemblage zone exposed at Banethi-Bagthan area, Shimla Hills, Himachal Pradesh. This zone has now been extended to Dharampur township and Chakki Ka Mor. This zone can be differentiated from the underlying and overlying zones in having and high percentage of *Thalassiphora velata, Thalassiphora pelegica* and *Thalassiphora* sp.

***Cleistosphaeridium* spp. Assemblage zon** Singh *et al.* (1978)

Type locality. Subathu section, Subathu town area, Shimla Hills.

Present locality. Patte Ka Mor section, Dharampur, Himachal Pradesh.

Lithology. Splintery shales.

Lower contact. Grey sandstone whose base is not exposed.

Upper contact. Shelly limestone.

Characteristic species of the zone. *Cleistosphaeridium diversispinosum, C. cephalum, C. flexosum, C. disjunction, Cyclonephelium compactum, C. indicum, Achmosphaera multifurcata, Adnatosphaeridium multispinosum, A. vittatum, Areoligera sentosa, Homotryblium abbriviatum, H. pallidium, Trichinidium hirstum, Thalassiphora velata, Th. pelegica, Polysphaeridium subtile, P. zoharyi, P. congregatum, Fromea amphora, Glaphyrocysta exuberans, G. intricate, G. pastieli, G.* sp., *Kopilisphaera sp., Muratodinium fimbriatum, Operculodinium centrocarpum, Pediastrum compactum, P. radiatus, P. angulatus, Spiniferites membraneous, S. monilis, S. ramous, Scriniodinium crystalline, Cycathidites minor, Laricoidites himalayensis, Nymphacidites sphericus, Palmidites intrafoveolatus, P. noviculatus, Podocarpidites couperi, P. decorus, P. kumarhatti, P.* sp., *Todisporites minor* and *T. rarus.*

Remarks: This zone was established by Singh *et al.* (1978) at Subathu town and was extended by Sarkar and Singh (1988) to Banethi-Bagthan area in Himachal Pradesh. In present studies the zone is also recorded at Patte Ka Mor. The zone can be differentiated from the overlying and underlying zones in having high frequency of *Cleistosphaeridium diverspinosum, C. flexosum,* and *C. disjunctum, algalcyst type* and *Nymphacidites sphericus.*

Further, the position of *Cleistosphaeridium* spp. Assemblage zone considered older than *Homotryblium* spp. Assemblage zone by Singh *et al.* (1978), Sarkar and Singh (1988) in the Subathu type area and Baneth-Bagthan has been found to be younger than *Thalassiphora* spp. Assemblage zone in the present area of investigation, in the light of FAD and LAD comparison of Dino taxa with the William *et al.* (1993).

***Pediastrum* spp. Assemblage zon. nov.**

Type locality. Dharampur township section at Dharampur in Shimla Hills.

Lithology. Pale yellowish splintery shale.

Lower contact. The lower part of *Pediastrum* complex assemblage zon. nov. is characterized by splintery shale.

Upper contact. Splintery shale.

Characteristic species of the zone. *Pediastrum radiatus, P. compactum, P. diffuses, P. angulatus, P. wilsonii, P. magnus, Operculodinium centrocarpum, Cleistosphaeridium flexosum, Cordosphaeridium exilimurum, C. inodes, Gonyaulcysta,* sp., *Homotryblium abbreviatum, H. tenuispinosum, Impletosphaeridium* sp., *Kopilisphaera eocaenica, Lingulodinium machaerophorum, Polysphaeridium subtile, Spiniferites ramous, S. supparus, Thalassiphora pelegica, Dictophyllidites* sp., *Eximspora meghalayensis, Graminidites* sp., *Laricoidites himalayensis, Lycopodiumsporites barogensis, L. enigmaticus, L. problematicus, Magnamonocolpites* sp., *Monoporopollites* sp., *Neocouperipollis brevispinosum, Palmidites* sp., *Podocarpidites kumarhatti, P. couperi, P. decorus, P. marginatus,* P. sp., *Trisyncolpites* sp., *Tricolpites levis, Todisporites* sp., *Retirecolpites* sp., *Franscritetrus* sp., *Multicellaesporites* sp. and *Paramthyrites robustus.*

Remarks. The *Pediastrum* spp. Assemblage zon. nov. is comparable with *Todisporites* spp. Assemblage zone of Singh *et al.* (1978) in Subathu town locality and Sarkar and Singh (1988) established in the Subathu sediments of Banethi-Bagthan area of Himachal Pradesh. The *Pediastrum* spp. Assemblage zon. nov. is characterized by the dominance of different species of *Pediastrum* and a sharp decline in the dinoflagellates population. The zone also exhibits the dominance of terrestrial mioflora.

The basal part of the Subathu Formation in the Subathu *Stratotype* which is represented by *Cyclonephelium* and *Hexagonifera* spp. Assemblage zones (Singh *et al.*, 1978) is not traceable to the present areas of investigation. Further the stratigraphic position of *Cleistosphaeridium* spp. Assemblage zone considered older than *Homotryblium* spp. Assemblage zone by Singh *et al.* (1978) and Sarkar and Singh (1988) in the Subathu Type Area and Baneth-Bagthan has been found to be younger than *Thalassiphora* spp. Assemblage zone in the present area of investigation, in the light of FAD and LAD comparison of Dino taxa with the Williams *et al.* (1993).

Age And Environment

On the basis of palynological assemblages from the Subathu Formation ,Singh (2002) and Singh *et al.* (2003) have considered this litho-unit to be ranging in age from Ypresian to Post Lutetian in the area. Singh *et al.*, (2003) on the basis of detailed percentage counts, frequency variation patterns of palynotaxa and precise correlation of index palynotaxa with the global dinoflagellates markers of Williams *et al.*, (1993) have assigned a precise age to the five palynostratigraphic zones as given below, in the ascending order of the stratigraphy:-

Palynostratigrphic zone	**Age and Duration**
5. *Pediastrum* spp. Assemblage zon. nov.	Post Lutetian
4. *Cleistosphaeridium* spp. Assemblage zone	Ypresian-Lutetian (51.0-44.0 ma)
3. *Thalassiphora* spp. Assemblage zone	Ypresian-Lutetian (51.0 ma - 49.0 ma)
2. *Cordosphaeridium* spp. Assemblage zone	Ypresian (52.0 -51.0 ma)
1. *Homotryblium* spp. Assemblage zone	Ypresian (52.0-51.0 ma)

The environmental interpretation based on palynoflora from the Subathu Formation of the area have been dealt in detail by Singh *et al.*, (2003). The palynological data favour shallow marine through marine to marginal marine environment of deposition from *Homotryblium-Cordosphaeridium* Assemblages through *Cleistosphaeridium* to *Thalassiphora* zones respectively in the ascending order. The topmost *Pediastrum* Assemblage zon. nov. has been considered to have been deposited in freshwater to marginal marine conditions.

Conclusions

1. The Subathu Formation of Kalka-Shimla section is palynostratigraphically differentiated into five palynozones namely *Homotryblium* spp. Assemblage zone, *Cordosphaeridium* spp. Assemblage zone, *Thalassiphora* spp. Assemblage zone, *Cleistosphaeridium* spp. Assemblage zone and *Pediastrum* spp. Assemblage zon. nov. in the ascending order.
2. These five palynozones when equated with the global dinoflagellate markers of Williams *et al.*, (1993) have been found to be ranging in age as *Homotryblium* spp. Assemblage zone and *Cordosphaeridium* spp. Assemblage zone, 52.0-51.0 ma; *Thalassiphora* spp. Assemblage zone, 51.0 ma - 49.0 ma; *Cleistosphaeridium* spp. Assemblage zone, 51.0-44.0 ma and *Pediastrum* spp. Assemblage zon. nov. as Post Lutetian
3. The *Homotryblium* and *Cordosphaeridium* spp. Assemblage zones have been deposited under shallow marine conditions, *Thalassiphora* spp. Assemblage zone has been deposited under brackish water to marine, *Cleistosphaeridium* spp. Assemblage zone has been deposited under marine condition and *Pediastrum* spp. Assemblage zone has been inferred to have been deposited under marginal marine to freshwater conditions.

References

BHATIA, S.B. and MATHUR, N.S. 1964. On occurrence of Pulmonate gastropods in the Subathu-Dagshai passage beds near Dharampur-Simla Hills. *Journal of the Geological Society of India,* **2:** 33-36.

CHAUDHRI, R.S. 1968. Stratigraphy of the Lower Tertiary Formations of Panjab Himalayas. *Geological Magazine of London,* **105:** 421-430.

CHAUDHRI, R.S. 1975. Sedimentology and genesis of the Cenozoic sediments of northwestern Himalayas (India), *Geol. Rund.,* **64:** 958-977.

CHAUDHRI, R.S. 1991. Sedimentation model of the Cenozoic sediments of Himalaya. *Publ. Cent. Adv. Stdy. Geol. Pb. Univ. Chd.,.* **4(N.S):** 1-32.

KARUNA KARAN, C. and RANGA RAO, A. 1979. Status of exploration for hydrocarbon in the Himalayan region-contribution to stratigraphy and structure. *Him. Geol. Surv. Misc. Publ. Sem.,* **41:** 1-67.

KRISHNAN, M.S. 1982. Geology of India and Burma. *Higginbothams (pvt) Ltd. Madras. 5th edition,*:1-535.

MATHUR, N.S. 1969. Biostratigraphy of the Subathu 'Bed' Simla Hills. *Unpublished. Ph.D. Thesis, Panjab University, Chandigarh.*

MEDLICOTT, H.B. 1879. Manual of the geology of India. *Geol. Surv. India, Culcutta,* **2:** 8-23.

MEDLICOTT, H.B. 1964. On the geological structures and relations of the southern portion of the Himalayan ranges between the rivers Ganges and Ravee. *Mem. Geol. Surv India,* **3:** 1-206.

PASCOE, E.H. 1964. A manual of geology of India and India and Burma. *Govt. of India.* **II:** 485-1344.

PILGRIM, G.E. and WEST, W.D. 1928. The structure and correlation of Simla rocks. *Mem. Geol. Surv. India,* **53:** 1-140.

RAIVERMAN, V. 1979. Stratigraphy and facies distribution of Subathu sediments, Simla Hills, northwestern Himalaya. *Geol. Surv. Ind. Misc. Publ.,* **41:** 111-126.

RAIVERMAN, V. and SESHVATARAM, B.T.V. 1965. On mode of deposition of Subathu-Dharamsala sediments in the Himalayan foothills in Punjab and Himachal Pradesh. *D.N. Wadia Comm.,*: 456-471.

SARKAR, S. and SINGH, H.P. 1988. Palynological investigation of the Subathu Formation (Eocene) in the Banethi-Bagthan area of Himachal Pradesh. India. *Palaeontographica Abt. B.,* **209 (1-3):** 29-109.

SINGH, H.P., KHANNA, A.K. and SAH, S.C.D. 1978. Palynological zonation of the Subathu Formation in the Kalka-Simla area of Himachal Pradesh. *Him. Geol.,* **8:** 33-46.

SINGH, Y.R. 2002. Palynostratigraphy of the Palaeogene rocks exposed around Kalka and Dharampur, Northwestern Himalaya, India. *Unpublished Ph. D. Thesis, Kurukshetra University, Kurukshetra.*

SINGH, Y. R., DOGRA, N.N. and SINGH, R.Y. 2003. Age and palaeonenvironmental constraints of Subathu Formation of Dharampur and Koti areas of Solan District, Himachal Pradesh-a palynological approach. Gond. *Geol. Magz., Spl.* **6:** 195-205.

SRIKANTIA, S.V. and BHARGAVA, O.N. 1998. Geology of Himachal Pradesh. *Publ. Geol. Soc. India, Banglore, 1st edition,*: 1-408.

WILLIAM, G.L., STOVER, L.E. and KIDSON, E.J. 1993. Morphology and stratigraphic ranges of selected Mesozoic-Cenozoic dinoflagellate taxa in the Northern Hemisphere. *Geol. Surv. Canada,* **20:** 1-137.

Occurrence and age significance of *Planorotalites palmerae* (Cushman and Bermudez) in the Dillai Parbat area of Assam, NE India

Kezhakielie Whiso[1], R. Venkatachalapathy[1*], P. Ramesh[2] and R.P. Kachhara[1]

[1]Department of Geology, Nagaland University, Kohima 797002, India
[2]Paleontology Laboratories KDMIPE, ONGC, Dehra Dun 248195, India

Introduction

The discovery of oil in Upper Assam has accelerated many geological studies in the region. The Paleogene shelf sediments of the Assam-Arakan region extending from the Garo hills (Meghalaya) in the west to the Mikir hill in the east adjoining the Naga hills are known for their rich assemblage of foraminifera. Sub-surface samples from Upper Assam Platform Nagaland and Dhansiri Valley also yield similarly rich assemblage of foraminifera. The present study has been carried out in Dillai Parbat area, which is located on the southeastern periphery of the Mikir Hills, Assam (Fig. 1). Foraminifera were separated using standard micropaleontological techniques. The authors followed the classification given by Loeblich and Tappan (1988) and the 'Plankton Stratigraphy' edited by Bolli *et al.* (1985). In this paper the occurrence of *Planorotalites palmerae* (Cushman and Bermudez) is reported along with its taxonomy and significance.

SYSTEMATIC PALEONTOLOGY
Order FORAMIIFERIDA, Eichwald, 1830
Suborder GLOBIGERININA, Delage and Herouard, 1896
Superfamily GLOBOROTALIACEA, Cushman, 1927
Genus Planorotalites, Morozova, 1957
Phanorotalites palmerae (Cushman and Bermudez)
(Plate-1)

Globorotalia palmerae Cushman and Bermudez, 1937, p. 26, p1.2, figs. 51-53.

Globorotalia palmerae Cushman and Bermudez, Raju, 1968, p.1.1, figs. 13a, b.

Globorotalia palmerae (Cushman and Bermudez), Toumarkine and Luterbacher, 1985, p. 117, fig. 20 (14-29).

Description

Test medium sized, compressed and has a stellate (star-shaped) axial periphery. This easily recognisable species generally has 5 to 7 chambers in the final whorl, which increase fairly rapidly in size. The chambers in the last whorl are radially elongated and prolonged by a spine formed by the extension of the faint keel bordering them. The early chambers are subglobular, whereas the later chambers are nearly conical in shape. Wall calcareous, perforate, surface slightly hispid. The sutures are radial, straight and slightly depressed. Aperture, a forward directed interiomarginal slit or arch, umbilical-extra-umbilical and bordered by a lip.

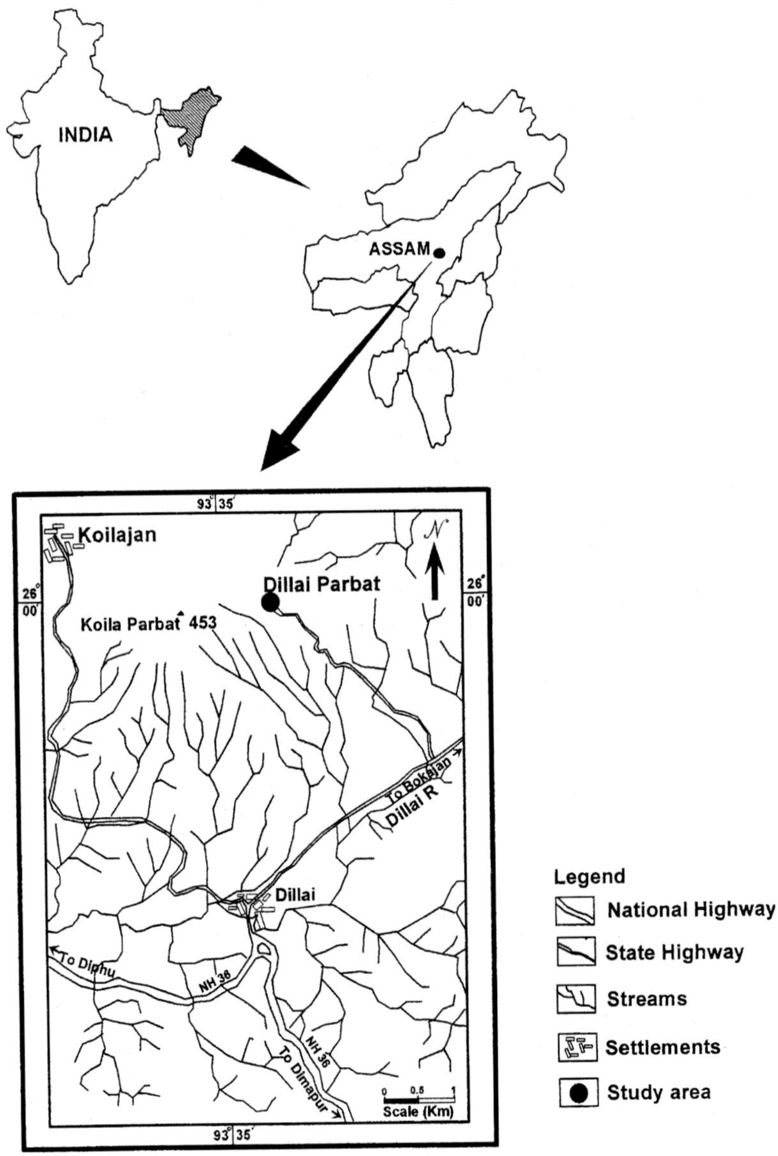

Fig. 1. Location map of Dillai Parbat.

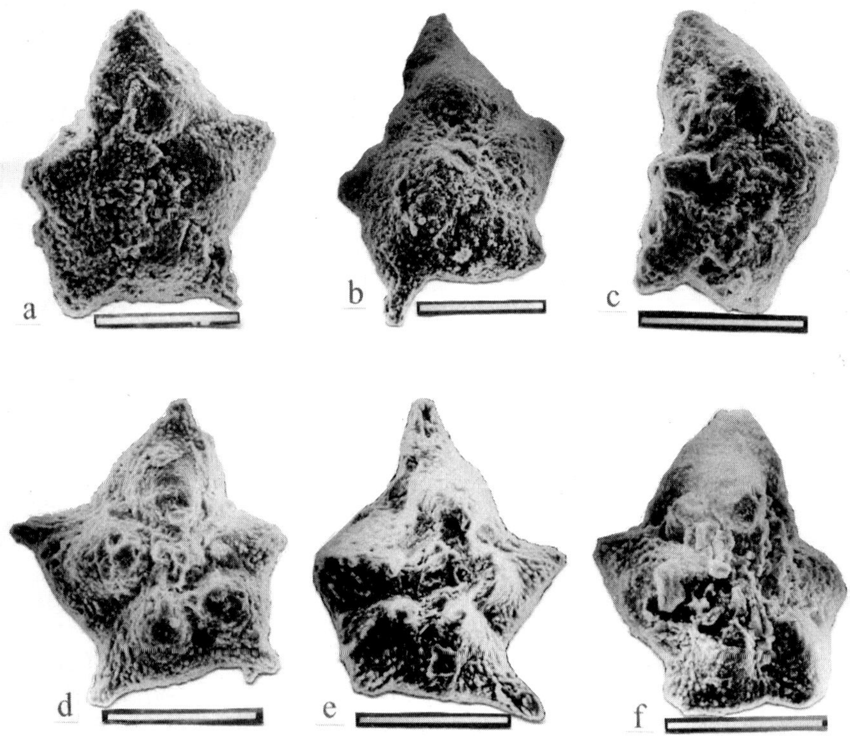

Plate 1. Planorotalites palmerae (Cushman and Bermudez). Scale bars represent 10 μm.

Remarks

This species was originally described by Cushman and Bermudez as *Globorotalia palmerae* for the specimens from Arroyo Naranjo, Havana Province, Cuba.

Planorotalites palmerae a significant Paleogene planktonic foraminiferal species has been confused with the genus *Pararotalia*. Schimdt and Raju (1973) clearly demonstrated the difference between these stellate representatives. The genus *Pararotalia* has a prominent ture umbilical knob/plug and an areal aperture. The *Planorotalites palmerae* sometimes may have a 'plug', which is composed of secondary incrustations and not related to the true umbilical knob of *Pararotalia*.

The occurrence of *P. palmerae* was reported from Cuba, Trinidad, Jamaica, Caribbean Sea and Spain. In India this species was first described by Raju (1968) from the subcrops of Cauvery Basin, South India. In North East India, *P. palmerae* is reported from the out-crops of Assam.

Bolli (1957) and Raju (1968) have used *Globorotalia palmerae* as a zonal marker for Early Eocene in Trinidad and India respectively. Because of the rarity of *Planorotalites palmerae*, the name *Acarinina pentacamerata* Zone is used as a substitute for the *Globoratalia palmerae* Zone of the original Trinidad zonation (Toumarkine and Luterbacher, 1985). Raju *et al* (1991) have proposed *Planorotalites palmerae* as a subzone within *Acarinina pentacamerata* Zone and correlated them with Zone P9, representing Lower Eocene.

Many authors including Berggren and Miller (1988) and Raju *et al* (1991) believe that the initial appearance of *P. palmerae* can be used to denote the base of Zone P9. The geographically restricted *P. palmerae* has a very patchy distribution. The species *P. palmerae* is found to be restricted within the *Acarinina pentacamerata* Zone of Toumarkine and Luterbacher (1985) and Raju *et al.*(1991).

The occurrence of *Planorotalites palmerae* becomes significant in the study area because its first appearance datum (FAD) denotes the base of Zone P9. In the present study area *P. palmerae* occurs in the base of F-block of Dillai Parbat Limestone mines of Cement Corporation of India, karbi Anglong District, Assam.

Hypotype

Maximum diameter: 0.25 mm., thickness: 0.03 mm.

Type locality

Approximately 18 km and 28 km from Bokajan in Assam and Dimapur in Nagaland respectively.

Acknowledgements

The authors are thankful to the authorities of Nagaland University for the facilities made available to carry out this work. The authors are grateful to Dr. D.S.N. Raju and Prof. M.S. Srinivasan for their constant encouragement during the course of this work. K.W. acknowledges the CSIR HRD Group, New Delhi for the fellowship (JRF) provided to him.

References

BERGGREN, W.A. and MILLER, KENNETH G. 1988. Paleogene tropical planktonic foraminiferal biosratigraphy and magnetobiochronology. *Micropaleontology*, 34(4): 362-380.

BOLLI, H.M. 1957a. The genera *Globigerina* and *Globorotalia* in the Paleocene-Lower Eocene Lizard Springs Formation of Trinidad, B.W.I. *Bulletin of the U.S. National Museum*, 215: 61-81.

BOLLI, HANS M., SAUNDERS, JOHN B. and NIELSEN, KATHARINA PERCH. 1985. *Plankton Stratigraphy*. Cambridge University Press, Cambridge.

CUSHMAN, J.A. AND BERMUDEZ, P.J. 1937. Further new species of foraminifera from the Eocene of Cuba. *Contrib. Cushman Lab. Foramin. Res.*, 13: 1-29.

LOEBLICH, ALFRED R. and TAPPAN, HELEN. 1988. *Foraminiferal genera and their classification*. Van Nostrand and Reinhold Company, New York.

RAJU, D.S.N. 1968. Eocene-Oligocene Planktonic Foraminiferal Biostratigraphy of the Cauvery Basin, South India. *Memoir Geol. Soc. India,* 2: 286-299.

RAJU, D.S.N., RAVINDRAN, C.N., MISHRA, P.K. and SINGH, JAGMOHAN. 1991. Cretaceous and Cenozoic Foraminiferal zonal framework for East Coast Sedimentary Basins of India. *Geoscience Journal*, XII(2): 155-175.

SCHMIDT, R.R. and RAJU, D.S.N. 1973. *Globorotalia palmerae* Cushman and Bermudez and closely related species from the Lower Eocene, Cauvery Basin, South India. *Proceedings, Koninks, Nederland Akademie van Wetenschappen, Series B*, 76(2): 167-187.

TOUMARKINE, M. and LUTERBACHER, H.P. 1985. Paleocene and Eocene planktonic foraminifera. In Bolli, H.M., Saunders, J.B. and Perch-Neilsen, K.D., Eds., *Plankton Stratigraphy*. Cambridge University Press, Cambridge, 117.

Stratigraphic, paleobiogeographic and paleoenvironmental significance of *Mesophyllum*, a Nongeniculate Coralline Alga from western Kachchh (Middle Eocene to Oligocene), India

P. Kundal and Sumedh K. Humane

Postgraduate Department of Geology, Nagpur University, Law College Campus,
Nagpur 440 001, India

Introduction

The sediments of Kachchh Basin of Gujarat ranges in age from Mesozoic to Cenozoic and the Cenozoic sediments of the basin are exposed as a more or less complete sequence from Paleocene to Pliocene (Biswas, 1992; Fig. 1). A glance on the work done on the Calcareous Algae from Tertiary rocks of Kachchh, district Gujarat (Table 1) divulges that 69 species of coralline, dasycladacean and udoteacean algae are known (Pal and Ghosh, 1974; Tandon, Gupta and Saxena, 1978; Kar, 1979; Misra, Jauhri, Singh, Kishore and Chowdhury, 2001; Singh and Kishore, 2001; Ghosh, 2002; Singh, Kishore, Misra and Jauhri, 2002; Kundal and Humane, 2002; Kundal and Humane, 2003a; Kundal and Humane, 2003 b and Kundal and Humane, 2005). *Mesophyllum*, is a genus of nongeniculate coralline algae belonging to family Corallinaceae and subfamily Melobesioideae. *Mesophyllum* occurs dominantly in two formations namely Fulra Limestone Formation (Middle Eocene) and Maniyara Fort Formation (Oligocene) (Fig. 2) and exhibits almost all the characters.

Only one named species of *Mesophyllum commune* from Khari Nadi Formation (Miocene) and two unnamed species of *Mesophyllum* from Maniyara Fort Formation (Oligocene) were known prior to commencement of present work (Pal and Ghosh, 1974; Misra *et al.*, 2001 and Ghosh, 2002). In the present paper six species of *Mesophyllum,* namely, *Mesophyllum commune* Lemoine, *Mesophyllum contii* Ishijima, *Mesophyllum curtum* Aguirre and Braga, *Mesophyllum koritzae* Lemoine, , *Mesophyllum roveretoi* Contii, *Mesophyllum* Sp. are identified on the basis of growth habit, thallus architecture, conceptacles and cell fusions. The specimens recovered in the present study are lodged in Museum of Department of Geology, Nagpur University, Nagpur- 440001.

Taxonomy

Division	:	Rhodophyta Wittstein, 1901
Class	:	Rhodophyceae Rabenhorst, 1863
Order	:	Corallinales Silva and Johansen, 1986
Family	:	Corallinaceae Lamouroux , 1816
Subfamily	:	Melobesioideae Bizzozero, 1885
Genus	:	***Mesophyllum*** Lemoine, 1928

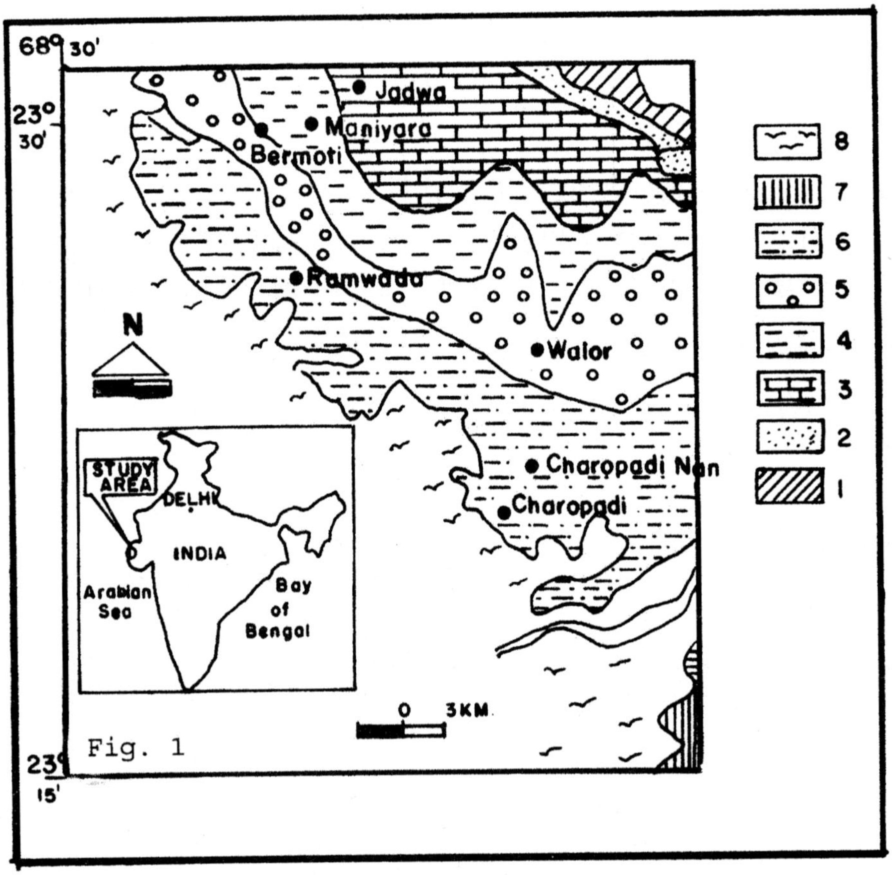

Fig. 1. Location and Geological Map of Kachchh, Gujarat, India (after Biswas 1992): 1. Naredi Formation (Upper Paleocene to Lower Eocene /Upper Thanetian to Ypresian), 2. Harudi Formation (Middle Eocene / Middle Lutetian), 3. Fulra Limestone Formation (Late Middle Eocene /Upper Lutetian to Lower Bartonian), 4. Maniyara Fort Formation (Oligocene /Upper Rupelian to Chattian), 5. Khari Nadi Formation (Early Lower Miocene / Aquitanian), 6. Chhasra Formation (Late Lower Miocene / Burdigalian) and 7. Sandhan Formation (Pliocene '& 8. Recent.

Mesophyllum commune Lemoine
Plate 1, figures 8,9
Mesophyllum commune Lemoine: Lemoine, 1939, p. 86, text figs. 54, 56 and 57
Mesophyllum commune Lemoine: Fravega, Giammarino, Vannucci, 1984, p.511, fig.7.
Material: PGTDG/MF/SCA/183
Morphology: Thallus encrusting with protuberances.

Vegetative Anatomy: Monomerous, thin coaxial core and multi layered peripheral region. Thickness of core region 100-125μm and cells of core region exhibit polygonal outlines. Cells of peripheral region rectangular and polygonal in oblique section. Height of cells of peripheral region found to be 10-15μm and 10-12μm in width.

Reproductive Organs: Multiporate conceptacle, rounded to crescent in shape. Diameter of conceptacle found to be 300-325μm and height is 80-160μm. Conceptacles unfertilized.

Remarks: The growth habit, thallus architecture, diameter of conceptacles and dimensions of cells of core and peripheral region of the present material more or less fit under the description of *Mesophyllum commune* Lemoine.

Stratigraphic and Geographic Distribution: Miocene of Algeria (Lemoine, 1939); Dolomitised micrite and biomicrite of Middle Miocene Benghazi Limestone of Rajmah Formation, Western Cyrenaica (Fravega, Giammarino and Vannucci, 1984) and Miocene of Malta (Bosence, 1993) .

Horizon and Locality: Maniyara Fort Formation (Oligocene) at Bermoti Hill Nadi Section.

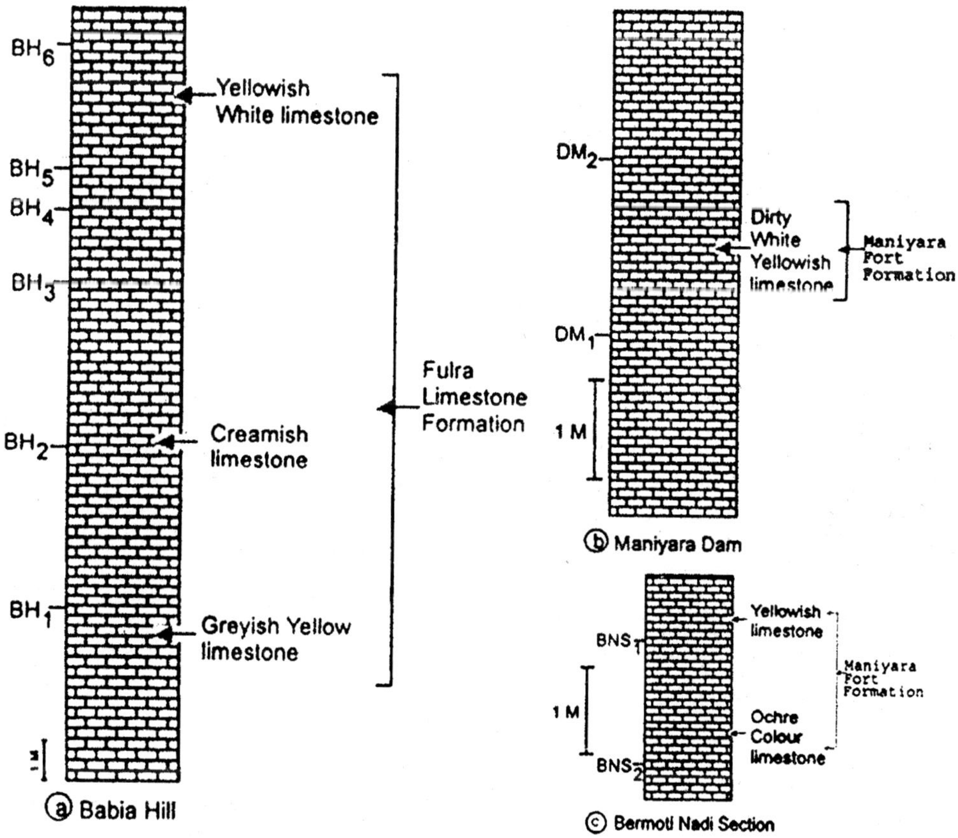

Fig. 2. Lithosections examined in the present study.

Mesophyllum contii Ishijima

Plate 1, figures 3 and 6

Mesophyllum contii Ishijima: Ishijima 1954, p. 33.

Material: PGTDG/MF/SCA/ 179

Morphology: Thallus monomerous, palm shaped, mutlilayered with crescentic core region.

Vegetative Anatomy: Monomerous thallus made up of coaxial and radiating core region 75-500µm in thickness. Cells of core filaments rectangular to squarish (12-25µm x 12-15µm). The peripheral region very thick representing several thalli growing one upon another. The cells of peripheral region subrectangular and elongated (25-30µm x 15-18µm). Cell fusion is clearly seen in the studied material. Secondary pit connection absent.

Reproductive Organs: Conceptacle (175µm in diameter and height is 85-125µm) multiporate, elliptical and with two to three pores at the roof.

Remarks: Mesophyllum contii Ishijima and the present specimen under study have similar appearance, cell dimensions, multiporate conceptacle and therefore the present specimen is described under *Mesophyllum contii* Ishijima

Stratigraphic and Geographic Distribution: Miocene limestone of Kawaguchi Formation, Western Pacific (Ishijima, 1954)

Horizon and Locality: Fulra Limestone Formation (Late Middle Eocene) exposed at Babia Hill.

Mesophyllum curtum Aguirre and Braga

Plate 1, figures 1 and 10

Mesophyllum curtum Aguirre and Braga: Aguirre and Braga 1998, p. 497, pl. 2, figs. 5 and 7.

Mesophyllum curtum Aguirre and Braga: Lemoine, 1939, p. 92, pl. 2, figs. 13, text. fig. 61.

Material: PGTDG/MF/SCA/ 180

Morphology: Thallus encrusting, semicircular in appearance.

Vegetative Anatomy: Thallus monomerous with coaxial core (150-275µm). Cells of the core region rectangular and at places flexuous in oblique section. The height of the cells of core region measures about 25 µm and 12 µm in width. At some places (oblique section) the cells are trapezoidal. Peripheral region (60 - 410µm) clearly demarcated by cells having rectangular, squarish and polygonal shape. The height of the cells of the pripheral region ranging from 8-10 µm and 4-5 µm in width.

Reproductive Organs: The multiporate bi/tetrasporangial conceptacles varying in dimensions disposed in peripheral region. The diameter of larger conceptacle ranging 425-450µm in diameter and 40-100µm in height while smaller conceptacle have diameter 250-275 µm and 30-80 µm in height .

Remarks: The present specimen possesses coaxial core, similar dimensions of peripheral filaments and multiporate bi/tetrasporangial conceptacle as that of *Mesophyllum curtum* Aguirre and Braga. Hence, it is described as *Mesophyllum curtum* Aguirre and Braga.

Stratigraphic and Geographic Distribution: Tortonian Limestone (Middle Miocene) of Algeria (Lemoine, 1939 and Aguirre and Braga, 1998)

Horizon and Locality: Maniyara Fort Formation (Oligocene) exposed at Maniyara Dam.

Mesophyllum koritzae Lemoine

Plate 1, figures 7

Mesophyllum koritzae Lemoine: Vannucci, 1970, p. 452, pl. 5, Fig. 1.

Material: PGTDG/MF/SCA/ 182

Morphology: Thallus encrusting having hook like appearance.

Vegetative Anatomy: Thallus monomerous with coaxial core in relation with peripheral region. Thickness of core region varies from 100 -150µm. Cells of core region 22-25µm in length and 10-

12μm in width. Peripheral region thin (22-75μm). Cells of the peripheral region rectangular varying from 8-10μm in height and 4-6μm in width.

Remarks: Dimensions of core and peripheral region, multiporate conceptacle, gross appearance and coaxial core in the present material are of similar nature as found in *Mesophyllum koritzae* Lemoine. Therefore, it is kept under *Mesophyllum koritzae* Lemoine.

Stratigraphic and Geographic Distribution: Miocene rocks of Algeria and Oligocene of Nullipore Formation, Morena Garda (Vannucci, 1970).

Horizon and Locality: Maniyara Fort Formation (Oligocene) exposed at Bermoti Hill Nadi Section.

Mesophyllum roveretoi Contii

Plate1, figures 2 and 5

Mesophyllum roveretoi Contii: Fravega, Piazza and Vannucci, 1993, p. 203.

Mesophyllum roveretoi Contii: Vannucci, Piazza, Fravega, Arnera, 1994, p. 102, table-2; p.115.

Mesophyllum roveretoi Contii: Vannucci, Piazza, Fravega and Abate, 1996, p.73, table. 1; p. 86, fig.c.

Mesophyllum roveretoi Contii: Vannucci, Piazza, Postarino and Fravega, 1997, p.16, table. 1.

Mesophyllum roveretoi Contii: Fravega, Giammarino, Piazza, Russo and Vannucci, 1987, p.52.

Material: PGTDG/MF/SCA/ 185

Morphology: Thallus an irregular crust.

Vegetative Anatomy: Monomerous thallus. Core region coaxial, thin with diameter 120-125μm and cells of core region ranges from 15-18μm while height falls in between 6-9μm. Peripheral region 250-500μm thick and cells of peripheral region 20-24μm in height and 15-20μm in width. Size of peripheral cells is rectangular.

Reproductive Organs: Tetra/bisporangial multiporate conceptacle. Diameter of conceptacles ranges in between 180-200μm while 80-90 μm in height.

Remarks: The dimensions of cells of coaxial core region and cells of peripheral region, the growth habit, thallus structure and reproductive organs of the present specimens resemble *Mesophyllum roveretoi* Contii. Therefore, the present material is placed under *Mesophyllum roveretoi* Contii.

Plate 1.

1, 10. *Mesophyllum curtum* Aguirre and Braga, Specimen No. PGTDG/MF/SCA/180 **1.** Thallus exhibiting semicircular appearance with coaxial core and peripheral region containing two multiporate fertile conceptacles. **10.** Same specimen showing multiporate fertile conceptacle. **2, 5.** *Mesophyllum roveretoi* Contii, Specimen No. PGTDG/MF/SCA/185 **2.** Thallus exhibiting coaxial core, peripheral region and two multiporate conceptacles. **5.** Same specimen showing coaxial core region. **3, 6.** *Mesophyllum contii* Ishijima, Specimen No. PGTDG/MF/SCA/179 3. Thallus exhibiting palm shaped appearance with thick core and layers of peripheral region containing multiporate conceptacle. **6.** Same specimen showing multiporate conceptacle. 4. *Mesophyllum* Sp., Specimen No. PGTDG/MF/SCA/187 Thallus showing multiporate conceptacle with coaxial core. **7.** *Mesophyllum koritzae* Lemoine, Specimen No. PGTDG/MF/SCA/186 Thallus showing elongated nature, coaxial core, peripheral region and conceptacle. **8, 9.** *Mesophyllum commune* Lemoine, Specimen No. PGTDG/MF/SCA/183 **8.** Thallus showing multistratose appearance with multiporate conceptacle. **9.** Same specimen showing multiporate conceptacle.

Plate 1.

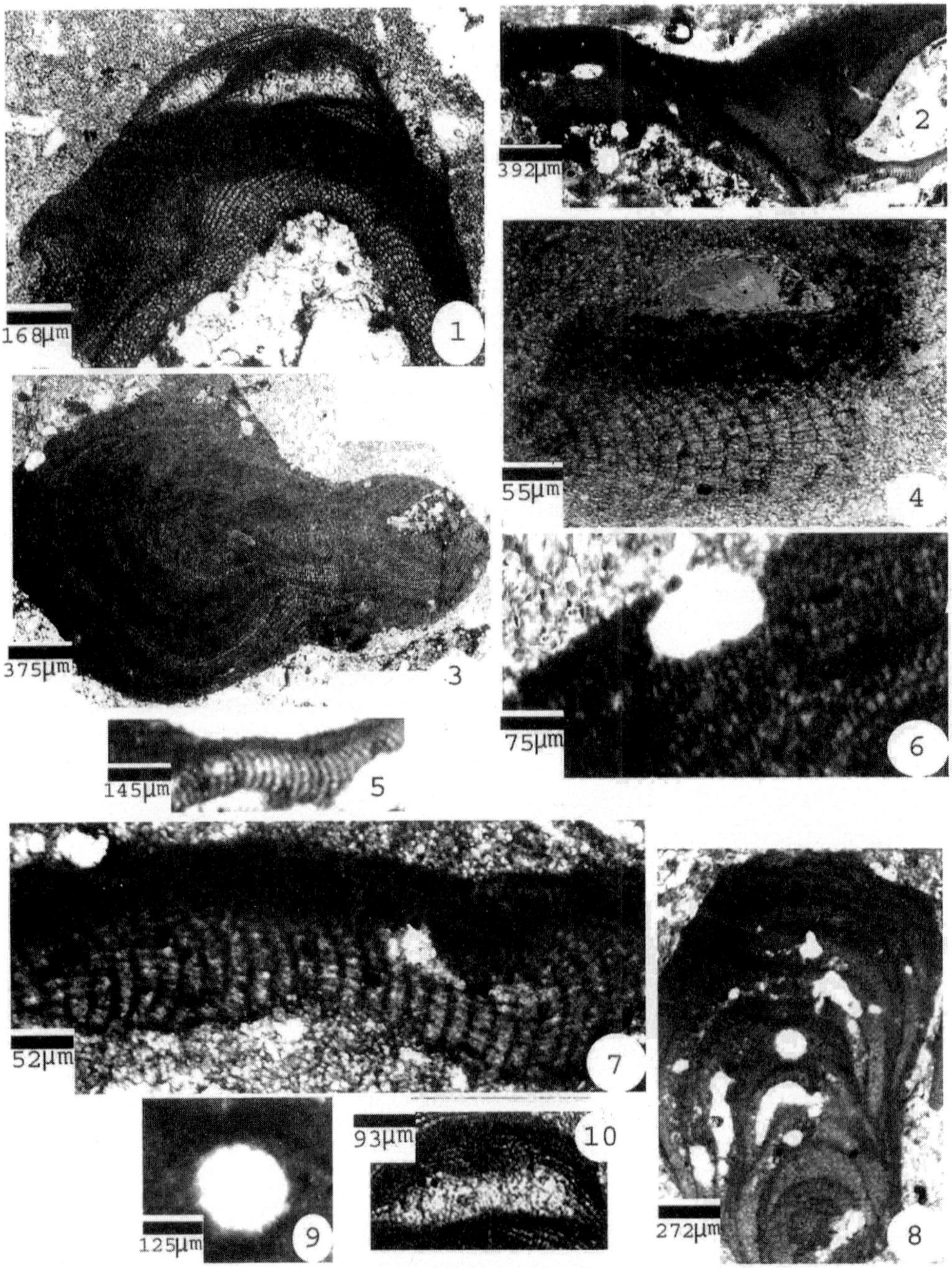

168µm
1
392µm
2
375µm
3
55µm
4
145µm
5
75µm
6
52µm
7
272µm
8
125µm
9
93µm
10

Table 1. Work on Calcareous Algae from Cenozoic of Kachchh

Authors	Algal Assemblage	Formation
Pal and Ghosh, 1974	*Lithophyllum* aff. *L. kladdosum, Mesophyllum commune,Aesthesolithon problematicum, A. cutchensis* & *Archaeoporolithon miocenicum*	Khari Nadi Formation (Miocene)
Tandon, Gupta and Saxena,1978	*Lithophyllum bermotiensis*	Maniyara Fort Formation (Oligocene)
Kar, 1979	*Lithothamnion cf. L. validum, L. cf. L. bofilli and Lithophyllum sp.*	Fulra Limestone Formation (Middle Eocene)
Misra, Jauhri, Singh, Kishore and Chowdhury, 2001	*Sporolithon* sp., *Lithoporella* sp., *Neogoniolithon* sp., *Spongites* sp. *Lithothamnion manni, L.*(?)*iorii, Lithothamnion* sp.1, *Lithothamnion* p.2, *Lithothamnion* sp. 3, *Mesophyllum* sp.1, *Mesophyllum* sp.2,*Corallina* sp. 1, *Corallina* sp. 2, *Arthrocardia* sp. 1 and *Amphiroa* sp.	Maniyara Fort Formation (Oligocene)
Singh and Kishore, 2001	*Cymopolia* sp., *Neomeris* sp., *Salpingoporella* sp., *Acicularia* sp., *Goniolithon* sp. and *Halimeda* sp.	Fulra Limestone (Middle Eocene) and Maniyara Fort Formation (Oligocene)
Ghosh, 2002	*Lithophyllum* sp. A, *Lithophyllum* sp. B, *Lithophyllum* sp. C, *Mesophyllum* sp.A & *Mesophyllum* sp.B	Maniyara Fort Formation (Oligocene)
Singh, Kishore, Misra and Jauhri,, 2002	*Lithothamnion* cf. *L. lacroixi* Lemoine, *Mastophoroideae* gen. et spec. indet., *Spongites* sp. 1, *Spongites* sp. 2	Maniyara Fort Formation (Oligocene)
Kundal and Humane, 2002	*Amphiroa anchiverricosa* Johnson and Ferris, *Arthrocardia cretacica* Raineri, *A. indica* Kundal and Humane, , *A. konitaensis* Ishijima and *Calliarthron antiquum* Johnson	Fulra Limestone Formation (Middle Eocene),Maniyara Fort Formation (Oligocene), Khari Nadi Formation (Lower Miocene), Chhasra Formation (Lower Miocene)
Kundal and Humane, 2003 a	Corallina elliptica, C. grandis , C. hayasakai, C. kachchhensis ,C. marshallensis, C. matansa , C. otsukiensis, C. prisca , *C. raoi* and *C. typica*	Fulra Limestone (Middle Eocene), Maniyara Fort (Oligocene), Chhasra (Lower Miocene)

Contd...

Table 1 (contd.)

Authors	Algal Assemblage	Formation
Kundal and Humane, 2003 b	*Acroporella indica, A. kachchhensis, Broeckella jauhrii, B. media, Cymopolia delicata, Dactylopora maniyarafortensis, Dissocladella deserta, D. nana, Orioporella villattae* and *Pagodaporella minuta*	Maniyara Fort (Oligocene) and Chhasra (Lower Miocene)
Kundal and Humane, 2005	*Jania badvei, J. guamensis, J. indica, J. mayei, J. sripadaraoi & J. vetus*	Fulra Limestone (Middle Eocene), Maniyara Fort (Oligocene) and Khari Nadi (Lower Miocene)

Stratigraphic and Geographic Distribution: Oligocene of Tertiary Piedmont Basin, NW Italy (Vannucci, Piazza, Postarino and Fravega, 1997); Miocene Rhodoliths of Tertiary Piedmont Basin, Mon Ferrato Alessandria, NW Italy (Vannucci, Piazza, Fravega, Arnera, 1994); Oligocene rocks of Sessello Basin, Tertiary Piedmont Basin, NW Italy (Vannucci, Piazza, Fravega and Abate, 1996) and Miocene rocks of tertiary Piedmont Basin, NW Italy (Fravega, Piazza and Vannucci, 1993).

Horizon and Locality: Maniyara Fort Formation (Oligocene) at Bermoti Hill Nadi Section.

Mesophyllum Sp.

Plate 1, figure 4

Material: PGTDG/MF/SCA/ 187

Morphology: Thallus encrusting, flattened.

Vegetative Anatomy: Thallus monomerous comprised of perfectly preserved coaxial core filaments and badly preserved coaxial core filament. Thickness of core region 80-100μm and cells 15-20μm in length and 5-12μm in width. Peripheral region thin ranging from 125-175μm in thickness. Cells of peripheral region are not preserved.

Reproductive Organs: Multiporate conceptacle with diameter 200-210μm and 50-60μm in height.

Remarks: The imperfect preservation of the present specimen does not permit to assign any formal name as the peripheral region and conceptacles are not properly preserved.

Horizon and Locality: Maniyara Fort Formation (Oligocene) at Bermoti Hill Nadi Section.

Stratigraphic, paleobiogeographic and paleoenvironmental significance

Mesophyllum commune occurs during Middle Eocene and Miocene in other parts of the World whereas this species has been documented in this paper from Fulra Limestone Formation (Middle Eocene) of Kachchh; *M. concretum* occurs in Maniyara Fort Formation (Oligocene) while it has been documented during Pliocene of Italy; *M. contii* occurs during Cenozoic in Western pacific while in Kachchh it is present in Fulra Limestone Formation (Middle Eocene) ; *M. curtum* occurs during Middle Eocene in Kachchh but earlier it has been documented from Tortonian (Middle Miocene) of Algeria and *M. koritzae* occurs in Maniyara Fort Formation (Oligocene) while earlier it has been documented from Miocene rocks of Algeria and Oligocene of Nullipore Formation, Morena Garda. Therefore, all the five species of *Mesophyllum*, broadly supports the ages of the formations in which they are present.

Mesophyllum commune is cosmopolitan as it occurs in Italy, Malta, Guatemala and now India. The other four species of *Mesophyllum* occur in Tethyan realm. Thereby, the species of *Mesophyllum* shows Tethyan affinities.

Documenting *Mesophyllum* from Maniyara Fort Formation (Oligocene) from Kachchh, Ghosh (2002) suggested that the dominance of *Mesophyllum* in association with *Lithophyllum* prevailed in water depth down to 40 m. Kundal and Humane (2002 and 2003a) interpreted that the limestone ranging in age from Middle Eocene to Lower Miocene of Kachchh was deposited at a depth of 60-80 m below low tide level in tropical water based on geniculate coralline algal species. However, Kundal and Humane (2003b) have documented dasycladacean species from Maniyara Fort Formation (Oligocene) and concluded that the dasycladacean species are suggestive of deposition of limestone at a depth of 10 m. While Fulra Limestone Formation also do contain dasycladalean algae. Hence, it is concluded that the limestone of Fulra Limestone Formation (Middle Eocene) and Maniyara Fort Formation (Oligocene) was deposited at a depth of 10-15 m while it could sometimes formed at 10-40m depth along the continental slope where the genus *Mesophyllum* with other nongeniculate algae are dominated.

Acknowledgement

We are grateful to the Department of Science and Technology (DST), Govt. of India for sponsoring the present research project (ESS / 23 / VES / 096 / 2000). One of us (SKH) is indebted to DST for award of JRF under the same project.

References

AGUIRRE, J. and BRAGA, J. C.,1998. Redescription of Lemoine's (1939) types of coralline algal species from Algeria. Paleontology, 41, 3: 489-507, pls. 3.

BISWAS, S. K., 1992. Tertiary Stratigraphy of Kutch. Journal of Palaeontological Society of India, 37: 1-29.

BOSENCE, D. W. J., 1983. Coralline algae from the Miocene of Malta. Paleontology, 26, 1: 147-173.

FRAVEGA, P., GIAMMARINO, S. and VANNUCCI, G., 1984. Algal remains and their correlation with microfaunas in the Benghazi limestone (Middle Miocene of Western Cyrenaica). Bolletine ofGeological Society of Italy, 103: 503-514, 7 ff.

FRAVEGA, P., GIAMMARINO, S., PIAZZA, M., RUSSO, A. and VANNUCCI, G. 1987. Significato Paleoecologico degli episodi coralgali a nord di sassello. Nuovi dati per una ricostruzione Paleogeografico-Evolutiva del margine meridionale del bacino Terziario del piemonte. Atti Society Tosc. Natural Science, Memoir Anno, 19-76, tab. 1.

FRAVEGA, P., PIAZZA, M. and VANNUCCI, G., 1994. Nongeniculate coralline algae association from the Calcare di Rosignano Formation , Lower Messinian, Tuscany (Italy). Studies on Ecology and Paleoecology of Benthic Communities. R. Matteucci *et al.* (eds.). Bulletin of Paleontological Society of Italy. 2, Mucchi, Modena: 127-140, text-figs 6.

GHOSH, A. K., 2002. Cenozoic coralline algal assemblage from southwestern Kutch and its importance in paleoenvironments and paleobathymetry. Current Science., 83 (2): 153-158.

ISHIJIMA, W., 1954. Cenozoic coralline algae from western Pacific, Tokyo. Privately Printed: 1-87.

JOHNSON, J. H., 1964. Eocene Algae from Ishigaki – Shima Ryūkyū – rettō. Geological Survey Professional Paper 399- C: C 1- C 13.

KAR, R. K., 1979. Fossil Algae from Fulra Limestone (Middle Eocene) Kutch. Gujarat. Geophytology, 9: 88-90.

KUNDAL, P. and HUMANE, S. K., 2002. Geniculate Coralline Algae from Middle Eocene to Lower Miocene of Kachchh, Gujarat, India. Gondwana Geological Magazine, 17(2): 89-101.

KUNDAL, P. and HUMANE, S. K., 2003a. *Corallina*, a geniculate coralline alga from Middle Eocene to Lower Miocene of Kachchh, Gujarat, India. Recent Developments in Micropaleontology , Gondwana Geological Magazine, (Ed. P. Kundal), 6: 261-275.

KUNDAL, P. and HUMANE, S. K., 2003 b. Dasycladalean algae from Middle Eocene to Lower Miocene of Kachchh, Gujarat, India. 8th International Symposium on Fossil Algae, Granada, Spain, 36.

KUNDAL, P. and HUMANE, S. K., 2005. *Jania* Lamouroux, a calcareous alga from Middle Eocene to Lower Miocene of Kachchh, Gujarat, India. Journal of Geological Society of India (In Press).

LEMOINE, M. P., 1939. Algues calcaires fossils de l'Algerie. Materiaux pour la Carte Geologique de l' Algerie. 1re Serie Paleontologie, 9: 1-128, pls. 1-15.

MISRA, P. K., JAUHRI, A. K., SINGH, S. K., KISHORE, S. and CHOWDHURY, A., 2001. Coralline Algae from the Oligocene of Kachchh, Gujarat, India. Journal of Palaeontological Society of India, 46: 59-76.

PAL, A. K. and GHOSH, R. N., 1974. Fossil Algae from the Miocene of Kutch, India. Palaeobotanist, 21(2): 189-192.

SINGH, S. K. and KISHORE, S., 2001. Chlorophycean Algae (Dasycladaceans and Udoteaceans) from the Eocene and Oligocene of Kachchh (= Kutch), Gujarat, India. Biological Memoirs, 27(1): 38-45.

SINGH, S. K., KISHORE, S., MISRA, P. K. and JAUHRI, A. K., 2002. Coralline algae from the Maniyara Fort Formation (Oligocene), Southwestern Kachchh, Gujarat, Western India. Biological Memoirs, 28(2): 51-60.

TANDON, K. K., GUPTA, S. K. and SAXENA, R. K., 1978. A new species of *Lithophyllum* from Oligocene of south western Kachchh. Journal of Palaeontological Society of India. 21- 22: 74-77.

VANNUCCI, G., PIAZZA, M., FRAVEGA, P., and CHIESA, I., 1994. Calcareous Rhodophyceae from the Astian facies (Pliocene), Valle San Bartolomeo, Alessandria, NW Italy. Studies on Ecology and Paleoecology of Benthic Communities. R. Matteucci *et al.* (eds.), Bulletin of Paleontological Society of Italy, 2,: 351-364, text-figs 4.

Holocene Sea Level Changes and neotectonism inferred from Beach rocks of South Andaman, Bay of Bengal

C. Rajshekhar, R.K. Sukhtankar and P.P. Reddy

Geology & Palaeontology Group, Agharkar Research Institute, Pune 411 004, India

Introduction

Beach rock is a good indicator of the past sea levels, as it is considered to have been formed within the range of intertidal zone (Kunio, 2001). Of late, it is realized that the precise dating of the beach deposits is necessary for the meaningful interpretations of the coastal evolution, neotectonism, sea level changes and associated climatic conditions (Agarwal and Guzder 1972; Gupta and Amin 1974; Gupta 1977; Guzder 1980; Kale and Rajaguru 1985; Bruckner 1989; Baskaran 1995; Nambiar, *et al.* 1991; Pirazzoli 1991; Nigam *et al.* 1992; Juyal *et al.* 1995). The application of various dating methods in reent years have enabled to gain a better insight into the Quaternary climatic events such as Last Glacial Maxima (LGM, around 18000 years B.P.), Younger Dryas (12,700-11,650 yrs. B.P.), Little ice age (1300-1850 A.D.) and medieval warming (1000-1300 A.D.) .

The aim of the present paper is to document the Late Holocene sea-level changes based on the elevation history and ^{14}C dates of the beach rocks and briefly review the geology and the tectonic events of the Pre-Quaternary sediments of Andaman Islands. An attempt has also been made to compare the sea level change record inferred from the beach rocks with those from the contemporaneous deposits of mainland. This study is also significant in the light of latest Tsunami of 26th December, 2004, as a result of which there have been considerable changes along the beaches of Andaman- Nicobar Islands.

Material and methods

This paper is the outcome of a field tour to South Andaman Island in 1999 under a sponsored scheme of Department of Science and Technology, New Delhi, Govt. of India. During the field work, systematic collection of rock samples and measuring of litho-columns from the distinct and workable outcrops from Wandoor and Chidyatapu (South Andaman) were made. The samples were processed using H_2O_2 method for micropaleontological analyses and petrographic studies were carried out with the help of thin sections. The ^{14}C dates of two beach rocks were estimated at ^{14}C laboratory in Birbal Sahni Institute of Paleobotany, Lucknow. The Wandoor beach rock – BS – 1634 yielded an age of

1580 ± 80 yrs. B.P. while an age of 3760 ± 150 yrs B.P was estimated for the beach rock (BS – 1630)of the Chidyatapu .

Geology

The Andaman- Nicobar Islands consists of chain of Islands between the parallels (6°45' N to 13°43' N) and longitudes (92°15' E to 94°00' E) in the Bay of Bengal. This region is geologically very interesting as the thick pile of Late Mesozoic to Holocene sedimentary sequences hold the important and useful clues to the past vertical sea floor tectonism, paleo sea level changes and related plate movements (Karunakaran, et al 1968; Krishnan 1982; Rajshekhar and Reddy, 2003, Srinivasan, 1968, 1977, 1978a,b, 1986). The stratigraphic classification proposed by Srinivasan (1986) is followed here.

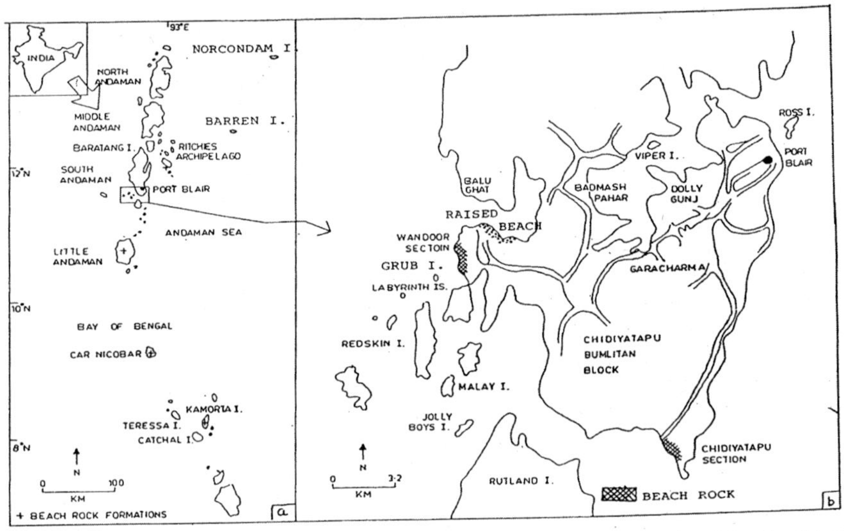

Fig. 1 Location Map (a) Distribution of beach rocks on the Andaman- Nicobar Islands; (b) Ritchie's Archipelago showing surface sections of Neill Limestone Member and Chidyatapu Member at Neill Island, (c) Position of Wandoor and Chidyatapu in the South Andaman Island.

Holocene Deposits

Holocene deposits of Andaman consist of the clays, beach sand, beach rocks, coral rags and mangrove clays. These deposits occur as local caps over the older sediments (Srinivasan and Azmi, 1976). As these sedimentary sequences hold clues to the past strand lines they have been studied and interpreted for sea level fluctuations for over a century (Buist 1851; Carter 1852; Guzder 1980).

The Andaman coastline is uneven and is characterized by wide and large beaches, rocky shores, mudflats, creeks and bays. The raised beach deposits are common across the chain of Islands indicative of neotectonic activities. Most of the shore deposits in this region occur as caps over the older sediments. The beach rock formations are mostly represented by moderately hard, buff coloured, loose to compact, medium to coarse grained, highly calcareous rock. Most of these rocks commonly contain gastropod shells - Nerita. Recently, Rajshekhar and Reddy (2003) proposed the stratigraphic classification to accommodate these beach rock formations of Andaman Archipelago and the same is followed here.

Stratigraphic Classification

Group	Formation	Member	Epoch
ARCHIPELAGO GROUP	NEILL FORMATION	Chidyatapu Member (Beach Rocks)	Holocene
		~~Unconformity~~~	
		Neill Limestone Member	Pleistocene
		~~Disconformity~~~	
		Silty Mudstone Member	Late Pliocene

The outcrops at Wandoor and Chidyatapu localities in South Andaman Island (Fig.1) are distinct during low tide. The beach rock of Wandoor is coarse grained, loosely cemented, moderately hard calcareous rock and it is youngest (1580 ± 80 years B.P.) of all the beach rocks known from this region. The beach outcrops of Chidyatapu are large in extent as compared to those of the Wandoor and its estimated ^{14}C date is 3760 ± 150 years B.P. In addition, Bruckner (1989) has also reported beach rocks from the Grub Island located in the vicinity of south Andaman coast (Fig.1b) and it is oldest (4460 ± 160 years B.P.) beach rock known so far from this region.

Fauna

Microfossils are common in the beach rock. The foraminiferal taxa consist of shallow water smaller benthic foraminifera predominantly represented by species of *Pararotalia, Neorotalia* and *Elphidium.* The larger benthic foraminifera are moderately preserved and reflect the reefoidal environment of deposition. benthic foraminifera including *Amphistegina lessoni, Sorites orbicularis, Amphisorus, borelis* and *Peneroplis* sp. are common. The report of Late Quaternary foraminifera from the Invisible Bank off the Andaman coast by Bhattacharjee and Ghosh (2000) reveals similarity between Indo-Pacific and Andaman Sea Holocene fauna.

Fossil Algae

Algal content of beach rock of Wandoor is rich as compared to the Chidiyatapu rocks (Rajshekhar and Reddy , 2003). The calcareous algae comprise species of four genera viz., *Messophyllum, Arthrocardia, Corallina* and *Halimede*. Among these *Corralina* and *Halimeda* show their dominance.

Geomicrobiology

Khadkikar and Rajshekhar (2003) for the first time reported the occurrence of microbially mediated cement from the beach rocks of South Andaman. They identified Needle-fiber, fungal filaments, fungal borings and calcispheres. Further studies on the microbial cements reveal the dominance of Needle-fiber cement in Wandoor beach rocks.

Depositional Environment of Pre-Quaternary Deposits

Geologically, the Late Cenozoic moderately deep-water marine sediments consist of chalk, mudstone and limestone of Archipelago Group well exposed in Andaman-Nicobar Islands. (Srinivasan and Sharma 1973, and Srinivasan and Azmi 1976a). The micropaleontological studies of these well developed sedimentaries have made possible to assign definite ages to these marine sequences. Based on the integrated Neogene planktic foraminiferal datum levels and radiometric dates Srinivasan (1978a, 1986) identified four phases of uplift for the sedimentary sequence of Andaman- Nicobar Islands at 22.5 Ma.; 9.8 Ma.; 2.8 Ma. and 1.95 Ma, punctuated by subsidence events. Thus the sedimentation history of Late Cenozoic marine sequences of Andaman reflects the active vertical sea floor tectonism. The uplift phases correspond to major orogenic phases in Himalaya (Srinivasan 1986). During the sedimentation, the Andaman region also bears the evidence of explosive volcanic activity in the form of Ash Beds in the Early Miocene (Srinivasan, 1980).

The information on depositional history of the Holocene deposits though inadequate at present, warrants more age data for proper understanding of the possible link between paleoseismicity, neotectonism and sea level changes.

Holocene Depositional Environment and tectonics

Based on paleobathymetric oscillations and sea floor tectonism inferred from the Late Cenozoic benthic foraminifera, Srinivasan (1986) suggested positive connection between, Late Cenozoic uplift events of Andaman basin with the contemporaneous Neogene uplift phases of the Himalaya.

Srinivasan and Azmi (1976) noticed the shell limestone designated here as the Chidyatapu Member occurring about 15 m above the mean sea-level at Neill Island of the forearc belt. Bruckner (1989) reported the beach rock of Grub Island near the south Andaman occurring about 1 m above the low tide level.

The authors have observed that the beach rock deposits of south Andaman (Chidyatapu 3760 ± 150 yrs. B.P. and Wandoor 1580 ± 80 yrs B.P.) occur almost at the sea-level and the deposits are distinct during low tide. Thus the beach rocks of varying age occurring at different heights from the sea level clearly suggest active Neotectonic activity in the region during the Holocene.

The authors have also dated the *Nerita* shells from the raised beach sand of South Andaman (700 yrs. B.P.) However, we restrain any further comment on it because of inadequate information about the occurrence of shells in situ / transported? We need to examine more beach rock outcrops and gather more information on their dates to comment on positive link between paleoseismicity, neotectonism and sea level changes including climate.

Discussion

The beach rock formations in mainland India too, are common along the west coast, east coast and coasts of Islands of the Lakshadweep in Arabian Sea and Andaman-Nicobar Islands in the Bay of Bengal (Vaidyanadhan and Ghosh 1993; Merh and Chamyal 1993). However, the beach rock formations are extensive along the west coast as compared to the east coast (Bruckner 1989; Badve *et al* 1984). They are often being referred to as 'Karal' or 'littoral concrete' in the literature (Guzder 1980). Most of the beach rocks are poorly fossiliferous and contain shallow marine foraminiferal derivatives commonly represented by species of *Ammonia, Elphidium* and *Pararotalia* (Badave *et al* 1984; Rajshekhar and Reddy 2002). Further, there is a marked difference in gross foraminiferal composition between the beach rocks of mainland Indian coast and the Andaman-Nicobar Islands. The former in most of the cases has smaller benthic foraminifera while the latter contains both the larger as well as the smaller benthic foraminifera (Rajshekhar and Reddy 2001). These formations are of intertidal origin (Guzder 1980; Wagle *et al* 1994; Badve *et al* 1984; Rajshekhar and Reddy 2003) and the beach rock fauna of Andaman possesses Indo-Pacific affinity (Rajshekhar and Reddy 2003). Bruckner (1989) noticed that the beach rock near Manori on the west coast was secondarily cemented and hence it was not considered as inter-tidal formation. He also observed that the lower 5 cm of beach rock at Valvati near Aravi on the west coast is intertidal in origin while the upper part probably is of aeolian origin.

As mentioned earlier, the beach rock indicates ancient strandline and therefore emphasis was given to date these formations (Bruckner 1989; Guzder 1980; Badve *et al* 1984; Rajshekhar and Reddy 2003; Agarwal and Guzder 1972; Venkatachala *et al* 1992; Kumaran and Rajshekhar 1999; present work) to address precisely the much needed information on sea level and neotectonic activity.

Taking into account the overall ^{14}C dates available so far, it has been observed that the beach rocks of India were formed between 6670 years B.P. and the present day and are in conformity with global occurrence.

Petrographically the beach rocks are of grainstone-packstone category and belong to algal biocalcarenite facies. The study further reveals that the aragonitic cement is the most common among the beach rocks. Similarly, the petrography of the Chidyatapu beach rock (3760 ± 150 years B.P.) also indicates the presence of aragonite cement implying that the lithification and alteration of aragonite are not yet completed. It may be mentioned here that the alteration of aragonite into calcite is the common phenomenon in the carbonate rocks (Tucker and Wright 1992). The presence of aragonite cement in the youngest as well as in the oldest beach rock of India indicates the continuance of diagenetic processes. Tucker and Wright (1992) observed that the Quaternary rocks in general show the slow rate of diagenesis and in most cases contained high meta-stable aragonite and high Mg-calcite. These observations necessitate to take up detailed petrographic studies to examine the diagenetic processes and cementation to be used while interpreting the climatic conditions with special reference to monsoon. At places, for example, on the South Andaman coast rocks encrusted by bryozoan colonies indicate the development of hard grounds and it is related with non-depositional events (Tucker and Wright 1992).

Besides the surface outcrops, the submarine beach rocks/carbonate sediments are also common along the coastline of India (Hashimi *et al* 1995; Venkatachala *et al* 1992; Malik 1992; Nigam *et al* 1982). A comparison of ^{14}C dates of submerged terraces with their depths suggests sea level fluctuations during the recent past. (Hashimi *et al* 1995). Venkatachala *et al* (1992) dated the carbonate sediments from the depth of 1134 m off the Narcondum Island as 4500 ± 150 yrs B.P. This suggests submergence history of the region around Narcondam Island at mid Holocene. Further, no definite relation has been observed between the age of the beach rocks and their respective heights from the Average Mean Sea Level (AMSL). However, it is observed that the coast in and around Mumbai (Bombay), Miraya Bhatti and Devgad on the west coast is more active as compared to its mild behaviour in the rest of the west coast region. Around 1580 yrs B.P. and 2000 yrs B.P., the relative height of the beach rock outcrops at Erangal Bhatti near Mumbai was 8.5 m. and at Devgad near Ratnagiri was 6 m respectively. Majority of the vertical thickness of the beach rocks do not exceed 1.75 m from the AMSL and thus suggest mild behaviour of the west coast as compared to Andamans in terms of neotectonic activity and sea level changes. The occurrence of submarine carbonate terrace at 120 m has been related with the LGM event around 18,500 years B.P. (Hashimi *et al* 1995). Incidentally, the LGM around 18,500 was characterised by high salinity and reduced precipitation and the period between 10,500 and 12,500 yrs B.P. was related to reduced salinity and strengthening of Monsoon in the Indian subcontinent (Cullen 1981). Interestingly, the Holocene climate is milder than the LGM around 18,500 yrs. B.P. (Bowen 1978). However, the so-called milder climate was punctuated by several wet and dry spells (Ramesh 2000). Oxygen istotopic studies of the Speleothems from the tropical India reveal the following interesting features (Ramesh 2000):

1200-400 years B.P.	High rainfall
3000-2900 years B.P.	Reduced rainfall
3350-3200 years B.P.	Humid climate
3400-3000 years B.P.	Enhanced rainfall
3700-3350 years B.P.	Arid climate

Besides these studies an exhaustive account of pre-history civilization and agriculture, Radhakrishna (1998) outlined the following climatic events within the Holocene epoch.

2000 years B.P. - Intensive aridity, great famine and end of Indus Civilization
3000 years B.P. - Tectonic disturbance - shifting of Yamuna to join Ganga and Sutlej to join Indus
4000 years B.P. - End of wet climate - commencement of aridity
5000 years B.P. - Stimulating environment - foundation laid for stable agriculture
6000 years B.P. - Lakes with higher water levels
7000 years B.P. - Wet spell - Break through in Himalayan Drainage

A comparison of these radiocarbon dates with the dates of beach rock it is observed that the beach rock formations at Kolamb (1110 $\pm$ 105 yrs. B.P.) south of Ratnagiri, Uran(700 $\pm$ 70 yrs. B.P.) south of Mumbai and Vadrai ki khadi, Mahim (3310 $\pm$ 110 yrs.B.P.) north of Mumbai along the west coast and Wandoor along the south Andaman coast formed under the high rainfall conditions while these rocks at Harnai (1910 $\pm$ 90 yrs.B.P.) north of Ratnagiri near Shrivardhan, Narpad (3660 $\pm$ 90 yrs.B.P.) north of Mahim, on the west coast and Chidyatapu on the south Andaman coast had formed under arid conditions. The effect of wet and dry spells on the beach rock formations is yet to be examined carefully and in this respect, Sr^{87}/Sr^{86} analyses of the beach rock may provide significant clues.

A review of the radiocarbon dates reveals that in general the ages of the surface outcrops of the beach rocks are younger to the ages of the submarine sedimentary terraces comprising limestone, oolitic limestone, algal limestone, wood, peat etc. As mentioned earlier the beach rock formations are located between 0.05 m and 8.5 m above the Average Mean Sea Level (AMSL) all along the coastline indicating emergent character. Pirazzoli (1991) has examined some 262 sea level curves and observed that the 55% of the data favour emergence as compared to the 44% submergence cases. Consequently, the occurrence of the beach rock outcrops above the AMSL is contemplated here to be due to neotectonic activity. Our studies on the beach rock outcrops and their ^{14}C dates suggest the following:

In recent times the geomorphological manifestations in the form of rejuvenated river terraces and plains in the Himalayan region have been attributed to the neotectonic activity. Valdiya (1993) has highlighted the unstable conditions of the Quaternary platforms of the Himalayan chain. The dating of these events have been discussed with special reference to the ancient culture and civilization. The ^{14}C dates of the artifacts known so far indicating Late Holocene mobile character of the Himalayan region. Thus the above evidences distinctly indicate the mobile character of the so-called genetically linked Himalaya-Arakan-Yoma-Andaman-Java-Sumatra Ridge.

Further a closer examination of ^{14}C dates from various sources (Bruckner 1989; Nambiar et al. 1991; Juyal et al. 1995; Vaidyanadhan and Ghosh 1993; Guzder 1980; Present study) of the surface and subsurface gastropod and bivalve shells from the East Coast and West Coast of mainland together with their elevational variation indicate differential behavior of the Indian coast line. It also brings into light that the behavior of the west coast is different from the Andaman coast necessitating to consider the mainland coast and the Andaman coast into two different and distinct set ups. The faunal evidences of the Late Quaternary rocks of Andaman show the dominance of Indo-Pacific affinity to the Andaman Sea since Miocene (Srinivasan, 1986; Bhattacharjee and Ghosh, 2000; Rajshekhar and Reddy, 2002).

In this context the observations made by Dickinson (2001) on the Neotectonism are significant. Dickinson (2001) while dealing with the Holocene sea level variations and the Neotectonic activity of the Pacific region observed mid Holocene high stand over the Pacific. He also noticed that these high stands were subjected to modifications due to uplift and subsidence events particularly along the forearc and volcanic belts. The ^{14}C dates of beach rock (4500 $\pm$ 150 years BP; Bruckner, 1989, south Andaman) also indicate mid Holocene high stand in this region and the gravity core from the sea bed near Narcondam Island (4500 $\pm$ 150 years BP, Venkatachala et al. 1992) seems to be the modification of Holocene highstands and thus the observations made by Dickinson (2001) is also applicable to the Andaman region.

Acknowlegement: The authors are grateful to Dr. V.S. Rao, Director, A.R.I., for encouragement. The financial assistance given by DST (No. ESS/CA/A3-23/98) is gratefully acknowledged. The carbon dating of the beach rocks was done by Dr.G. Rajagopalan at Radio Carbon Laboratory, B.S.I.P., Lucknow. Thanks are also due to Andaman authorities, especially the conservator of Forest, Deputy Conservator of Forest and P.W.D Executive Engineer, South Andaman for providing necessary assistance during CR's visit to Andaman. P.P.R. is grateful to C.S.I.R for providing S.R.F.

References:

AGARWAL, D.P., GUZDER, S.J., 1972. Quaternary studies on the western coast of India: Preliminary observations. The Palaeobotanist 21, 216-222.

BASKARAN, M., 1995. The origin of Quaternary carbonate deposits (Miliolites); a theoritical perspective. In: Quaternary Environments and Geoarchaeology of India, Memoir of Geological Society of India. 32, 360-371.

BHATTACHARJEE, D., GHOSH, S.K., 2000. Seabed sediments and Late Quaternary sea level changes around the invisible bank, Andaman Sea. Proceedings of International. Seminar on INQUA shorelines, Indian Ocean subcommission, Thanjavur, India. Pp. 212-217.

BRUCKNER, H., 1989. Late Quaternary shorelines in India. In: LateQuaternary Sea Level Correlation and Applications. Scott, D.B., Pirazzoli, P.A, Honig, C.A. (Eds.), Kluwer Academic Publishers, Dordrecht. pp. 169-194.

BOWEN, D.Q., 1978. *Quaternary Geology*, Pergamon Press, Publisher: Robert Maxwell, M.C., Oxford.

BUIST, G., 1851. Geology of the Island of Bombay. Trans. Bombay geographic Society 10, 167-204.

CARTER, H. J., 1852. Geology of the Island of Bombay. Journal of Bombay Asiatic Society, 4, 116.

CHATTERJEE, A. K., 1964. The Tertairy fauna of Andamans. International Geological Congress Report, 22nd Session, New Dehli (Ed. Sudaram, R.K.) pp. 303-318.

DICKINSON, W.R., 2001. Paleoshoreline record of relative Holocene sea Levels on Pacific Islands. Earth Science Reviews 55, 191-234.

GUPTA, S.K., AMIN, B.S., 1974. Io/U ages of corals from Saurashtracoast. Marine Geology 16, 79-83.

GUPTA, S.K., 1977. Quaternary sea level changes on the Saurashtra coast. In: Ecology and Archaeology of Western India concept, Agarwal.

D.P. AND PANDE B.M, (Eds.) concept publishing co. Delhi, 181-193.

GUZDER, S.J., 1980. Quaternary Environment and Stone Age Cultures of the Konkan, Coastal Maharashtra: Deccan college P G and Research Institute Publication, Pune., p. 101.

JUYAL, N., PANT, R.K., BHUSHAN, R., SOMAYAJULU, B.L.K., 1995. Radiometric dating of the Late Quaternary sea levels of the Saurashtra caost, western India; an experiement with Oyester and Clam shells. In: Quaternary Environments and Geoarchaeology of India, Memoir of Geological Society of India. 32, 372-379.

KALE, V.S., RAJGURU, S.N., 1985. Neogene and Quaternary transgression and regression of west coast of India; An overview. Bulletin of Deccan College P G and Research Institute, Pune., 44, 153-165.

KARUNAKARAN, C., RAY, K.K., SAHA, S.S., 1968. Tertairy sedimentation in the Andaman-Nicobar Geosyncline. Journal of Geological society of India. 9, 32-39.

KHADKIKAR AS & RAJSHEKHAR C Microbial cements in Holocene beach rocks of south Andaman Island, Bayof Bengal. Current Science 2003, 84 (1), 933-936.

KRISHNAN, M.S., 1982. Geology of India and Burma, New Delhi. 536p.

KUNIO, O. (2001) , Radiocarbon ages of beach rocks and the Late Holocene Sea level changes in the southern part of the Nansei Islands , Southwest of Japan. Radiocarbon, vol.43, no.2, pp. 887-898.

NAMBIAR, A.R., RAJAGOPALAN, G., RAO, B.R.J., 1991. Radiocarbon dates of sediment cores from intercontinental shelf off Karwar, west coast of India. Current Science 61, 353-354.

NIGAM, R., HASHIMI, N.H., MENEZES, E.T., WAGH, A.B., 1992. Fluctuating sea levels off Bombay (India) between 14,500 and 10,000 years B.P. Current Science 62, 309-311.

PIRAZZOLI, P.A., 1991. World Atlas of Holocene Sea level Changes.Elsevier Science Publishers B.V., Amsterdam. 300 p.

RAJSHEKHAR, C. AND REDDY P.P. 2002. Ecology of the beach rock fauna of the South Andaman Island, Bay of Bengal.Current science 82(7), 881-885.

RAJSHEKHAR, C., REDDY, P.P., 2003. Quaternary Stratigraphy of Andaman-Nicobar Islands, Bay of Bengal. Journal of Geological Society of India, vol.62, pp 485-493.

ROY, T.K., 1983. Geology and Hydrocarbon Prospects of Andaman-Nicobar basin. Petroleum Asia journal. VI(4), 37-50.

SRINIVASAN, M.S., 1968. Andaman-Nicobar Islands : A future petroleum source for India. Oil and Gas News, New Delhi. pp. 19-21.

SRINIVASAN, M.S., 1977. Standard planktonic foraminiferal zones of the Andaman-Nicobar Late Cenozoic. Recent Researches in Geology, Hindustan Publishing corporation, Delhi, 3, 23-29.

SRINIVASAN, M.S., 1978a. New Chronostratigraphic division of the Andaman Nicobar Late Cenozoic. Recent Researches in Geology, Delhi. 4, 22-36.

SRINIVASAN, M.S., 1978b. Geology and Mineral resources of Andaman Nicobar Islands. Andaman-Nicobar Information Govt. Press, Port Blair. pp. 44 -52

SRINIVASAN, M.S. 1980. Early Neogene volcanism in Southeast Asia : Evidence of Ash Beds from Andaman-Nicobar. *GPSEA Symposium Ibaraki, Japan* (1978), Geol. Paleont. SE Asia, vol. 21, pp. 229-236.

SRINIVASAN, M.S., 1986. Neogene reference sections of Andaman-Nicobar: Their bearing on volcanism, sea-floor tectonism and global sea-level changes. Ophiolites and Indian Plate Margin. Ghosh, N.C., Vardarajan, S., {Eds.} pp. 295-308.

SRINIVASAN, M. S., AND AZMI, R.J., 1976. New development in the LateCenozoic lithostratigraphy of Andaman-Nicobar Islands, Bay of Bengal. Proceedings of VI Indian Colloquium on Micropaleontology and Stratigraphy, Varanasi (Ed. M.S., Srinivasan) pp. 302-327.

SRINIVASAN, M.S., AND SHARMA, V., 1973. Stratigraphy and microfauna of Car-Nicobar Island, Bay of Bengal. Journal of Geological Society of India. 14, 1-11.

TUCKER, M. E., AND WRIGHT, P. V., 1992. Carbonate sedimentology. Blackwell scientific publications.

VAIDYANADHAN, R., GHOSH, R.N., 1993. Quaternary of the East coast of India. Quaternary Periods in India . Special Issue, Current Science 64 (11 & 12), 804-816.

VALDIYA K.S., 1993. Uplift and geomorphic rejuvenation of the Himalaya in the Quaternary Period. Current Science. 64(11&12), 873-885.

VENKATACHALA, B.S., RAJAGOPALAN, G., KAR, R.K., RAJNIKANT, A, 1992. Palynological studies and ^{14}C dating of gravity core from the seabed, west coast of Norcondam Island in the Andaman Sea. Geological Survey of India, special publication 29, 107-110.

Progress in Cenozoic Radiolarian research: The Indian context

V. Sharma

Department of Geology, University of Delhi, Delhi 110 007, India

Introduction

The earliest publication on Radiolaria, though, is more than 160 years old by Meyen (1834), the initial progress in the radiolarian study in the world had been slow. Earlier publications dealt mainly with the description and illustration of radiolarians; the advancement in the application of Radiolaria is relatively recent. Applied aspect of this microfossil group in Geology has been recognized not more than five decades ago. This was because of the simple reason that these fossils were not studied systematically in respect of their phylogeny and critical examination of morphology. Consequently, the fossil species seemed similar to the modern radiolarians, with the result that they were considered to have long geological range rendering them stratigraphically unimportant. Inspite of being very abundant in many of the marine sequences radiolarians attracted a very few workers, while some other microfossil groups were already recognized as useful tools in stratigraphy.

With a few scattered references in the nineteen fifties and sixties, there seems to be rather a sudden increase in number of publications dealing with various aspects of radiolarian research in the nineteen seventies and afterwards. This is because the progress in radiolarian research is linked to the availability of material collected by the Deep Sea Drilling Project (DSDP) and Ocean Drilling Program (ODP), which made available undisturbed long cores of sediments from below the ocean floors of the world oceans since 1968 and 1985 respectively.

Advantage of oceanographic expeditions

DSDP, ODP and other oceanographic expeditions enhanced our knowledge on Radiolaria as well as other microfossil groups. These programmes provided excellent opportunities

(a) to examine microfossil groups ranging from Mesozoic to Recent from all latitudes of the world oceans,
(b) to recognize several new taxa,
(c) in phylogenetic studies,
(d) to trace precise stratigraphic ranges of taxa, and
(e) to identify water mass related species.

These advantages revolutionized our understanding in biostratigraphic, paleoclimatic and paleoceanographic studies. Radiolaria, in particular, had seen establishment of a more natural taxonomy based on phyletic relationship unlike the artificial ones of Ehrenberg and Haeckel, and establishment of stratigraphically important taxa.

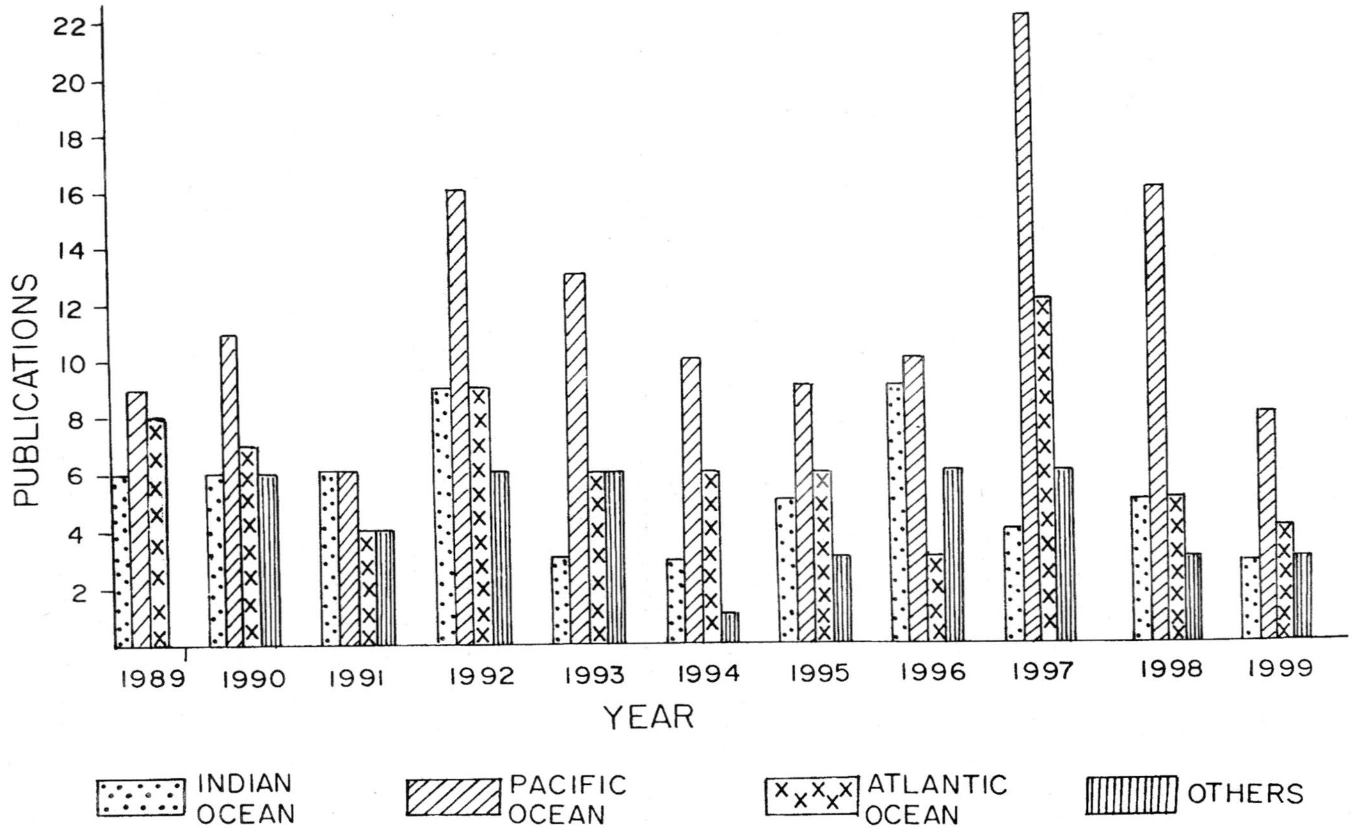

Fig. 1. Diagram showing number of publications on Cenozoic Radiolaria during 1989-1999 (Data prepared from the Newsletter of the International Association of Radiolarian Paleontologists).

Several other oceanographic expeditions also resulted in collection and study of radiolarians in plankton tows. These studies provided understanding of skeletal secretion, which proved significant in taxonomy.

Progress In Radiolarian Research

The world over, research on Radiolaria has been greatly influenced by the availability of deep-sea cores. In a period stretching from 1970 to 1978, maximum number of publications on radiolarian biostratigraphy from DSDP cores appeared in 1973-74, indicating increasing awareness of stratigraphic importance of radiolarian in dating and correlation of the DSDP sequences (Riedel, 1981). Number of publications on Cenozoic Radiolaria during the period 1989-1999 base on available data are shown in Fig. 1.

Not only radiolarians showed their potential in biostratigraphy, the studies on this microfossil group also demonstrated that they are equally useful to unravel both long and short term climatic and oceanographic changes. While occurrences of calcareous fossils in the marine sediments are controlled by the Carbonate Compensation Depth (CCD), limiting their usefulness in strata deposited above the CCD, radiolarians have no such a limiting factor and are usually found at all depths. Consequently, radiolarians (and other siliceous microfossils) proved much more useful in oceanic biostratigraphy where strata are deposited below the CCD and are devoid of calcareous microfossils. Role of Radiolaria in the study of sediments deposited at such great depths is very significant, where absence of calcareous microfossils may lead to erroneous interpretation. For example, there can be a false perception of presence of a hiatus based on absence of calcareous microfossils if the examined sediments lay below the CCD during a part of their deposition. Radiolarians, however, can be of immense value in such situations.

Oceanic Biostratigraphy

Biostratigraphic zonal scheme for the tropical Cenozoic (except Paleocene) is based on the work of Riedel and Sanfilippo (1970, 1971, 1978), Nigrini (1971), Moore (1971) and Foreman (1973, 1975), a summary of which was given by Sanfilippo *et al.* (1985). Research contributions by Nigrini (1974), Maurasse and Glass (1976), Saunders *et al.* (1985), Johnson *et al.* (1989), Nishimura (1992), Caulet *et al.* (1993) and Hollis (1993) resulted in minor amendments in the scheme. Nishimura's work (1987, 1992) on DSDP cores from the Atlantic led to the identification of zone and sub zones in the upper Paleocene and that of Hollis (1993) for the lower Paleocene on the basis of samples from New Zealand.

Though ranges of a few species of Radiolaria, foraminifera and diatoms were related to paleomagnetic record in order to compare their ranges (Hays *et al.*, 1969) applicability of magnetic polarity time scale for calibrating microfossil events was realized in the seventies (Foster and Opdyke, 1970; Opdyke, 1972; Opdyke *et al.,* 1974; Theyer and Hammond, 1974a, b; Theyer *et al.,* 1978). This allowed paleomagnetic calibration of the tropical radiolarian zonal boundaries for the duration from latest Oligocene to Recent. With the establishment of continuous magnetostratigraphy in the Neogene, calibration of microfossil events to the geomagnetic polarity time scale became possible (Berggren *et al.*, 1985).

Code numbers in Radiolarian biostratigraphy were first applied by Caulet (1979) who numbered six zones, from younger to older, identified in the late Pliocene to Pleistocene sediments in the Central Indian Ocean. Johnson *et al.* (1989) proposed a revised eleven-fold radiolarian zonation for the tropical Indian Ocean Pliocene- Pleistocene, the zones being defined by events synchronous in the tropical Indo-Pacific. By examining additional cores from the Central Indian Basin, Caulet *et al.* (1993) refined the ages of Plio-Pleistocene radiolarian datums. Sanfilippo and Nigrini (1998) updated the standard tropical Cenozoic zonation (Sanfilippo *et al.*, 1985) and presented 39 zones along with code numbers for each zone. The scheme, which is considered applicable for the tropical Indian, Pacific and Atlantic Oceans, incorporates 22 zones for the Paleogene and 17 zones for the Neogene. Identification of radiolarian events in different zones led to the establishment of their chronology in each zone (Sanfilippo and Nigrini, 1998).

In temperate and high latitude biostratigraphy, zonations have been proposed by Hays (1965), Hays and Opdype (1967), Chen (1975) and Caulet (1986). In recent years recovery of high quality hydraulic-piston cores from Falkland Plateaus (Weaver,1983) and Weddel Sea (Abelmann, 1990; Lazarus 1990) and study of cores from Kerguelen Plateau (Leg 119 and 120) (Caulet, 1991 and Lazarus, 1992) led to refinement in high latitude radiolarian zonation for the Cenozoic and their calibration with the geological time scale.

Lazarus (1990) discussed the history and inventory of Antarctic radiolarian zonation and proposed a new biozonation, based mainly on the Plio-Pleistocene zonation of Hays (1965) and Hays and Opdyke (1967) and on Miocene zonation of Chen (1975). Caulet (1991), based on the work of Chen (1975) and modified by Weaver (1983) and Caulet (1982, 1985, 1986), proposed biostarigraphic scheme for the middle Eocene to Pleistocene sequences in which he identified 16 zones.

Lazarus (1992) presented zonal scheme for the Neogene based on earlier work. For the Plio-Pleistocene interval, he used the scheme of Lazarus (1990), in which he slightly modified the ages of some of the datums based on the work of Caulet (1991). For the Miocene interval, Lazarus (1992) presented a revised zonation, in part based on the work of Chen (1975), Weaver (1983) and Lazarus (1990), and for earlier Miocene, he followed the scheme of Abelmann (1992). The zones are calibrated with magnetostratigraphy syntheses of Legs 119 and 120 by Barron *et al.* (1991). Work of Takemura (1992) on cores of Leg 120 established three zones for the interval from late Eocene to late Oligocene.

Paleoclimatic studies

Like other marine microfossils, radiolarian assemblages preserve signatures of oceanic and climatic changes. Relationships between watermass and radiolarian distribution in modern oceans are of prime importance in understanding paleoclimatic change and paleoceanographic history.

Paleotemperatures using radiolarians have been determined by many (Casey, 1972; Keany, 1973; Keany and Kennett, 1975; Moore and Lombari, 1981). Keany (1973) used percentage distribution of an Antarctic species, *Antarctica strelkovi*, across the examined sequence and found a cooler Matuyama than the Brunhes in the Antarctic and subantarctic waters (South of Australia and from central South Pacific). Moore and Lombari (1981) applied regression analysis to decipher paleotemperature in the North Pacific.

Paleoceanography

Radiolarians have been successfully used to understand past oceanographic changes, many of the studies incorporated transfer functions to find such changes (Casey, 1971; Moore, 1973; Sancetta, 1978; Morley, 1979, 1989; Moore *et al.*, 1980; Romine and Moore, 1981; Moore and Lombari, 1981; Hays *et al.*, 1969; Farrell and Janecek, 1991; Pisias and Moore, 1995). Paleogeographic changes since the beginning of Miocene brought about great changes in the oceanographic circulation, the record of which are found in the radiolarian assemblages of that period. Evolution of oceanic circulation in the equatorial and North Pacific during the Miocene (23 Ma – 5 Ma) was studied by Romine and Lombari (1985) using radiolarian assemblages at DSDP site 289 in western tropical Pacific.

Relative intensity of upwelling can be detected by recording fluctuations in percentage of upwelled species. Late Miocene – early Pliocene upwelling evidenced by radiolarians in the Monterey Formation, California, is one of the typical examples incorporating land-based sections (Mullineaux and Westberg-Smith, 1986). Categorising radiolarian taxa into a number of counting groups and using Principle Component Analysis on radiolarians from middle Miocene – early Pliocene Monterey Formation, Mullineaux and Westberg – Smith (1986) suggested increase in upwelling of cold, subsurface water during the deposition of Monterey Formation. Similar studies involving surface sediments and cores have been carried out by other investigators (Molina-Cruz, 1977, 1984; Nigrini, 1991, 1992; Nigrini and Caulet, 1992).

Radiolarian Research – Indian Context

Radiolarian studies on Cenozoic deposits in the Indian Ocean region, that is, in the region bordering the Indian Ocean and on the drill sites and surface sediments in the Indian Ocean, have not been as extensive as in the Pacific and Atlantic. Unlike the Pacific and the Atlantic, Indian Ocean is not open in the north and is relatively smaller in size. Somewhat restricted oceanographic conditions due to typical geographic configuration of the Indian Ocean and consequently, presence of not so varied faunal provinces, as are found in the Pacific and Atlantic Oceans, may be other factors for drawing attention of a fewer workers.

Occurrences of radiolarians in the rock exposures and surface sediments of the Indian Ocean have been known for about 150 years. *Rhapalostrum lagena* and *Stylosphaera hispida* were reported from the rocks of Kamorta Island and Nancowry Island (both in Nicobar group of islands) respectively by Ehrenberg (1854). Similarly *Schizoma quadrilobum* was reported by Ehrenberg (1861) from the surface sediments of Indian Ocean (and other oceans). Though scattered reports of radiolarian occurrences from this region have been made thereafter, it was not till the nineteen sixties that the true potential of radiolarians were recognized. On the radiolarian occurrence in the Ritchie's Archipelago, Andaman Sea, Jacob and Shrivastava (1952) commented, "it is generally recognized that even after the satisfactory identification of radiolarian remains, it is extremely difficult to assess their stratigraphic value as most of the known genera are believed to have long ranges".

As elsewhere, radiolarian studies in the Indian Ocean region too gained momentum after the advent of DSDP, ODP and other oceanographic programmes (Figs. 2 and 3). A number of publications appeared in the last 30 years on Radiolaria from Arabian Sea, Bay of Bengal, Banda Sea and rest of Indian Ocean. These dealt with biogeography, ocean-climate history, biostratigraphy, magnetostratigraphy and other studies connected with Radiolaria.

Studies on land based sections in the Indian Ocean region are, similarly, relatively recent. A few preliminary reports (Jacob and Shrivastava, 1952; Singh and Vimal, 1973; Singh and Jauhari, 1976; Singh, 1981) are from Indian subcontinent including Andaman-Nicobar. Detailed work documenting radiolarian taxa, biostratigraphy and other radiolarian based studies are those of Srinivasan *et al.* (1983); Sharma and Sharma (1988, 1989); Gupta and Srinivasan (1992); Sharma and Singh (1993); Sharma *et al.* (1993); Mahapatra and Sharma (1994); Ling *et al.* (1995); Sharma and Singh (1997a, b); Sharma and Daneshian (1998); Sharma *et al.* (1999); Singh *et al.* (2000); Sharma and Daneshian (2003) and Sharma and Ram (2003) in Andaman-Nicobar and by Ling and Anikouchine (1967); Ling and Nicholas (1991) and Ling and Samuel (1998) in Indonesia. Late cenozoic sequences of Andaman-Nicobar Islands have been studied in detail in the last few years. Results of these studies are presented in the subsequent sections.

Biostratigraphy and Events

Radiolarian zones for the tropical Cenozoic sequences (Sanfilippo and Nigrini, 1998) are applicable in the Indian Ocean as well; some zones, however, are exclusive to this region. High-resolution study by Johnson *et al.* (1989) in the tropical Indian Ocean resulted in establishing finer zones in the Plio-Pleistocene intervals and calibrating them with the geomagnetic polarity time scale. Four of these zones (Fig. 4) are recognizable in the Indian Ocean only (Sanfilippo and Nigrini, 1998). The zones are distinctly recognized in the Pliocene sediments of Car Nicobar Island in the northeast Indian Ocean (Sharma and Singh, 1993).

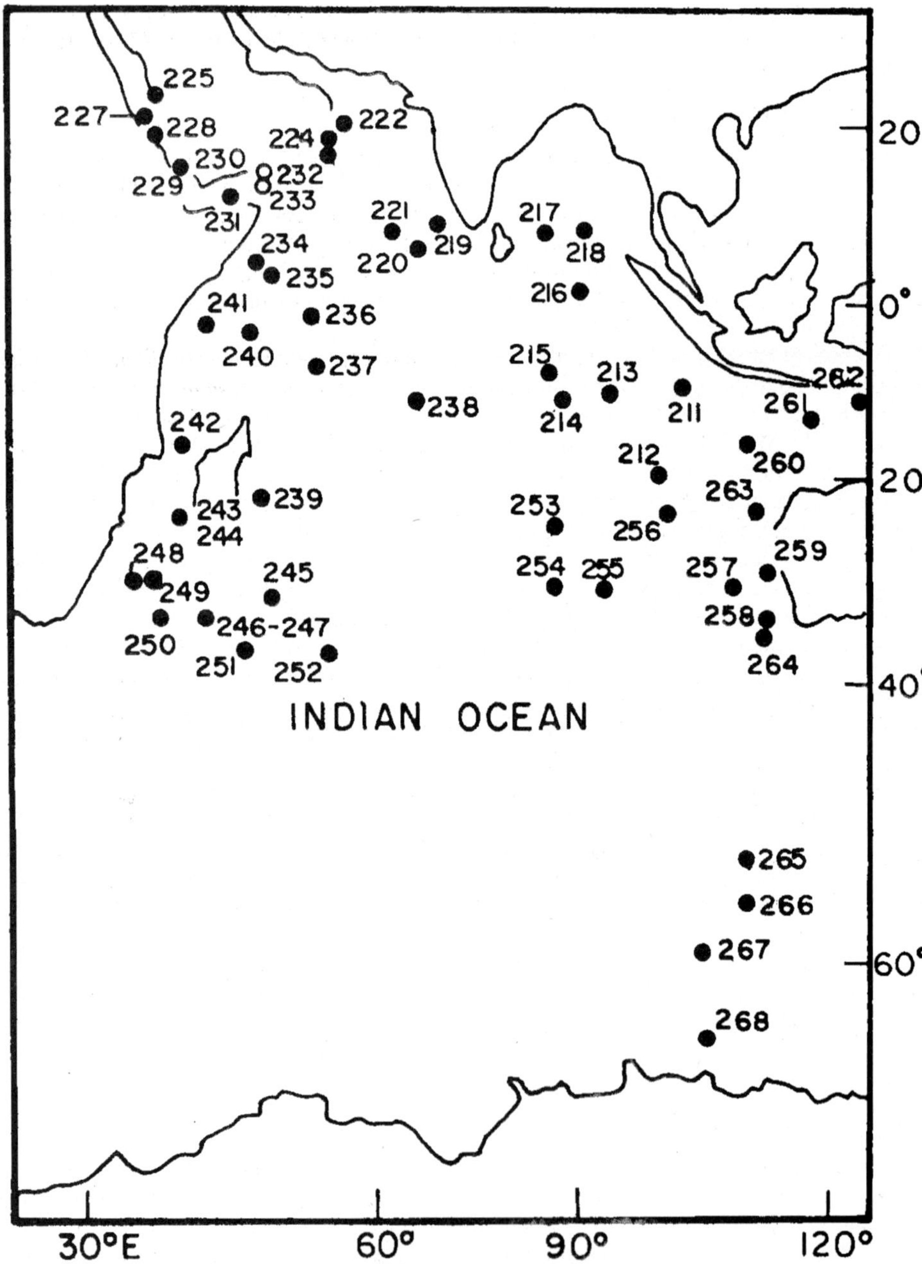

Fig. 2. Location of DSDP sites in the Indian Ocean and Indian Ocean sector of Southern Ocean.

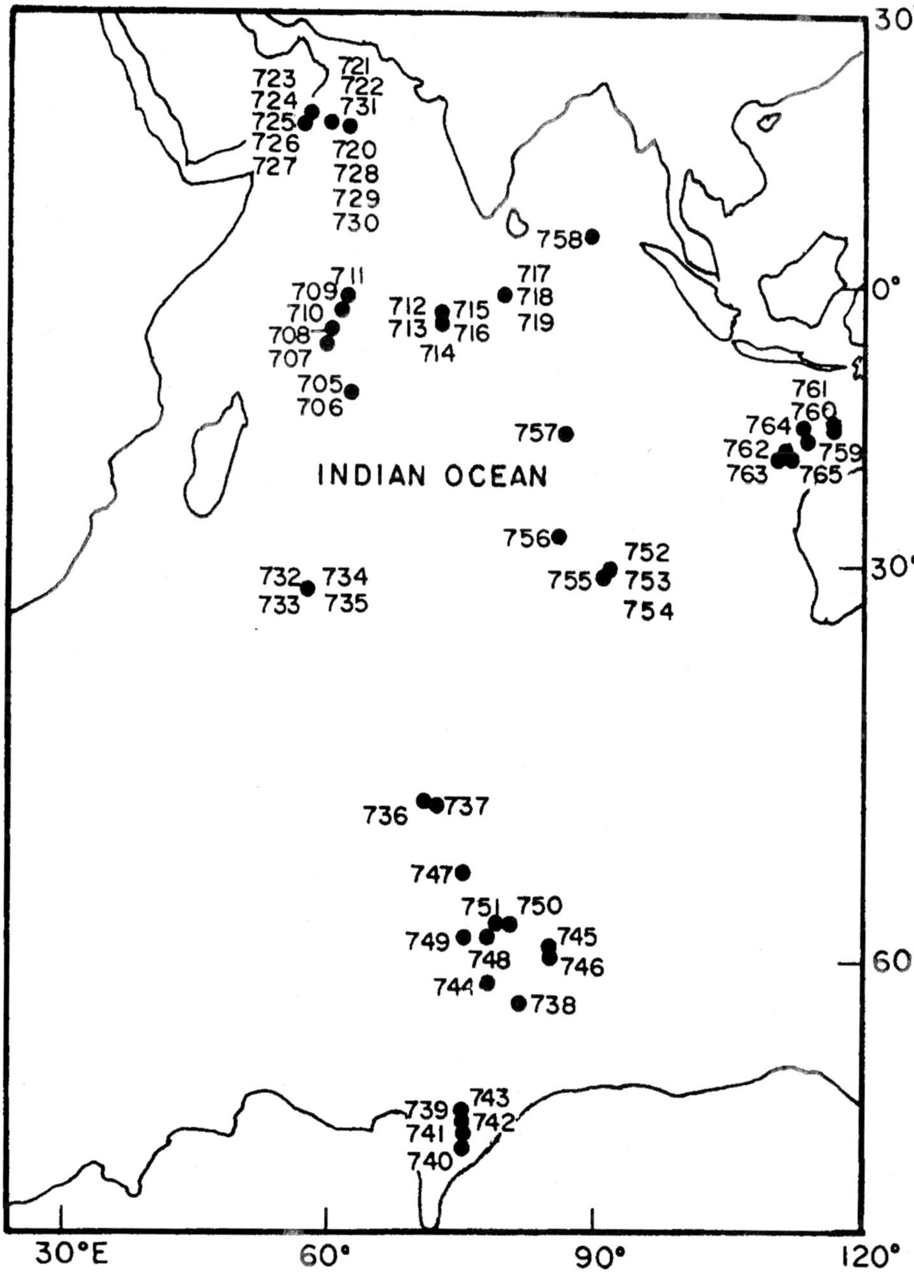

Fig. 3. Location of ODP sites in the Indian Ocean and Indian Ocean sector of Southern Ocean.

Sanfilippo et. al., 1985	Johnson et. al., 1989	Moore, 1995	Sanfilippo and Nigrini, 1998	
Zonal Name and boundary event	**Zonal Name and boundary event**	**Zonal Name and boundary event**	**Zonal Name, Code Number and boundary event**	
LO Pterocanium prismatium	LO Pterocanium prismatium	LO Pterocanium prismatium		LO Pterocanium prismatium **RN12b**
Pterocanium prismatium	**Pterocanium prismatium**	*Pterocanium prismatium*	**Pterocanium prismatium** **RN12**	
	LO Anthocyrtidium jenghisi			LO Anthocyrtidium jenghisi
	Anthocyrtidium jenghisi			**RN12a**
LO Stichocorys peregrina	LO Stichocorys peregrina	LO Stichocorys peregrina		LO Stichocorys peregrina
	Stichocorys peregrina		**Lychnodictyum audax**	**RN11b**
	LO Phormostichoartus fistula	**Anthocyrtidium jenghisi**		LO Phormostichoartus fistula
Spongaster pentas	**Phormostichoartus fistula**		**RN11**	**RN11a**
	LO Phormostichoartus doliolum	LO Phormostichoartus doliolum		LO Phormostichoartus doliolum
Spongaster berminghami-> S pentas	**Phormostichoartus doliolum**		**Phormostichoartus doliolum**	**RN10**
	LO Anthocyrtidium prolatum	**Phormostichoartus doliolum**	**RN10**	
		LO Didymocyrtis penultima		LO Didymocyrtis penultima
Stichocorys peregrina	**Anthocyrtidium prolatum**	**Stichocorys peregrina**	**Stichocorys peregrina** **RN9**	**RN9**
	……….?……………			
Stichocorys delmontensis ->S. peregrina		Stichocorys delmontensis ->S. peregrina		Stichocorys delmontensis ->S. peregrina
Didymocyrtis penultima		**Didymocyrtis penultima**	**Didymocyrtis penultima** **RN8**	**RN8**
LO Diartus hughesi		LO Diartus hughesi		LO Diartus hughesi

Fig 4. Upper Miocene and Pliocene tropical radiolarian zones in the Pacific and Indian oceans and their correlation. LO = Last morphotypic appearance; → = Evolutionary transition (Based on Fig. 1 Sanfilippo and Nigrini, 1998). Four zones (RN11a, RN11b, RN12a and RN12b) are recognized in the Indian Ocean only.

Biostratigraphic zones in the Central Indian Ocean drill sites of Leg 115 are recognized by Johnson (1990). He identified 55 radiolarian events in the middle Eocene to earliest Oligocene and 31 radiolarian events in the late Miocene to Holocene and found succession of events closely comparable with that of the tropical Pacific.

Study on time relationship between radiolarian events demonstrated that diachroneity exists between many events in the tropical Indian and Pacific Oceans. Many events are not found synchronous (range of age < 0.4 my) in these oceans (Johnson and Nigrini, 1985).

Establishment of Neogene radiolarian biostratigraphy and identification of events in a large number of sections have been carried out in Andaman–Nicobar Islands in the northeast Indian Ocean. As a result, radiolarian biostratigraphy of many islands is now known and events established (Srinivasan *et al.*, 1983; Sharma and Sharma, 1989; Sharma and Singh, 1993, 1997b; Sharma and Daneshian, 1998; Sharma *et al.,* 1999 and Sharma and Ram, 2003). Integrating biostratigraphic zones based on Radiolaria and other microfossils (Foraminifera and calcareous nannoplankton) multiple microfossil biostratigraphy of a number of sections has also been established for attaining enhanced biostratigraphic resolution. (Sharma *et al.*, 1993; Singh *et al.*, 2000, Sharma and Daneshian 2003).

Biogeographic Studies

Knowledge on the radiolarian distribution in the Indian Ocean based on their occurrence in water column and surface sediments has resulted in a fairly good understanding of their biogeography in this ocean. Noteworthy contributions in this regard are by Nigrini (1967), Petrushevskaya (1967, 1971), Johnson and Nigrini (1980, 1982) and Morley (1989) (Fig. 5). Based on the census data from core top sample of the world oceans by CLIMAP project members, Lombari and Boden (1985) presented maps of distribution of 70 recent radiolarian species in the world oceans (including Indian Ocean), which are useful to understand relationship between their distribution and oceanic parameters.

Paleoceanography

Oceanographic/paleoceanographic and paleoclimatic studies in the Indian Ocean have been attempted made using radiolarian transfer functions (Morley, 1989; Gupta and Fernandes, 1995; Gupta, 1996; Gupta *et al.*, 1996). Based on radiolarian distribution in surface sediment samples from Indian Ocean (about 21°S to 59°S latitudes), Morley (1989) identified four geographically distinct assemblages whose distributions coincide with present day patterns of sea-surface temperature and water masses. Incorporating sediment data with sea- surface temperature and surface salinity variations determined by transfer function, paleomonsoonal changes in the tropical Indian Ocean are estimated in the late Miocene and the Quaternary (Gupta and Fernandes, 1995, 1997). Usefulness of Pyloniid/Spongodiscid ratio is found in understanding paleomonsoonal fluctuation by studying radiolarian assemblages from surface sediments in the tropical Indian Ocean (Gupta, 1996). Based on radiolarian assemblages warmer and cooler conditions are indicated in the northeast Indian Ocean in the late Early Miocene (*Stichocorys wolffii* to early *Calocycletta costata* zones) and Middle Miocene (late *Calocycletta costata* and *Dorcadospyris alata* zones) respectively, with northward expansion of cool surface water during the Middle Miocene (Sharma and Daneshian, 2003) (Fig. 6).

Radiolarian distribution in the Central Indian Basin has been used to infer presence of Antarctic Bottom water (AABW) current (Sharma and Mahapatra, 1990; Gupta and Jauhari, 1994). Based on the radiolarian distribution in the surface sediments of stations located in a more or less east west transect in the tropical Indian Ocean, influence of AABW is seen in the form of displaced Antarctic taxa in the Central Indian Basin (Sharma and Mahapatra, 1990). A high ratio of two major groups of Radiolaria, Nassellaria/Spumellaria, was considered as an index for the oxygen content and thus a proxy for the presence of the AABW current in the eastern part of Central Indian Basin (Gupta and Jauhari, 1994) (Fig. 7). The existence of past upwelling was indicated in the Indian Ocean by Johnson and Nigrini (1980) when they identified presence of upwelling fauna in the surface sediments of the Western Indian Ocean. Late Neogene radiolarian assemblages recovered from the northwestern Indian Ocean (Oman Margin and Owen Ridge) have shown presence of upwelling and fluctuations in its intensity (Nigrini, 1991, 1992). Nigrini (1991, 1992) identified three categories of forms indicative of upwelling, beginning at about 11.9 Ma with a decrease in the strength of upwelling signal at about 9.6 Ma. A strengthening and weakening are recognized again at 4.7 Ma and 1.5 Ma respectively.

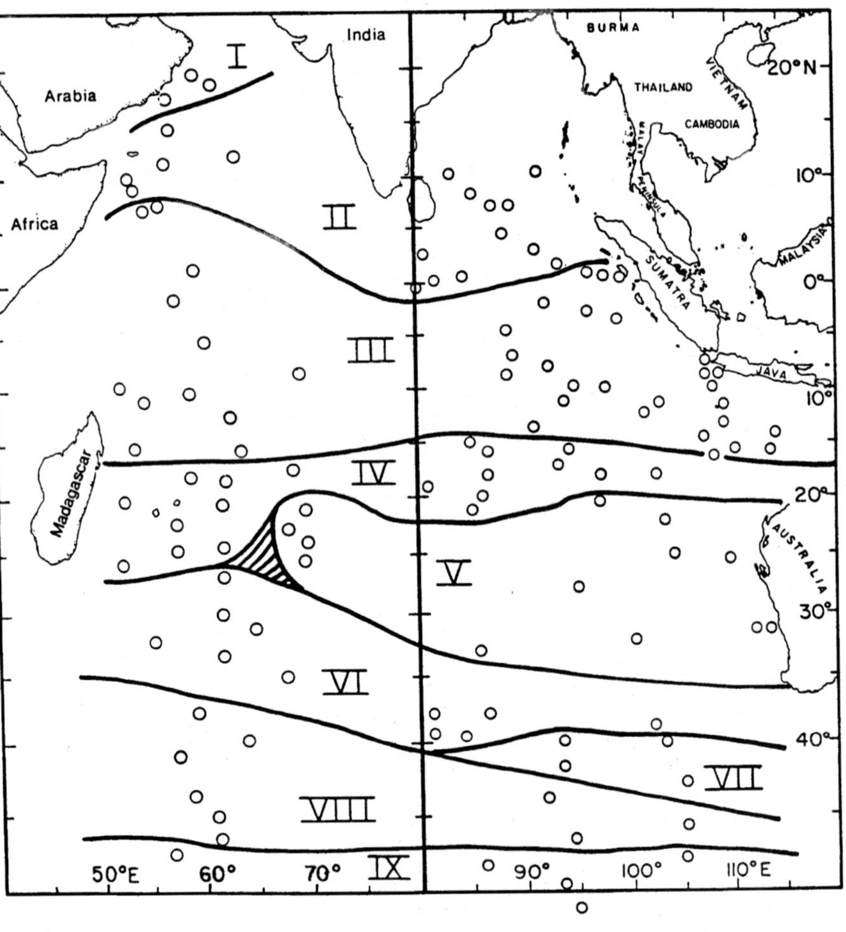

Fig. 5. Nine radiolarian assemblages identified in the Indian Ocean by Johnson and Nigrini (1980, 1982). Open circles mark the position of surface sediment samples. I = Arabian upwelling, II = Northern Indian Ocean, III = Tropical, IV = South Equatorial Current, V = Central, VI = Temperate, VII = Transitional, VIII = Subpolar, IX = Polar Front.

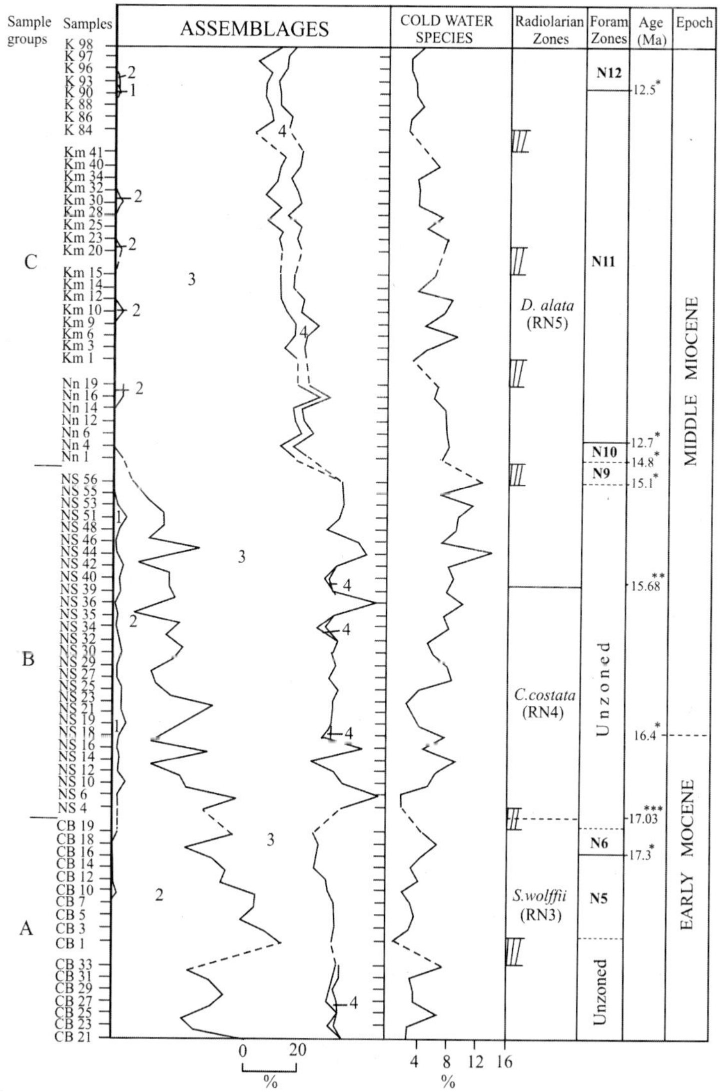

Fig. 6. Radiolarian assemblages (1-4) and percentage of cold water species in the Early to Middle Miocene sequences of Andaman-Nicobar Islands. Cooling is suggested from upper *C. costata* to lower *D. alata* Zones (Sharma and Daneshian, 2003). Planktonic foraminiferal zones of Blow (1969) are based on Srinivasan *et al.* (1983) and Srinivasan and Dave (1985). The Early/ Middle Miocene boundary, which is placed at 16.4 Ma by Berggren *et al.* (1995), is tentatively drawn below the first appearance (in sample NS21) of radiolarian species *Liriospyris parkerae* which is dated at 15.82 Ma (Shackelton *et al.*, 1995). Age: * = Berggren *et al.* 1995, ** = Shackelton *et al.* (1995), *** = Lazarus *et al. (1995).* (Based and modified after Fig. 6 of Sharma and Daneshian 2003).

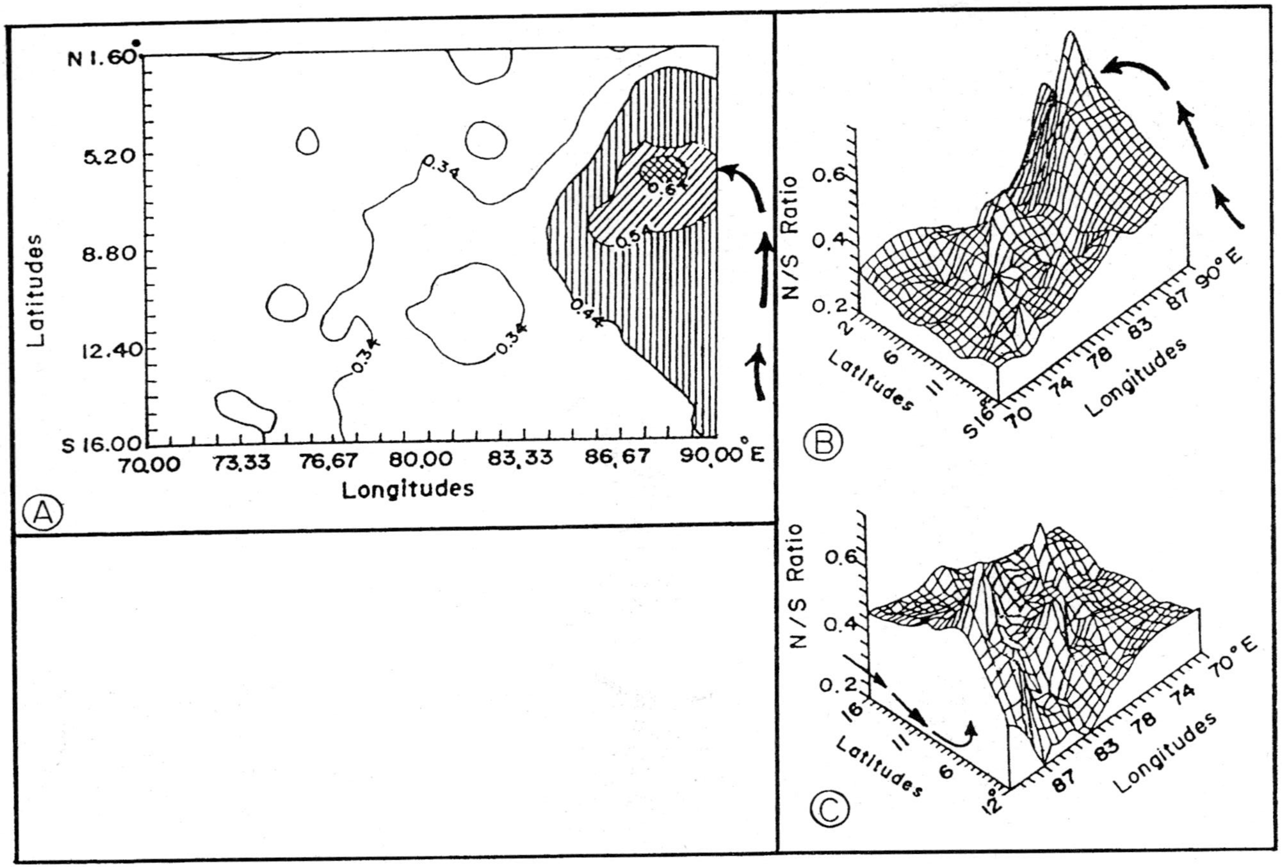

Fig. 7. Contour map of Nassellaria/Spumellaria (N/S) ratio in the Central Indian Basin (a), 3-D views of rotation of perspective axis of the basin (b) and (c), suggest movement of Antarctic Bottom Water along 90°E Ridge and its entry near 6°S into the basin (Gupta and Jauhari, 1994).

Radiolarian Research: Future Directions

The primary aims of paleontological studies is to find answers related to origin, evolution, diversification and its causes, migration, geographic distribution, etc. of a group under study and to unravel the past geological events, besides their application in dating and stratigraphic divisions. In biostratigraphic studies, an integrated biostratigraphic zonation is more meaningful in age assignment and correlation. Radiolarian zonation for example, can be combined with calcareous microfossil group. A high resolution multiple microfossil biostratigraphy with magnetostratigraphy will be of value in global stratigraphic correlation. Sustained efforts in phylogenetic studies of different radiolarian groups in future will help to refine biostratigraphy and hence in precise correlation.

Biogeographic distributions of Radiolaria in the surface waters have been known to some extent. However, there is paucity of data regarding their vertical distribution. The ongoing Joint Global Ocean Flux Studies (JGOFS) can help to a great extent to build this gap.

One of the important approaches warranted in radiolarian biostratigraphic studies is the use of graphic correlation method. With the availability of dated cores from different oceans, the method can be used to throw light on phylogeny, migration and evolution of radiolarians, which in turn would enable us to gain a better understanding of the Cenozoic paleoceanographic evolution.

Probably the most challenging problem which requires extensive studies, is to understand environmental control on growth and development of radiolorian skeletons. This includes work related to various parameters that influence morphologic variation in radiolorian skeletons, and this can be accomplished through laboratory cultures.

References

ABELMANN, A., 1990. Oligocene to middle Miocene radiolarian stratigraphy of southern high latitudes from Leg 113, sites 689-690, Maud Rise. Proc. ODP, Sci. Results, 113: 675-708.

ABELMANN, A., 1992. Early to middle Miocene radiolarian stratigraphy of the Kerguelen Plateau, Leg 120, Proc. ODP, Sci. Results, 120: 757-783.

BARRON, J., BAULDAUF, J.G., BARRERA, E., CAULET, J.P., HUBER, B.T., KEATING, B.H., LAZARUS, D., SAKAI, H., THIERSTEIN, H.R. and WEI, W., 1991. Biochronologic and magnetochronologic synthesis of ODP Leg 119 sediments from the Kerguelen Plateau and Prydz bay, Antarctica. Proc. ODP, Sci. Results, 119: 813-847.

BERGGREN, W.A., KENT, D.V. and VAN COUVERING, J.A., 1985. The Neogene: Part 2, Neogene geochronology and chronostratigraphy p.211-260. *In:* The Chronology of the Geological Record (Ed. N.J. Snelling), Geol. Soc. Mem. (London), vol. 10.

BERGGREN, W.A., KENT, D.V., SWISHER III, C.C. and AUBRY, M.P., 1995. A revised Cenozoic Geochronology and chronostratigraphy. Soc. Econ. Pal. Min. Spec. Publ. 54: 129-212.

BLOW, W.H. 1969. Late Middle Eocene to Recent planktonic foraminiteral biostratigraphy, p. 99-421. In: Proceeding of the First International Conference on Planktonic Microfossils (Eds. P. Bronnimann and H.H. Renz), Geneva 1967.

CASEY, R.E., 1971. Radiolarians as indicators of past and present water masses, p.33-341. *In:* Micropaleontology of Oceans (Eds. B.M. Funnell and W.R. Riedel), Cambridge Univ. Press.

CASEY, R.E., 1972. Neogene Radiolarian Biostratigraphy and Paleotemperatures: Southern California, the Experimental Mohole, Antarctic Core E 14-8. Paleogeogr. Paleocl. Paleoecol. 12: 115-130.

CAULET, J.P., 1979. Les depots a radiolaires d'age Pliocene superieur a Pleistocene dans locean Indien central: nouvelle zonation biostratigraphique. Mem. Mus. Nat. Hist. Nat. Ser. C, 43: 119-141.

CAULET, J.P., 1982. Faunes de radiolaires et fluctuations climatiques dans les sediments de l'Ocean Indien Austral : une nouvelle biozonation. Bull. Soc. Geol. Fr. 24: 555-562.

CAULET, J.P., 1985. Radiolarians from the Southwest Pacific. Initl. Repts. DSDP, 90: 835-861.

CAULET, J.P., 1986. A refined radiolarian biostratigraphy for the Pleistocene of the temperate Indian Ocean. Mar. Micropal. 11: 217-229.

CAULET, J.P., 1991. Radiolarians from the Kerguelen Plateau, Leg 119. Proc. ODP, Sci. Results, 119: 513-546.

CAULET, J.P., NIGRINI, C. and SCHNEIDER, D.A., 1993. High resolution Pliocene- Pleistocene radiolarian stratigraphy of the tropical Indian Ocean. Mar. Micropal. 22: 111-129.

CHEN, P.H., 1975. Antarctic Radiolaria. Intl. Rept. DSDP, 28: 437-513.

EHRENBERG, C.G., 1854. Mikrogeologie. Leipzig: Voss, XXVIII + 374 pp. Atlas, 31 pp., 41 pls. Fortsetzung (1856), 88p + 1p. errata.

EHRENBERG, C.G., 1861. Uber die Tiefgrund-Verhaltnisse des Oceans am Eingange der Davisstrasse und bei Island. Kgl. Preuss. Akad. Wiss. Berlin, Monatsber., Jahre, 1861: 275-315.

FERRELL, J.W. and JANECEK, T., 1991. Late Neogene Paleoceanography and Paleoclimatology of the Northeast Indian Ocean (site 758). Proc. ODP, Sci. Results, 121: 297-355.

FOREMAN, H.P., 1973. Radiolaria of Leg 10 with systematics and ranges for the families Amphipyndacidae, Artostrobiidae and Theoperidae. Intl. Rept. DSDP, 10: 407-474.

FOREMAN, H.P., 1975. Radiolaria from the North Pacific, Deep Sea Drilling Project. Intl. Rept. DSDP, 32: 579-676.

FOSTER, J.H. and OPDYKE, N.D., 1970. Upper Miocene to Recent magnetic stratigraphy in deep-sea sediments. Jour. Geophys. Res. 75: 464-473.

GUPTA, S.M. and FERNANDES, A.A., 1995. Supra-generic transfer function based late Miocene Paleomonsoonal changes in the northeastern Tropical Indian Ocean. Bull. Indian Geologist Assoc. 28(1): 29-51.

GUPTA, S.M. and FERNANDES, A.A., 1997. Quaternary radiolarian faunal changes in the central Indian Ocean: inferences to oscillating hydrographic front at 10 degree south of equator. Curr. Sci. 72: 965-972.

GUPTA, S.M. and JAUHARI, P., 1994. Radiolarian abundances and geochemistry of the surface sediments from the central Indian Basin: Inferences to Antarctic Bottom Water Current. Curr. Sci. 66(9): 659-663.

GUPTA, S.M. and SRINIVASAN, M.S., 1992. Late Miocene radiolarian biostratigraphy and paleoceanography of Sawai Bay Formation, Neill Island, Andamans, India. Micropal. 38(3): 209-235.

GUPTA, S.M., 1996. Quantitative radiolarian assemblages in surface sediments from the Central Indian Basin and their paleomonsoonal significance. Jour. Geol. Soc. India, 47: 339-354.

GUPTA, S.M., FERNANDES, A.A. and RAHUL MOHAN, 1996. Tropical sea surface temperatures and the Earth's Eccentricity cycles. Geophysical Res. Letters, 23(22): 3159-3162.

HAYS, J.D. and OPDYKE, N.D., 1967. Antarctic Radiolaria, magnetic reversals and climate change. Science, 158: 1001-1011.

HAYS, J.D., 1965. Radiolaria and late Tertiary and Quaternary history of Antarctic Seas. In: Biology of Antarctic Seas II (Ed. G.A. Llano). Am. Geophys. Union, Antarct. Res. Ser. 5: 125-184.

HAYS, J.D., SAITO, T., OPDYKE, N.D. and BURCKLE, L.H., 1969. Pliocene-Pleistocene sediments of the Equatorial Pacific : Their Paleomagnetic, Biostratigraphic, and Climatic Record. Geol. Soc. Am. Bull. 80: 1481-1514.

HOLLIS, C.J., 1993. Latest Cretaceous to Late Paleocene radiolarian biostratigraphy: A new zonation from the New Zealand region. Mar. Micropal. 21: 295-327.

JACOB, K. and SHRIVASTAVA, R.N., 1952. Fossil Radiolaria and silicoflagellata from the Tertiary clays of Colebrook Island and Ritchie's Archipelago, Andaman Islands. Sci & Cult. 17(8): 346-348.

JOHNSON, D.A. and NiGRINI, C., 1980. Radiolarian biogeography in surface sediments of the Western Indian Ocean. Mar. Micropal. 5: 111-152.

JOHNSON, D.A. and NIGRINI, C., 1982. Radiolarian biogeography in surface sediments of the Eastern Indian Ocean. Mar. Micropal. 7: 237-281.

JOHNSON, D.A. and NIGRINI, C., 1985. Synchronous and time-transgressive Neogene radiolarian datum levels in the equatorial Indian and Pacific Oceans. Mar. Micropal. 9: 489-523.

JOHNSON, D.A., 1990. Radiolarian biostratigraphy in the Central Indian Ocean, Leg 115. Proc. ODP, Sci. Results, 115: 395-409.

JOHNSON, D.A., SCHNEIDER, D.A., KENT, D.V., NIGRINI, C.A. and CAULET, J.P., 1989. Plio–Pleistocene radiolarian events and magnetostratigraphic calibrations for the tropical Indian Ocean. Mar. Micropal. 4: 33-66.

KEANY, J. and KENNETT, J.P., 1975. Pliocene – Pleistocene biostratigraphy and paleoclimatology at DSDP site 278 on the Antarctic convergence. Intl. Rept. DSDP, 9: 757-767.

KEANY, J., 1973. New radiolarian paleoclimatic index in the Plio – Pleistocene of Southern Ocean. Nature, 246: 139-141.

LAZARUS, D., 1990. Middle Miocene to Recent radiolarians from the Weddell Sea, Antarctica, ODP leg 113. Proc. ODP, Sci. Results, 113: 709-727.

LAZARUS, D., 1992. Antarctic Neogene radiolarians from the Kerguelen Plateau, Leg 119 and 120. Proc. ODP, Sci. Results, 120: 785-809.

LAZARUS, D., SPENCER-CERVATO, C., PIANKA-BIOLZI, M., BECKMANN, J.P., VON SALIS, K., HILBRECHT, H. and THIRSTEIN, H, 1995. Revised chronology of Neogene DSDP Holes from the World ocean. Ocean Drilling Program, Coolege Station, Texas, Tech. Note 24.

LING, H.Y. and ANIKOUCHINE, W.R., 1967. Some spumellarian radiolaria from the Java, Philippine and Mariana trenches. Jour. Pal. 41(6): 1481-1491.

LING, H.Y. and SAMUEL, M.A. 1998. Siliceous microfossil from Nias Island; their significance for the Tertiary paleoceanography of the Northeast Indian Ocean. Jour. Asian Earth Sci. 16: 407-417.

LING, H.Y., HALL, R. and NICHOLAS, G.J., 1991. Early Eocene Radiolaria from Waigeo island, Eastern Indonesia. Jour. Southeast Asian Earth Sci. 6(3-4): 299-305.

LING, H.Y., SHARMA, V., SINGH, S., MAZUMDAR, D. and MAHAPATRA, A.K. Cretaceous and Middle Eocene Radiolaria from ejected sediments of mud volcanoes of Baratang Island in Andaman sea of the Northeastern Indian Ocean. Jour. Geol. Soc. India, 45: 463-469.

LOMBARI, G. and BODEN, G., 1985. Modern radiolarian global distributions. Cushman Found. Foram. Res., Spec. Publ. 16A: 1-125.

MAHAPATRA, A. and SHARMA, V., 1994. Trissocyclid Radiolaria from the late Early Miocene sequences of Colebrook, North Passage and Great Nicobar Islands, Northeast Indian Ocean. Micropal. 40: 157-168.

MAURASSE, F. and GLASS, B.P., 1976. Radiolarian Stratigraphy and North American microtektites in Caribbean RC9-58: implications concerning Late Eocene radiolarian chronology and the age of the Eocene-Oligocene boundary, p.205-212. *In:* Trans. De la VII conf. Geol. Des Caraibes du 30 juin au 12 juillet 1974 (Ed. R. Causse), St. Francois, Guadeloupe.

MEYEN, F.J.F., 1834. Über das Leuchten des Meeres, und Beschreibung einiger Polypen und anderer niederer Thiere. *In:* Beiträge zur Zoologie, gesammelt auf einer Reise um die Erde. Kai Leopold. Carol. Akad. Naturforsch., Verh., 18 (suppl.1, abh.5): 125-216.

MOLINA-CRUZ, A., 1977. Radiolarian assemblages and their relationship to the Oceanography of the subtropical southeastern Pacific. Mar. Micropal. 2: 315-352.

MOLINA-CRUZ, A., 1984. Radiolaria as indicators of upwelling processes: The Peruvian connection. Mar. Micropal. 9: 53-75.

MOORE, T.C., Jr. and LOMBARI, G., 1981. Sea surface temperature changes in the North Pacific during the Late Miocene. Mar. Micropal. 6: 581-597.

MOORE, T.C., Jr., 1971. Radiolaria. Intl. Rept. DSDP, 8: 727-775.

MOORE, T.C., Jr., 1973. Late Pleistocene-Holocene oceanographic changes in the northeastern Pacific. Quat. Res. 3: 99-109.

MOORE, T.C., Jr., BURCKLE, L.H., GEITZENAURE, K., LUZ, B., MOLINA-CRUZ, A., ROBERTSON, J.H., SACHS, H., SANCETTA, C., THEIDE, J., THOMPSON, P. and WENKAM, C., 1980. The reconstruction of sea surface temperature in the Pacific Ocean of 18,000 years B.P. Marine Micropal. 5: 215-247.

MORLEY, J.J., 1979. A transfer function for estimating oceanographic conditions based on deep-sea surface sediment distribution of radiolarian assemblages in the South Atlantic. Quat. Res. 12: 381-395.

MORLEY, J.J., 1989. Radiolarian–based transfer functions for estimating paleoceanographic conditions in the South Indian Ocean. Mar. Micropal. 13: 293-307.

MULLINEAUX, L.S. and WESTBERG – SMITH, M.J., 1986. Radiolarians as paleoceanographic indicators in the Monterey Formation, Upper Newport Bay, California, Micropal. 32: 48-71.

NIGRINI, C. and CAULET, J.P., 1992. Late Neogene radiolarian assemblages characteristic of Indo-Pacific areas of upwelling. Micropal. 38(2): 139-164.

NIGRINI, C., 1967. Radiolaria in pelagic sediments from the Indian and Atlantic Oceans. Bull. Inst. Oceanogr., Univ. Calif., La Jolla, 11: 1-125.

NIGRINI, C., 1971. Radiolarian zones in the Quaternary of equatorial Pacific Ocean, p. 443-462. *In:* Micropal. of Oceans (Eds. B.M. Funnell and W.R. Riedel). Cambridge Univ. Press.

NIGRINI, C., 1974. Cenozoic Radiolaria from the Arabian Sea, DSDP Leg 23. Intl. Rept. DSDP, 26: 1051-1121.

NIGRINI, C., 1991. Composition and biostratigraphy of radiolarian assemblages from an area of upwelling (northwestern Arabian Sea, leg 117). Proc. ODP, Sci. Results, 117: 89-126.

NISHIMURA, A., 1987. Cenozoic Radiolaria in the Western North Atlantic, Site 603, Leg 93 of the Deep Sea Drilling Project. Intl. Rept. DSDP, 93: 713-737.

NISHIMURA, A., 1992. Paleocene radiolarian biostratigraphy in the Northwest Atlantic at Site 384, Leg 43, of the Deep Sea Drilling Project. Micropal. 38(4): 317-362.

OPDYKE, N.D., 1972. Paleomagnetism of deep-sea cores. Rev. Geophys. Space Phys. 10: 213-249.

OPDYKE, N.D., BURCKLE, L.H. and TODD, A., 1974. The extension of the magnetic time scale in sediments of the Central Pacific Ocean. Earth Planet. Sci. Lett. 22: 300-306.

PETRUSHEVSKAYA, M.G., 1967. Radiolyarii otryadov Spumellaria i Nassellaria antarkticheskoi ovlasti. Issledovaniya Fauny Morei 4 (12), Rez. biol. Issled. sov. antarkt. Exsped. 1955-58, pp. 1–186.

PETRUSHEVSKAYA, M.G., 1971. Radiolaria in the plankton and Recent Rediments from the Indian Ocean and Antarctic, p.318-329. *In:* Micropaleontology of Oceans (Eds. B.M. Funnell and W.R. Riedel), Cambridge Univ. Press.

PISIAS, N.G. and MOORE, T.C., Jr., 1995. Radiolarian response to oceanographic changes in the eastern equatorial Pacific at 2.3 and 4.8 Ma: Relationship between changing carbonate deposition and surface oceanography. Proc. ODP, Sci. Results, 138: 461-477.

RIEDEL, W.R. and SANFILIPPO, A., 1970. Radiolaria, Leg 4, Deep Sea Drilling Project. Intl. Repts. DSDP, 4: 503-575.

RIEDEL, W.R. and SANFILIPPO, A., 1971. Cenozoic Radiolaria from the Western tropical Pacific, Leg 7. Intl. Repts. DSDP, 7: 1529-1672.

RIEDEL, W.R. and SANFILIPPO, A., 1978. Stratigraphy and evolution of tropical Cenozoic radiolarians. Micropal. 24(1): 61-96.

RIEDEL, W.R., 1981. DSDP biostratigraphy in retrospect and prospect, p.253-261. *In:* The Deep Sea Drilling Project: A decade of Progress (Ed. J.E. Warme *et al.*). Spec. Publ. Soc. Econ. Paleont. Mineral. Tulsa, vol. 32.

ROMINE, K. and LOMBARI, G. 1985. Evolution of Pacific Circulation in the Miocene: Radiolarian evidence from DSDP site 289. Geol. Soc. Am. Mem. 163: 273-290

Romine, K. and Moore, T.C., Jr., 1981. Radiolarian assemblage distributions and paleoceanography of the eastern equatorial Pacific Ocean during the last 127000 years. Paleogeogr. Paleocl. Paleoecol. 35: 281-314.

SANCETTA, C.A., 1978. Neogene Pacific microfossils and Paleoceanography. Mar. Micropal. 3: 343-376.

SANFILIPPO, A. and NIGRINI, C., 1998. Code numbers for Cenozoic low latitude radiolarian biostratigraphic zones and GPTS conversion tables. Mar. Micropal. 33: 109-156.

SANFILIPPO, A., WESTBERG-SMITH, M.J. and RIEDEL, W.R., 1985. Cenozoic Radiolaria, p. 631-712. *In:* Plankton Stratigraphy (Eds. H.M. Bolli, *et al.*). Cambridge Univ. Press.

SAUNDERS, J.B., BERNOULLI, D., MÜLLER-MERZ, E., OBERHÄNSLI, H., PERCH-NIELSEN, K., RIEDEL, W.R., SANFILIPPO, A. and TORRINI, R., Jr., 1985. Stratigraphy of the late middle

Eocene to early Oligocene in the Bath Cliff Section, Barbados, West Indies. Micropal. 30(4): 390-425.

SHAKLETON, N.J., BALDUF, J.G., FLORES, J.A., IWAI, M., MOORE, T.C., Jr., RAFFI, I., VINCENT, E. (1995). Biostratigraphic summary for leg 138, Ocean Drilling Program, Sci Results, 138: 517-536.

SHARMA, V. and DANESHIAN, J., 1998. Miocene Radiolaria from Nicholson and John Lawrence Islands, Andaman Sea. Jour. Geol. Soc. India, 52: 695-707.

SHARMA, V. and DANESHIAN, J., 2003. Early Neogene radiolarian changes in the Northeast Indian Ocean and paleoceanographic implications: Evidence from the Andaman–Nicobar Islands. Proc. 8th Int. Cong. Pac. Neogene Strat., Chiang Mai, Thailand, p.126-158.

SHARMA, V. and MAHAPATRA, A., 1990. Radiolarian Occurrences in surface sediments of the Indian Ocean in relation to bottom water circulation. Jour. Geol. Soc. India, 35: 251-261.

SHARMA, V. and RAM, M.P., 2003. Early to Middle Miocene radiolarian assemblages and biostratigraphy, Andaman Islands, Northeast Indian Ocean. Jour. Pal. Soc. India (accepted for publication).

SHARMA, V. and SHARMA, G.K., 1988. Radiolaria from Neill Island, Andaman Sea and their distributional characteristics. Jour. Pal. Soc. India, 33: 7-19.

SHARMA, V. and SHARMA, G.K., 1989. Late Miocene to Early Pliocene radiolarian biostratigraphy of Neill Island, Andaman Sea. Jour. Geol. Soc. India, 34: 76-82.

SHARMA, V. and SINGH, S., 1993. Radiolarian biostratigraphy of Early Pliocene sequences, Car Nicobar Island, Northeast Indian Ocean. Jour. Geol. Soc. India, 41: 199-213.

SHARMA, V. and SINGH, S., 1997a. Early Neogene radiolaria from Anadman-Nicobar Islands, Indian Ocean : paleoclimatic implications, p.80-102. Proc. Seminar on Neogene Evolution of Pacific Ocean Gateways IGCP-355, Japan.

SHARMA, V. and SINGH, S., 1997b. Late Neogene Radiolarian events in Andaman-Nicobar Islands, Northeast Indian Ocean. Micropal. 43(1): 41-50.

SHARMA, V., SRINIVASAN, M.S. and MAHAPATRA, A.K., 1993. Early Miocene radiolarian and planktonic foraminiferal biostratigraphy, North Passage Island, Andaman Sea. Jour. Geol. Soc. India, 42: 154-162.

SHARMA, V., SURENDER SINGH and NEERU RAWAL, 1999. Early Middle Miocene Radiolaria from Nicobar Islands, Northeast Indian Ocean. Micropal. 45(3): 251-277.

SINGH, O.P., SRINIVASAN, M.S. and SHARMA, V., 2000. Early Neogene multiple microfossil biostratigraphy, John Lawrence Island, Andaman Sea. Micropal. 46(4): 343-352.

SINGH, P., 1981. A note on the Early Pliocene Radiolaria from Neill Island, South Andaman, India. Geoscience Jour., 2 (1): 97-107.

SINGH, P. and JAUHARI, A.K., 1976. A preliminary report on the occurrence of Radiolaria in the middle Eocene of the Vinjhan-Miani area, Kutch, Gujarat. Curr. Sci. 45 (3): 100.

SINGH, P. and VIMAL, K.P., 1973. A note on the geology and micropaleontology of Neill Island, South Andaman. Cur. Sci. 42(7): 239-241.

SRINIVASAN, M.S. and DAVE, A. 1985. Middle Miocene sediments of Nancowry and Kamorta Islands: Their biostratigraphic and climatic record. Bull. Geol. Min. Met. Soc. India, 52: 319-334.

SRINIVASAN, M.S., LOMBARI, G. and DAVE, A., 1983. Early Miocene Planktonic Foraminiferal and Radiolarian zonation, Colebrook Island, Andaman Sea. Jour. Geol. Soc. India, 24: 1-18.

TAKEMURA, A., 1992. Radiolarian Paleogene biostratigraphy in the Southern Indian Ocean, Leg 120. Proc. ODP, Sci. Results, 120: 735-756.

THEYER, F. and HAMMOND, S.R., 1974a. Paleomagnetic polarity sequence and radiolarian zones, Brunhes to Epoch 20. Earth Planet. Sci. Lett. 22: 307-319.

THEYER, F. and HAMMOND, S.R., 1974b. Ccnozoic magnetic time scale in deep-sea sediments: Completion of the Neogene. Geology, 2: 487-492.

THEYER, F., MATO, Y. and HAMMOND, S.R., 1978. Paleomagnetic and geochronologic calibration of lastest Oligocene to Pliocene radiolarian events, equatorial Pacific. Mar. Micropal. 3: 377-395.

WEAVER, F.M., 1983. Cenozoic radiolarians from the southwest Atlantic, Falkland Plateau region, Deep Sea Drilling Project Leg 71. Intl. Repts. DSDP, 71(2): 667-686.

Patterns of Radiolarian species diversity in the Early Neogene of Andaman – Nicobar Islands

V. Sharma[1] and J. Daneshian[2]

[1]Department of Geology, University of Delhi, Delhi 110007, India.
[2]Department of Geology, Teachers' Training University of Tehran, Mofateh Avenue, Tehran, Iran.

Introduction

Good preservation and abundance of diverse microfossils in the Neogene rocks of Andaman – Nicobar islands in the northeast Indian Ocean (Fig. 1) have attracted a number of researchers. As a result, a large number of stratigraphic sections have been investigated for foraminifera and Radiolaria, besides other microfossil groups. The radiolarian studies are mainly on biostratigraphy, biochronology and paleoclimatology (Singh, 1981; Srinivasan *et al.,* 1983; Sharma and Sharma, 1989; Sharma and Singh, 1993, 1997a,b; Sharma *et al.,* 1993,1999; Singh *et al* ., 2000; Sharma and Daneshian, 2003; Sharma and Ram, 2003).

The episode of ice sheet formation and its expansion in the Antarctica in the early Neogene resulted in climatic and oceanographic changes which can be recognized in marine microfaunas. The changes gave rise to appearances and disappearances of species and also affected their abundances. In the northern Indian Ocean, though signatures of such climatic changes are not so clear as are found in high latitude areas, they still can be recognized and interpreted. Most Neogene sections of Andaman – Nicobar contain abundant Radiolaria and are suited for understanding diversity trends and their possible causes.

Though a large number of publications on species diversity exist, work on radiolarian diversity is relatively few. In the last about forty years, biogeographic distributions of Radiolaria in relation to water mass have been known to a great extent. Attempts to understand radiolarian diversity have also been made by a number of workers (Casey and Sloan, 1971; Casey, 1977; McMillen and Casey, 1978; Casey *et al.,* 1979,1981, 1990; Swanberg and Bjorklund, 1987; Cheng, 1989; Blueford *et al.,* 1990; Vishnevskaya, 1992; Yamamoto, 1992; Chen and Tan, 1997; Nishimura *et al.*, 1997; Vishnesvskaya, 1997; Li and Xu, 1998; Bartolini *et al.*, 1999; Kiessling, 1999; Sashida and Kurihara, 1999). Some of these publications deal with simple diversity, in terms of number of species, observed in studied stratigraphic sequences or in surface sediment samples without assigning any reason for the change in their values. A few, however, have linked diversity with the change in temperature and in water masses.

In this work, the authors determined radiolarian species diversity in samples (77 in number) from sections spanning from *Stichocorys wolffii* to *Dorcadospyris alata* zones of Early to Middle Miocene. Possible causes responsible for variations in diversity are discussed.

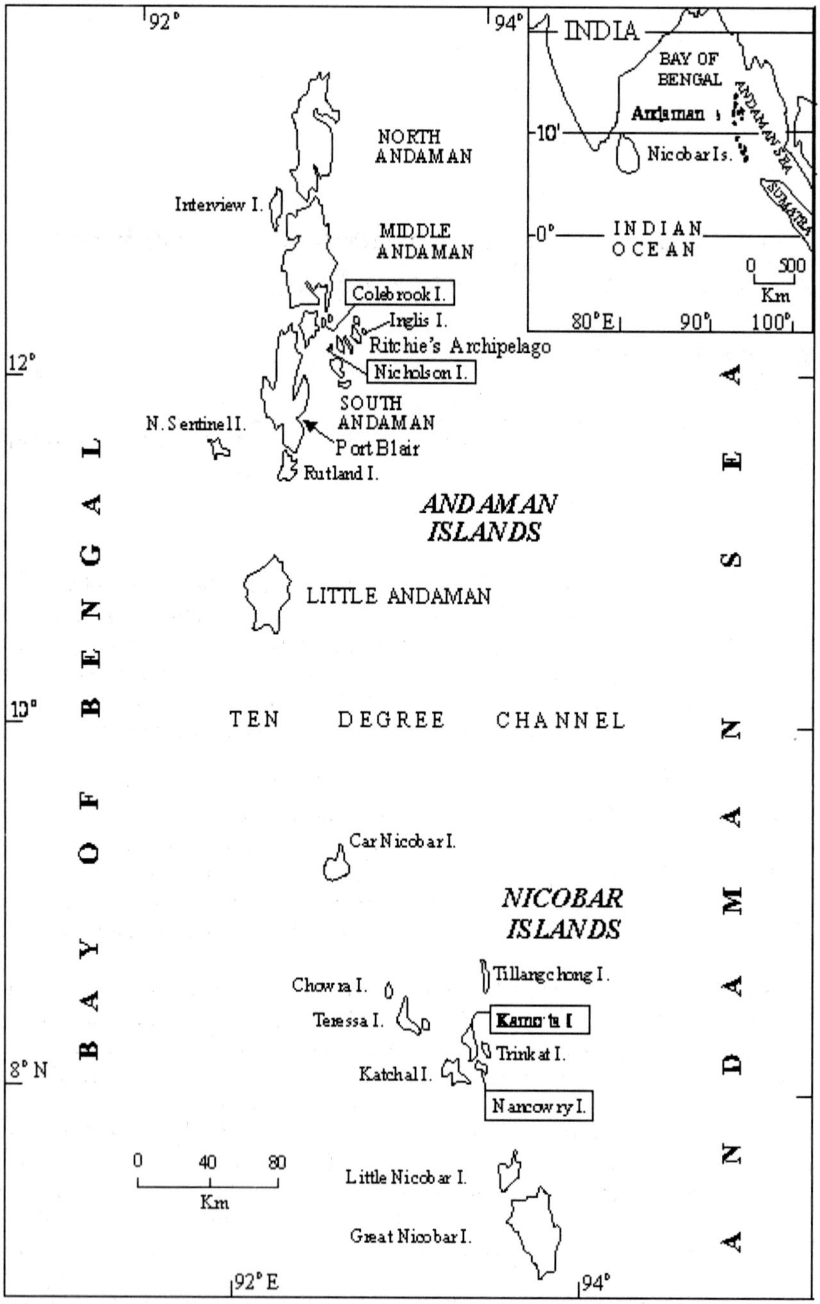

Fig. 1. Location map. Names of islands exposing studied sections are shown within boxes.

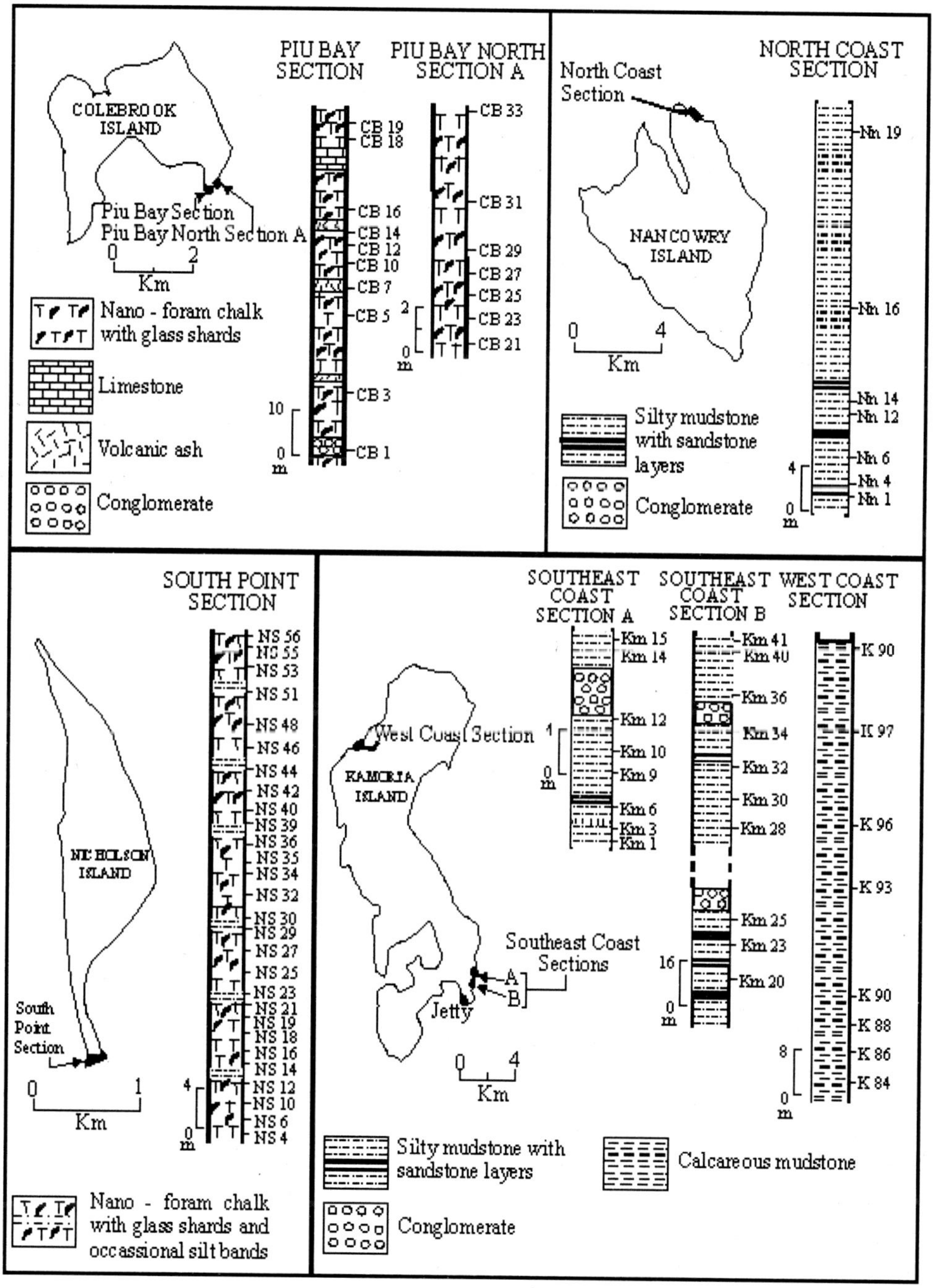

Fig. 2. Lithologs and locations of studied sections.

Material and methods

Radiolarians of seventy seven samples from seven sections are studied (Fig. 2), representing a span from about 17.3 Ma to 12.5 Ma. Within this span, however, a few minor intervals remain unrepresented in the sequence.

The samples are disaggregated by the usual method (Sanfilippo *et al.*, 1985) and fractions of washed residue coarser than 53 μm are used in this study. Radiolarian specimens, about 600 to 700, were identified in a part of quantitative slides prepared by random distribution method of Moore (1973). The random distribution method allows the assemblage of a sample to be accurately represented by counting a portion of that slide (Romine and Moore, 1981). From the distributional data of specimens, species diversity of Radiolaria in each sample is determined. Reworked species are not considered in determination of diversity.

Measurement of species diversity

Species diversity of a taxon in a recent organism can be determined by counting the total number of species of the organism inhabiting an area. In case of fossils, it can be measured by counting number of species present in a sample. In diversity measurements, relative abundance of species (= equitability) is an important aspect and simple counting of number of species in a sample does not take into account the abundance of a species. Further, diversities in different samples determined by simple counting can not be compared with each other because of variable sample size. To minimize such problems a number of mathematical expressions are in use. The most commonly used is Shannon – Wiener index, which is utilized here. The species diversity index is expressed as:

$$H(S) = \Sigma_{i=1}^{s} pi \ln pi$$

Where, H(S) is the diversity index
pi is the proportion of the ith species, and
s is the number of species

Other components in the study of diversity, like equitability (species abundance) and total number of species in each sample, are also determined (Fig.3). Buzas and Gibson (1969) suggested that equitability, E, can be measured by

$$E = \frac{e^{H(S)}}{S}$$

Where, H(S) is species diversity
and S is number of species

Equitability influences diversity H(S), for, higher (or, lower) diversity may result not only due to increase (or, decrease) in number of species, but also due to high (or, low) equitability (Gibson and Buzas, 1973). For example, high diversity in samples CB 18 and NS 16 is due to high equitability in these samples as compared to that in other samples having same number of species (Fig. 3). Likewise, low equitability in sample K 93 is responsible for low species diversity.

Radiolarian assemblages, paleoceanographic and paleoclimatic implications

Sharma and Daneshain (2003) subjected the distributional data of radiolarians from the same set of samples to cluster analysis. Fifty three taxa with an abundance of 2% or more in any one sample were used in the analysis. They identified four assemblages (Table 1) and three sample groups, A, B, and C.

The data, in the form of a diagram on radiolarian assemblage, cold water taxa and biostratigraphy by Sharma and Daneshian (2003) is reproduced after modification, supplemented by data on species diversity (Fig. 4).

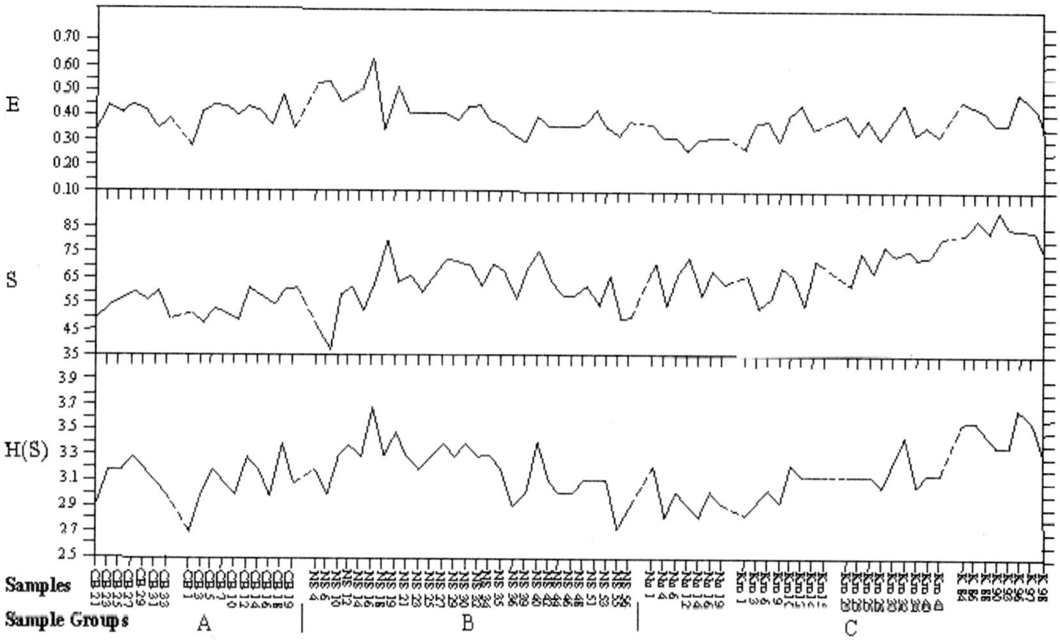

Fig. 3. Radiolarian species diversity, H (S), number of species, S and equitability, E, in each sample.

The paleoceanographic interpretations were mainly based on distributional pattern of different assemblages and percent distribution of water mass related species in the sequence including cold water forms, like *Spongopyle osculosa, Lithelius minor* and *Spongotrochus* (?) *venustum,* which are characteristic of temperate/transitional waters (Sharma and Daneshian, 2003). The cold water species show fluctuations in their percentage occurrence in the radiolarian assemblages which are characteristically warm water. A warmer condition during the interval represented by *Stichocorys wolffii* and *C. costata* zones and in upper *D. alata* Zone, and a colder condition in the interval from upper *Calocycletta costata* to lower *Dorcadospyris alata* zones were identified (Sharma and Daneshian, 2003). The climatic cooling recognized in the upper *Calocycletta costata* and lower *Dorcadospyris alata* zones is responsible for increase in percentage frequency of cold water (temperate) species in an otherwise warmer assemblage as a result of their northward transport by cooler surface waters. The climatic fluctuations recorded in the study (Sharma and Daneshian, 2003) are found to correspond to similar fluctuations determined by oxygen isotopic data of planktic foraminifera during the same interval at tropical pacific DSDP sites 574 and 575 and Indian Ocean sites 216, 237, 238 and 758 (Vincent and Killingley, 1985; Vincent *et al.*, 1985; Vincent *et al.*, 1991).

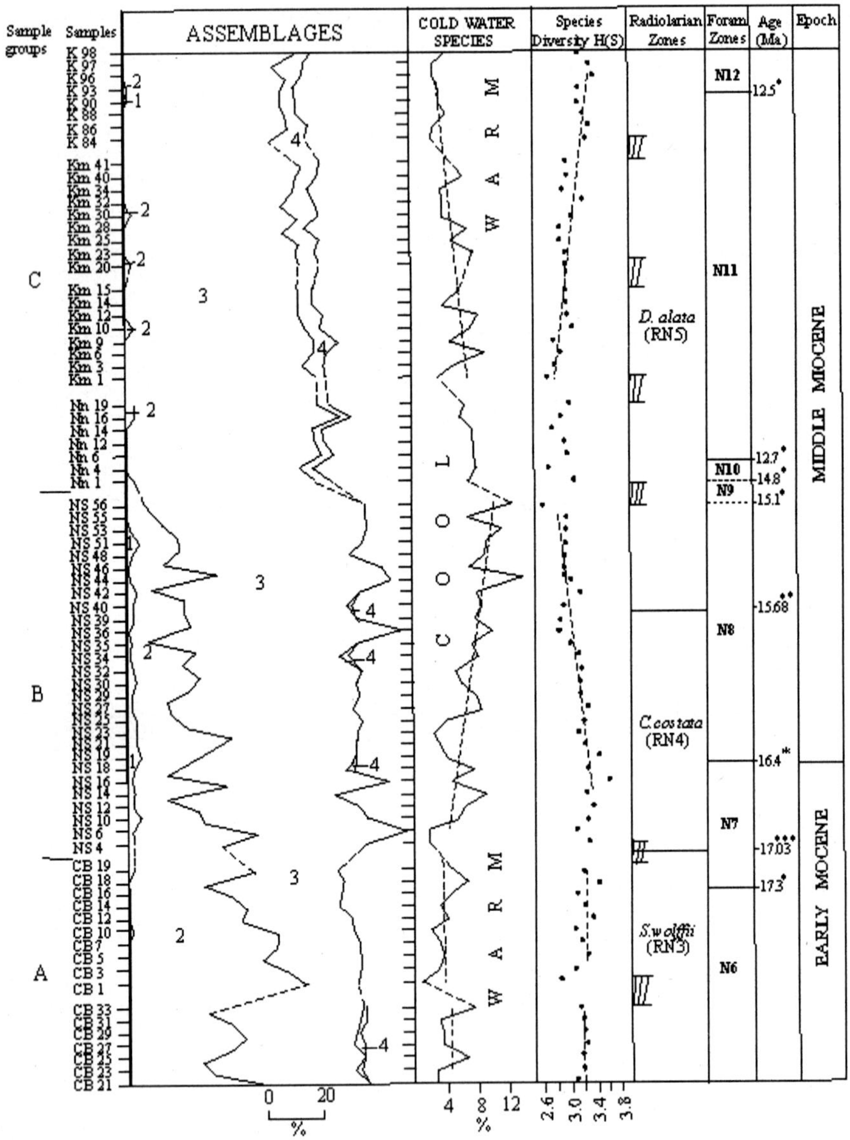

Fig. 4. Commulative percentage of constituent species of four identified assemblages, percent occurrence of cold water species and radiolarian species diversity. Foraminiferal zones of Blow (1969) recognized in the sections are after Srinivasan *et. al.* (1983), Srinivasan and Dave (1985) and Daneshian (1998). Intervals unrepresented by strata are shown in the column of radiolarian zones. Dash lines in the columns of cold water species and species diversity are drawn to show gradients in their values. * = Berggren *et. al.* (1995), ** = Shackleton *et. al.* (1995), *** = Lazarus *et. al.* (1995). (Modified after Sharma and Daneshian, 2003).

Table 1. Radiolarian assemblages identified by cluster analysis (Sharma and Daneshian, 2003).

Assemblage 1 1. *Calocycletta costata* 2. *Liriospyris stauropora* 3. *Spongaster* spp. **Assemblage 2** 1. *Calocycletta virginis* 2. *Carpocanarium* spp. 3. *Carpocanopsis bramlettei* 4. *Carpocanopsis cingulata* 5. *Dendrospyris bursa* 6. *Didymocyrtis mammifera* 7. *Gorgospyris schizopodia* 8. *Rhodospyris* (?) spp. De 1 group 9. *Stichocorys wolffii* 10. *Tholospyris anthophora* 11. *Tholospyris mammillaris*	13. *Phormostichoartus marylandicus* 14. *Phorticium polycladum* 15. *Phorticium pylonium* 16. *Porodiscus* sp. A 17. *Porodiscus* (?) sp. B 18. *Spongodiscus* spp. 19. *Spongopyle osculosa* 20. *Spongotrochus glacialis* group 21. *Spongotrochus* (?) *venustum* 22. *Spongurus* (?) spp. 23. *Stichocorys delmontensis* 24. *Stylochlamydium asteriscus* 25. *Tetrapyle octacantha* 26. *Theocorys redondoensis* 27. *Tholospyris kantiana* 28. *Trisolenia magalactis magalactis*
Assemblage 3 1. *Botryocyrtis* spp. 2. *Carpocanistrum* spp. 3. *Circodiscus microporus* 4. *Circodiscus* sp. 5. *Cubotholus* sp. 6. *Hymeniastrum* spp. 7. *Liriospyris geniculosa* 8. *Liriospyris mutuaria* 9. *Liriospyris* sp. "L2" 10. *Lithelius minor* 11. *Lophophaena* spp. 12. *Phacodiscus* sp.	**Assemblage 4** 1. *Albatrossidium* spp. 2. *Calimitra atavia* 3. *Ceratocyrtis histricosa* 4. *Cycladophora bicornis* 5. *Dictyocoryne truncatum* 6. *Didymocyrtis laticonus* 7. *Eucecryphalus* spp. 8. *Eucyrtidium hexagonatum* 9. *Larcospira moschkovskii* 10. *Lophospyris pentagona pentagona* 11. *Peridium* spp.

Species diversity: Observation and discussion

An overall high diversity values are in the *Stichocorys wolffii* Zone and lower part of *Calcoycletta costata* Zone. This interval is characterised by a warmer condition. Lower diversity values in the upper *Calocycletta costata* and lower *Dorcadospyris alata* zones correspond to colder condition. Higher diversity values are seen again in the interval represented by upper *D. alata* Zone, which marks a warmer period. Also, a general declining / rising trend in diversity values in the sequence corresponds to an increasing/decreasing trend in percentage abundance of cold water species. A fall in diversity values in the lower part of *Dorcadospyris alata* Zone (represented by the samples of Nicholson Island) is due to the extinction of many species, particularly of those belonging to Assemblage 2. This interval is marked by cooling that seems responsible for extinction of many taxa of Assemblage 2, which is comprised of typical warm water species. It is of interest to note that while number of species of Assemblage 2 largely remains unaltered in samples belonging to *Calocycletta costata* and early *Dorcadospyris alata* zones (Fig. 5) their percentage abundances decrease appreciably (Fig. 4). This is the interval marked by cooling. The increase in radiolarian diversity in the upper *D. alata* Zone (Fig. 4) presence is largely due to taxa of Assemblage 4 in greater number (Fig. 5). Percent abundance of these taxa, too, increased during this interval.

Marine plankton diversity is found affected by oceanographic circulation, primary productivity, salinity of surface water and temperature (Stehli, 1965; Ruddiman, 1968; Williams and Johnson, 1975, Cline and Hays, 1976; Balsam and Flessa, 1978).

Diversity and abundance of Radiolaria are found low in brackish water produced due to mixing of fresh water from rivers with the sea surface waters and high in open Oceans (Casey *et al.,* 1981). Abundance of Radiolaria and lithology of the studied sections suggest deposition at great depths, quite away from shores. Though influx of terrigenous clastics is seen in North Coast Section (Nancowry Island) and Southeast Coast Sections A and B (Kamorta Island), the associated benthic foraminiferal fauna indicates their deposition in deeper waters (Srinivasan and Dave, 1985; Srinivasan 1988). Role of primary productivity in influencing diversity is not known in the present case. Rutherford *et al.* (1999), found poor correlation between primary production and zooplankton (planktic foraminifera) diversity.

Temperature change seems to be one of the major causes to affect zooplankton diversity (Berggren, 1969; Casey and Sloan, 1971; Balsam and Flessa, 1978, McGowan and Walker, 1985; Krasheninnikov *et al.*, 1992; Sashida and Kurihara, 1999). Radiolarian diversity is higher in warm waters as compared to its diversity in cold waters (Casey and Sloan, 1971). Stehli *et al.* (1969) observed a typical diversity gradient sloping poleward in Maestrichtian planktic forminifera in the northern hemisphere and considered climatic change as the primary cause of such a gradient. Based on planktic foraminifera, Krasheninnikov *et al.* (1992) found late Paleocene and early Eocene as the warmest interval within the Paleogene and planktic foraminiferal species diversity being highest during this interval. Sashida and Kurihara (1999) observed higher radiolaraian diversity in samples of warm water than that in cool and moderate water condition, in the surface water off the cost of Shimoda, Japan. It is interesting to note that in other microfossil groups, like nannoplankton and discoasters, a positive correlation between paleotemperature and diversity has also been observed during the Cenozoic (Haq, 1973).

The Middle Miocene cooling as recorded in the examined sequence was restricted not only to this region but was global in nature. It was a period of intense cooling and Antartic ice accumulation. The cooling not only had a direct effect in lowering sea surface temperature, the ice build up during this interval resulted in roughly 50m drop in sea level, beyond the general regression from uplift in Asia and elsewhere (Seibold and Berger, 1993). Haq (1973) considered transgression and regression responsible for influencing diversity while explaining the diversity of Jurassic to Pleistocne nannoplankton. Large continental areas exposed during regression result in extreme 'continental' climate with larger difference in maximum and minimum temperature, as opposed to more equable 'maritime' climate having lesser difference in maximum and minimum temperature during transgression, giving rise to narrow ecospace and reduction in nannoplankton taxa (Haq, 1973).

As outlined in the previous lines, diversity values have a positive correlation with temperature – high diversity in warmer and low diversity in cooler intervals, indicating that climate (temperature) has a vital role in affecting diversity. Further, the decreasing trend in percentage of cold water species is indicative of increasing trend in temperature and vice versa and hence, gradients in diversity too, correspond with the gradients in temperature. Climatic cooling, which began around early Middle Miocene (upper *C. costata* Zone) not only affected constituent species of Assemblage 2 by gradually reducing their abundances, it also seems responsible for extinction of many species. Extinction of several taxa belonging to Assemblage 2, in the lower *D. alata* Zone, is most likely due to climatic deterioration.

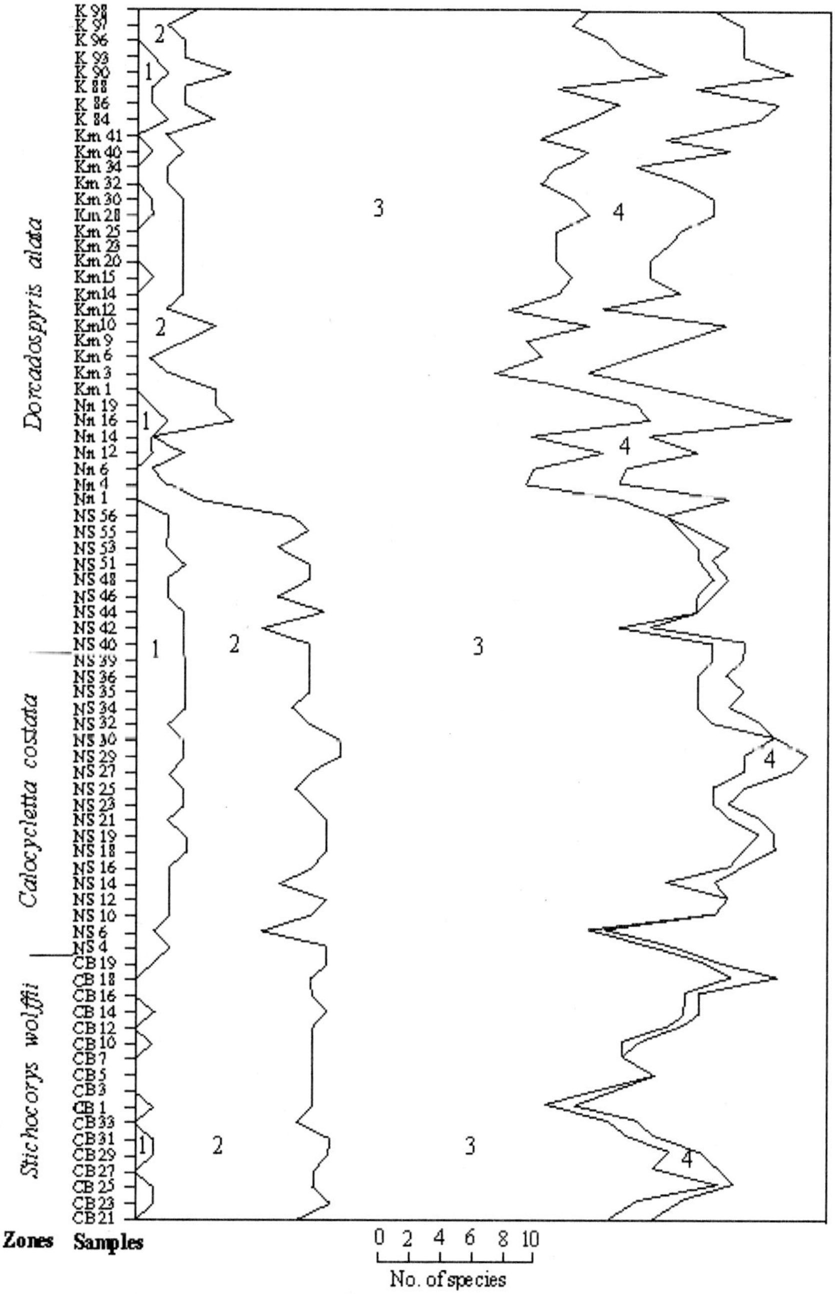

Fig. 5. Number of species of each assemblage present in the samples. Unrepresented stratigraphic intervals (as marked in Fig. 4) are not shown.

Acknowledgements

It is a great pleasure for the authors to submit this paper to the volume dedicated in honour of Professor M.S. Srinivasan for his outstanding contributions in micropaleontology. The work was supported by grants from the Ministry of Culture and Higher Education of I.R. Iran and by the Council of Scientific and Industrial Research, New Delhi, grant no. 24 (23)/ 97/ EMR – II.

References

BALSAM, W.L. and FLESSA, K.W., 1978. Patterns of planktonic foraminiferal abundance and diversity in surface sediments of the Western North Atlantic. Mar. Micropal., 3 : 279-294.

BARTOLINI, A., BAUMGARTNER, P.O. and Guex, J., 1999. Middle and Late Jurassic radiolarian paleoecology versus carbon-isotope stratigraphy. Paleogar. Paleocl. Paleoecol., 145 : 43-60.

BERGGREN, W.A., 1969. Rates of Evolution in some Cennozoic planktonic Foraminifera. Micropal., 15 : 351-365.

BERGGREN, W.A., KENT, D.V., SWISHER, C.C.III and AUBRY, M.P., 1995. A revised Cenozoic geochronology and chronostratigraphy. SEPM Spec. Publ., 54 : 129-212.

BLOW, W.H., 1969. Late Middle Eocene to Recent planktonic foraminiferal biostratigraphy. In: P. Bronnimann, H.H. Renz (eds.), Proceedings of the First International Conference on Planktonic Microfossils (Geneva, 1967), pp. 199-421.

BLUEFORD, J.R., GONZALES, J.J. and SCOY, K.V., 1990. Comparing Radiolarian and Diatom Diversity and Abundance from the Northeast Pacific. Mar. Micropal., 15 : 219-232.

BUZAS, M.A. and GIBSON, T.G., 1969. Species diversity: Benthonic foraminifera in Western North Atlantic. Science, 163 : 72-75.

CASEY, R.E., 1977. The Ecology and Distribution of Recent Radiolaria. In: Ramsay, A.T.S. (ed), Oceanic Micropaleontology, Academic Press (London) Ltd., 2 : 809-845.

CASEY, R.E., GUST, L., LEAVESLEY, A., WILLIAMS, D., REYNOLDS, R., DUIS, T. and SPAW, J.M., 1979. Ecological niches of radiolarians, planktonic foraminiferans and pteropods inferred from studies on living forms in the Gulf of Mexico and adjacent waters. Trans. Gulf Coast Assoc. Geol. Soc., 29 : 216-223.

CASEY, R.E., LEAVESLEY, A., SPAW, J.M., MCMILLEN, K. and SLOAN, J., 1981. Radiolarian species composition, density and diversity as indicators of water structure and circulation on the South Texas Shelf. Trans. Gulf Coast Assoc. Geol. Soc., 31 : 257-263.

CASEY, R.E. and SLOAN, J.R., 1971. Possible causes of diversities and extinctions of radiolarians and other microplanktons. Geol. Soc. Am. Ann. Meeting, 1971, Abstract, 763-765.

CHEN, M.H. and TAN, Z.Y., 1997. Radiolarian distribution in surface sediments of the Northern and Central South China Sea. Mar. Micropal. 32 (1-2) : 173-194.

CHENG, Y.N., 1989. Upper Paleozoic and Lower Mesozoic radiolarian assemblages from the Busuanga Islands, North Palawan Block, Philippines. Bull. Natl. Mus. Nat. Sci., Taiwan, 1 : 129-176.

CLINE, R.M. and HAYS, J.D., 1976. Investigation of late Quaternary paleoceanography and paleoclimatology. Geol. Soc. Am. Mem., 145 : 464 pp.

DANESHIAN, J., 1998. Miocene Radiolaria from Andaman–Nicobar Islands, Northeast Indian Ocean. Unpublished Ph. D. thesis, University of Delhi, pp. 1-290.

GIBSON, T.G. and BUZAS, M.A., 1973. Species diversity: Patterns in modern and Miocene foraminifera of the eastern margin of North America. Geol. Soc. Amer. Bull., 84 : 217-238.

HAQ, B.U., 1973. Transgressions, climatic change and the diversity of calcareous nannoplankton. Mar. Geol., 15 : M25-M30.

KISSLING, W., 1999. Late Jurassic radiolarians from the Antarctic Peninsula. Micropal., 45 (suppl.): 1-96.

KRASHENINNIKOV, V.A., SERNOVA, M.YA. and BASOV, I.A., 1992. Paleogene zonal scales based on planktonic foraminifers and their significance for elaboration of the paleogene stratigraphic schemes of the Pacific high latitudes. In: Centenary of Japanese micropaleontology, Ishizaki, K. and Saito. T. (eds.) Terra Scientific Publishing Company, Japan, p. 129-141.

LAZARUS, D., SPENCER – CERVATO, C., PIANKA – BIOLZI, M., BECKMANN, J.P., VON SALIS, K., HILBRECHT, H. and THIERSTEIN, H., 1995. Revised chronology of Neogene DSDP Holes from the world ocean. Ocean Drilling Program, College Station, Texas, Tech. Note 24.

LI, J. and XU, Y.L., 1998. Late Eocene – Quaternary Radiolaria from the northeastern Philippine. Sea and their significance for paleoceanography. Xiandai Dizhi = Geoscience, 12 : 336-343.

McGOWAN, J.A. and WALKER, P.W., 1985. Dominance and diversity maintenance in an oceanic ecosystem, Ecol. Monogr., 55 : 103-118.

McMILLEN, K.J. and CASEY, R.E., 1978. Distribution of living polycystine radiolarians in the Gulf of Mexico and Caribbean Sea, and comparison with the sedimentary record. Mar. Micropal., 3 : 121-145.

NISHIMURA, A., NAKASEKO, K. and OKUDA, Y. 1997. A few costal water radiolarian assemblage recovered from sediment samples from the Antartic Ocean. Mar. Micropal., 30 (1-3) : 29-44.

ROMINE, K. and MOORE, T.C., Jr., 1981. Radiolarian assemblage distributions and paleoceanography of the eastern equatorial Pacific Ocean during the last 127,000 years. Paleogr. Paleocl. Paleoecol., 35 : 281-314.

RUDDIMAN, W.F., 1968. Recent planktonic foraminifera : Dominance and Diversity in North Atlantic surface Sediments. Science, 164 : 1164-1167.

RUTHERFORD, S., D'HONDT, S. and PRELL, W., 1999. Environmental controls on the geographic distribution of zooplankton diversity. Nature, 400 : 749-753.

SANFILIPPO, A., WESTBERG – SMITH, M.J. and RIEDEL, W.R., 1985. Cenozoic radiolaria. In : Bolli, H.M. *et al.* (Eds.), Plankton Stratigraphy : Cambridge, Cambridge University Press, p. 631-712.

SASHIDA, K. and KURIHARA, T., 1999. Recent radiolarian faunas in the surface water off the coast of Shimoda, Izu Peninsula, Japan. Sci. Repts. Univ. Tsukuba Inst. Geosci. Section B : Geol. Sci., 20 : 115-144.

SEIBOLD, E. and BERGER, W.H., 1993. The sea floor. Springer–Verlag, Berlin, 335pp.

SHACKLETON, N.J., BALDAUF, J.G., FLORES, J.A., IWAI, M., MOORE, T.C., Jr., RAFFI, I., VINCENT, E., 1995. Biostratigraphic summary for Leg 138, Ocean Drilling Program, Sci. Results 138 : 517-536.

SHARMA, V. and DANESHIAN, J., 2003. Early Neogene radiolarian changes in the Northeast Indian Ocean and their paleoceanographic implications : Evidence from the Andaman – Nicobar Islands, p. 126-158. Proc. 8th Int. Cong. Pac. Neogene Strat., Chiang Mai, Thailand.

SHARMA, V. and RAM, M.P., 2003. Early to Middle Miocene radiolarian assemblages and biostratigraphy, Andaman Islands, Northeast Indian Ocean. Jour. Pal. Soc. India, 48 : 1-39.

SHARMA, V. and SHARMA, G.K. 1989. Late Miocene to Early Pliocene radiolarian biostratigraphy of Neill Island, Andaman Sea. Jour. Geol. Soc. India, 34 : 76-82.

SHARMA, V. and SINGH, S., 1993. Radiolarian biostratigraphy of Early Pliocene sequences, Car Nicobar Island, Northeast Indian Ocean. Jour. Geol. Soc. India, 41 : 199-213.

SHARMA, V. and SINGH, S., 1997 a. Early Neogene Radiolaria from Andaman–Nicobar Islands, Indian Ocean : paleoclimatic implications, p. 80-102. Proc. Int. seminar on Neogene Evolution of Pacific Ocean Gateways, IGCP – 355, Japan.

SHARMA, V. and SINGH, S. 1997 b. Late Neogene Radiolarian events in Andaman Nicobar Islands, Northeast Indian Ocean. Micropal. 43(1) : 41-50.

SHARMA, V., SINGH, SURENDER and RAWAL, NEERU., 1999. Early Middle Miocene Radiolaria from Nicobar Islands, Northeast Indian Ocean. Micropal. 45(3) : 251-277.

SHARMA, V., SRINIVASAN, M.S., and MAHAPATRA, A.K., 1993. Early Miocene radiolarian and planktonic foraminiferal biostratigraphy, North Passage Island, Andaman Sea. Jour. Geol. Soc. India, 42 : 154-162.

SINGH, O.P., SRINIVASAN, M.S. and SHARMA, V., 2000. Early Neogene multiple microfossil biostratigraphy John Lawrence Island, Andaman Sea. Micropal. 46(4) : 343-352.

SINGH, P., 1981. A note on the Early Pliocene Radiolaria from Neill Island, South Andaman, India. Geoscience Jour., 2(1) : 97-107.

SRINIVASAN, M.S. 1988. Late Cenozoic sequences of Andaman Nicobar Islands: their regional significance and correlation. Ind. Jour. Geol., 60(1) : 11-34.

SRINIVASAN, M.S. and DAVE, A., 1985. Middle Miocene sediments of Nancowry and Kamorta Islands : their biostratigraphic and climatic record. Bull. Geol. Min. Met. Soc. India., 52 : 319-334.

SRINIVASAN, M.S., LOMBARI, G. and DAVE, A., 1983. Early Miocene Planktonic Foraminiferal and Radiolarian zonation, Colebrook Island, Andaman Sea. Jour. Geol. Soc. India, 24: 1-18.

STEHLI, F.G., 1965. Paleontologic technique for defining ancient ocean currents. Science, 148 : 943-946.

STEHLI, F.G., DOUGLAS, R.G. and NEWELL, N.D., 1969. Generation and maintenance of Gradients in Taxonomic Diversity. Science, 164 : 947-949.

SWANBERG, N.R. and BJORKLUND, K.R., 1987. Radiolaria in the plankton of some fjords in western and northern Norway : the distribution of species. Sarsia, 72 : 231-244.

VINCENT, E. and KILLINGLEY, J.S., 1985. Oxygen and carbon isotope record for the Early and Middle Miocene in the central equatorial Pacific, DSDP leg 85. In : Mayer, L. *et al* . (eds.), Init. Rept. DSDP 85, Washington, U.S. Govt. Printing Office, pp. 749-769.

VINCENT, E., KILLINGLEY, J.S. and BERGER, W.H., 1985. Miocene Oxygen and Carbon isotope stratigraphy of the tropical Indian Ocean. In : Kennett, J.P. (ed.), The Miocene Ocean : Paleoceanography and biogeography. Geol. Soc. Amer. Mem. 163, pp. 103 -130.

VINCENT, E., SHACKLETON, N.J. and HALL, M.A., 1991. Miocene Oxygen and carbon isotope stratigraphy of planktonic foraminifers at sites 709 and 758, tropical Indian Ocean. In : Weissel, J. *et al.* (eds.), Ocean Drilling Program, Sci. Results 121, College Station, Texas, pp. 241-252.

VISHNEVSKAYA, V., 1992. Significance of Mesozoic radiolarians for tectonostratigraphy in Pacific Rim terranes of the former USSR. In : Aitchison, J.C. and Murchey, B.L. (eds.), Significance and application of Radiolaria to terrane analysis. Paleogr. Paleocl. Paleoecol. Special Issue, pp. 23-39.

VISHNEVSKAYA, V.S., 1997. Development of Paleozoic – Mesozoic Radiolaria in the Northwestern Pacific Rim. Mar. Micropal., 30 (1-3): 79-95.

WILLIAMS, D.F. and JOHNSON, W., 1975. Diversity of recent planktonic foraminifera in the southern Indian Ocean and late Pleistocene paleotemperatures. Quaternary Res., 5 : 237-250.

YAMAMOTO, K., 1992. Composition and diversity of radiolarian fossil assemblages : Case study on Upper Cretaceous Futakawa Formation, Sotoizumi Group. In : Sakai T. and Aita, Y. (eds.), Proceedings of the third Radiolarian Symposium, News of Osaka Micropaleontologists. Spec. vol. 8 : 7- 88 (in Japanese with abstract in English).

Planktic marine Diatom events of the Oligocene and earliest Miocene in the Equatorial Pacific Ocean

John A. Barron

U.S. Geological Survey, MS 910, Menlo Park, CA 94025, USA.

Introduction

Over 35 years of deep ocean coring by the Deep Sea Drilling Project and its successor, the Ocean Drilling Program, has resulted in the collection of hundreds of diatom-bearing sediment cores from virtually all ocean regions. Detailed reports on the occurrences of diatom taxa have been completed for many of these cores, a good portion of which had been well dated by magnetostratigraphy and/or by correlation with other microfossil groups. As reviewed by Barron (2003), compilations of North Pacific, equatorial Pacific, and Southern Ocean biochronolgies for the past 18 million years have allowed the times of appearance and disappearance of numerous diatom taxa to be accurately compared between regions with an accuracy of 100,000 years or less. In this manner, the age of evolutionary appearance and the region of origin of various taxa can be distinguished from migrational appearances.

Oligocene and earliest Miocene diatom biostratigraphy have lacked such refinement. While calibration of Southern Ocean diatom events with magnetostratigraphy has been established for much of the Oligocene (Baldauf and Barron, 1991; Harwood and Maruyama, 1992; Ramsay and Baldauf, 1999), no direct correlation of low latitude diatom biostratigraphies for the Oligocene and earliest Miocene have been made. Refinement of Oligocene and earliest Miocene diatom biostratigraphy in the North Pacific is even more severely limited by the lack of complete stratigraphic sequences (Gladenkov and Barron, 1995).

The pioneering paper of Jousé (1974) compiled the ranges of key diatom taxa in Oligocene and Miocene deep sea sediments and proposed a series of assemblages zones based on distinctive species.

Fenner (1984a) combined the results of diatom biostratigraphy in nine low latitude Atlantic and Pacific DSDP sites and proposed a much more refined late middle Eocene through Oligocene low-latitude diatom zonation, emphasizing stratigraphic markers that were easily identifiable, somewhat resistant to dissolution, and generally consistent in their occurrences. Subsequent studies (Fenner, 1984b, 1985; Barron, 1985a; Fenner and Mikkelsen, 1990; Fourtanier, 1991) have proven Fenner's (1984a) diatom zonation to be widely applicable in low latitude regions of the world's oceans. Prior to ODP Leg 199, the primary correlation of these diatom zones was first with calcareous nannofossil zones (Fenner, 1984 a,b; Fenner and Mikkelsen, 1990), with secondary calibration of calcareous nannofossil zones with magnetostratigraphy (Barron and Baldauf, 1995).

This paper will review biostratigraphically useful diatom events in the Oligocene and lowermost Miocene sediments (*circa* 33.5-21.5 Ma) of the equatorial Pacific. Calibration of Oligocene and earliest Miocene diatom biostratigraphy with magnetostratigraphy at ODP Site 1220 (10°10.601'N, 142°45.491'W, water depth 5217.9 m) by Barron *et al.* (in press) allows accurate ages to be assigned to numerous diatom datum levels. Compilations by Fenner (1984a, 1985) and Barron (1983) of Oligocene and early Miocene diatom biostratigraphy provide a framework for regional applications.

Diatom Zonation

The diatom zonation used by Barron *et al.* (in press) for the latest Eocene, Oligocene, and earliest Miocene of equatorial ODP Site 1220 is reviewed below. Ranges of stratigraphically important diatoms and this diatom zonation are correlated with the paleomagnetic stratigraphy of ODP Site 1220 in Figure 1.

Baxteriopsis brunii Partial Range Zone
Author: Fenner (1984a)
Base: first continuous occurrence of specimens of *Baxteriopsis brunii* (van Huerck) Karsten that contain 4 to 7 areolae in 10 μm.
Top: first occurrence of *Coscinodiscus excavatus*

Coscinodiscus excavatus Range Zone
Author: Barron (1985a)
Base: first occurrence of *Coscinodiscus excavatus*
Top: last occurrence of *Coscinodiscus excavatus*
Jousé (1974) proposed a *Coscinodiscus excavatus* var. *quadriocellata* Assemblage Zone which is essentially equivalent to this zone. Fenner (1984a) chose to separate this early Oligocene interval into two zones: the *Coscinodiscus excavatus* Partial Range Zone and the *Cestodiscus reticulatus* Interval Zone (incorrectly termed a range zone). This separation was based on the recognition of *Cestodiscus reticulatus* Fenner. Barron (1985a) remarked that it was very difficult to distinguish *C reticulatus* from *C. robustus* Jousé (= *C. gemmifer* Castracane) and chose to not to recognize the *C. reticulatus* zone of Fenner (1984a).

Cestodiscus trochus Interval Zone
Author: Barron *et al.* (in press)
Base: last occurrence of *Coscinodiscus excavatus*
Top: first occurrence of *Rocella vigilans*
The *Cestodiscus reticulatus* Interval Zone of Fenner (1984a) extends above the last occurrence of *Coscinodiscus excavatus* to the first occurrence of *Rocella vigilans.* Although, Barron (1985a) chose to include this interval in his *Rocella vigilans* Zone, it seems more appropriate to recognize the first occurrence of *R. vigilans* as the base of a *R. vigilans* Partial Range Zone, in the manner used by Fenner (1984a). Because *Cestodiscus trochus* (= *C. mukinae* Jousé, 1974) is consistently present in this interval, it is appropriate to name this the *Cestodiscus trochus* Interval Zone.

The first occurrence of *Cavitatus miocenicus* appears to coincide with the last occurrence of *Coscinodiscus excavatus* and the base of the *Cestodiscus trochus* Interval Zone at ODP Site 1220, a relationship that is seemingly supported at DSDP 77 by Fenner's (1984b) occurrence chart. However, relatively poor preservation in these sections makes this relationship uncertain. *Skeletonemopsis barbadense* and *Cestodiscus robustus* have their last occurrences in the lower part of the *C. trochus* Zone.

Rocella vigilans Partial Range Zone
Author: Fenner (1984a)
Base: first occurrence of *Rocella vigilans*

Top: first occurrence of *Bogorovia veniamini*

As proposed by Fenner (1984a) the top of Subzone a and the base of Subzone b is recognized by the first occurrence of *Rossiella symmetrica.*

At ODP Site 1220, *Cavitatus jouseanus, C. rectus, Coscinodiscus rhombicus* and *Kozloviella minor* all have their first occurrences in the upper part of Subzone b of Fenner (1984a). Barron *et al.* (in press) found it appropriate to split Subzone b into an additional subzone. The base of their new Subzone c and the top of their modified Subzone b is placed at the first occurrence of *Cavitatus jouseanus. Cestodiscus trochus* (= *C. mukinae* of Jousé,1974) has its last occurrence in Subzone c.

Bogorovia veniamini Partial Range Zone

Author: Fenner (1984a)

Base: first occurrence of *Bogorovia veniamini*

Top: first occurrence of *Rocella gelida*

Cestodiscus kugleri is restricted to the *B. veniamini* Partial Range Zone at ODP Site 1220. *Kozloviella minor* has its last occurrence in the lower part of the zone, and *Rossiella symmetrica* has its last occurrence in the upper part of the zone.

Rocella gelida Range Zone

Author: sense used by Gombos (1983), essentially equivalent to the *Rocella gelida* Partial Range Zone of Barron (1983).

Base: first occurrence of *Rocella gelida*

Top: last occurrence of *Rocella gelida*

Coscinodiscus lewisianus var. *rhomboides*, which has been mistakenly called *C. lewisianus* var. *similis* (see Barron 1985a), has basically the same range as *R. gelida* and is a convenient secondary marker for the *R. gelida* Zone *Rocella schraderi* is restricted to the middle part of the zone. *Craspedodiscus barronii* (= *C. coscinodiscus* of Jousé, 1974 and Gombos,1983) has its first occurrence in the lowermost part of the *R. gelida* Partial Range Zone. *Rocella vigilans* last occurs in the upper part of the zone.

Rossiella fennerae Interval Zone.

Author: Barron *et al.* (in press), essentially equivalent to the *Rossiella paleacea* Zone of Barron (1983).

Base: last occurrence of *Rocella gelida.*

Top: last occurrence of *Bogorovia veniamini*

Yanagisawa (1995a) proposed that the base of the *Rossiella paleacea* Concurrent Zone of Barron (1983) should be redefined at the first abundant occurrence of *Rossiella fennerae* Yanagisawa, arguing that Barron's (1983) identification of *Rossiella paleacea* actually included *R. fennerae.* Previously, Barron (1983, 1985a) had remarked that the last occurrence of *Rocella gelida* coincided with the base of his *Rossiella paleacea* Zone at DSDP Sites 71 and 574. It is proposed here to redefine the base of Barron's (1983) *Rossiella paleacea* Zone at the last occurrence of *Rocella gelida.* Taxonomic changes by Yanagisawa (1995a) to Barron's (1983) concept of *Rossiella paleacea*, however, that such a modified *Rossiella paleacea* Zone of Barron (1983) should be renamed the *Rossiella fennerae* Interval Zone.

The last occurrence of *Thalassiosira primalabiata* was used by Barron (1983) as the marker for the top of Subzone a and base of his Subzone b of his *R. paleacea* Zone. Barron (1985a) argued that the first occurrence of *Actinocyclus hajosiae* was isochronus between DSDP Sites 574 and 575 and suggested that it would be a more consistent marker for the Subzone a/Subzone b boundary. At ODP Site 1220 the first occurrence of *A. hajosiae* also coincides with the last sporadic occurrence of

T. primalabiata; therefore, the first occurrence of *A. hajosiae* has been used by Barron *et al.* (in press) to redefine the top of Subzone a and the base of Subzone b of the *Rossiella fennerae* Interval Zone.

The first occurrences of *Rossiella fourtanierae*, *Thalassiosira praefraga*, and *T. spumellaroides* are present in Subzone a at ODP Site 1220. However, Barron (1983) reports that *T. praefraga* (identified as *T. spinosa* Schrader) ranges into the lower part of the *Rocella gelida* Zone at DSDP Site 71.

The top of Subzone b and the base of Subzone c of the *Rossiella paleacea* Zone of Barron (1983) (and therefore, the *Rossiella fennerae* Interval Zone of Barron *et al.* (in press)) is defined by last occurrence of *Azpeitia oligocenica.* At ODP Site 1220, however, the last occurrence of *A. oligocenica* falls within an interval of dissolution (Fig. 1).

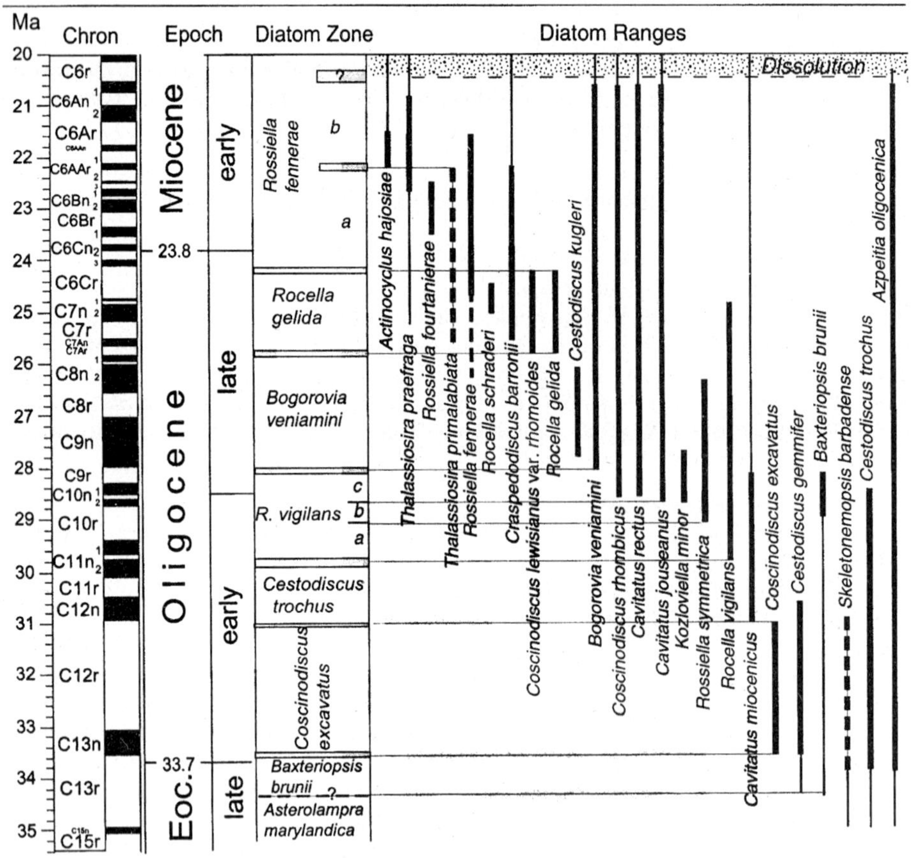

Fig. 1. Diatom zones and ranges of selected diatoms in the Oligocene and earliest Miocene of the equatorial Pacific correlated to paleomagnetic stratigraphy after Barron *et al.* (in press). Thick range lines depict ranges at ODP Site 1220; thick, dashed line refers to sporadic occurrences; thin lines show extended ranges of these diatom taxa at other sites after Fenner (1984a), Barron (1983, 1985b) and Yanagisawa (1995 a,b). Paleomagnetic chrons and ages after Cande *et al.* (1995); geologic time scale after Berggren *et al.* (1995).

Taxonomy & age Ranges of Diatoms

The taxonomy of Oligocene and earliest Miocene diatoms can be confusing, as new taxonomic studies such as those of Yanagisawa (1995 a,b) and Strelnikova *et al.* (2001) must be considered along with an increasing number of papers by various authors on sediments collected by Ocean Drilling Program. Consequently, an important part of this study will be to refer to updated diatom taxonomy, reference good illustrations of species concepts, and suggest synonyms of various taxa. The following section lists important Oligocene and earliest Miocene diatoms, reference to illustrations and their taxonomy, and their updated age ranges according to Barron *et al.* (in press) and other studies. Ages followed by a question mark (?) are suspect due to sporadic occurrences and/or dissolution. The ranges of many of these diatoms in terms of paleomagnetic stratigraphy at ODP Site 1220 are shown on Fig. 1. Paleomagnetic chron designations and ages are after Cande *et al.* (1995) and are consistent with the Berggren *et al.* (1995) time scale.

Actinocyclus hajosiae Barron, 1983, p. 504, pl. I, figs. 3, 4; pl. V, fig. 13; pl. VI, fig. 3.
First Occurrence: 22.06-22.31 Ma (C6AAn.1r)(Barron *et al.*, in press)
Last Occurrence: *circa* 19.3 Ma (age updated from Barron, 1985a)

Azpeitia oligocenica (Jousé) Sims in Sims *et al.*, 1989, p. 302, pl. 2, figs. 1-3; pl. 3, figs. 8, 9. Synonym: *Coscinodiscus oligocenicus* Jousé sensu Barron, 1983, p. 512, pl. 3, figs. 8 and 11.
First Occurrence:>33.79 Ma (Barron *et al.*, in press)
Last Occurrence: >20.72 Ma? (C6n.1n) (Barron *et al.*, in press); 20.3 Ma (Barron and Baldauf, 1995)

Azpeitia praenodulifera (Barron) Sims et Fryxell in Sims *et al.*, 1989, p. 298, pl. 1, figs. 8-13, pl. 3, fig. 11. Synonym: *Coscinodiscus praenodulifer* Barron, 1983, p. 511, pl. III, figs. 9, 10; pl. I, fig. 8.
First Occurrence: 21.12-21.44 Ma (C6An.2n-C6An.2r) (Barron *et al.*, in press)
Last Occurrence: 15.4 Ma (Barron, in press)

Baxteriopsis brunii (van Heurck) Karsten; Fenner. 1985, p. 727, pl. 10, figs. 7, 8.
First Occurrence: *circa* 34.4 Ma (Barron and Baldauf. 1995)
Last Occurrence: 28.12-28.35 Ma (C9n.1r-C10n.1n) (Barron *et al.*, in press)

Bogorovia barronii Yanagisawa 1995b, p. 31, figures 4, 11-14; 5, 3-5.
First Occurrence: 22.40-22.67 Ma (C6AAr.1r-C6AAr.2r) (Barron *et al.*, in press)
Last Occurrence: 21.12-21.44 Ma (C6An.2n-C6An.2r) (Barron *et al.*, in press)

Bogorovia veniamini Jousé, 1974 emend. Yanagisawa. 1995b, p. 25, figures 4, 4-10; 8, 1-9.
First Occurrence: 28.12-28.35 Ma (Barron *et al*,. in press)
Last Occurrence: >20.72 Ma? (C6n.1n) (Barron *et al.*, in press); 19.7 Ma (Barron and Baldauf 1995)

Cavitatus jouseanus (Sheshukova-Poretzkaya) Williams; Akiba *et al.*, 1993, p. 20, figs. 6 and 19-20; Gladenkov and Barron, 1995, p. 31, pl. 5, figs. 3, 15, 18, 19, 22, 23.
First Occurrence: 28.65-28.82 Ma (C10n.1r-C10n.2r) (Barron *et al.*, in press)
Last Occurrence: 11.41 Ma (Barron,. in press)
Remarks: see remarks for *Cavitatus miocenicus*

Cavitatus rectus Akiba et Hiramatsu in Akiba *et al.*, 1993, p. 28, pl. 6, figs. 7-15; Gladenkov and Barron, 1995, p. 31, pl. 5, figs.7-10. Synonym: *Thalassionema hirosakiensis* (Kanaya) Schrader sensu Schrader, 1976, p. 636, pl. 1, figs. 14-16.

First Occurrence: 28.51-28.65 Ma (C10n.1n- C10n.1r) (Barron *et al.*, in press)
Last Occurrence: >20.72 Ma? (C6n.1n) (Barron *et al.*, in press)

Cavitatus miocenicus. (Schrader) Akiba et Yanagisawa; Gladenkov and Barron (1995), pl. 5, figs. 1, 2, 27. Synonym: *Synedra miocenica* Schrader, 1976, p. 636, pl. 1, fig. 1; *Cavitatus jouseanus* (Sheshukova-Poretzkaya) Williams sensu Harwood and Bohaty, 2001, p. 324, pl. 1, fig. 1, 2?; *Synedra* sp. of Jousé, 1974, pl. 4, fig. 23.

First Occurrence: 30.99-31.07 Ma (C12r) (Barron *et al.*, in press); Harwood and Maruyma's (1992) age of 30.9 Ma for the first occurrence of *C. jouseanus* (*S. jouseana*) in the Southern Ocean probably refers to the first occurrence of this taxon (see remarks below).

Last Occurrence: 16.3 Ma (Barron, in press)

Remarks: Harwood and Bohaty (2001) remark that "Specimens of *Cavitatus jouseanus* present in CRP-3 differ from *C. jouseanus* s.s. in that they are smaller and more lightly silicified with narrow, tapered ends, rather than broadly-rounded ends. This morphology appears to be characteristic of 'early forms' of *C. jouseanus*." Schrader (1976) shows *Synedra miocenica* ranges below *S. jouseana* at Southern Ocean Site 278. Gladenkov and Barron (1995) show a similar relationship between *C. miocenicus* and *C. jouseanus* at ODP Site 884 in the North Pacific.

Cestodiscus convexus Castracane, 1886, p. 128, pl. 7, fig. 6; Strelnikova *et al.*, 2001, pl. VII, fig. 43; pl. X, figs. 49, 50; pl. XI, figs. 56-59.
First Occurrence: 33.27-33.39 Ma (C13n) (Barron *et al.*, in press)
Last Occurrence: 30.73-30.99 Ma (C12n-C12r) (Barron *et al.*, in press)

Cestodiscus robustus Jousé, 1974, p. 345, pl. 1, figs. 14-15; Barron, 1985b, pl. 1, fig. 1. Synonyms: *Coscinodiscus superbus* Hardman sensu Fenner, 1978, p. 516, pl. 13, figs. 1-5, pl. 14, figs. 1-4; sensu Gombos and Ciesielski, 1983, p. 601, pl. 4, figs. 1-8. *Cestodiscus gemmifer* Castracane sensu Fenner and Mikkelsen, 1990, p. 441, pl. 1, fig. 3. *Cestodiscus trochus* Castracane sensu Strelnikova *et al.*, 2001, p. 79, figs. 51-54, 60-63; *Cestodiscus parmula* Castracane sensu Radionova, 1987, p. 149, pl. 2, figs. 6-8. Remarks: The marginal spines are often not preserved.

First Occurrence: 33.50-33.60 Ma (C13n-C13r)(Barron *et al.*, in press); ranges to base of *Baxteriopsis brunii* Partial Range Zone (*circa* 34.4 Ma) according to Fenner (1984b).

Last Occurrence: 30.56-30.73 Ma (C12n) (Barron *et al.*, in press)

Cestodiscus kugleri Lohman, 1974, p. 340, pl. 4, figs. 4, 5 and 8; Fourtanier, 1991, pl. 1, fig. 6; Gladenkov and Barron, 1995, p. 31, pl. 3, Figs. 2-6, 7?. Synonym: *Cestodiscus* aff. *trochus* Castracane sensu Fenner, 1984b, p. 1270, pl. 1, fig. 4; *Cestodiscus* sp. 6 of Schrader, 1976, pl. 12, fig. 4.
First Occurrence: 27.60-27.92 Ma (C9n) (Barron *et al.*, in press)
Last Occurrence: 26.01-26.10 Ma (C8n.2n) (Barron *et al.*, in press)

Cestodiscus reticulatus Fenner 1984a, p. 331, pl. 1, fig. 10. Remarks: These forms closely resemble *C. gemmifer* (or *C. robustus* of Jousé, 1974) except for the reticulate pattern in the valve's center, and it is not clear that they should be separated.

Cestodiscus trochus Castracane, 1886, p. 123, pl. 7, figs. 1 and 3. Synonym: *Cestodiscus mukhinae* Jousé, 1974, p. pl. 1, figs. 1, 2, 3-5? Both Fenner (1985) and Harwood and Maruyama (1992) suggested this synonymy. Remarks: Appears to include flat valve forms tabulated as *C. antarcticus* Fenner by Strelnikova *et al.* (2001), p. 79, fig. 55, 64-67.
First Occurrence: >33.79 Ma (Barron *et al.*, in press)
Last occurrence: 28.40-28.51 Ma (C10n.1n) (Barron *et al.*, in press)

Cestodiscus trochus with marginal spines: Synonym: *Cestodiscus pulchellus* Greville sensu Schrader, 1976, pl. 13, fig. 5.
First Occurrence:33.16-33.27 Ma (C13n) (Barron *et al.*, in press)
Last Occurrence: 29.33-29.44 Ma (C10n.2r-C11n.1n) (Barron *et al.*, in press)

Cestodiscus sp. 2 sensu Fenner 1981, p. 89, pl. 14, figs. 3-4; Strelnikova *et al.*, 2001, pl. VI, fig. 34; pl. VIII, figs. 39-42; pl. IX, figs. 44-48. Synonym: *Cestodiscus pulchellus* var. *novazealandica* Grove sensu Harwood and Maruyama, 1992, p. 701, pl. 1, fig. 4. Remarks: does not include the flat forms figured as *C. pulchellus* var. *novazealandica* Grove (1890) in Schmidt, 1875-1959, pl. 148, fig. 7, pl. 163, figs. 8, 10.
First Occurrence: >33.79 Ma
Last Occurrence: 33.05-33.16 (C12r-C13n) (Barron *et al.*, in press)

Coscinodiscus excavatus Greville in Pritchard; Fenner, 1985, p. 728, pl. 7, figs. 26, 28; Barron, 1985a, pl. 1, fig. 8; Strelnikova *et al.*, 2001, plate XVII, figs. 92-97.
First Occurrence: 33.50-33.60 Ma (C13n-C13r) (Barron *et al.*, in press); 33.6 Ma (Barron and Baldauf, 1995)
Last Occurrence: 30.99-31.07 Ma (C12r) (Barron *et al.*, in press)

Coscinodiscus excavatus var. *semilunaris* Grunow; Fenner, 1978, p. 514, pl. 10, figs. 13,1 4.
First Occurrence: 32.30-32.37 Ma (C12r) (Barron *et al.*, in press)
Last Occurrence: 31.53-31.69 Ma (C12r) (Barron *et al.*, in press):

Coscinodiscus hajosiae Fenner 1984a, p. 331, pl. 2, fig. 1. Synonym: *C. spiralis* sensu Hajós 1976, p. 826, pl. 7, figs. 1-3.
First Occurrence: >33.79 Ma (Barron *et al.*, in press)
Last Occurrence: 32.30-32.37 Ma? (C12r) (Barron *et al.*, in press)

Coscinodiscus lewisianus Greville; Barron, 1985b, p. 781, pl. 9, fig. 10.
First Occurrence: 27.12-27.44 Ma? (C9n) (Barron *et al.*, in press)
Last Occurrence: 12.93 Ma (Barron, in press).

Coscinodiscus lewisanus var. *levis* (Jousé) Harwood et Maruyama, 1992, p. 702, pl. 3, figs. 12-15; Gladenkov and Barron, 1995, pl. 1, fig. 14. Synonym: *Actinocyclus levis* Jousé, 1974, p. 353, pl. 5, fig. 2.
First Occurrence: 28.51-28.65 Ma (C10n.1n-C10n.2n) (Barron *et al.*, in press)
Last Occurrence: 25.70-26.01 Ma (C7An.1r-C8n.2n) (Barron *et al.*, in press)

Coscinodiscus lewisianus var. *rhomboides* Barron, 1985b, p. 443, pl. 7, figs. 2, 3. Synonym: *Coscinodiscus lewisianus* var. *similis* Rattray sensu Jousé 1974, pl. 3, figs. 6, 7; sensu Barron (1983), p. 512, pl. 1, fig. 8; sensu Gombos (1983), pl. 3, figs. 3-6.
First Occurrence: 25.70-26.01 Ma (C7An.1r-C8n.2n) (Barron *et al.*, in press)
Last Occurrence: 24.13-24.32 Ma (C6Cn.3r) (Barron *et al.*, in press)
Remarks: Barron (1985a) noted that *C. lewisianus* var. *similis* Rattray is restricted to the later part of the early Miocene and is distinguished from the late Oligocene diatom *C. lewisianus* var. *rhomboides* by its being less rhombic in shape and its possession of a broader, finely areolated margin.

Coscinodiscus rhombicus Castracane; Gombos, 1983, pl. 3, fig. 1; Fenner 1985, p. 729, pl. 7, figs. 1-4.
First Occurrence: 28.51-28.65 Ma (C10n.1n-C10n.2n) (Barron *et al.*, in press)

Last Occurrence: *circa* 18.5 Ma (age updated from Barron, 1983)

Craspedodiscus barronii Bukry, 1984, p. 464, pl. 2, figs. 13-15, pl. 3, figs. 1-5; Fourtanier, 1991, pl.3, fig. 7. Synonym: Oligocene and early Miocene forms have been recorded as *C. coscinodiscus;* Gombos, 1983, pl. 4, figs. 1-3.

First Occurrence: 25.53-25.70 Ma (C7A.1n-C7An.1r) (Barron *et al.*, in press)

Last Occurrence: *circa* 16.4 Ma (Barron, unpublished notes)

Kozloviella minor Jousé, 1974, p. 352, pl. 4, fig. 18; Gladenkov and Barron, 1995, pl. 1, figs. 13, 20.

First Occurrence: 28.65-28.82 Ma ((C10n.2n-C10n.2r) (Barron *et al.*, in press)

Last Occurrence: 27.60-27.92 Ma (C9n) (Barron *et al.*, in press)

Kozloviella sp. cf. *K. pacifica* Jousé, 1974, p. 352, pl. 4, fig. 17; Barron *et al.*, in press, pl. 1, fig. 9.

First Occurrence: 28.40-28.51 Ma (C10n.1n) (Barron *et al.*, in press)

Last Occurrence: 28.12-28.35 Ma (C9r-C10n.1n) (Barron *et al.*, in press)

Lisitzinia ornata Jousé; Fenner 1985, p. 734, pl. 10, fig. 11; Gladenkov and Barron, 1995, pl. 1, figs. 15-18.

First Occurrence: 26.10-26.19 Ma? ((C8n.2n) (Barron *et al.*, in press)

Last Occurrence: 25.53-25.70 Ma? (C7A.1n-C7An.1r) (Barron *et al.*, in press)

Remarks: This high latitude taxon apparently evolved in the Southern Ocean (First occurrence = *circa* 28.0 Ma), migrated across the equator, and persisted in the North Pacific into the earliest Miocene (Gladenkov and Barron, 1995).

Rocella gelida (Mann) Bukry; Gombos, 1983, pl. 1, figs. 1-6; Barron, 1985b, pl. 12, fig. 16; Gladenkov and Barron, 1995, pl. 1, figs. 3, 7.

First Occurrence: 25.70-26.01 Ma (C7A.1r-C8n.2n) (Barron *et al.*, in press)

Last Occurrence: 24.13-24.32 Ma (C6Cn.3r) (Barron *et al.*, in press)

Rocella schraderi Bukry, Gombos, 1983, pl. 1, figs. 13-16. Synonym: *R. gelida* var. *schraderi* (Bukry) Barron; Barron, 1985b, pl. 12, fig. 15.

First Occurrence: 24.96-25.05 Ma (C7n.2n) (Barron *et al.*, in press)

Last Occurrence: 24.13-24.32 Ma (C6Cn.3r) (Barron *et al.*, in press)

Rocella princeps (Jousé) Fenner 1984a, p. 335; Gombos 1983, pl. 2, figs. 13-16.

First Occurrence: 26.01-26.10 Ma (C8n.2n) (Barron *et al.*, in press)

Last Occurrence: 24.80-24.84 Ma (C7n.1r-C7n.2n) (Barron *et al.*, in press)

Rocella vigilans Fenner, 1984a, p. 333, pl. 1, fig. 11; Gombos, 1983, pl. 1, figs. 7-12; Fenner, 1985, pl. 7, figs. 14 and15.

First Occurrence: 29.74-29.98 Ma (C11n.1r-C11n.2n) (Barron *et al.*, in press)

Last Occurrence: 24.74-24.80 Ma (C7n.1n-C7n.1r) (Barron *et al.*, in press)

Remarks: Although Harwood and Maruyama (1992) recognize a small and a large form, specimens from ODP Site 1220 display a range in sizes and morphology and are not clearly separable.

Rossiella fennerae Yanagisawa 1995a, p. 8, pl. 4, figs. 7-10; pl. 7, figs. 1-8. Synonyms: *Rossiella symmetrica* Fenner, 1984a, pl. 1, figs. 3, 4 (non fig. 2); *Rossiella paleacea* (Grunow) Desikachary et Maheswari sensu Barron (1983), pl. 5, fig. 4.

First Occurrence: 26.29-26.39 Ma (C8n.2n) (Barron *et al.*, in press)

First Consistent Occurrence: 24.62-24.74 Ma (C6Cn.3r-C7n.1n) (Barron *et al.*, in press)

Last Occurrence: 21.44-21.81 Ma (C6An.3r-C6AAn) (Barron *et al.*, in press)

Rossiella fourtanierae Yanagisawa, 1995a, p. 9, pl. 4, figs. 11-14; pl. 9, figs. 1-6. Synonym: *Rossiella paleacea* (Grunow) Desikachary et Maheswari, robust form of Barron (1983), p. 512, pl. 5, fig. 6.
First Occurrence: 23.40-23.64 Ma (C6Cn.1n-C6Cn.1r) (Barron *et al.*, in press)
Last Occurrence: 22.40-22.67 Ma? (C6AAr.1r-C6Bn.1n) (Barron *et al.*, in press)

Rossiella paleacea (Grunow) Desikachary & Maheshwari; Yanagisawa 1995a, p. 11, pl. 4, figs. 15-20; pl. 12, figs. 1-6.
First Occurrence: 21.44-21.81 Ma (Barron *et al.*, in press)
Last Occurrence: 7.6 Ma (Barron, in press)

Rossiella symmetrica Fenner 1984a, p. 333, pl. 2, figs. 2 (non pl. 1, figs. 3, 4; Fenner 1985, p. 737, pl. 7, figs. 5-7; Yanagisawa, 1995a, p. 5, pl. 4, figs. 1-6.
First Occurrence: 29.05-29.11 Ma (C10n.2r) (Barron *et al.*, in press); 29.0 Ma (Barron and Baldauf, 1995)
Last Occurrence: 26.29-26.39 Ma (C8n.2n) (Barron *et al.*, in press)

Skeletonemopsis barbadense (Greville) Sims, 1994. Synonym: *Skeletonema barbadense* Greville; Fenner, 1985, pl. 9, figs. 1, 2; Fenner and Mikkelsen, 1990, p. 447, pl. 1, fig. 5.
First Occurrence: >33.79 Ma (Barron *et al.*, in press)
Last Occurrence: 30.73-30.99 Ma (C12n-C12r) (Barron *et al.*, in press)

Thalassionema sp. cf. *T. nitzschioides*
First Occurrence: *circa* 31.30-31.45 Ma (Barron *et al.*, in press)
Last Occurrence: living

Thalassiosira praefraga Gladenkov et Barron, 1995, p. 30, pl. 2, figs. 3-6 and 9. Synonym: *Thalassiosira spinosa* Schrader sensu Barron, 1983, p. 512, pl. IV, fig. 8; sensu Barron, 1985b, p. 793, pl. 11, fig. 15; sensu Fourtanier, 1991, pl. 2, figs. 9, 10.
First Occurrence: 22.67-22.94 Ma? (C6Bn.1n-C6Bn.2n) (Barron *et al.*, in press); Barron (1983) indicates this taxon (identified as *T. spinosa*) ranges down to the first occurrence of *Rocella schraderi* (*circa* 25 Ma) at DSDP Site 71.
Last Occurrence: 17.6 Ma (age updated from Barron, 1985a).

Thalassiosira primalabiata Gombos in Gombos and Ciesielski, 1983, pl. 9, figs. 1-8; Barron, 1983, p. 512, pl. III, figs. 7, 12.
First Occurrence: *25.53-25.70 Ma (C7An.1n-C7An.1r)* (Barron *et al.*, in press)
Last Occurrence: *22.06-22.31 Ma* (C6AAr.1r-C6AAr.2r) (Barron *et al.*, in press). 22.2 Ma (age updated from Harwood and Maruyama 1992).

Thalassiosira spumellaroides Schrader 1976, p. 636, pl. 6, figs. 1, 2; Barron, 1983, p. 513, pl. IV, fig. 7; Barron, 1985a, pl. 5, fig. 4; Fourtanier 1991, pl. 2, fig. 12.
First Occurrence: 22.40-22.67 Ma (C6AAr.1r-C6AAr.2r) (Barron *et al.*, in press); 22.6 Ma (age updated from Harwood and Maruyama 1992)
Last Occurrence: 17.20 Ma (age updated from Barron, 1985a).

Thalassiothrix sp. cf. *T. primitiva* Schrader, 1976, p. 637, pl. 1, fig. 3.
First Occurrence: 32.97-33.01 Ma (C12r) (Barron *et al.*, in press)
Last Occurrence: ?

Acknowledgements

I am happy to write this paper for the Volume dedicated in honour of Professor M.S. Srinivasan, a dedicated researcher who helped shape my thinking about Miocene paleoceanography and biostratigraphy. I am grateful to the Ocean Drilling Program for providing the samples. Scott Starratt and Dave Bukry reviewed the manuscript and offered very useful suggestions.

References

AKIBA, F., HIRAMATSU, C., and YANAGISAWA, Y., 1993. A Cenozoic diatom genus *Cavitatus* Williams; an emended description and two new biostratigraphically useful species, *C. lanceolatus* and *C. rectus* from Japan. Bulletin of the Natural Science Museum of Tokyo, Series. C., 19, 11-39.

BALDAUF, J.G. and BARRON, J.A., 1991. Diatom biostratigraphy: Kerguelen Plateau and Prydz Bay regions of the Southern Ocean. In: Barron, J.A., Larsen, B., *et al.* (Eds.), Proceedings of Ocean Drilling Program, Scientific Results,119, 547-588.

BARRON, J.A., 1983. Latest Oligocene through early middle Miocene diatom biostratigraphy of the eastern tropical Pacific. Marine Micropaleontology, 7, 487-515.

BARRON, J.A., 1985a. Late Eocene to Holocene diatom biostratigraphy of the equatorial Pacific Ocean, Deep Sea Drilling Project Leg 85. In: Mayer, L., Theyer, F., *et al.* (Eds.), Initial Reports of the Deep Sea Drilling Project, 85, 413-456.

BARRON, J. A., 1985b. Miocene to Holocene planktic diatom stratigraphy. In: Bolli, H.M., Saunders, J.B. and Perch-Nielsen, K. (Eds.), Plankton Stratigraphy. Cambridge University Press. pp. 763-805.

BARRON, J.A., 2003. Planktonic marine diatom record of the past 18 m.y.: appearances and extinctions in the Pacific and Southern Oceans. Diatom Research, 14(2), 203-224.

BARRON, J.A. and BALDAUF, J.G., 1995. Cenozoic marine diatom biostratigraphy and applications to paleoclimatology and paleoceanography. In: Blome, C.D., Whalen, P.M. and Reed, K.M. (Convenors), Siliceous Microfossils. Paleontological Society Short Courses in Paleontology, 8, 107-118.

BARRON, J.A., FOURTANIER, E., and BOHATY, S.M., in press. Oligocene and earliest Miocene diatom biostratigraphy of Site 1220, ODP Leg 199, equatorial Pacific Ocean. In: Lyle, M.W., Wilson, P.A. and Janecek, T.R. (Eds.), Scientific Results, Ocean Drilling Program, 199, 000-000.

BERGGREN, W.A., KENT, D.V., SWISHER, C.C. and AUBRY, M.-P., 1995. A revised Cenozoic geochronology and chronostratigraphy. In: Berggren, W.A., Kent, D.V. and Hardenbol, J. (Eds.), Geochronology, Time Scales and Global Stratigraphic Correlation. SEPM Special Publication, 54, 129-212.

BUKRY, D., 1984. Neogene silicoflagellates from Deep Sea Drilling Project Site 543, western tropical Atlantic Ocean. In: Biju-Duval, B., Moore, J.C., *et al.* (Eds.), Initial Reports of the Deep Sea Drilling Project, 78, 463-468.

CANDE, S.C., and KENT, D.V., 1995. Revised calibration of the geomagnetic polarity time scale for the Late Cretaceous and Cenozoic. Journal of Geophysical Research, 100, 6093-6095.

CASTRACANE CONTE A.F. ANTELMINELLI, 1886. Report on the scientific results of the voyage of the H.M.S. Challenger during the years 1873-76. Botany, 2, 1-178.

FENNER, J., 1978. Cenozoic diatom biostratigraphy of the equatorial and southern Atlantic Ocean. In: Supko, P.R., Perch-Nielsen, K., *et al.* (Eds.), Initial Reports of the Deep Sea Drilling Project, 39, 491-623.

FENNER, J., 1981. Diatoms in the Eocene and Oligocene sediments off NW- Africa. Their stratigraphic and paleogeographic occurrences. Doctoral Dissertation, University of Kiel, Germany, 231 pp.

FENNER, J., 1984a. Eocene-Oligocene planktic diatom stratigraphy in the low latitudes and the high southern latitudes. Micropaleontology, 30, 319-342.

FENNER, J., 1984b. Middle Eocene to Oligocene planktonic diatom stratigraphy from Deep Sea Drilling Sites in the South Atlantic, Equatorial Pacific, and Indian oceans. In: Supko, P.R., Perch-Nielsen, K., *et al.* (Eds.), Initial Reports of the Deep Sea Drilling Project, 75, 1245-1271.

FENNER, J., 1985. Late Cretaceous to Oligocene planktic diatoms. – In: Bolli, H.B., Saunders, J.B. and Perch-Nielsen, K. (Eds.), Plankton Stratigraphy. Cambridge University Press. pp. 713-763.

FENNER, J. and MIKKELSEN, N., 1990. Eocene-Oligocene diatoms in the Western Indian Ocean: Taxonomy, stratigraphy, and paleoecology. In: Duncan, R.A., Backmann, J., Peterson, L.C., *et al.* (Eds.), Proceedings of the Ocean Drilling Program, Scientific Results, 115, 433-463.

FOURTANIER, E., 1991. Diatom biostratigraphy of equatorial Indian Ocean Site 758, ODP Leg 121. In: Peirce, J. Weissel, J., *et al.* (Eds.), Proceedings of Ocean Drilling Program, Scientific Results, 121, 189-208.

GLADENKOV, A. Y. and BARRON, J.A., 1995. Oligocene and early middle Miocene diatom biostratigraphy of Hole 884B. In: Rea, D.K, Basov, I.A., Scholl, D.M. and Allan, J.F., (Eds.), Proceedings of the Ocean Drilling Program, Scientific Results, 145, 21-41.

GOMBOS, A.M., Jr., 1983. Survey of diatom in the upper Oligocene and lower Miocene in Holes 515B and 516F. In: Barker, P.F., Carlson, R.L. and Johnson, D.A. (Eds.), Initial Reports of the Deep Sea Drilling Project, 72, 793-804.

GOMBOS, A.M., Jr. and CIESIELSKI, P.F., 1983. Late Eocene to early Miocene diatoms from the southwest Atlantic. In: Ludwig, W.J., Krasheninnikov, V.A., *et al.* (Eds.). Initial Reports of the Deep Sea Drilling Project, 71, 583-632.

HAJÓS, M., 1976. Upper Eocene and lower Oligocene Diatomaceae, Archaeomonadaceae, and Silicoflagellatae in southwestern Pacific sediments, DSDP Leg 29. In: Hollister, C.D., Craddock, C. *et al.* (Eds.), Initial Reports of the Deep Sea Drilling Project, 35, 817-883.

HARWOOD, D. M. and BOHATY, S. M., 2001. Early Oligocene siliceous microfossil biostratigraphy of Cape Roberts Project Core CRP-3, Victoria Land Basin, Antarctica. Terra Antarctica, 8, 315-338.

HARWOOD, D.M. and MARUYAMA, T., 1992. Middle Eocene to Pleistocene diatom biostratigraphy of Southern Ocean sediments from the Kerguelen Plateau, Leg 120. In: Wise, S.W., Jr., Schlich, R., *et al.* (Eds.), Proceedings of Ocean Drilling Program, Scientific Results, 120, 683-733.

JOUSÉ, A.P., 1974. Diatoms in the Oligocene-Miocene biostratigraphic zones of the tropical areas of the Pacific Ocean. Nova Hedwigia Beihefte, 45, 333-364.

LOHMAN, K.E., 1974. Lower middle Miocene marine diatoms from Trinidad. Verhandlungen der Naturforschenden Gesellschaft im Basel, 84, 326-360.

RADIONOVA, E.P., 1987. Diatom morphology of genus *Cestodiscus* from lower middle Miocene depositions of the tropical zone of the Pacific Ocean. In: Rauser-Chernoyusov, D.M. (Ed.), Methods of Zonal Stratigraphic Work-up According to Microorganisms. Micropaleontology Edition, 29. Akademia Nauk SSSR, p. 141-154. (in Russian).

RAMSAY, A.T.S. and BALDAUF, J.G., 1999. A reassessment of the Southern Ocean biochronology. Geological Society of London Memoir, 18, 1-122.

SCHRADER, H.-J., 1976. Cenozoic planktonic diatom biostratigraphy of the southern Pacific Ocean. In: Hollister, C.D., Craddock, C., *et al.* (Eds.), Initial Reports of the Deep Sea Drilling Project, 35, 605-671

SCHMIDT, A. (FORTGESETZT VON SCHMIDT, M., FRICKE, F., HEIDEN, H., MÜLLER, O. and HUSTEDT, F.), 1875-1959. Atlas der Diatomaceenkunde. Tafel 1-480. Leipzig.

SIMS, P. A., 1994. *Skeletonemopsis*, a new genus based on the fossil species of the genus *Skeletonema* Grev. Diatom Research, 9, 387-410.

SIMS, P.A., FRYXELL, G.A., and BALDAUF, J.G., 1989. Critical examination of the diatom genus *Azpeitia*. Species useful as stratigraphic markers for the Oligocene and Miocene Epochs. Micropaleontology, 35(4), 293-307.

STRELNIKOVA, N I., FOURTANIER, E., KOCIOLEK, J.P., and BARRON, J.A., 2001. Ultrastructure studies of *Coscinodiscus* and *Cestodisccus* species from the Eocene and Oligocene. In: Jahn, R., Kociolek, J.P, Witkowski, A., and Compère, P. (Eds.), Lange-Bertalot-Festschrift: Studies on Diatoms. Koeltz Scientific Books, Koenigstein, pp. 63-96.

YANAGISAWA, Y., 1995a. Cenozoic diatom genus *Rossiella* Desikachary et Maheswari: An emended description. Transactions of the Proceedings of the Paleontological Society of Japan, New Series, 177, 1-20.

YANAGISAWA, Y., 1995b. Cenozoic diatom genus *Bogorovia* Jousé: An emended description. Transactions of the Proceedings of the Paleontological Society of Japan, New Series, 177, 21-42.

Guembelitria Blooms –Environmental Catastrophcs Index

Gerta Keller and Alfonso Pardo

Department of Geosciences, Princeton University, Princeton, NJ 08544, USA
Fundación San Valero, c/ Violeta Parra 9, 50015 Zaragoza, Spain

Introduction

The small triserial planktic foraminiferal *Guembelitria* species evolved in the middle Cretaceous and are sporadically recorded in the sedimentary record to the Recent (Kroon and Nederbragt, 1990; Leckie, 1987; Keller *et al.*, 2001). They thus form one of the longest ranging plankton groups. Their stratigraphic record, however, is poorly documented and their paleoecology little understood, largely due to the very small species size (38-100 μm), which is not routinely examined in foraminiferal analysis. *Guembelitria* populations are best known from the aftermath of the Cretaceous-Tertiary (K-T) mass extinction where no species larger than 100 microns in size survived and the immediate survivors and newly evolved species are frequently smaller than 63 microns (Keller *et al.*, 1995, 2001). *Guembelitria* species are the sole Cretaceous long-term survivors, and the sole group that thrived in the aftermath of the K-T mass extinction, when they formed major plankton blooms that earned them the label of disaster opportunist species.

During the last decade, systematic statistical documentation of the small-sized species in late Maastrichtian and early Danian sediments in the world's oceans across latitudes and from shallow marginal seas to the open ocean has yielded an extensive dataset for *Guembelitria* and its environmental associations along with the small low oxygen tolerant biserial species (e.g., *Heterohelix globulosa, H. dentate, H. navarroensis, Zeauvigerina waiparaensis, Woodringina hornerstownensis, W. claytonensis, Chiloguembelina midwayensis*). We review here the *Guembelitria* blooms and low oxygen tolerant small biserial heterohelicid species and species richness from sections in the Indian Ocean (Ninetyeast Ridge DSDP Site 216 and Madagascar) and correlate these to sequences in the eastern Tethys (Israel and Egypt; Fig. 1) to show the geographic distribution of *Guembelitria* blooms during the late Maastrichtian and early Danian and its relationship to mantle plume volcanism in the Indian Ocean.

Methods

Guembelitria species are very small (<100 μm) and their blooms can only be identified by analysis of the small size fractions (>38 μm or > 63 μm). Statistical counts of planktic foraminiferal species are based on random sample splits (using an Otto microsplitter) of about 150-250 specimens (average of 5-10 species), or 300 specimens for more diverse assemblages (Buzas, 1990). Bias introduced by preservational effects (dissolution, breakage) can usually be identified based on the number of broken specimens. Since *Guembelitria* tend to be among the most thin-walled and therefore fragile species, they are usually present in well preserved assemblages. Reliable statistical data can also be obtained from more indurated sediments, where foraminiferal shells are recrystallized and infilled with micritic carbonate, which makes them more resistant to breakage and solution effects. Species used in this study are shown in Plate 1.

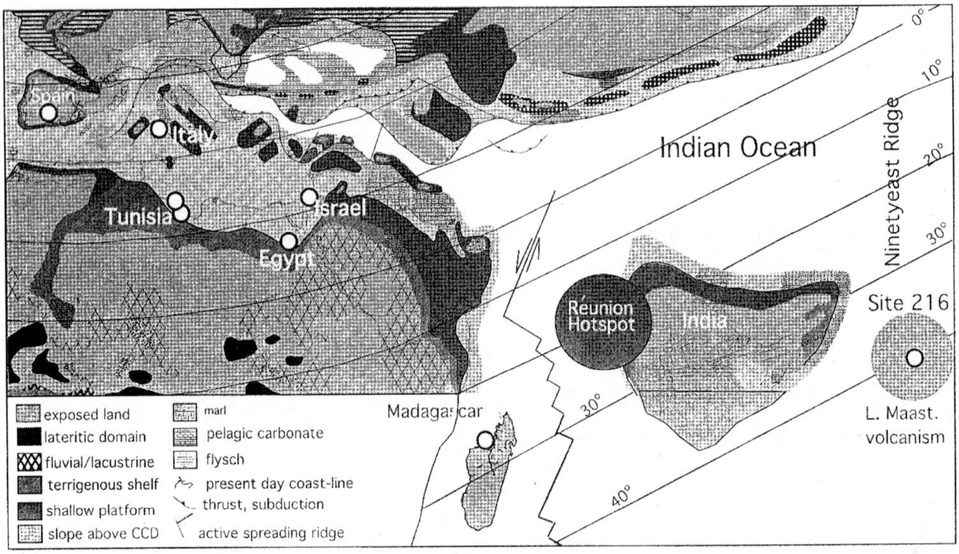

Fig. 1. Paleoenvironmental map of the Indian Ocean and eastern Tethys modified after Dercourt *et al.* (1993) with paleolocations of studied sections. Note the biotic effects of volcanism decreased westward from the Indian Ocean to the eastern Tethys with no biotic effects west of Tunisia.

Plate 1.

Ecological generalists that thrive at times of major environmental perturbations caused by either volcanism or impacts. The first stage of the faunal succession after the perturbation is characterized by the disaster opportunists *Guembelitria* blooms (Figs. 1-8). This second stage is characterized by low oxygen tolerant small heterohelicids (Fig. 11-20). All morphotypes of these species are small, thin-walled and lack surface ornamentation.

1, 2. *Guembelitria cretacea* (Cushman), Qreyia section, Egypt. Scale bar = 50 μm **3.** *Guembelitria cretacea* (Cushman), Amboanio section, Madagascar. Scale bar = 50 μm **4-6.** *Guembelitria trifolia* (Morozova), Qreyia section, Egypt. Scale bar = 50 μm **7, 8.** *Guembelitria dammula* (Voloshina), Amboanio, Madagascar. Scale bar = 50 μm **9.** *Heterohelix globulosa* (Ehrenberg). Qreyia section, Egypt. Scale bar = 100 μm **10.** *Heterohelix globulosa* (Ehrenberg). Amboanio, Madagascar. Scale bar = 100 μm **11.** *Chiloguembelina midwayensis* (Cushman), Qreiya section, Egypt. Scale bar = 50 μm **12.** *Chiloguembelina claytonensis*, Loeblich & Tappan, Qreiya, Egypt. Scale bar = 50 μm **13.** *Woodringina hornerstownensis* Olson,Qreiya section, Egypt. Scale bar = 50 μm **14.** *Heterohelix dentata* Stenestad, Amboanio section, Madagascar. Scale bar = 50 μm **15.** *Heterohelix dentata* Stenestad, Qreiya section, Egypt. Scale bar = 50 μm **16.** *Heterohelix navarroensis* Loeblich, Qreiya section, Egypt. Scale bar = 50 μm

Plate 1.

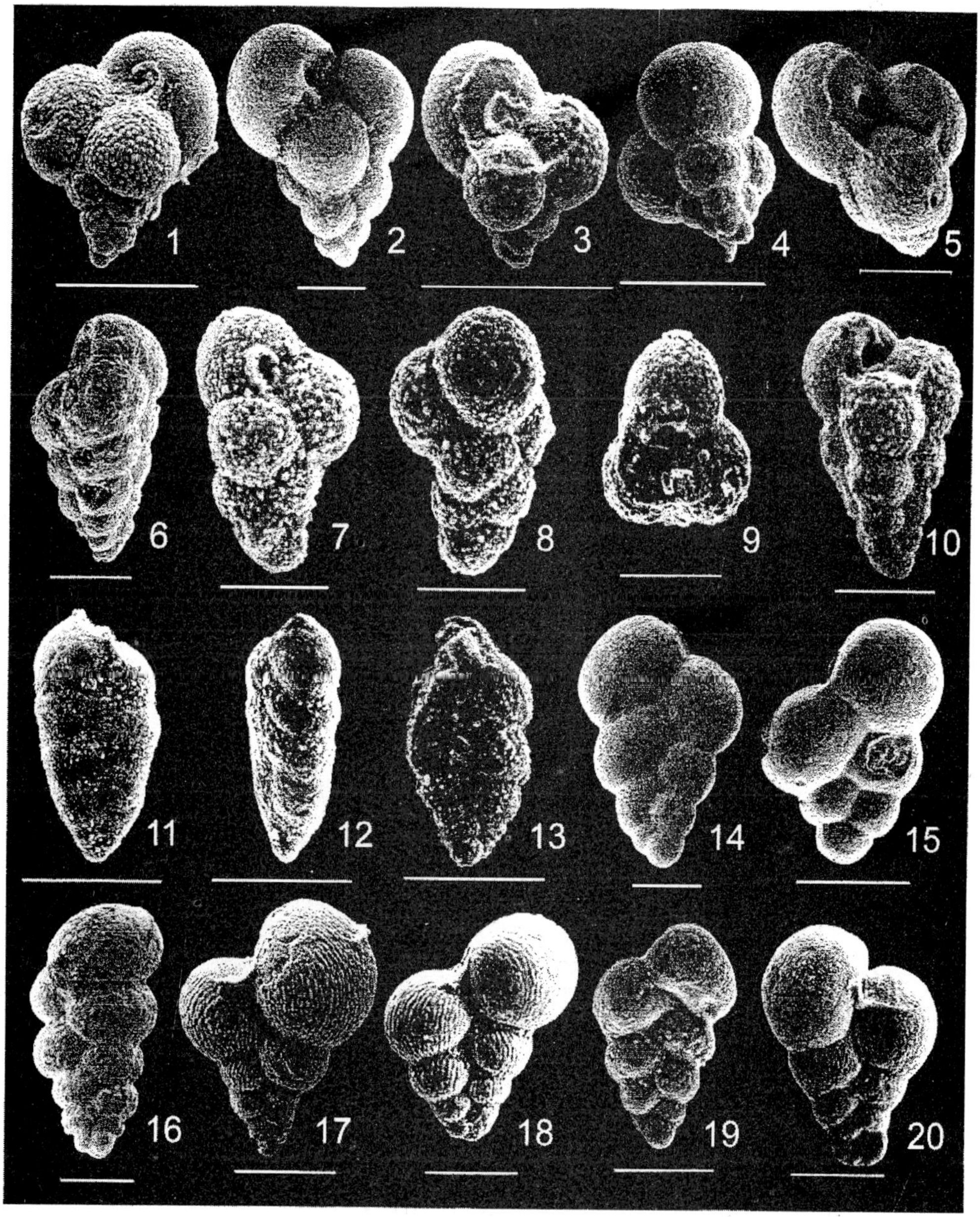

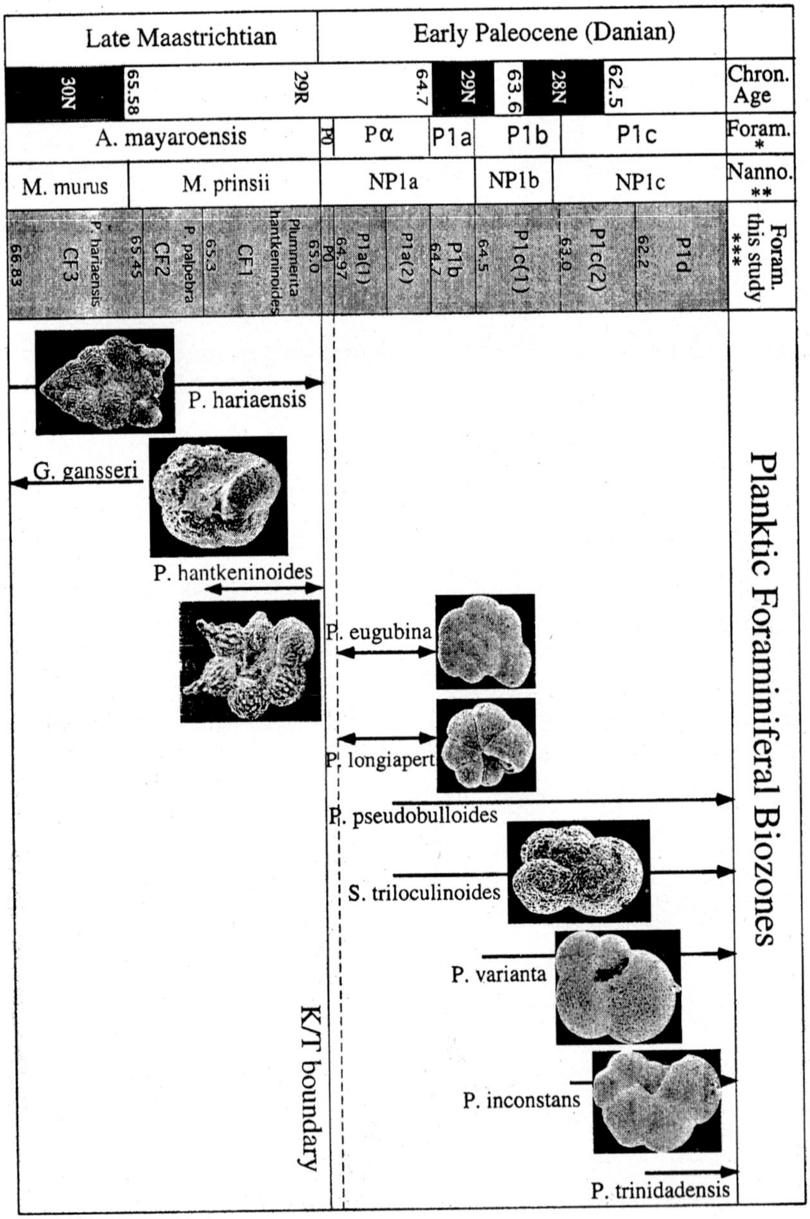

Fig. 2. High resolution biozonation of Keller *et al.* (1995) and Li and Keller (1998b) used in this study and illustration of planktic foraminiferal index species for biozones. The biozonation scheme by Berggren *et al.* (1995) is given for comparison. Calcareous nannofossil zones by Tantawy (2002). CF= Cretaceous Foraminifera)

The high resolution biozonation used in this study is from Keller *et al.* (1995) and Pardo *et al.* (1996) and Li and Keller (1998a) as shown in Figure 2. There are multiple criteria for identification of the K-T boundary in addition to the mass extinction of planktic foraminifera, including the evolution of the first Tertiary species, a lithological change usually marked by organic-rich clay deposition, the presence of a thin (3-4 mm) red clay layer enriched in Ir and sometimes altered spherules, and a negative excursion in the $\delta^{13}C$ record.

Ninetyeast RidgeDSDP Site 216

During the late Maastrichtian DSDP Site 216 was located at a paleolatitude of about 40°S on the crest of Ninetyeast Ridge, which consisted of islands built to sea level as the oceanic plate moved over a mantle plume resulting in lithospheric uplift and volcanic deposition (Thompson *et al.*, 1974; Moore *et al.*, 1974). After passage over the mantle plume the site rapidly subsided and deepened from shallow to deeper marine below the photic zone as Site 216 passed beyond the influence of mantle plume volcanism and chalk deposition resumed during the last 500 ky of the Maastrichtian. Microfossils associated with these volcanic-rich sediments reveal unusually low diversity and small size, which was originally attributed to a high latitude location and dissolution effects (McGovern, 1974), but restudy of the small size fraction reveals a high stress fauna related to mantle plume volcanism (Keller, in press).

The K-T boundary is marked by a hiatus that spans the early Danian zones P0 to the lower part of Plc and possibly the uppermost part of the Maastrichtian. The chalks below the K-T boundary contain planktic foraminiferal and nannofossil assemblages characteristic of zones CF1-CF2 and *Micula prinsii*. The abrupt transition from chalk to volcanic-rich deposition is marked by a hiatus and correlates with the sea level lowstand at 65.5 Ma. Widespread erosion at this sea level lowstand is also observed in Madagascar, Israel and Egypt (Abramovich *et al.*, 2002; Keller, 2002, in press a).

Late Maastrichtian planktic foraminiferal assemblages in the volcanic-rich sediments at Site 216 are characterized by unusually low diversity (6-10 species), very small adult species sizes (<100 µm) and low abundances compared with assemblages at similar middle to high latitudes, except for the unusually high abundances (>80%) of the disaster opportunist *Guembelitria cretacea*. More diverse species assemblages with normal adult species sizes first appear in the chalk near the top of the section and above the volcanic-rich sediments (Fig. 3). Low oxygen tolerant small heterohelicids (mostly *H. dentata, H. globulosa*) dominate at times of low *Guembelitria* abundances. There are few other species (hedbergellids, globigerinellids, rugoglobigerinids) and no large complex species (globotruncanids, pseudotextularids etc). Preservational effects cannot account for these unusual assemblages since foraminiferal tests are generally well preserved, with minor recrystallization and calcite infilling. Dissolution effects are minor (some breakage and holes in tests) as evident in the high abundance of small delicate species. Moreover, the absence of other more dissolution resistant species indicates that these assemblages are not the result of preservational bias. These unusual high-stress assemblages thus appear to be directly related to the passage of the oceanic plate over a mantle plume and consequent volcanic activity.

Madagascar

The Amboanio section from the Mahajanga Basin of northwestern Madagascar was deposited in an outer neritic environment, which shallowed to middle neritic depths near the end of the Maastrichtian (Abramovich *et al.*, 2002). Major hiatuses are present between zones CF4/CF3 and CF3/CF1 with most of zones CF3 and CF2 missing. These hiatuses correspond to sea level lowstands at 65.5 Ma and 67 Ma. During the late Maastrichtian zone CF4 *Guembelitria* species dominate (*G. cretacea* 40%, *G. dammula* 10-15%) along with *Heterohelix* species (Fig. 4). In zone CF3

Heterohelix species reach 80% (mainly *H. dentata)* and *Guembelitria* average 10%. Near the end of the Maastrichtian in zone CF1 *Guembelitria* blooms reach 40%. Other ecologic generalist and specialist species are relatively minor components of the assemblage. A hiatus marks the K-T boundary.

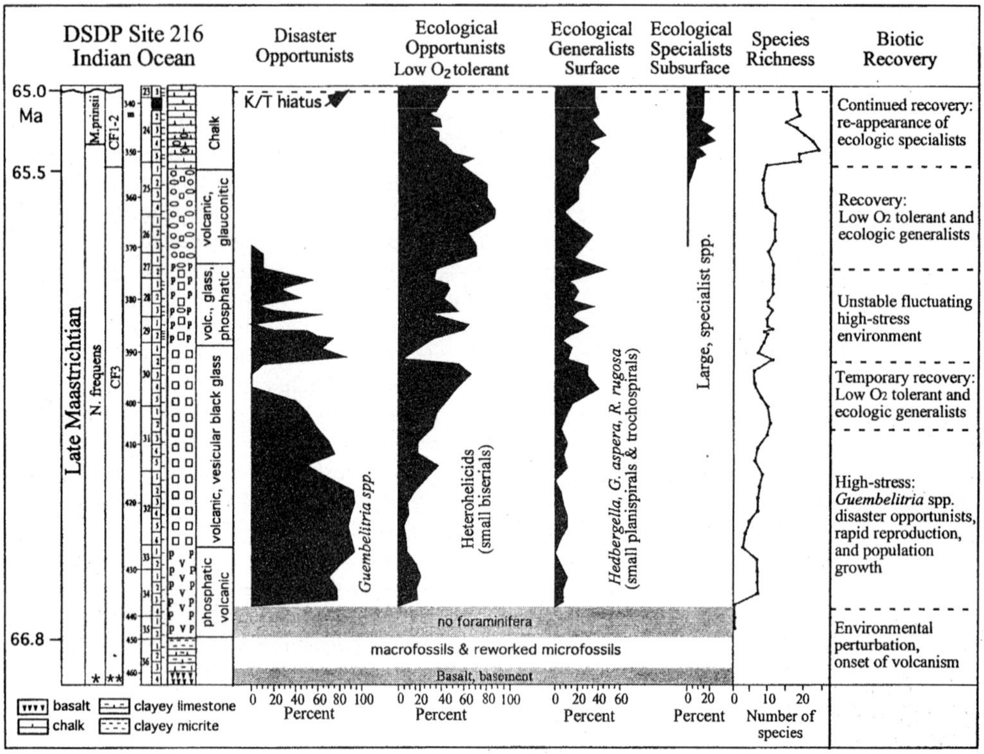

Fig. 3. Relative abundances of disaster opportunists, low oxygen tolerant ecological generalists and species richness at DSDP Site 216 on Ninetyeast Ridge, Indian Ocean. Faunal assemblages reflect biotic stress associated with mantle plum volcanism. *Guembelitria* blooms are frequently the sole survivors in this volcanic environment, Heterohelicids appear at times of reduced biotic stress. Biotic effects mimic the effects of the K-T mass extinction.

DSDP Site 525 on Walvis Ridge in the South Atlantic was deposited at a similar paleodepth as Madagascar and shares the same globotruncanid fauna (Abramovich and Keller, 2003). However, *Guembelitria* are generally absent and only rarely appear at times of blooms in Madagascar, although many of the globotruncanids and other larger sized species are dwarfed. This attests to the different oceanographic conditions in the South Atlantic and Indian Ocean, with maximum biotic stress in the latter.

Egypt

The Qreiya section in central Egypt was deposited in a middle neritic environment (Keller, 2002; Keller *et al.*, 2002; Tantawy, 2003). The K-T boundary is marked by a thin clay layer and Ir anomaly above a bioturbated marly shale with an erosional surface and hiatuses also reduced early Danian zones Pla, Plb and Plc (Fig. 5). *Guembelitria* blooms are present in zones Pla and Plc, similar to other Tethys sections. But major *Guembelitria* blooms (50-70%) also occurred in the late Maastrichtian zones CF3 and CF1, similar to the Indian Ocean localities. At times of low *Guembelitria* abundances, the small *Heterohelix navarroensis* dominates along with increased abundances of the small trochospiral and planispiral generalists. Large specialized species are rare throughout the section. Species richness is also very low (25-30 species) compared with similar paleodepths at Elles, Tunisia (40-45 species).

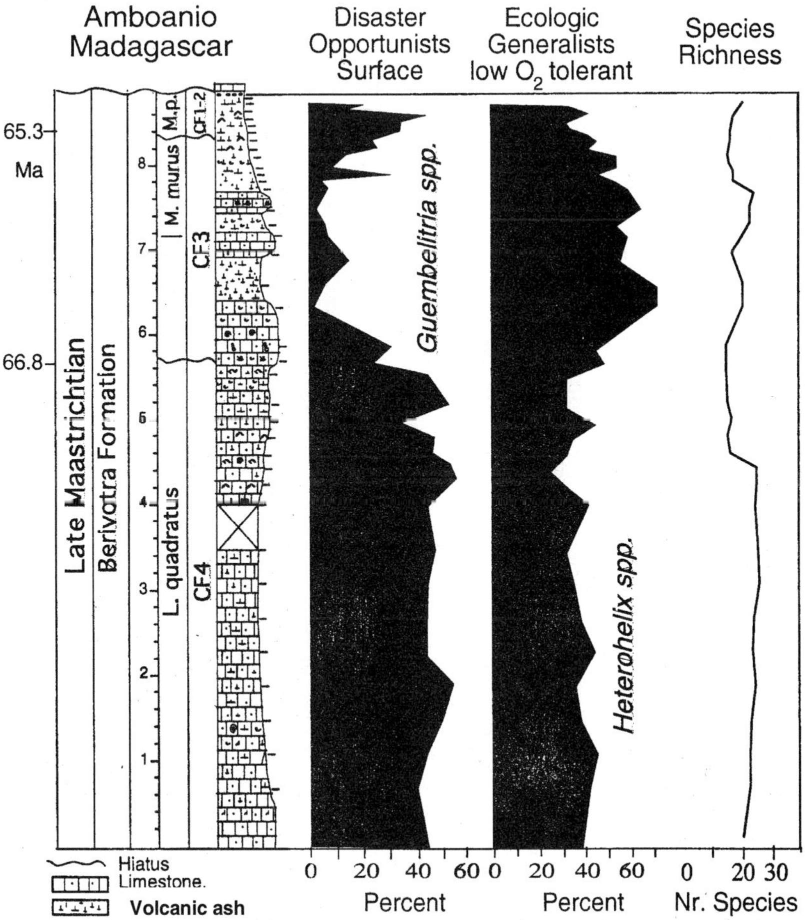

Fig. 4. Relative abundances of disaster opportunists and low oxygen tolerant ecological generalists, species richness and stable isotopes at Amboanio, Mahajanga Basin, Madagascar. Sedimentation occurred in an outer neritic environment that shallowed to middle neritic depths near the top of the Maastrichtian. *Guembelitria* species dominate during the late Maastrichtian zone CF4 and again in Zone CF1, which contains ash layers. These *Guembelitria* blooms may reflect volcanism. Data from Abramovich *et al.*, 2002.

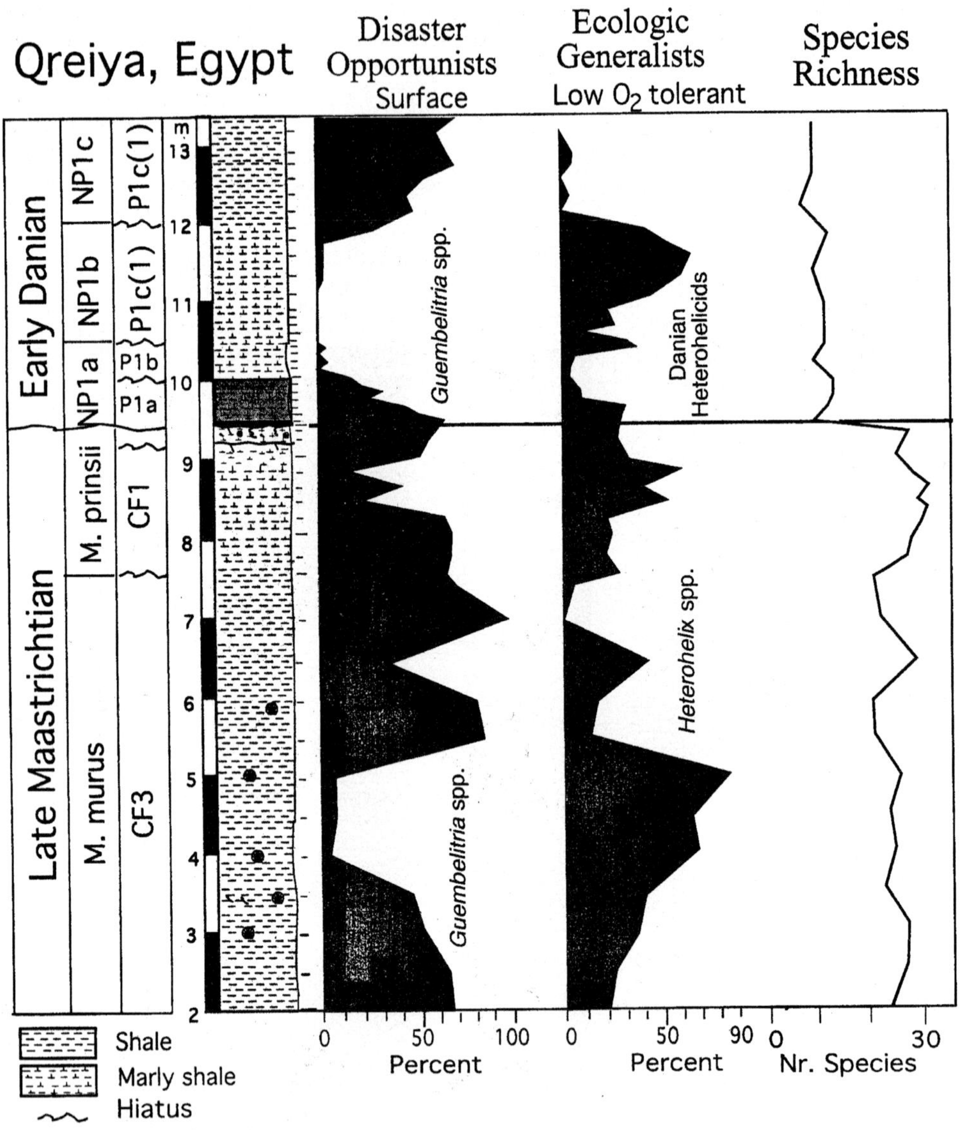

Fig. 5. Relative abundances of disaster opportunists and low oxygen tolerant ecological generalists, species richness and stable isotopes at Qreiya in central Egypt where deposition occurred in a middle to outer neritic environment. *Guembelitria* blooms indicate that intermittent high stress conditions similar to the K-T boundary event prevailed through the late Maastrichtian and early Danian as also indicated by stable isotopes. Data from Keller, 2002; Keller *et al.*, 2002.

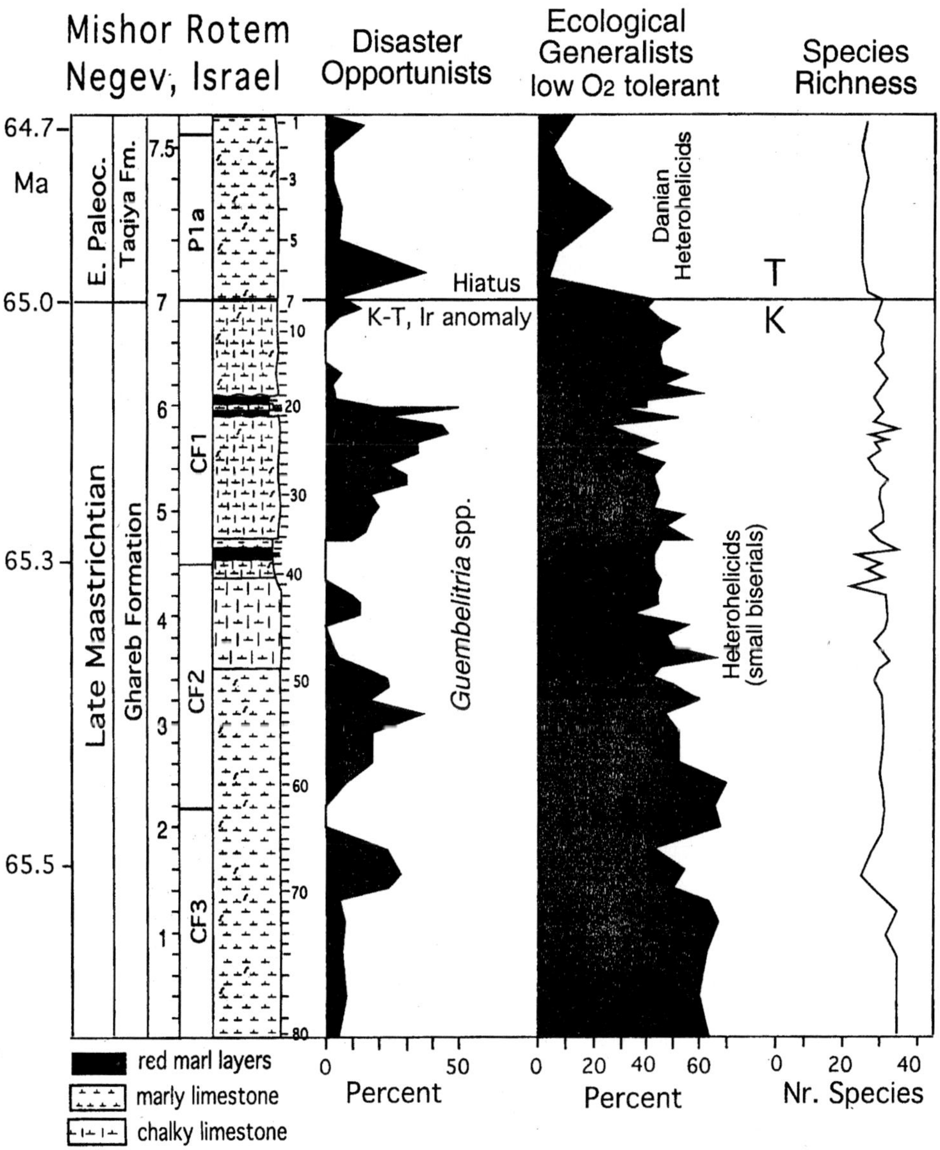

Fig. 6. Relative abundances of disaster opportunists, low oxygen tolerant ecological generalists and species richness at the Mishor Rotem section of the Negev, Israel where sedimentation occurred in an upper bathyal environment. *Guembelitria* blooms indicate intermittent high stress conditions similar to the K-T boundary event prevailed through the late Maastrichtian and early Danian similar to Egypt, though with less severe biotic effects. Data from Keller, in press a.

Israel

The high stress pattern observed in central Egypt is also recognized in sections from the Negev of Israel, which were deposited in deeper upper slope environments (Abramovich *et al.*, 1998; Keller, in press a). At the Mishor Rotem section in the Negev, the K-T boundary is marked by the characteristic thin (3-4 mm) red clay layer enriched in iridium. A reduced zone Pla overlies the boundary clay (Fig. 6) with a reduced *Guembelitria* bloom possibly due to preservational bias. In the late Maastrichtian, three distinct *Guembelitria* blooms are present, though they are less intense than at Qreiya as suggested by the lower percent abundance. This is also indicated by the high (50-60%, though fluctuating) abundance of small *Heterohelix* species (mostly *H. globulosa*), which is more similar to Elles, Tunisia, than Qreiya (Abramovich and Keller, 2002). Small planispiral and trochospiral generalists average 10% during the late Maastrichtian, but increase to 25% during the last 150 ky of the Maastrichtian and reach 40% during the early Danian zone Pla. Large specialist species are rare. Species richness is relatively low at 30-35 species, but still higher than at Qreiya. Israel thus experienced increased stress during the late Maastrichtian, but not as severe as Egypt.

Paleogeography of *Guembelitria* blooms

Blooms of the disaster opportunist *Guembelitria* species are best known from the aftermath of the K-T boundary mass extinction where they characterize the early Danian zones P0 and Pla in all oceans and across latitudes as shown in Figure 7. However, there are significant geographic differences in the intensity of the blooms (relative abundance with respect to other species) and the temporal distribution during the Danian, as well as the late Maastrichtian, which indicate recurring biotic crises conditions.

The post-K-T *Guembelitria* bloom (Pla and P0 zones) shows a clear geographic demarcation between dominant blooms, here identified as abundances greater than 40%, and lesser blooms of 10-20%. The dominant blooms occur in low to middle latitudes, whereas the lesser blooms are restricted to northern and southern high latitudes (Fig. 7). This indicates that low and middle latitudes experienced the most intense biotic stress in the post-K-T mass extinction. This is independently supported by the mass extinction pattern, which shows all tropical and subtropical species extinct, but the survival of ecologic generalists adapted to environmental changes (review in Keller, 2001). This pattern is also independently supported by $\delta^{13}C$ records, which show a 2-3 ‰ negative shift in low to middle latitudes, but only 0.5 ‰, or even a positive shift, in high latitudes (e.g. Denmark, Kazakstan, Bulgaria, ODP Site 738; Schmitz *et al.*, 1992, Keller *et al.*, 1993; Barrera and Keller, 1994; Pardo, 1999; Pardo and Keller, 1999; Adatte *et al.*, 2002). Low latitudes thus experienced the major productivity crash and the most severe biotic consequences. In contrast, high latitudes experienced relatively little change in bioproductivity and highly reduced biotic stress.

The most intense (>60%) *Guembelitria* blooms appear to be associated with particular paleoenvironments, such as the shallow continental shelf, continental slope margins, and areas of intense volcanism (Fig. 7). All of these environments have one common factor – high nutrient influx either from continental runoff, upwelling, or volcanic input. A large meteorite impact could temporarily have the same effect causing increased nutrient influx from ejecta fallout and erosion associated with post-impact climate cooling. Planktic foraminiferal diversity, which is at the maximum in oligotrophic and well-stratified oceans (Lipps, 1979), significantly decreases in such nutrient-rich environments.

Guembelitria thus exhibits a well-developed r-strategy that allows these organisms to take full advantage from stressed and disturbed habitats. The chain of events after a major disturbance in the habitat has been widely studied and modeled in recent ecosystems (Margalef, 1996; Begon *et al.*, 1998), and may be extrapolated to paleoecological studies. Thus, in this biotic crisis scenario following an environmental disturbance, their rapid population increase allows r-strategists to take full advantage of the high nutrient content in marine waters, which is characteristic of the aftermath of a biotic crisis. In this way, guembelitrids are the starter fauna of a new ecological succession in highly disturbed areas where they help recover the oligotrophic balance that allows the opening of a variety of ecological niches which support more diverse or polytaxic assemblages. The upturn of the ecological crisis is marked by the recovery of low oxygen tolerant small heterohelicids, which herald the initial improvement of environmental conditions.

In this respect, modern representatives of *Guembelitria* appear to have retained this same affinity for high nutrient environments, as they are most abundant in upwelling areas with highly variable conditions where few other species thrive (Kroon and Nederbragt, 1990). Times of biotic crises may thus be related to a number of different factors all of which lead to increased nutrient influx, including impacts, volcanism, restricted shelf environments, and upwelling along continental margins.

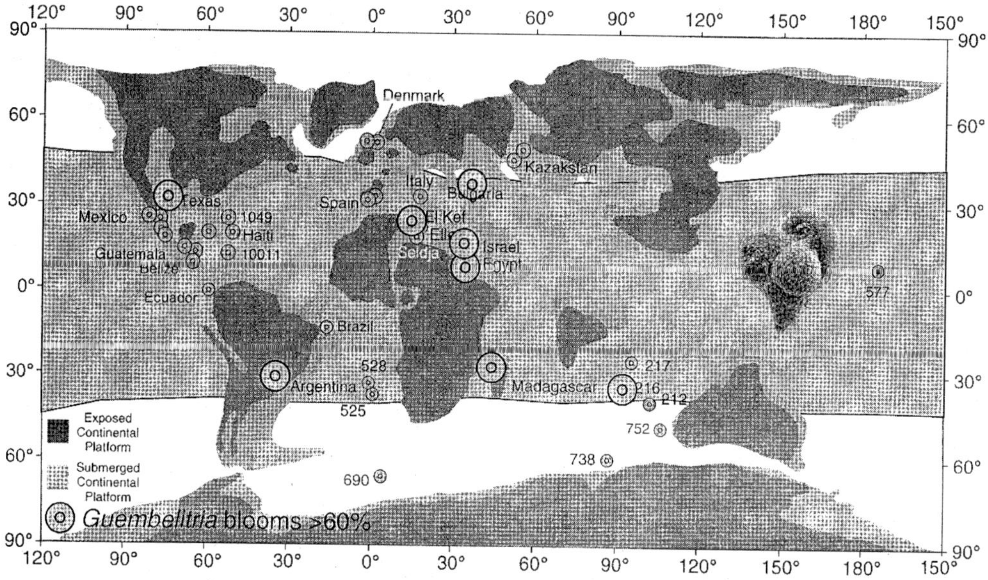

Fig. 7. Biogeographic distribution of Guembelitria blooms. Guembelitria blooms dominate (40-80%) low to middle latitudes during the early Danian after the K-T mass extinction. The highest dominance of these disaster opportunists is found in shallow continental seas, continental margins and volcanic provinces. These environments are characterized by high nutrient influx either due to terrestrial runoff , upwelling, or volcanic input.

Guembelitria blooms and volcanism

Ninetyeast Ridge DSDP Site 216 provides a clear example of the biotic effects of mantle plume volcanism from inception to cessation of volcanic influx. The onset of this environmental perturbation eliminated nearly all planktic foraminifeal species. The intitial assemblage in the volcanic-rich sediments is dominated (80-90%) by the disaster opportunist *Guembelitria* species, which represent high stress conditions. *Guembelitria* are characterized by small size, rapid reproduction rate and fast population growth. They thrived in eutrophic waters, where few other species survived. When population growth exceeds nutrient availability, they cease and are replaced by low oxygen tolerant species, which further deplete nutrients leading to mesotrophic conditions and opening niches for other ecological generalists. At Site 216 the intial recovery is followed by a period of instability with fluctuating *Guembelitria* and heterohelicid populations (Fig. 3). This period is followed by low oxygen tolerant heterohelicids until the end of volcanic influx. Environmental recovery marked by increased species diversity, larger size and the return of specialized k-strategists occurred only when volcanic influx ceased.

In Madagascar the unusual prolonged faunal dominance by *Guembelitria* followed by low oxygen tolerant heterohelicids is likely related to volcanism also. Ash layers are present near the top of the Maastrichtian associated with a *Guembelitria* bloom (Abramovich *et al.*, 2002). No ash layers are preserved in the lower part of the section, though they may have altered to clay. There is ample evidence of prolonged periods of volcanism in the Indian Ocean Ninetyeast Ridge mantle plume and Reunion hotspot during the late Maastrichtian and earlier in Madagascar.

There is no evidence of volcanism in sections from Israel and Egypt. However, the *Guembelitria* blooms and alternating low oxygen tolerant biserial peak abundances can be correlated stratigraphically with Madagascar and Site 216, suggesting that they are a response to the same environmental perturbation. The late Maastrichtian *Guembelitria* blooms can be traced into Tunisia, where they represent minor, but significant increases of 10%. No late Maastrichtian *Guembelitria* blooms have been observed in Italy or Spain. This suggests that the biotic effects of volcanism in the Indian Ocean may have been relatively limited.

Biotic effects of impacts and volcanism

If major biotic catastrophes can be attributed to volcanism, what biotic effects can be attributed to major impacts? In fact, the biotic effects would be substantially the same. The biotic response to any major environmental catastrophe appears to be universal. The first stage is the elimination of k-strategists, or highly specialized species restricted to narrow ecological niches with little opportunity to adjust to changing environmental conditions. This happened at the K-T boundary where all specialized tropical and subtropical species went extinct along with most ecological generalists (review in Keller, 2001). In the Indian Ocean Site 216, the same group disappeared with the onset of volcanism on Ninetyeast Ridge during the late Maastrichtian, though this group returned after volcanism ceased.

The second stage is the invasion of r-strategists, or ecological generalists. The wide fundamental niche of these organisms enables adaptation to variable environmental conditions and hence permits the niche drift that assures survival of these stress tolerant species (Begon *et al.*, 1998). Their adaptation to high stress disturbed environments includes dwarfing, which is an ecological response for taking advantage of the nutrient supply by means of heterochronic acceleration (i.e. early sexual maturation and faster reproduction, MacLeod *et al.*, 2000). The result of this intense reproduction is two-fold. The sudden bloom of disaster opportunists (*Guembelitria*) depletes the environment of excess nutrients, which re-opens the pre-crisis ecological niches initially for other r-strategists.

The third stage is the rise of low oxygen tolerant heterohelicids following the decline of the disaster opportunists. Other small generalist species (r-strategists) also occupy ecologic niches (e.g. hedbergellids and globigerinellids). In the early Danian, this stage also provides opportunity for the evolution of new species.

The fourth stage represents the recovery of the ecosystem, which eventually leads to the return of k-strategists. Since the recovering k-strategy species are better adapted to the intense ecological niche competition of the latter stages of an ecological succession, r-strategy species are displaced to the background assemblage. The re-appearance of the k-strategist community usually marks restoration of the ecosystem to a pre-crisis oligotrophic environment. During the late Maastrichtian at Site 216 this happened only after volcanic influx ceased. After the K-T mass extinction, where no k-strategists survived, the evolution of new k-strategists took several million years.

Conclusions and suggestions for further studies

Guembelitria blooms are reliable proxies for environmental catastrophes during the late Maastrichtian and post-K-T early Danian, particularly in low and middle latitudes. Quantifying these easily identifiable small species thus provides a simple method for locating times of high environmental stress. The most intense *Guembelitria* blooms (>60% *Guembelitria*) occurred in shallow continental shelf areas, slope/shelf margins, and volcanic provinces of the Indian Ocean, though their temporal and geographic distributions are still poorly known. A more complete dataset of *Guembelitria* blooms through time will help identify times of biotic catastrophes whether due to impact or volcanism.

Investigating the temporal and geographic distribution of *Guembelitria* blooms during the late Maastrichtian and K-T boundary and its association with flood basalts and/or impacts, particularly in the Indian Ocean are necessary to determine the biotic effects associated the Reunion and Ninetyeast Ridge mantle plumes and Deccan flood basalts. There is also the possibility that India may have been the K-T boundary impact location. The Shiva crater has been proposed as the K-T impact crater (Chatterjee *et al.*, 2003). The combination of a major impact superimposed over ongoing Deccan volcanism would produce a powerful biotic catastrophe.

Little is known of the environmental conditions that lead to *Guembelitria* blooms. Preliminary $\delta^{13}C$ species ranking suggests that *Guembelitria* species are among the isotopically lightest species, along with *Z. waiparaensis* in high latitudes (Barrera and Keller, 1994; Pardo *et al.*, 1999), and both groups lived and thrived in eutrophic surface waters. Further geochemical and paleoecological studies are needed to determine the paleoecology of these species, including (1) stable isotope ranking of *Guembelitria* species relative to other species in order to determine depth habitat, productivity and temperature conditions, (2) trace element analysis of benthic and planktic foraminiferal tests to determine toxicity levels that may have proved lethal for specialized species, (3) evaluate nutrient, salinity, temperature and oxygen variations within crises intervals, and (4) evaluate the importance of the *Guembelitria* blooms in terms of biomass production as compared to the K-strategist species for a better understanding of the organic matter cycles in the oceans.

Acknowledgments

We dedicate this paper to Professor M.S. Srinivasan in honor of his 65th birthday and in recognition of his lifelong contributions to planktic foraminiferal micropaleontology and in particular his major contributions to marine paleoceanography. This study was supported by NSF Grant EAR-0207407. DSDP and ODP samples were provided by the Ocean Drilling Program.

References

ABRAMOVICH, S., ALMOGI, L.A. and BENJAMINI, C., 1998. Decline of the Maastrichtian pelagic ecosystem based on planktic Foraminifera assemblage change; implication for the terminal Cretaceous faunal crisis. Geology 26(1), 63-66.

ABRAMOVICH, S. and KELLER, G., 2002. High stress late Maastrichtian paleoenvironment: Inferences from planktic foraminifera in Tunisia. Palaeogeogr., Palaeoclimatol., Palaeoecol., 178: 145-164.

ABRAMOVICH, S. and KELLER, G., 2003. Planktic foraminiferal response to the latest Maastrichtian abrupt warm event: a case study from South Atlantic DSDP Site 525A. Marine Micropaleontology, 48, 225-249.

ABRAMOVICH, S., KELLER, G., ADATTE, T., STINNESBECK, W., HOTTINGER, L., STUEBEN, D., BERNER, Z., RAMANIVOSOA, B., RANDRIAMANANTENASOA, A., 2002. Age and paleoenvironment of the Maastrichtian-Paleocene of the Mahajanga basin, Madagascar: a multidisciplinary approach, Marine Micropaleontology, 47,17-70.

ADATTE, T., KELLER, G., STINNESBECK, W., 2002. Late Cretaceous to Early Paleocene climate and sea-level fluctuations. Paleogeogr., Paleoclimatol., Paleoecol. 178, 165-196.

BARRERA, E. and KELLER, G., 1994. Productivity across the Cretaceous-Tertiary boundary in high latitudes. *Geol. Soc. Amer. Bull.*, 106: 1254-1266.

BEGON, M., HARPER, J. L., TOWNSEND, C.R. 1998. Ecology: Individuals, Populations and Communities. 3rd ed. Blackwell Science, Boston, MA, 1068 pp.

BERGGREN, W. A., KENT, D. V., SWISHER III, C. C. and MILLER, K. G., 1995. A revised Paleogene geochronology and chronostratigraphy. In Berggren, W. A., Kent, D. V. and Handervol, J., eds. Society of Economic Geology and Paleontology, Special Publication, 54: 129-213.

BUZAS, M. A., 1990. Another look at confidencial limits for species proportions. *J. Paleont.*, 64: 842-843.

CHATTERJEE, S., GUVEN, N, YOSHINOBU, A., DONOFRIO, R., 2003. the Shiva crater: implications for Deccan volcanism, India-Seychelles rifting, dinosaur extinction, and petroleum entrapment at the KT boundary. 2003 GSA Seattle Annual Meeting (November 2-5, 2003), Abstract. # 58126.

DERCOURT, J., RICOU, L.E., and VRIELYNCK, B., 1993. Atlas: Tethyc Paleoenvironmental Maps. Gauthier-Villars, Paris, 307pp.

KELLER, G., 2002. *Guembelitria* dominated late Maastrichtian planktic foraminiferal assemblages mimic early Danian in Central Egypt, Mar. Micropaleontology, 47, 71-99.

KELLER, G. in press a. Low diversity late Maastrichtian and early Danian planktic foraminiferal assemblages of the eastern Tethys, Journal of Foraminiferal Research.

KELLER, G., in press b. Biotic effects of volcanism and impacts. Earth and Planetary Science Letters

KELLER, G., BARRERA, E., SCHMITZ, B., and MATSSON, E., 1993. Gradual mass extinction, species survivorship, and long term environmental changes across the Cretaceous-Tertiary boundary in high latitudes. Geol. Soc. Am. Bull. 105, 979-997.

KELLER, G., LI, L., and MacLEOD, N., 1995. The Cretaceous-Tertiary boundary stratotype section at El Kef, Tunisia: how catastrophic was the mass extinction? Paleogeogr., Paleoclimatol., Paleoecol. 119, 221-254.

KELLER, G., HAN, Q., ADATTE, T., BURNS, S. J., 2001. Paleoenvironment of the Cenomanian Turonian transition at Eastbourne, England. Cretac. Res. 22, 391-422.

KELLER, G., ADATTE, T., STINNESBECK, W., LUCIANI, V., KAROUI-YAAKOUB, N. and ZAGHBIB-TURKI, D., 2002. Paleoecology of the Cretaceous-Tertiary mass extinction in planktonic foraminifera. Paleogeogr., Palaeoclimatol., Paleoecol. 178, 257-297.

KROON, D. and NEDERBRAGT, A. J., 1990. Ecology and Paleoecology of triserial planktic foraminifera. Marine Micropaleontology 16, 25-38.

LECKIE, R. M., 1987. Paleoecology of mid-Cretaceous planktonic foraminifera: A comparison of open ocean and epicontinental sea assemblages. Micropaleontology, 33, 264-276.

LI, L. and KELLER, G., 1998a. Maastrichtian climate, productivity and faunal turnovers in planktic foraminifera in South Atlantic DSDP Sites 525 and 21. Mar. Micropaleontol. 33, 55-86.

LIPPS, J. H., 1979. Ecology and Paleoecology of planktic foraminifera. In Foraminiferal Ecology and Paleoecology, SEPM Short Course No. 6, Houston, Texas, 62- 104

MacLEOD, N., ORTIZ, N., FEFFERMAN, N., CLYDE, W., SCHULTER, C., MACLEAN, J., 2000. Phenotypic response of foraminifera to episodes of global environmental change. In Biotic response to global environmental change: the last 145 million years, Culver, S.J., Rawson, P. (eds.), Cambridge University Press, Cambridge, pp. 51-78.

MARGALEF, R., 1996. Ecología. Editorial Omega. Barcelona. 951 pp.

McGOWRAN, B., 1974. Foraminifera. In von der Borch, C.C., Slater, J. G., *et al.*, Initial Reports of the Deep Sea Drilling Project, Vo. 22, Washington (U.S. Government Printing Office) p. 609-628.

MOORE, D.G., CURRAY, J.R., RAITT, R.W., and EMMEL, F.J., 1974. Stratigraphic-seismic sections correlations and implications to Bengal fan history. In von der Borch, C.C., Slater, J. G., *et al.*, Initial Reports of the Deep Sea Drilling Project, Vo. 22, Washington (U.S. Government Printing Office) p. 403-412.

PARDO, A., 1999. Planktic foraminiferal turnovers across the K-T and P-E transitions in the northern Tethys realm. Anubar Ediciones, Zaragoza, 177 pp.

PARDO, A. and KELLER, G., 1999. Aspectos paleoceanográficos y paleoecológicos del límite Cretácico/Terciario en la Península de Mangyshlak (Kazakstan): inferencias a partir de foraminíferos planctónicos. Rev. Esp. Micropal. 31 (2), pp. 265-278.

PARDO, A., ORTIZ, N. and KELLER, G., 1996. Latest Maastrichtian and K/T boundary foraminiferal turnover and environmental changes at Agost, Spain. In McLeod, N. and Keller, G. eds: *Biotic and Environmental Events across the Cretaceous/Tertiary Boundary. Norton Press, New York*: 139-171.

PARDO, A., ADATTE, T., KELLER, G. and OBERHÄNSLI, H., 1999. Paleoenvironmental changes across the K-T transition in Koshak (Kazakstan) based on planktic foraminifera and clay minerals. Palaeogeogr., Palaeoclimat., Palaeoecol. 154, pp. 247-273.

SCHMITZ, B., KELLER, G. and STENVALL, O., 1992. Stable isotope and foraminiferal changes across the Cretaceous-Tertiary boundary at Stevns Klint, Denmark: Arguments for long-term oceanic instability before and after bolide impact. Paleogeogr., Paleoclimatol., Paleoecol., 96, 233-260.

TANTAWY A. A. A. 2002. Calcareous nannofossil biostratigraphy and paleoecology of the Cretaceous-Tertiary transition in the western desert of Egypt. Marine Micropaloentology 47, 323-356.

THOMPSON, G., BRYAN, W.B., FREY, F.A. and SUNG, A.C., 1974. Petrology and geochemistry of basalts and related rocks from sites 214, 215, 217 DSDP Leg 22, Indian Ocean. In von der Borch, C.C., Slater, J. G., *et al.*, Initial Reports of the Deep Sea Drilling Project, Vol. 22, Washington (U.S. Government Printing Office) p. 459-468.

Discovery of Early Miocene Microtektite horizons in the northern Indian Ocean DSDP Site 216: Geological implications

M.S. Srinivasan and Devesh K. Sinha

Department of Geology, Banaras Hindu University, Varanasi 221005, India

Introduction

Microtektites are microscopic (< 1mm) glassy objects ranging in size from pin –head down to about 40 microns (μm), known so far from oceanic sediments, although possibility of their occurrence in ice cores from polar regions can not be ruled out. These microforms of glass with essentially the same splash- form shapes as land tektites have same age and composition and provide evidence for a possible extraterrestrial bolidc impact.

It was Glass and Heezen in 1967 who first discovered the small glassy objects known as microtektites, stratigraphically associated with the Brunhes/ Matuyama paleomagnetic boundary (0.7 Ma) in Indian Ocean sediments (Glass and Heezen, 1967; Glass, 1967).

The DSDP site 216 from where the present microtektites were discovered was drilled on the crest of the Ninetyeast Ridge near the equator (Lat. 127^0.73'N and Long $90^0$12E) in the water depth of 2247 meter (Fig. 1) The Neogene at this site consists of nanno- foram ooze with abundant and well preserved calcareous microfossils. Thus, the site is quite good for the study of Neogene planktic foraminiferal Zones (Srinivasan and Chaturvedi, 1992). Nevertheless, discontinuous coring at this site preludes a detailed Neogene planktic foraminiferal biostratigraphy (Fig. 2).

It was during the investigation of Neogene planktic foraminifera from site 216, that the discovery was made of the two microtektite horizons in the Early Miocene, hitherto unrecognized. Further, it is shear coincidence that the present paper is written at a time when Dr. B.P. Radhakrishna (2004) made a call for conducting studies for locating and dating extraterrestrial impact events, and a need for specific search to be made for tektites, spherules and dust particles of extraterrestrial origin.

Material and Methods

Microtektites, as illustrated in Plate 1 were recovered in sufficient numbers from the two horizons in the Early Miocene nanno- foram chalk at DSDP site 216. These two horizons (22-216,6-5, 44-46 cm and 22-216,7-2,40-42 cm) occur above the first appearance datum of *Globoquadrina dehiscens* (23.2 Ma) which delineates the base of the Miocene (Top of Zone N4A) and before the last appearance datum of *Globorotalia kugleri* (21.8 Ma) which marks the top of Zone N4B in the Early Miocene (Fig.2). Based on the planktic foraminiferal time scale, an age of 23.2 to 21.8 Ma has been assigned to these microtektite horizons.

The microtektites found in these horizons are microscopic (1mm) glassy objects, which are spherical, flask shaped or elongate and range from bottle green to translucent black. The surface of microtektite is generally smooth while some exhibit impact structures and conchoidal fractures. Some flask shaped microtektites have small tubercles on their surfaces (Plate1). In general, the characteristics of the glassy spherules and elongate microtektites discovered in the deep sea sediments of Early Miocene age from northern Indian Ocean DSDP site 216 are similar to the microtektites reported from the late Eocene to Middle Oligocene marine sediments by Keller *et al.* (1983).

The newly discovered microtektites also resemble glassy objects reported from the late Pliocene marine sediments by Margolis *et al.* (1991) and 0.7 my old microtektite reported from Indian Ocean sediments (Glass and Heezen, 1967; Glass, 1967; Glass and Perch Nielsen, 1986; Prasad and Sudhakar, 1996a,b).

Stratigraphic record of microtektite horizons in the Cenozoic:

Growing realization of possible connection between impact events, microtektite occurrences and geochemical anomalies, led to search for microtektite horizons in the stratigraphic record, where they could serve as datums for improved correlation. So far, only a few occurrences of microtektites are known from oceanic sediments, and this is primarily due to incomplete search for microtektites.

Stratigraphic occurrences of microtektites in the Cenozoic are limited and in some cases they are known to be accompanied by geochemical anomalies. The discovery of geochemical anomalies accompanied by the occurrences of tektites in Cretaceous/ Paleocene boundary sequence is now well documented on a global scale (Ryder *et al.,* 1996). It is now well established that the extraterrestrial impact caused these anomalies and also led to mass extinction (Alvarez *et al.* 1980; Ganapathy, 1980, 1982; Smit and Hertozen, 1980; Grieve, 1997).

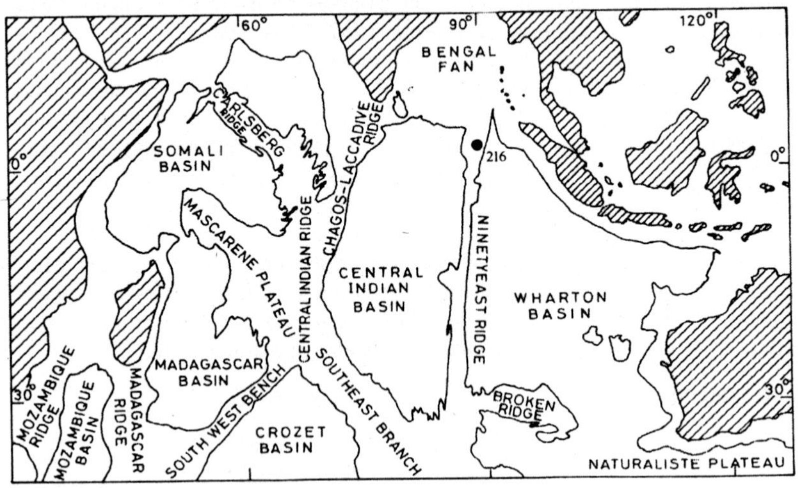

Fig. 1. Physiographic features of the Indian Ocean as defined by 4000m contour and the location of the DSDP site 216 from where Microtektites have been discovered.

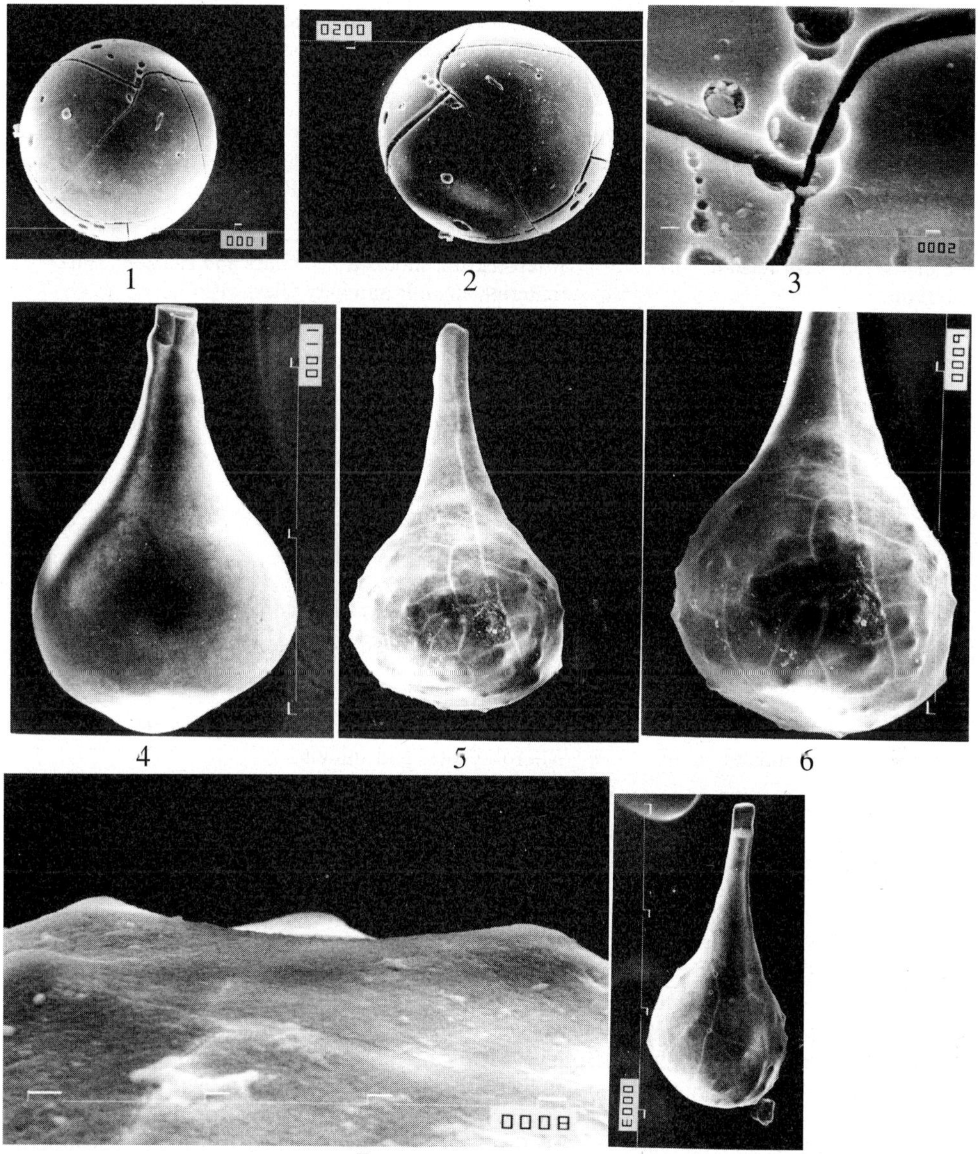

Plate 1

1. Microtektite (Spherical) – DSDP site 216-6-5, 44-46 cm, × 420; **2.** Microtektite (Spherical) - DSDP site 216-6-5,44-46 cm, × 420; **3.** Enlarged view of fig.2 - DSDP site 216-6-5,44-46 cm ×4100; **4-6** and **8-** Funnel shaped Microtektite , DSDP site 216,7-2,40-42cm ×300; **7.** Enlarged view of surface of Microtektite in fig.5, ×2700

Unlike the terminal Cretaceous event, the Cenozoic microtektite occurrences cover a small area. So, the following distinct microtektite horizons (four in the Neogene and three in the Paleogene) are known including the present discovery in the Early Miocene (Table-1)

Table 1. Microtektite horizons in the Cenozoic Deep sea Record

Stratigraphic Horizon	Age (Ma)	Evidence	Reference
Late Pleistocene Australasian Microtektites	0.7	Microtektite, shocked material, small Ir anomaly	Glass and Heezen (1967) Glass (1967) Shyam Prasad and Rao (1990) and Shyam Prasad and Sudhakar (1996a,b)
Early Pleistocene Ivory Coast Microtektites	1.1	Microtektite, Ir anomaly, $^{187}Os/^{188}Os$ ratio	Koeberl and Shirey (1993)
Late Pliocene Microtektites, North Pacific	2.2	Microtektite, Siderophile anomaly	Peng (1994)
Late Pliocene Microtektites, South Pacific	2.3	Microtektite	Kyte *et al* (1988), Margolis *et al* (1991)
Early Miocene Microtektites, Northern Indian Ocean	22	Microtektite	Present Work
Middle Miocene Microtektites, Pacific, Atlantic and Indian Oceans	31-32	Microtektite, Microkrystites and spinels	Keller *et al* (1983)
Late Eocene Microtektites, Pacific, Atlantic and Indian Oceans	37.5-38.0	Microtektite	Keller *et al* (1983)
Middle Eocene Microtektites Atlantic and Indian Oceans	38.5-39.5	Microtektite	Keller *et al* (1983)

i) 0.7 my old microtektite horizon was first discovered by Glass and Heezen (1967) and Glass (1967) associated with the Brunhes/Matuyama paleomagnetic boundary in the Indian Ocean. By examining a large number of deep sea cores, Glass (1972) and Glass *et al.* (1979) identified more than 40 occurrences in the Australasian region. This youngest microtektite horizon is also the largest strewn field covering almost the entire Indian Ocean and part of the western Pacific (Australasian Microtektite Strewn field). Studies on geochemical anomalies and shocked minerals associated with Australasian microtektites, have firmly established the impact origin of strewn field (Prasad and Rao, 1990, Prasad and Sudhakar, 1996a, 1996b).

ii) 1.1 my old microtektite horizon has been found in the equatorial Atlantic (Ivory Coast Microtektite Strewn field). The Ivory Coast microtektites are accompanied by small Ir anomaly, low $^{187}Os/^{188}Os$ ratio and associated with Bosumtwi structure, Ghana (Koeberl and Shirey, 1993).

iii) ~ 2.3 my old microtektite occurrence in the southern Ocean was first discovered by Margolis *et al.* (1991). Microtektite bearing sediments are enriched in Ir and Au and contain impact debris. According to Margolis *et al* (1991) about 2.3 million years ago an asteroid ~ 0.5 km in diameter impacted the south Pacific Ocean and produced these microtektites and microkrystites. 2.2 my old microtektite horizon reported by Peng (1994) from the north Pacific appears to be linked to the late Pliocene impact event reported by Margolis *et al.* (1991).

iv) ~22 my old microtektite occurrence is reported here for the first time from the northern Indian Ocean DSDP site 216 located on the crest of the Ninetyeast Ridge. Since the microtektite horizons are within the *Globoquadrina dehisces* Zone, and immediately below the last appearance datum of *Globorotalia kugleri* (21.8 Ma), an age of 22 million years has been assigned to this new microtektite horizon.

Keller *et al.* (1983) recognized the following three microtektite horizons in the Paleogene:

a) 31-32 my old (Middle Oligocene) horizon recorded in the DSDP site 292 in the Pacific and site SSQ in the Atlantic and DSDP site 242 in the Indian Ocean.

b) 37.5-38.0 my old (Late Eocene) occurrence in the DSDP sites 292 and 167 in the Pacific, site 94, E67-128 and SSQ in the Atlantic and site 242 in the Indian Ocean. Occurrence of coeval microtektite horizons is also known from Barbados in the Caribbean Sea and Gulf of Mexico representing north American microtektite strewn field (Saunders *et al*, 1984; Keller *et al*, 1983). The microtektite horizon reported from the marine outcrop of Tapti River section in Gujarat by Shukla and Bhandari (1997) appears to be linked with the late Eocene bolide impact that occurred at about 37.5 to 38.0 million years ago (Glass *et al*, 1985).

c) 38.5-39.5 my old (Middle Eocene) horizon recorded in the DSDP site 149 and RC9-58 in the Atlantic and DSDP site 242 in the Indian Ocean.

Of the three Paleogene multiple microtektite horizons, the one occurring at the Late Eocene with paleomagnetic age of 37.5-38.0 Ma (Keller *et al*, 1983) appear to be more widespread similar to the youngest Cenozoic microtektite horizon recorded at 0.7 Ma.

Origin of Microtektites

The origin of microtektites and tektites has still remained an unresolved issue. Glass's initial (1967, 1972) studies involved a search for microtektites in oceanic sediments and to trace their geographic and stratigraphic distributions in order to date their formation. Although, the origin of microtektites is still unsolved, the principle hypothesis of their origin involves collision of asteroids, meteorites or comets with the Earth or the Moon. Thus, occurrence of microtektites in the stratigraphic record present evidence of extraterrestrial impact events.

There have been some speculations about lightening or meteorite being the cause of their formation. Still recently, there was a great deal of credibility given to the theory of the lunar origin of microtektites/ tektites propagated mainly by O' Keefe (1969) and King (1977). The analyses of lunar samples brought back to Earth however have thrown some cold water upon this theory, as these were very different from the composition of tektites. Nevertheless, some feel that the lunar theory can not yet be rejected completely.

It is clear from the published data that many more microtektites remain to be discovered and a detailed study of geochemical anomalies and physical characteristics related to impact events need to be carried out to resolve the issue of origin of microtektites.

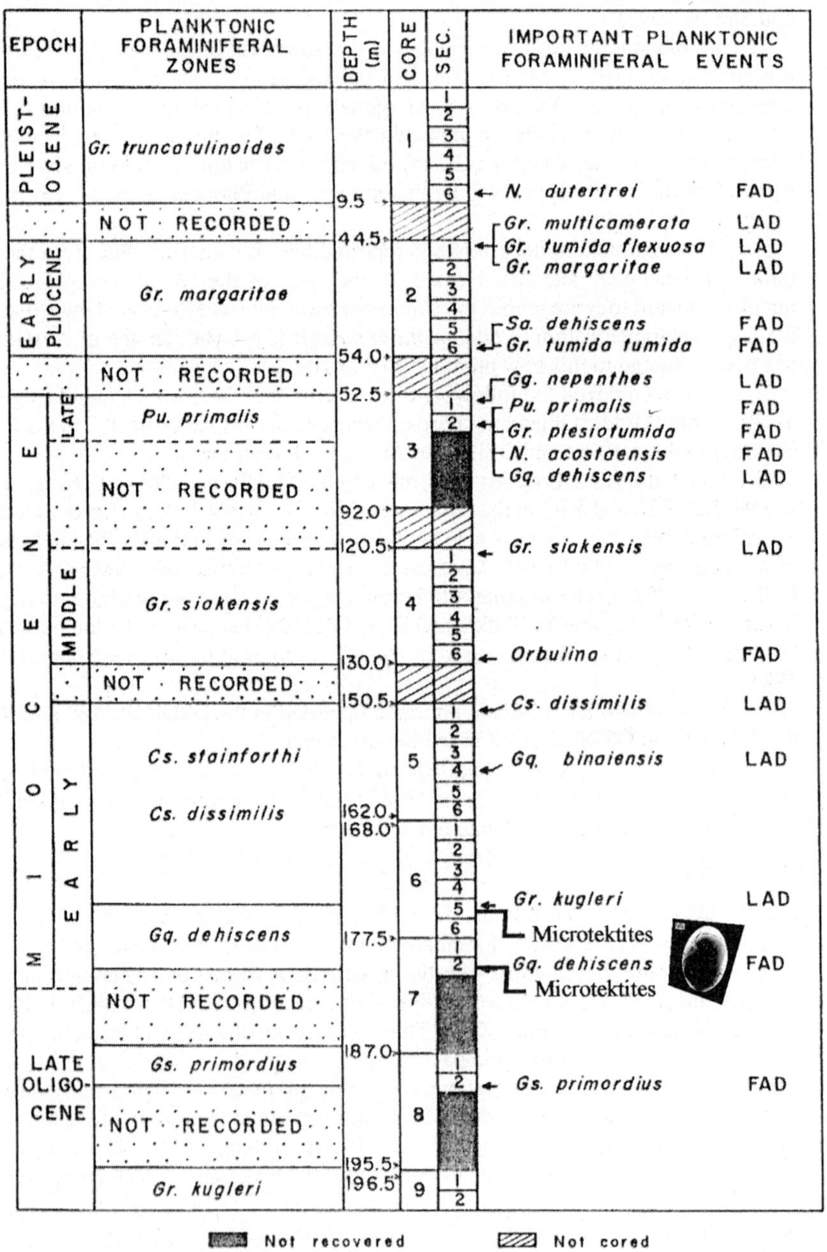

Fig. 2. Neogene Planktic foraminiferal zones, important planktic foraminiferal events and microtektite horizons at DSDP site 216. (after Srinivasan and Chaturvedi, 1992).

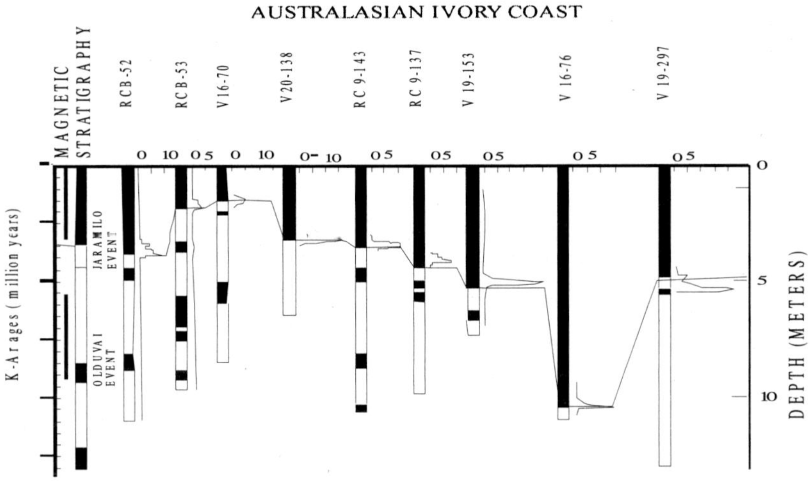

Fig. 3. Correlation of the microtektite horizons with the paleomagnetic stratigraphy (after Gentner *et al*, 1970)

Microtektites and Geomagnetic reversals

Soon after the discovery of magnetostratigraphy and its application to deep sea sequences, some intriguing relations between geomagnetic reversals and other geological processes recorded in the deep sea emerged (Kennett, 1980). The earliest of these was by Glass and Heezen (1967) and Glass (1967) of the Australasian Microtektite horizon which coincided with Brunches/ Matuyama paleomagnetic boundary (0.7 my BP) suggesting that a common cause was responsible for both the phenomena.

Glass and Heezen (1967a, b) speculated that the magnetic reversals may be at times linked to collisions of cosmic bodies with the Earth. This still remains as speculative, as this assumption is based on single stratigraphic observation.

An examination of magnetostratigraphy of the other known microtektite horizons for example, the ivory coast microtektite (1.1 my old), the Early Miocene (22 Ma) and the multiple microtektite horizons in the Paleogene (Keller *et al*, 1983) do not seem to be associated with any magnetic reversals. Nevertheless, paleomagnetostratigraphy has been important for dating microtektite horizons in the deep sea cores. Gentner *et al* (1970) based on paleomagnetic dating demonstrated that the Australasian and Ivory Coast tektite fields are of different age (Fig.3) in contrast to the earlier view of Chapman and Schreiber (1969) who considered both of them as part of the same strewn field on geochemical evidences.

Thus there is still not sufficient data linking microtektites with magnetic reversals. Therefore, an intensive search for microtektites in sediments deposited during geomagnetic reversals is necessary to ascertain cosmic encounters as cause of geomagnetic reversals.

Microtektites and faunal extinctions

Evidence of extraterrestrial impact events in stratigraphic record occurs most commonly as microtektite or tektite horizons which in some cases are accompanied by geochemical anomalies, but there is no confirmatory evidence that they are impact related.

Since the discovery of major faunal extinction associated with bolide impact at the Cretaceous / Paleocene boundary, there is a general assumption that the impact events are linked to the large scale biotic extinctions of great geological significance. Nevertheless, how theses bolide impacts might have affected organisms in the past remains a matter of speculation at this time. According to Glass and Heezen (1967a, b), collisions of cosmic bodies with the Earth appears to be linked with the geomagnetic reversals. Kennett and Watkins (1970) on the other hand observed coincidences of geomagnetic polarity change, volcanic maxima and faunal extinctions in the south Pacific deep sea cores. The faunal extinction according to Kennett and Watkins (ibid) is due to volcanically induced climatic change. However no compelling evidence has yet been presented to demonstrate that paleoclimatic change of large magnitude during geomagnetic reversals could have caused extinctions as suggested by Harrison (1968).

The reported occurrence of microtektite horizons in the Atlantic (RC9-58 and DSDP site 149) and the suggested extinction of five radiolarian species and the major mammal extinctions .Alvarez *et al.* (1980), Ganpathy (1982), and Sanfilipo *et al*, 1985) had to speculate that these horizons also mark major extinction event caused by a bolide impact at the Eocene / Oligocene boundary, similar to the terminal Cretaceous impact event.

However Keller *et al* (1983) observed that the extinctions of radiolarian species coinciding with the microtektite horizon is due to a hiatus and that the mammalian extinctions which Alvarez *et al* (1980) reported as mass extinctions occur in the middle Oligocene and not at the Eocene / Oligocene boundary, atleast 5 million years after the supposed bolide impact.

Further, quantitative nannofossils and planktic foraminiferal data also do not reveal any mass extinction in oceanic record at the Eocene –Oligocene transition. However, climatic cooling induced by the growth of the Antarctic ice sheet during middle Eocene to Middle Oligocene triggered by separation of Australia from Antarctica and the subsequent development of the Circum Antarctic Current appears to have caused gradual faunal turnover during this interval.

In the present study there is no evidence of major faunal extinctions associated with any of the microtektite horizons in the Early Miocene. Hence, it is difficult to ascertain any link between microtektite occurrence and faunal extinctions. On the other hand these microtektite horizons preceded planktic foraminiferal radiation and diversification that occurred during the Early Miocene. Thus our findings are in agreement with Keller *et al.*(1983) who earlier found no evidence for mass extinctions associated with any of the microtektite horizons in the upper Eocene marine sediments.

It is believed that microtektite horizons of different ages exist in the deep sea stratigraphic record. Large number of micropaleontologists who are now engaged in examining thousands of core samples recovered from the Deep Sea Drilling Project and Ocean Drilling Program can search and discover new microtektite horizons. Similarly, paleoclimatologists who are examining ice cores from the Arctic and Antarctic regions should also look for microtektites as these are the ideal locations for the preservation of microtektites and other impact related material of extraterrestrial origin.

Discovery of new microtektite horizons will provide excellent tool for improved chronostratigraphic correlation on a regional scale and may also provide evidences of past catastrophic events that the Earth has witnessed hitherto unknown. This would enable us further to accept the extra-terrestrial impact events as just another potential force changing the face of the planet.

Conclusions

i) Examination of DSDP cores from site 216 in the northern Indian Ocean resulted in the discovery of two new early Miocene microtektite horizons, hitherto unknown.

ii) The present study does not reveal any evidence of major faunal extinctions coinciding with the microtektite horizons. On the other hand, these horizons precede the planktic foraminiferal radiation and diversification that occurred during the Early Miocene.

iii) Also there is no indication that these microtektite horizons are stratigraphically associated with geomagnetic reversal change

iv) So far only a few occurrences of microtektites are known from the deep sea record and this is primarily due to incomplete search for microtektites. Therefore much new data still needs to be accumulated with regard to the microtektite horizons in the deep sea stratigraphic record to understand how far geomagnetic reversals and faunal extinctions are linked to bolide impact.

v) Discovery of new microtektite horizons will provide excellent tool for improved correlation on a regional scale and will also furnish information regarding hitherto unknown impact events and their consequences.

Acknowledgements

The Deep Sea Drilling Project (DSDP) is thankfully acknowledged for providing core samples for the present study. One of the authors (MSS) acknowledges the financial support provided by the Indian National Science Academy, New Delhi, in the form of INSA Senior Scientist Scheme. DKS thanks the Department of Science and Technology, New Delhi for financial support.

References

ALVAREZ, L.W., ALVAREZ, W., ASARO, F., and MICHEL, H.V. 1980. Extra-terrestrial cause for the Cretaceous Tertiary extinctions. Science 208, . 1095-1108.

CHAPMAN, D.R. and SCHREIBER, L.C. 1969. Chemical investigations of Australasian Tektites. Jour. Geophys. Research .74, .6737-6776.

GANAPATHY, R. 1980. A major meteorite impact on the Earh 65 million years ago. Evidence from the Cretaceous-tertiary boundary clay. Science, 209, 921-923.

GANAPATHY, R.1982. Evidence for a major meteorite impact on the Earth 34 million years ago. Implications for Eocene extinctions. Science, 216, 885-886.

GENTNER, W., GLASS, B.P., STORZER, D. and WAGNER, G.A. 1970. Fission Track ages and ages of deposition of deep sea microtektites. Science 168: 359-361.

GLASS, B.P. 1967. Microtektites in deep sea sediments. Nature 214, 372-374.

GLASS, B.P. 1968. Microtektites and the origin of Australasian tektite Strewn field. Publication No.6 by the center for Meteorite studies, tempe, Arizona State University. 22 p.

GLASS, B.P.1972. Australasian Microtektites in deep sea sediments. In : Antarctic Oceanology II, The Australian-New Zealand sector, Antarctic Research Series, vol.19, 335-348.

GLASS, B.P. and HEEZEN,B.C.1967a. Tektites and geomagnetic reversals. Nature 214, 372.

GLASS, B.P. and HEEZEN, B.C. 1967b. Tektites and geomagnetic reversals. Scientififc American, vol.217,No.1, 33-39.

GLASS, B.P. and PERCH-NIELSEN, K. 1986. Microtektite in Micropaleontological samples. Micropaleontology, vol.32, no.1, 46-47.

GLASS, B.P. SWINCKI, M.B. and ZWART, P.A. 1979 Australaisan, Ivory coast and north American Tektite Strewn fields: size, mass and correlation with geomagnetic reversals and other Earth events. Proc. 10th Lunar and Palnetary Sci. Conf. 2535-2545.

GLASS, B.P., BURNS, C.A., CROSBIE, J.R., DU. and BOIS, D.L. 1985. Late Eocene north American Microtektites and clinopyroxene bearing spherules. In : proc. 16th Linar planet. Sci. Conf.. Jour. Geophyic. Reasearch Support 90, D 175-D196.

GRIEVE, R.A.F.1997. Extraterresrial impact events : The record in the rocks and the Stratigraphic coloumn. Paleogr. Paleoclimat. Paleoecol. 132, 5-23.

HARRISON, C.G.A. 1968. Evolutionary processes and and reversals of the Earth's magnetic field. Nature. 217, 46-47

KELLER.G., D' HONDT, S. and VALLIER, T.L. 1983. Multiple Microtektite horizons in Upper Eocene marine sediments: No evidence for mars extinctions Science, 221, 150-152.

KENNETT, J.P. 1980. Magnetic Stratigraphy of sediments (Editor). Bench mark papers in Geology/ 54. Dowden, Hutchingon & Ross, Inc. Pennsylvania, 438p.

KENNETT, J.P. and WATKINS, N.D. 1970. Geomagnetic polarity change, volcanic maxima, and faunal extinctions in the South Pacific. Nature. vol. 227, 930-934.

KING, E.A. 1977. The origin of tektites. A brief review. American Scientist 65, 212-218.

KOEBERL. C. and SHIREY, S.B. 1993 detection of a meteorite component in Ivory Coast tektites with Rhenium-Osmium isotopes. Science 261, 595-598.

KYTE, F.T., ZHOU, Z. and WASSON, J.T. 1988. New evidence on the size and possible effects of a late Pliocene oceanic asteroid impact. Science 241, 63-65.

MARGOLIS, S.V., CLAEYS and KYTE. 1991. Microtektite, Microkrystites and Spinels from a late Pliocene Asteroid impact in the southern Ocean. Science 251, 1594-1597.

O' KEEFE, J.A..1969. The Microtektite data: implications for the hypothesis of the lunar origin of tektites. Jour. Gephysic. Research. 74. 6795-6804.

PENG, H. 1994. An extraterrestrial event at the Tertiary Quaternary boundary (Abst) In: New Delopments regarding the KT events and other catastrophies in Earth's Hiostory. Lunar Palnet Inst. Contrib. 825, 88-89.

PRASAD M.S. and RAO, P.S.1990. Tektites far and wised. Nature, vol347, 340.

PRASAD M.S. and SUDHAKAR, S. 1996b. Microcrators on an Australasian Microtektites. Meteorite and Planetary Science, 31, 46-49.

PRASAD M.S. and SUDHAKAR, S.1996a. Australasian Microtektites from the Central Indian basin: Implications for ejecta distribution patterns. Current cience vol.70, No.1, 74-77.

RADHAKRISHNA, B.P.2004. Extra terrestrial impact in the geological past some reflections, Jour. Geol.Soc. India. vol.64 no.2 127-132.

RYOLER,G., FASTOVSKY,D. and GARDNER, D. (Eds.) 1996. New Deleopments regarding the KT event and other catasptrophies in Earth History. Geol. Soc. Ameri. Spec. Pap. 306, 569p

SANFLIPPO, A. RIEDEL, W.R., GLASS, B.P. and KYTE, F.T.1985. Late Eocene Microtektites and radiolarian extinctions on Barbados. Nature 314, 613-615.

SAUNDERS, J.B., BERNOWLLI, D., MIILLER-MERZ, E, OBERHANSLI, H., PERCH, NIELSEN, K., RIEDEL, W.K. SANFLIPPO, A. and TORRINI, R. Jr. 1984. Stratigraphy of the late Middle Eocene to early Oligocene in the bath Clif Section, Barbados, West Indies, Micropaleontology, vol. 30, No.4, 390-425.

SHUKLA, P.N. and BHANDARI, N. 1997. Physical chemical and biological changes at geological boundaries: causes, consequences and clues based on the study of Indian sections. Paleobotanists, 46, 41-62.

SMIT, J. and HERTOGEN, J. 1980. An extraterrestrial event at the Cretaceous- tertiary boundary. Nature, 285, 198-200.

SRINIVASAN, M.S. and CHATURVEDI, S.N.1992. Neogene Planktonic foraminiferal Biochronology of the DSDP sites along the Ninetyeast Ridge, Northern Indian Ocean. In: centenary of Japaneses Micropal. (eds. Ishizaki and Saito), 175-188.

Astronomic and Oceanographic Influences on Global Carbon Cycling across the Oligocene/Miocene Boundary

Benjamin P. Flower[1], James C. Zachos[2] and Paul N. Pearson[3]

[1]College of Marine Science, University of South Florida, St. Petersburg, FL 33701
[2]Earth Sciences Department, University of California, Santa Cruz CA 95064
[3]Department of Earth Sciences, University of Bristol, Bristol BS8 1RJ, UK

Introduction

Oligocene/Miocene organic carbon burial

The $\delta^{13}C$ composition of marine inorganic carbon on geologic time scales has been used to assess the balance between burial and weathering of organic and inorganic carbon (e.g., Shackleton, 1984, 1985; Kump and Garrels, 1986). This is based on the observation that storage of light organic carbon in marine sedimentary reservoirs (~-20 to -25‰) or continental reservoirs (~-25‰) leaves the residual marine dissolved inorganic carbon (DIC) pool enriched in $\delta^{13}C$ (e.g., Broecker and Peng, 1982). For example, a sustained increase in marine organic carbon burial of $2.1*10^{12}$ tons C would enrich ocean $\delta^{13}C$ by ~1‰ (e.g., Vincent and Berger, 1985). Shackleton (1985) produced a Cenozoic $\delta^{13}C$ record based on bulk carbonate sediment that documented four major episodes of increased relative organic carbon burial, during the middle Eocene, early Oligocene, Oligocene/Miocene (O/M) boundary, and early to middle Miocene. These increases are also recorded in foraminiferal calcite, although other processes modify their $\delta^{13}C$ composition.

A distinct positive excursion in benthic and planktic foraminiferal $\delta^{13}C$ of ~1‰ was confirmed near the O/M boundary (Miller *et al.*, 1991a; Wright *et al.*, 1992; Hodell and Woodruff, 1994; Flower *et al.*, 1997a,b) ca. 23 Ma on the timescale of Cande and Kent, 1992, 1995). The excursion began near the top of Subchron C6Cr and ended near the Chron C6AA/C6Bn boundary (Hodell and Woodruff, 1994). In fact, this $\delta^{13}C$ maximum has since been utilized as a distinct stratigraphic marker for the O/M boundary (Hodell and Woodruff, 1994). The $\delta^{13}C$ excursion is composed of several individual carbon isotope maxima that have a periodicity of approximately 400 k.y. (Zachos *et al.*, 1997; 2001b; Paul *et al.*, 2000). Carbonate preservation is good during the maxima and poor in between, suggesting control by ocean circulation and chemistry changes. A similarrelationship occurred during the "Monterey Carbon Isotope Excursion" in the early to middle Miocene (Vincent and Killingley, 1985; Vincent and Berger, 1985; Miller *et al.*, 1991a,b; Woodruff and Savin, 1991; Flower *et al.* 1993a,b).

Trade wind induced upwelling

One hypothesis for increased organic carbon burial near the O/M boundary is essentially an application of the "Monterey Hypothesis" originally developed for the early to middle Miocene $\delta^{13}C$ excursion (Vincent and Berger, 1985). The "Monterey Hypothesis" holds that polar cooling increased the planetary temperature gradient, which increased trade wind intensity and in turn upwelling, particularly in coastal regions around the Pacific marked by marginal basins (Vincent and Berger, 1985). A similar mechanism may have operated near the O/M boundary. High-resolution stable isotope records demonstrate that $\delta^{13}C$ maxima within the excursion are associated with $\delta^{18}O$ maxima, confirming a relation to deep sea cooling and ice growth (Flower *et al.*, 1997a; Zachos *et al.*, 1997). A key prediction of this mechanism is increases in trade wind strength, at least during the individual $\delta^{13}C$ maxima. Such increases may be necessary to enhance coastal upwelling and associated organic carbon deposition. However, there is as yet no evidence for trade wind strength increases during individual $\delta^{13}C$ maxima (Hovan and Rea, 1992). Indeed, eolian dust flux near the O/M boundary is not particularly high relative to the rest of the Cenozoic (Rea, 1994).

More recently, proxy estimates of latest Oligocene to Miocene pCO_2 indicated surprisingly low levels throughout the Miocene (Pagani *et al.*, 1999; Pearson and Palmer, 2000). Although these data hint at orbital-scale variations in Miocene pCO_2, mean levels (200-300 ppmv) appear similar to preindustrial levels. Thus it is possible that the O/M boundary and early to middle Miocene $\delta^{13}C$ excursions were closely associated with atmospheric CO_2 drawdown (Pagani *et al.*, 1999; Flower, 1999; Pearson and Palmer, 2000).

Nutrient delivery

Raymo (1994) advanced the hypothesis that Miocene organic carbon deposition increased because of greater nutrient delivery to the oceans related to Himalayan uplift. This hypothesis does not invoke trade wind variability. She used the strontium isotope composition of marine carbonate (Richter *et al.*, 1992) as a proxy of continental weathering, and in combination with Shackleton's (1985) estimate of organic carbon vs. carbonate burial, suggested a pronounced increase in organic carbon burial during the early to middle Miocene (Raymo, 1994). Increased continental nutrient supply could also have been derived from other regions (Ravizza, 1993; Raymo, 1994; Reusch *et al.*, 1998). The nutrient delivery hypothesis is consistent with the observation that carbonate accumulation also increased during the mid-Miocene (Woodruff and Savin, 1991), although this trend could alternatively be due to regional shifts in carbonate preservation. This model can be applied to the O/M boundary. Increased nutrient supply should enhance the "biologic pump" removing nutrients from surface waters and returning them to sub-thermocline waters via remineralization (e.g., Broecker and Peng, 1982). Such remineralization should also enhance development of the oxygen minimum zone (OMZ) in certain areas.

Sea-level-controlled upwelling

One other hypothesis that deserves mention is sea-level-controlled upwelling on shallow carbonate banks (Compton *et al.*, 1990). Widespread phosphorite deposits in the southeastern U.S. have been associated with $\delta^{13}C$ maxima from the O/M boundary to the middle Miocene (Compton *et al.*, 1990; 1993; Mallinson and Compton, 1997). Originally rich in organic carbon, these deposits are inferred to have contributed to a mean ocean $\delta^{13}C$ increase, and subsequently diagenetically altered to phosphorite. Noting that these deposits roughly coincide with Miocene $\delta^{13}C$ maxima during an interval of generally rising sea level on the >1 m.y. scale, these workers proposed that topographically induced upwelling in this region may have resulted in organic carbon and phosphate accumulation. However, the association of $\delta^{13}C$ maxima with $\delta^{18}O$ maxima (sea-level lowering) on the <1 m.y. scale indicates that different processes control organic carbon burial on orbital time scales (Mallinson and Compton, 1997). Nevertheless, it is possible that sea-level-controlled upwelling played a significant role in organic carbon burial on the super-orbital scale during the O/M $\delta^{13}C$ maximum.

Approach

Western Equatorial Atlantic hydrography as a monitor of global change

A central tenet of the Monterey Hypothesis is that increased planetary thermal gradients enhanced trade wind strength, and hence coastal upwelling (Vincent and Berger, 1985). Increased trade winds would also have two important effects in equatorial settings: (1) Open ocean upwelling would increase in central and eastern regions, and (2) warm surface waters would "pile up" in western equatorial areas, leading to a thicker mixed layer (and a deeper thermocline; Merle, 1980). These effects should have been partially reduced in the Miocene ocean because the Isthmus of Panama was open to at least intermediate depths (Duque-Caro *et al.*, 1990) and therefore the bathymetric barrier that allowed piling up of warm surface waters in the Caribbean region was diminished. However, the southeasterly trade winds and the South Equatorial Current (SEC) are robust features of the climate system (Philander and Pacanowski, 1986) and probably controlled mixed layer depth near Ceara Rise, as in the present day ocean (Fig. 1). Southeasterly trade wind strength is greatest during the late boreal summer when the intertropical convergence zone (ITCZ) is at its northernmost extent, leading to a thick mixed layer and increased strength of the SEC off the northeast coast of South America (Philander and Pacanowski, 1986). Conversely, decreased trade wind strength during boreal spring when the ITCZ is further south leads to a shallower thermocline and reduced SEC strength (Fig. 1). Therefore the Ceara Rise region should be quite sensitive to mixed layer variability related to strength of the SEC and associated southeasterly trade winds.

Correspondingly, the nutrient delivery hypothesis predicts that $\delta^{13}C$ maxima should be associated with increased carbonate and/or organic carbon productivity (Raymo, 1994). If increased nutrient delivery came from increased global weathering of continental rocks and sediments, it may be associated with increased bioproductivity across the low-latitude regions. Alternatively, it is possible that increased bioproductivity was limited to coastal and upwelling regions. In either case, increased continental weathering should also have been accompanied by increased radiogenic contributions (e.g., $^{87}Sr/^{86}Sr$ and $^{187}Os/^{186}Os$; Richter *et al.*, 1992; Ravizza, 1993).

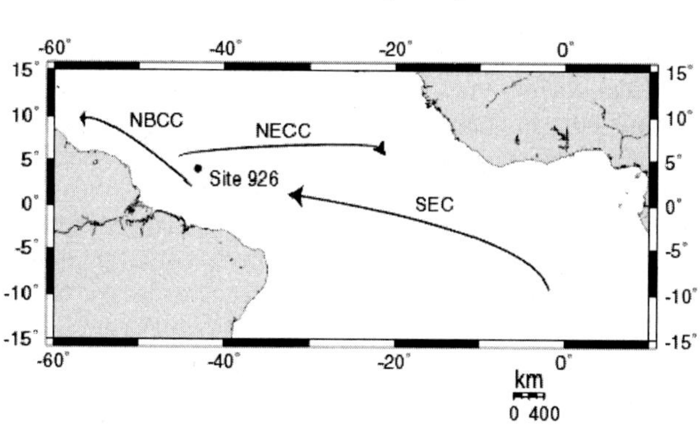

Fig. 1. Location map showing position of Site 926 on Ceara Rise in the western equatorial Atlantic (3°43.148'N, 42°54.507W; 3598.3 mwd). Also shown are average boreal summer position and relative strength of the Southeast Equatorial Current (SEC), North Brazil Coastal Current (NBCC), and North Equatorial Counter Current (NECC).

Foraminifera as recorders of mixed layer hydrography and productivity

The relative abundance of mixed layer species in a foraminiferal assemblage has been widely used as a proxy for mixed layer thickness in the western equatorial Atlantic (e.g., Ravelo *et al.*, 1990; Cullen and Curry, 1997; Chaisson and Ravelo, 1997). Mixed layer species thrive at the expense of thermocline species when the mixed layer is deep, because of increased niche availability in the photic zone. When the thermocline shoals into the photic zone, the relative abundance of thermocline species increases (Leckie, 1989).

The stable isotope geochemistry of planktic foraminiferal tests is also an excellent recorder of the environments in which foraminifera live. Surface-dwelling forms typically exhibit low $\delta^{18}O$ values reflecting warm SSTs and high $\delta^{13}C$ values reflecting nutrient-depleted surface water DIC (e.g., Fairbanks *et al.*, 1982; Ravelo *et al.*, 1990; Ravelo and Fairbanks, 1992; 1995). Thermocline-dwelling forms exhibit higher $\delta^{18}O$ values reflecting colder temperatures and lower $\delta^{13}C$ values reflecting remineralization of organic matter. Differences in $\delta^{18}O$ and $\delta^{13}C$ from coeval samples can be considered to reflect temperature gradients and nutrient gradients from mixed layer to thermocline depths (e.g., Broecker and Peng, 1982). In the modern tropical Atlantic, *Globigerinoides sacculifer* (without sac) is commonly used as a recorder of near-surface conditions, and *Menardella tumida* is often used to record lower mixed layer conditions.

Planktic foraminiferal $\delta^{18}O$ is also influenced by evaporation/precipitation balance. However, salinity variations are expected to be small in this open equatorial ocean setting. This is because the slope of the $\delta^{18}O$seawater/S relationship is quite low in the equatorial Atlantic (Fairbanks *et al.*, 1992). Secondary calcification at depth is an acknowledged complication (Lohmann, 1995). This along with partial shell recrystallization during early burial may also alter primary isotope values. However, while diagenetic influences shift the absolute values, because the majority of primary calcite is retained, the original isotope gradients are largely preserved (Pearson *et al.*,1993, Crowley and Zachos, 1999; Zachos *et al.*, 2002).

Overall, variability in mixed layer depth and bioproductivity should be manifested in assemblage and isotopic records based on planktic foraminifera that inhabit mixed layer to thermocline depths (e.g., Ravelo *et al.*, 1990; Ravelo and Fairbanks, 1992). In particular, a thick mixed layer that extends below the photic zone should be reflected by (a) dominance of mixed layer species at the expense of thermocline-dwellers, and (b) increased $\delta^{18}O$ gradients between mixed-layer species and thermocline species (Ravelo *et al.*, 1990; Ravelo and Fairbanks, 1992; Pearson *et al.*, 1997; Chaisson and Ravelo, 1997; 2000; Faul *et al.*, 2000). In contrast, a shallower thermocline within the photic zone will (a) reduce the relative abundance of mixed layer foraminifera and (b) decrease the $\delta^{18}O$ gradient between mixed-layer and thermocline species. Similarly, increased productivity in a low-latitude setting should increase the surface-to-thermocline $\delta^{13}C$ gradient as added nutrients enhance the "biologic pump." Such an increase should be recorded as an increase in the $\delta^{13}C$ gradient between surface and thermocline planktic foraminifera. Increased productivity should also increase abundances of planktic foraminifera and coccolithophorids relative to terrigenous clays in Ceara Rise sediments.

For the O/M boundary interval, we combine full foraminiferal assemblage counts with stable isotope data on *Globigerina praebulloides* and *Paragloborotalia mayeri* and investigate their interrelationships with published benthic foraminiferal stable isotope data. These species are chosen because they are interpreted to reflect mixed layer and upper thermocline conditions, respectively (Pearson *et al.*, 1997). Our main goal is to investigate the relation of inferred changes in surface hydrography to global carbon cycling across the O/M boundary.

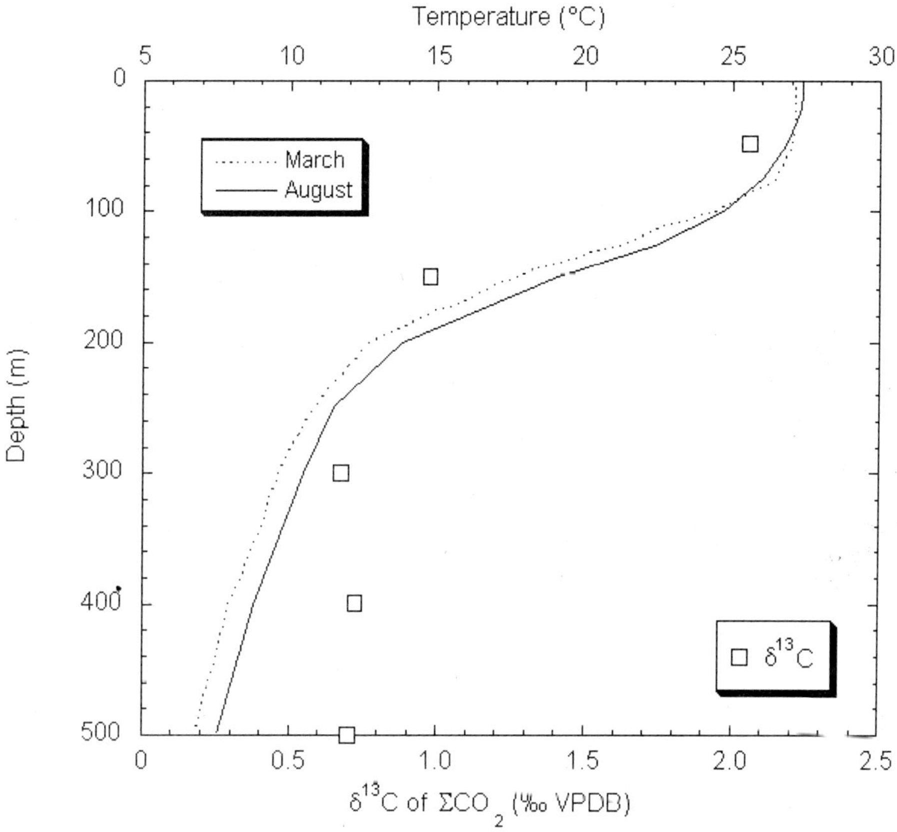

Fig. 2. Hydrographic profile in the Ceara Rise region from surface to 500 m (after Levitus and Boyer, 1994). Marked by the 18°C isotherm, mixed layer thickness is greatest during the late boreal summer (August shown) in response to increased southeasterly trade winds. Also shown is the δ^{13}C composition of σCO_2 at GEOSECS Stations 39-46 in the western equatorial Atlantic (Kroopnick, 1985), exhibiting high values in surface waters and low values in the thermocline related to bioproductivity.

Methods

Sediment samples from Hole 926B cores 50X and 51X were taken at 10 cm intervals, washed over a 63 micron sieve, oven-dried at <50°C, and split for assemblage and isotopic work. Full assemblage counts were conducted on quantitative splits of ~300 specimens in the >150 μm size fraction, identified to the species level.

Full planktic foraminiferal assemblage counts were conducted to trace the relative dominance of mixed-layer species vs. thermocline species, which is an index of mixed layer thickness. Mixed layer species are considered to include *G. primordius, G. praebulloides*, *Globigerinita* spp., *P. kugleri/pseudokugleri*, *Tenuitellinata angustiumbilicata*, and *Zeaglobigerina woodi*. Thermocline species include *Globoquadrina venezuelana, G. globularis* and related forms, and *P. mayeri*, while *Catapsydrax dissimilis* is a sub-thermocline species (Pearson *et al.*, 1997; 2001). Species composition is influenced by dissolution, because mixed-layer species are generally more susceptible to dissolution in sediments. Accordingly, the ratio of foraminiferal fragments to fragments plus whole tests was used as an assessment of test breakage, an excellent but labor-intensive dissolution monitor (e.g., Berger *et al.*, 1982; Peterson and Prell, 1985).

G. praebulloides and *P. mayeri* from Cores 50X and 51X were analyzed on a Finnigan MAT DeltaPlus XL stable isotope ratio mass spectrometer (SIRMS) equipped with a Kiel III automated carbonate preparation device in the College of Marine Science at the University of South Florida. External precision based on over 900 NBS-19 standards run on this instrument since July 2000 is better than +/-0.04 ‰ for $\delta^{13}C$ and +/-0.08 ‰ for $\delta^{18}O$. Data on the above two species, plus *P. kugleri/pseudokugleri*, *C. dissimilis*, and *G. venezuelana* from Core 50X were generated on a VG Prism SIRMS equipped with an automated carousel preparation device in the Godwin Laboratory at the University of Cambridge.

To investigate the coherence and phasing among our proxy records, Blackman-Tukey cross-spectral analysis was conducted using the Analyseries package (Paillard *et al.*, 1996). We elected not to similarly analyze some of our isotope gradient records, because of a lack of significant orbital-scale variations observed and a poor signal-to-noise ratio in these time series. Each other series was interpolated to a constant 5 k.y. time step, and a cross-correlation function was calculated using 1/3 lags and a Bartlett window with a bandwidth of 6.8 cycles/m.y. This bandwidth encompassed spectral peaks in the eccentricity (125 and 95 k.y.), obliquity (41 k.y.), and precession (23 k.y.) bands.

Stratigraphy and chronology

The development of an orbitally tuned chronostratigraphy based on magnetic susceptibility back into the Oligocene (Weedon *et al.*, 1997; Shackleton *et al.*, 1999; Shackleton *et al.*, 2000) allows precise dating in the studied interval (ca. 23.43-22.7 Ma). High-resolution benthic isotope stratigraphy and biostratigraphy have refined the chronology during the late Oligocene to early Miocene (Flower *et al.*, 1997a; Zachos *et al.*, 1997; 2001b), and shown that the O/M boundary coincides with the onset of the Mi1 glaciation ca. 23.0 Ma (Shackleton *et al.*, 2000). An average sedimentation rate of ~2.6 cm/k.y. and 10 cm sampling allows temporal resolution better than 5 k.y.

Results and discussion

Global carbon cycling and deep-sea temperatures

Stable isotope data on the planktic foraminifers *G. praebulloides* and *P. mayeri* reveal substantial orbital-scale variability across the O/M boundary. All three species $\delta^{13}C$ records closely covary during the 22.43-22.7 Ma interval (Fig. 3). Indeed, Pearson *et al.* (1997) demonstrated that consistent $\delta^{13}C$ gradients were maintained between five species of planktic foraminifera during the earliest Miocene. Overall, the strong covariance suggests that the $\delta^{13}C$ records are primarily recording entire water column variations in the $\delta^{13}C$ composition of DIC, which is best explained by reservoir changes in carbon cycling. In particular, this covariance indicates substantial orbital-scale variability in global organic carbon burial across the O/M boundary. Therefore, we can use any of these records as a first-order proxy for global carbon cycling in this interval. Hodell and Woodruff (1994) also demonstrated that deep-sea *Cibicidoides* $\delta^{13}C$ values were within 0.45 ‰ during the O/M $\delta^{13}C$ maximum, indicating minimal regional differences due to deep-circulation. In succeeding sections, we use the *Cibicidoides* $\delta^{13}C$ record as a proxy for global carbon cycling.

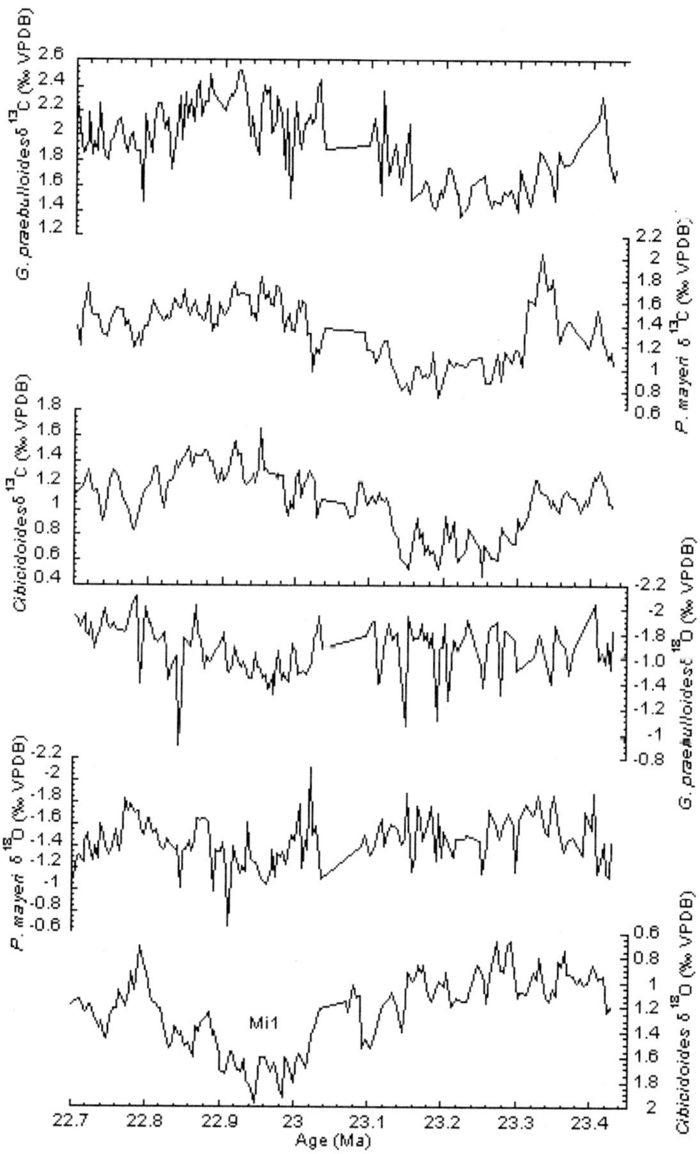

Fig. 3. Stable isotope records from Hole 926B vs. age. Planktic foraminiferal $\delta^{13}C$ records based on *Globigerina praebulloides* and *Paragloborotalia mayeri* covary with benthic $\delta^{13}C$ data based on *Cibicidoides*, reflecting control by global carbon cycling across the O/M boundary. Planktic $\delta^{18}O$ records exhibit reduced amplitude relative to *Cibicidoides* $\delta^{18}O$ during the Mi1 $\delta^{18}O$ increase, suggesting deep-sea cooling accompanied polar ice growth during Mi1 (Paul *et al.*, 2000; this study).

Covariance between the oxygen isotope records is weaker, suggesting important deep-sea temperature change during the Mi1 $\delta^{18}O$ increase of ~1‰ at this site (Fig. 3). *G. praebulloides* $\delta^{18}O$ values range from –2.1 to –0.9 ‰ (mean –1.7 ‰), while *P. mayeri* values range from –2.1 to –0.6 ‰ (mean –1.4 ‰). *Cibicidoides* $\delta^{18}O$ exhibits a somewhat greater range (0.7 to 2.0 ‰), including an interval of distinctly higher values from 23.02-22.9 Ma (Mi1) that represents some combination of deep-sea cooling and Antarctic ice sheet growth (Miller *et al.*, 1991b; Flower *et al.*, 1997a; Zachos *et al.*, 1997; 2001b; Billups and Schrag, 2002). Indeed, the amplitude of Mi1 in the planktic records is half that in the *Cibicidoides* record, as found previously based on lower resolution data (Paul *et al.*, 2000). This observation may indicate that ~0.5‰ of the Mi1 benthic $\delta^{18}O$ increase was due to deep-sea cooling of ~2°C, or that the ice growth effect in planktic $\delta^{18}O$ was offset by warmer tropical sea surface temperatures (Paul *et al.*, 2000). However, existing benthic Mg/Ca data indicate minimal temperature across the O/M boundary (Billups and Schrag, 2002).

Mixed layer variability

Percent mixed layer species (%MLS) data exhibit significant variations from 10-76%, but do not show a clear relation to *Cibicidoides* $\delta^{13}C$ and therefore inferred global carbon burial (Fig. 4). A pair of distinct %MLS increases from ca. 23.33-23.26 Ma and ca. 22.94-22.81 Ma do not coincide with *Cibicidoides* $\delta^{13}C$ increases. The earlier %MLS increase is associated with a 0.6‰ decrease in *Cibicidoides* $\delta^{13}C$, and the later %MLS increase is associated with the O/M $\delta^{13}C$ peak. Furthermore, no major increase in %MLS occurs during the initiation of the O/M $\delta^{13}C$ maximum ca. 23.14 Ma. Similarly, our oxygen isotope proxy for mixed layer thickness reveals no significant change associated with the O/M $\delta^{13}C$ maximum or Mi1 (Fig. 4). The mixed layer to thermocline $\delta^{18}O$ gradient (*G. praebulloides – P. mayeri*) shows little mean change over the interval studied, suggesting minimal changes in temperature gradients that might be associated with mixed layer depth.

Significantly, each *Cibicidoides* $\delta^{13}C$ maximum (on the 10^5 year scale) appears to be followed by an increase in %MLS (Fig. 4). Cross-spectral analysis confirms that these records are coherent in the eccentricity band, and that %MLS lags *Cibicidoides* $\delta^{13}C$ by 22.6 ± 8.1 k.y. (Table 1). This relationship indicates that mixed layer thickness changed about a quarter cycle (on average) after *Cibicidoides* $\delta^{13}C$. The coherence and consistent phase suggests that trade wind variability was probably not a major driver of global carbon burial in this band, but may have provided a reinforcing feedback through a sustaining influence on coastal upwelling. In contrast, in the precession band %MLS shows a small lead (5.5 ± 3.4 k.y.) vs. *Cibicidoides* $\delta^{13}C$, implying a stronger sensitivity of global carbon cycling to trade wind variability in this band.

Bioproductivity

Our use of *Cibicidoides* $\delta^{13}C$ as a monitor of global carbon burial is supported by strong covariance between $\delta^{13}C$ records from throughout the water column (Pearson *et al.*, 1997; this study). In particular, the mixed layer to deep $\delta^{13}C$ gradient is maintained; our reconstruction based on *G. praebulloides – Cibicidoides* $\delta^{13}C$ records indicates no substantial changes across the O/M boundary (Fig. 5). However, reconstruction of mixed layer to thermocline $\delta^{13}C$ gradients (*G. praebulloides – P. mayeri*) shows considerable signal. In particular, this gradient (planktic-planktic $\delta^{13}C$ in Fig. 5) exhibits seven maxima in ~700 k.y., plus a small increase associated with the initiation of the O/M $\delta^{13}C$ maximum ca. 23.14 Ma (Fig. 5). Another distinctive feature is a minimum in $\delta^{13}C$ from ca. 23.35-23.31 Ma that corresponds in part with relatively high *Cibicidoides* values. We have no ready explanation for this enigmatic event, which is controlled by a marked $\delta^{13}C$ increase in *P. mayeri* (Fig. 3). The primary cause of increased mixed layer to thermocline $\delta^{13}C$ gradients is enhanced extraction of nutrients from surface waters and remineralization at depth (e.g., Broecker and Peng, 1982). An associated increase in bioproductivity can be accomplished with or without addition of nutrients to surface waters by upwelling or riverine delivery.

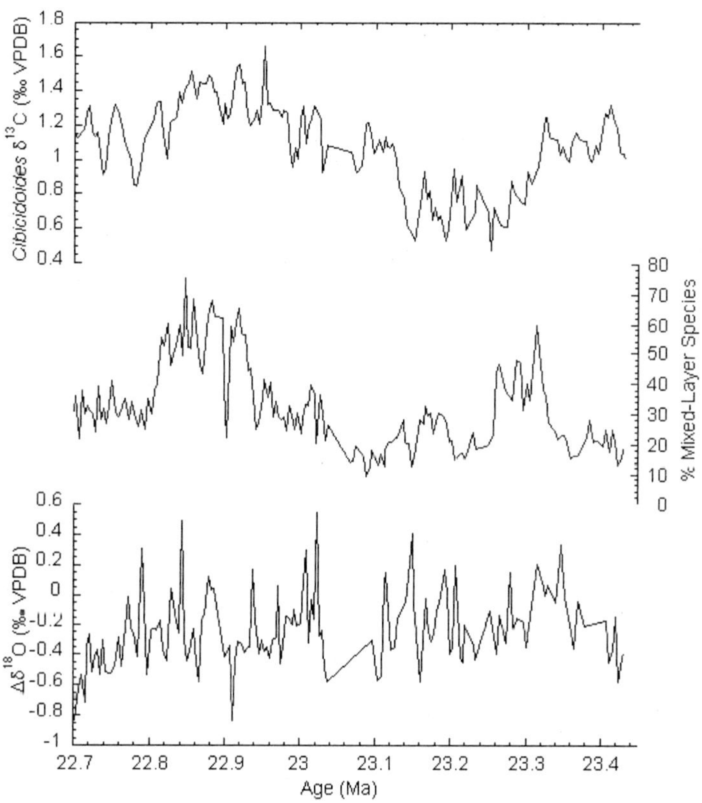

Fig. 4. Proxies for mixed layer variability and *Cibicidoides* $\delta^{13}C$ across the O/M boundary. Neither percent mixed layer species (%MLS) nor the mixed layer to thermocline $\delta^{18}O$ gradient (*G. praebulloides* – *P. mayeri* $\delta^{18}O$) shows a major increase at the onset of the O/M $\delta^{13}C$ maximum ca. 23.13 Ma. However, %MLS is coherent with *Cibicidoides* $\delta^{13}C$, and exhibits a small lead, in the precession band (Table 1), suggesting modulation of global carbon cycling by trade wind strength in this band.

Cross-spectral analysis confirms coherence between planktic-planktic $\delta^{13}C$ and *Cibicidoides* $\delta^{13}C$ in the eccentricity (100 k.y.) and precession (23 k.y.) bands (Table 1). Therefore we find some support for the hypothesis that global organic carbon cycling (not necessarily in the western equatorial Atlantic) was accompanied by orbital-scale variations in bioproductivity in this region. Furthermore, a phase lead of 5.8 $\pm$ 2.7 k.y. vs. *Cibicidoides* $\delta^{13}C$ in the precession band suggest that Ceara Rise bioproductivity changed somewhat earlier (on average) than global carbon cycling in this band. We find no significant coherence in the obliquity band. Overall, our proxies for both mixed layer thickness and bioproductivity are coherent with and lead *Cibicidoides* $\delta^{13}C$ in the precession band, consistent with evidence from the late Pleistocene for trade wind influence on equatorial Atlantic Ocean dynamics in this band (McIntyre *et al.*, 1989; Molfino and McIntyre, 1990). Control of trade wind variability by low-latitude insolation is strong and is related to African monsoon variability (Manabe and Broccoli, 1985; McIntyre *et al.*, 1989; Molfino and McIntyre, 1990). Our results suggest that similar mechanisms were operating across the O/M boundary.

Table 1. Results of Cross-Spectral Analysis

	Band (k.y.)					
	100		41		23	
□δ^{13}C vs. Cibδ^{13}C						
Coherence	0.41		(0.34)		0.51	
Phase (°)	-129.30	±86.90	-69.78	±91.54	-90.41	±42.53
k.y.	-35.92	±24.14	-7.95	±10.43	-5.78	±2.72
%MSL vs. Cibδ^{13}C						
Coherence	0.59		(0.31)		0.46	
Phase (°)	81.47	±29.29	-35.42	±99.74	-85.59	±52.52
k.y.	22.63	±8.14	-4.03	±11.36	-5.47	±3.36
%fragments vs. Cibδ^{13}C						
Coherence	(0.39)		0.63		(0.31)	
Phase (°)	36.46	±66.99	155.81	±32.17	-32.10	±99.72
k.y.	10.13	±18.61	17.74	±3.66	-2.05	±6.37
%reflectance vs. Cibδ^{13}C						
Coherence	0.45		0.86		0.55	
Phase (°)	-52.33	±48.67	142.35	±12.50	-68.31	±37.01
k.y.	-14.54	±13.52	16.21	±1.42	-4.36	±2.36
%CF vs. Cibδ^{13}C						
Coherence	(0.19)		0.83		(0.38)	
Phase (°)	83.28	±142.52	-56.50	±14.46	104.70	±77.51
k.y.	23.13	±39.59	-6.44	±1.65	6.69	±4.95
%CF vs. %reflectance						
Coherence	0.41		0.83		0.78	
Phase (°)	-27.52	±52.26	-7.26	±14.46	26.20	±17.74
k.y.	-7.64	±14.52	-0.83	±1.65	1.67	±1.13
%fragments vs. %reflectance						
Coherence	0.59		0.64		0.49	
Phase (°)	90.92	±29.75	3.47	±31.99	14.99	±38.97
k.y.	25.26	±8.26	0.40	±3.64	0.96	±2.49

Blackman-Tukey cross spectral analysis was conducted with the Analyseries program (Paillard *et al.*, 1996) using 1/3 lags and a Bartlett window with a bandwidth of 6.8 cycles/m.y. Coherence is significant at the 80% confidence level (k=0.40). Negative phase (given in degrees and kiloyears; k.y.) indicates the given parameter leads the reference variable.

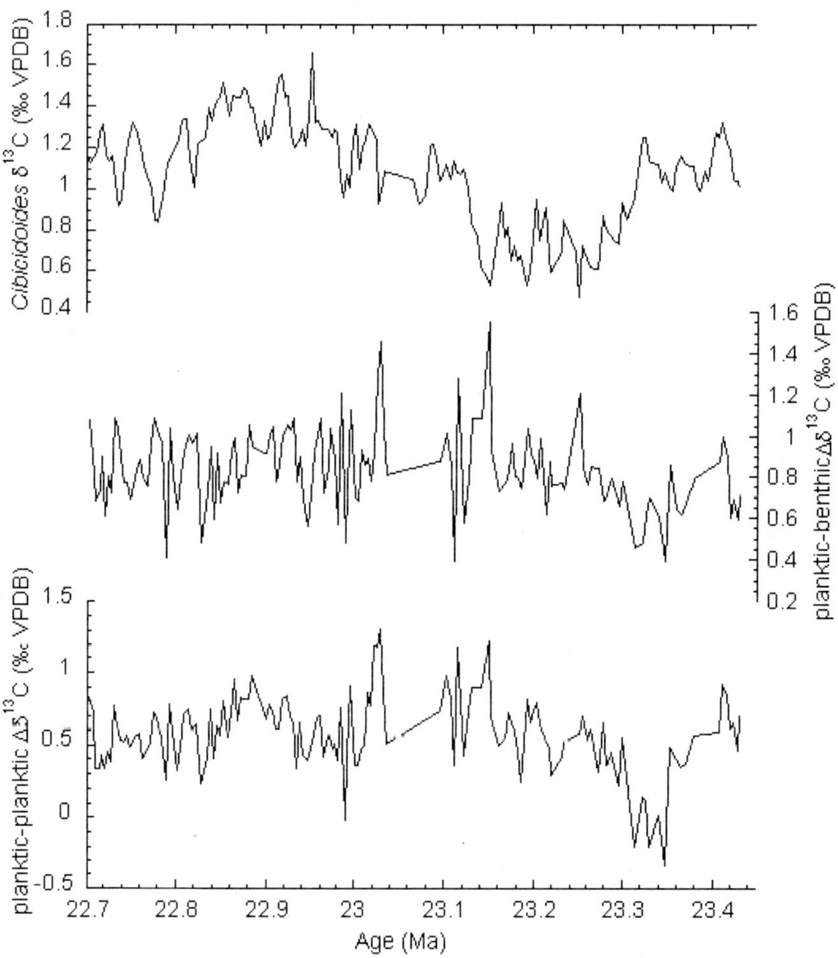

Fig. 5. Proxies for bioproductivity and *Cibicidoides* $\delta^{13}C$ across the O/M boundary. Neither the mixed layer to deep-water $\delta^{13}C$ gradient (*G. praebulloides – Cibicidoides* $\Delta\delta^{13}C$) nor the mixed layer to thermocline $\delta^{13}C$ gradient (*G. praebulloides – P. mayeri* $\Delta\delta^{13}C$) shows a substantial increase associated with the O/M $\delta^{13}C$ maximum ca. 23.13 Ma. However, the latter is coherent with *Cibicidoides* $\delta^{13}C$, and exhibits a small lead, in the precession band (Table 1), suggesting modulation of global carbon cycling by bioproductivity in this band.

Nutrient delivery

Previous work has indicated no particular increase in nutrient accumulation in Ceara Rise sediments associated with the O/M $\delta^{13}C$ maximum (Delaney and Anderson, 1997). Increased nutrient supply should be accompanied by increased surface-to-deep water $\delta^{13}C$ gradients (e.g., Miller and

Fairbanks, 1985; Vincent and Berger, 1985; Delaney, 1990), in contrast to our findings. Raymo (1994) suggested a pronounced increase in organic carbon burial during the early to middle Miocene related to Himalayan uplift. Her study found no significant increase near the O/M boundary, but did not attempt to examine organic carbon burial at the orbital scale. Another possible source region is South America. However, Ceara Rise records indicate that sediment delivery by the Amazon was low until ca. 8 Ma, after substantial uplift of the Andes during the middle to late Miocene (Dobson *et al.*, 1997). In contrast, Himalayan uplift and weathering apparently were well underway by the O/M boundary (Uddin and Lundberg, 1993; Najman *et al.*, 1997). Finally, radiogenic isotope proxies for global continental weathering show little increase across the O/M boundary, based either on $^{87}Sr/^{86}Sr$ (Richter *et al.*, 1992; Hodell and Woodruff, 1994; Oslick *et al.*, 1994) or $^{187}Os/^{186}Os$ (Ravizza, 1993; Reusch *et al.*, 1998).

On the other hand, our results may be consistent with a corollary of the nutrient delivery hypothesis, that increased nutrient supply during glaciations enhanced surface bioproductivity and thereby the mixed layer to thermocline $\delta^{13}C$ gradient in the western equatorial Atlantic. We cannot rule out orbital-scale increases in nutrient delivery controlling bioproductivity. Our suggestion is that nutrient supply (and possibly weathering-induced CO_2 drawdown) may have been an important feedback mechanism on the orbital scale for O/M organic carbon burial. An associated decrease in temperature during the Mi1 glaciation may have provided a negative feedback on weathering that led to minimal changes in radiogenic isotope proxies for global continental weathering (Berner *et al.*, 1983). This hypothesis can be tested further by examining orbital-scale variability in bioproductivity throughout Himalayan weathering history. Similarly, the relationships between western tropical hydrography and global carbon burial can be compared in other intervals of enhanced global carbon burial such as the early to middle Miocene. Exploring how orbital-scale changes in wind-driven upwelling and productivity are manifested in non-upwelling areas is a fruitful avenue for research aimed at understanding organic carbon burial.

Dissolution and deep-water variability

To monitor the potential effect of secondary processes on our foraminiferal assemblage and stable isotope proxies, we compare proxies for lithologic variability and carbonate dissolution (Fig. 6). Controlled by the relative contribution of carbonate sediments and terrigenous clays, lithologic cycles in cores 50X and 51X (based on %reflectance in the 550 nm band; Fig. 6) are dominated by cyclicity in the 41 k.y. band (Shipboard Scientific Party, 1995). Based on coherent cyclicity in Ba/Al, P/Al, Si/Al, and %$CaCO_3$, Weedon and Shackleton (1997) attributed these cycles to a combination of (1) carbonate bioproductivity, (2) carbonate dissolution, and (3) terrigenous supply. Percent coarse fraction data also indicate a strong influence of dissolution on Ceara Rise sediments, especially in the obliquity band (Flower *et al.*, 1997a; Zachos *et al.*, 1997, 2001; Paul *et al.*, 2000). Fig. 6 shows coherent variations with percent reflectance. However, both percent reflectance and percent coarse fraction data are also influenced by carbonate bioproductivity and terrigenous supply.

Foraminiferal fragmentation provides an independent tracer of carbonate dissolution that has advantages over percent reflectance and percent coarse fraction records employed in earlier studies (Flower *et al.*, 1997a; Zachos *et al.*, 1997; 2001; Pearson *et al.*, 1997; Paul *et al.*, 2000). Our foraminiferal fragmentation proxy (fragments / fragments plus whole planktic foraminifers) overcomes these limitations because it should be influenced only by dissolution. In particular, this proxy is largely controlled by the CO_3^{-2} concentration in bottom waters (e.g., Broecker and Peng, 1982). Carbonate-corrosive water bathing Ceara Rise near the O/M boundary is inferred to be Southern Ocean Water (SOW) (Wright *et al.*, 1992; Flower *et al.*, 1997b; Paul *et al.*, 2000).

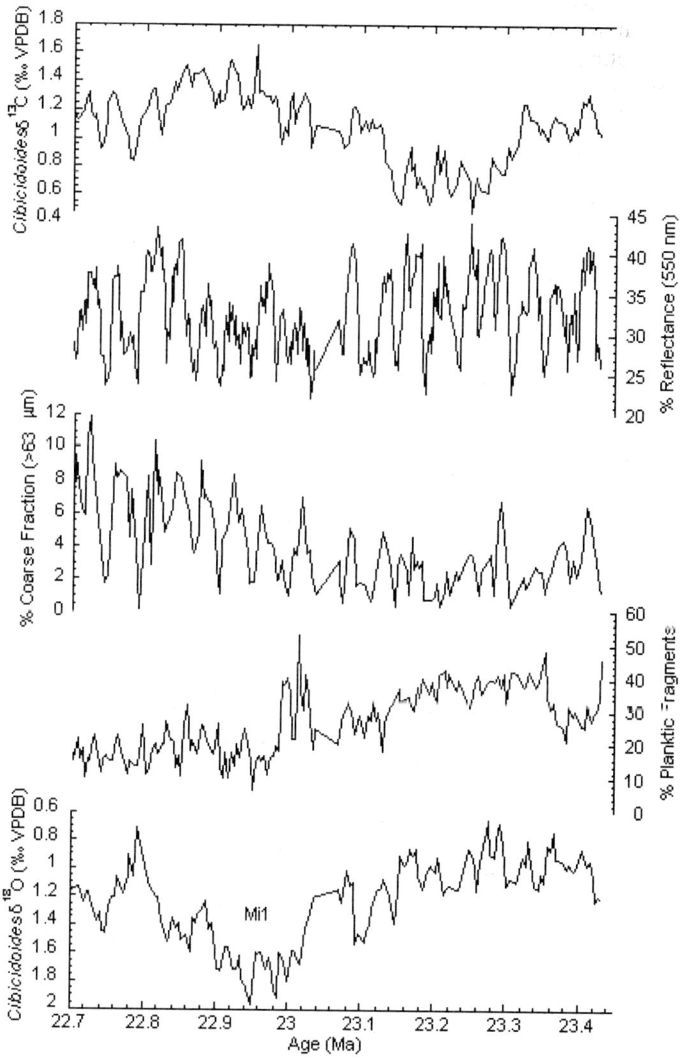

Fig. 6. Proxies for lithologic variability (percent reflectance in the 550 nm band), percent coarse fraction (>63 µm), dissolution (percent planktic fragments/fragments plus whole foraminifera) shown vs. *Cibicidoides* $\delta^{13}C$ and $\delta^{18}O$. A long-term increase in percent coarse fraction and percent fragments after 23.1 Ma suggests enhanced carbonate preservation. All three proxies are coherent and in phase in the obliquity band (~41 k.y.), indicating a common control by dissolution (Table 1). However, percent fragments and percent reflectance lag *Cibicidoides* $\delta^{13}C$, suggesting global carbon cycling was decoupled from deep-water circulation in this band.

Our fragmentation proxy (Fig. 6) exhibits trends that are independent of both the %MLS and the planktic-planktic $\Delta\delta^{13}C$ records. This proxy is not coherent with either record in any orbital band. This implies that our mixed layer thickness and bioproductivity records are not fundamentally controlled by dissolution. Pearson *et al.* (1997) presented evidence for changes in carbonate bioproductivity based on a $\Delta\delta^{13}C$ proxy similar to that used here, and suggested that its variability was not associated with lithologic cycles. These authors found consistency in the planktic isotope data they generated across lithologic cycles interpreted to reflect dissolution variability. These findings are consistent with the suggestion that although dissolution is significant, it does not appreciably affect either planktic isotope data (Pearson *et al.*, 1997; this study) or the assemblage data (this study).

The percent fragments proxy is coherent and in phase with percent reflectance and percent coarse fraction the obliquity band (Table 1), suggesting a common control by carbonate dissolution. However, percent fragments and percent reflectance also exhibit a significant lag relative to *Cibicidoides* $\delta^{13}C$ of ~15-21 k.y. This consistent lag underscores that factors other than SOW strongly influenced *Cibicidoides* $\delta^{13}C$ in this interval, because SOW variability should be accompanied by coeval dissolution changes. Indeed, this relationship is consistent with a phase sequence in which global carbon cycling leads deep circulation changes and dissolution cycles in the deep Atlantic Ocean, at least in the obliquity band.

Finally, comparison of our dissolution index with benthic $\delta^{13}C$ and $\delta^{18}O$ reveals important long-term trends (Fig. 6). In particular, a transition from high to low percent fragments from ca. 23.36 to 22.9 Ma indicates decreasing dissolution across the O/M $\delta^{13}C$ maximum. This finding may turn indicate that the long-term increase in percent coarse fraction beginning during Mi1 was also caused by enhanced preservation more than either increased carbonate productivity or reduced terrigenous input. Furthermore, a distinct dissolution pulse associated with Mi1 at ca. 23 Ma suggests that the initiation of Mi1 was marked by increased SOW at Ceara Rise.

Overall, these results suggest a dominant control of lithologic cycles by carbonate dissolution, tied to SOW variability driven by high latitude forcing. However, *Cibicidoides* $\delta^{13}C$ appears to be decoupled from SOW variability, at least in the obliquity band. Indeed, if *Cibicidoides* $\delta^{13}C$ primarily reflects global carbon cycling, then the carbon cycle must have a complex relationship to Atlantic deep circulation across the O/M boundary.

Conclusions

In Oligocene/Miocene upwelling-dominated settings, it is difficult to ascertain whether organic carbon burial increased because of increased trade wind strength and associated upwelling, increased nutrient supply from continents and associated bioproductivity, or some other mechanism. Therefore it is advantageous to examine limited- or non-upwelling settings that are sensitive to these processes. Here we attempt to reconstruct western tropical Atlantic hydrography across the O/M boundary to assess the relative importance of trade wind strength and nutrient supply in controlling global organic carbon burial.

Covariance between planktic and benthic $\delta^{13}C$ indicates that global carbon cycling controlled water column $\delta^{13}C$ in this interval, which can be approximated by *Cibicidoides* $\delta^{13}C$. Foraminiferal assemblage and isotopic data indicate limited variation in mixed layer thickness and surface bioproductivity associated with $\delta^{13}C$ maxima that comprise the Oligocene/Miocene $\delta^{13}C$ excursion. However, in the precession band both percent mixed layer species and the surface to thermocline $\delta^{13}C$ gradient are coherent with *Cibicidoides* $\delta^{13}C$, and exhibit a small lead of ~6 $\pm$ 3 k.y. These results are consistent with evidence from the late Pleistocene for trade wind influence on equatorial Atlantic Ocean dynamics in this band (McIntyre *et al.*, 1989; Molfino and McIntyre, 1990).

Use of the percent mixed layer species (%MLS) proxy rests on the assumption that mixed layer species did not change their depth preferences due to evolutionary processes in this interval. Testing this assumption is beyond the scope of this paper. However, we speculate that episodic expansion and contraction of the mixed layer may have contributed to the diversification of mixed layer species known from the late Oligocene to early Miocene (e.g., Cifelli, 1969; Kennett and Srinivasan, 1982; Srinivasan and Kennett, 1983; Spezzaferri, 1995).

In the obliquity band, coherence among lithologic variability (based on percent reflectance in the 550 nm band), percent coarse fraction, and a dissolution index (fragments/fragments plus whole planktic foraminifers) indicates a common control by carbonate dissolution. However, none of these is coherent with our surface water proxies, suggesting minimal overprinting by dissolution. Furthermore, percent fragments and percent reflectance exhibit a significant lag relative to *Cibicidoides* $\delta^{13}C$ of ~15-21 k.y. Because Southern Ocean Water (SOW) variability should be accompanied by coeval dissolution changes, this observation implies that other factors strongly influenced *Cibicidoides* $\delta^{13}C$ across the O/M boundary. We speculate that global carbon cycling led deep circulation changes and dissolution cycles in the deep Atlantic Ocean, at least in the obliquity band. Our results suggest that SOW variability was decoupled from *Cibicidoides* $\delta^{13}C$, and underscore the importance of global carbon cycling in Oligocene / Miocene climate change.

Acknowledgments

We would like to thank Al Hine for helpful discussions, and Kirsten Eliot and Amy Wright assistance in generating the foraminiferal assemblage and stable isotope data at the University of South Florida. This research used samples provided by the Ocean Drilling Program (ODP). The ODP is sponsored by the U.S. National Science Foundation (NSF) and participating countries under the management of Joint Oceanographic Institutions (JOI), Inc. This work was partially supported by NSF grants EAR-9725311 to BPF and JCZ and EAR-9814883 to JCZ. Acknowledgment is also made to the Donors of The Petroleum Research Fund, administered by the American Chemical Society, for partial support of this research (ACS PRF#38056-AC2 to BPF).

References

BERGER, W.,H., BONNEAU, M.-C., PARKER, F.L., 1982. Foraminifera on the deep-sea floor: lysocline and dissolution rate. Oceanol. Acta, 5, 249-258.

BERNER, R.A., LASAGA, A.C., and GARRELS, R.M., 1983. The carbonate-silicate geochemical cycle and its effect on atmospheric carbon dioxide over the past 100 million years. Am. J. Science, 283, 641-683.

BILLUPS K., SCHRAG D.P., 2002. Paleotemperatures and ice volume of the past 27 Myr revisited with paired Mg/Ca and O-18/O-16 measurements on benthic foraminifera. Paleoceanography, 17, Art. No. 1003.

BROECKER, W.S., PENG. T.-H., 1982. Tracers in the Sea, Eldigio Press, New York.

CANDE, S.C., KENT, D.V.,1992. A new geomagnetic polarity time scale for the late Cretaceous and Cenozoic. Journal of Geophysical Research. 97, 13,917-19,951.

CANDE, S.C., KENT, D.V.,1995. Revised calibration of the Geomagnetic Polarity Timescale for the late cretaceous and Cenozoic. Journal of Geophysical Research, 100, 6093-6095.

CHAISSON, W.P., RAVELO, A.C., 2000. Pliocene development of the east-west hydrographic gradient in the equatorial pacific. Paleoceanography, 15, 497-505.

CIFELLI, R., 1969. Radiation of Cenozoic planktonic Foraminifera. Systemic Zoology, 18, 154-168.

COMPTON, J. S., SNYDER, S. W., HODELL, D. A., 1990. Phosphogenesis and weathering of shelf sediments from the southeastern United States: implications for Miocene $\delta^{13}C$ excursions and global cooling. Geology, 18, 1227-1230.

COMPTON, J. S., HODELL, D. A., GARRIDO, J. R., MALLINSON, D. J., 1993. Origin and age of phosphorite from the south-central Florida Platform: relation of phosphogenesis to sea-level fluctuations and $\delta^{13}C$ excursions. Geochimica et Cosmochimica Acta, 57, 131-146.

CHAISSON, W.P., RAVELO, A.C., 1997. Changes in upper water-column structure at Site 925, Late Miocene – Pleistocene: Planktonic foraminifer assemblage and isotopic evidence. In: Shackleton, N.J., Curry, W.B., Richter, C., Bralower, T.J. (Eds.), Proceedings of the Ocean Drilling Program, Scientific Results, 154: College Station, TX (Ocean Drilling Program), pp. 255-268.

CHAISSON, W.P., RAVELO, A.C., 2000. Pliocene development of the east-west hydrographic gradient in the equatorial pacific. Paleoceanography, 15, 497-505.

CROWLEY, T., ZACHOS, J.C., 1999. Comparison of zonal temperature profiles for past warm time periods. In: Huber, B.T., MacLeod, K.G., Wing, S.L. (Eds.), Warm Climates in Earth History, Cambridge Univ. Press, Cambridge, pp. 50-76.

CULLEN, J.L., CURRY, W.B., 1997. Variations in planktonic foraminifer faunas and carbonate preservation in Site 927: Evidence for changing surface water conditions in the western tropical Atlantic Ocean during the middle Pleistocene, In: Shackleton, N.J., Curry, W.B., Richter, C., Bralower, T.J. (Eds.), Proceedings of the Ocean Drilling Program, Scientific Results, 154: College Station, TX (Ocean Drilling Program), pp. 207-228.

DELANEY, M.L., 1990. Miocene benthic foraminiferal Cd/Ca records: South Atlantic and Western Equatorial Pacific. Paleoceanography, 5, 743-760.

DELANEY, M.L. and ANDERSON, L.D., 1997. Phosphorus geochemistry in Ceara Rise sediments. In: Shackleton, N.J., Curry, W.B., Richter, C., Bralower, T.J. (Eds.), Proceedings of the Ocean Drilling Program, Scientific Results, 154: College Station, TX (Ocean Drilling Program), pp. 475-482.

DePAOLO, D.J., FINGER, K.L., 1991. High-resolution strontium-isotope stratigraphy and biostratigraphy of the Miocene-Monterey-formation, central California. Geological Society of America Bulletin, 103, 112-124.

DOBSON, D.M., DICKENS, G.R., REA, D.K., 1997. Terrigenous sedimentation at Ceara Rise. In: Shackleton, N.J., Curry, W.B., Richter, C., and Bralower, T.J. (Eds.), Proceedings of the Ocean Drilling Program, Scientific Results, 154: College Station, TX (Ocean Drilling Program), pp. 465-473.

DUQUE-CARO, H., 1990. Neogene stratigraphy, paleoceanography and palebiogeography in northwest South America and the evolution of the Panama Seaway. Palaeogeogr., Palaeoclimatol., Palaeoecol., 77, 203-234.

FAIRBANKS, R.G, SVERDLOVE, M., FREE, R., WIEBE, P.H., BÉ, A.W. H., 1982. Vertical distribution and isotopic composition of living planktonic foraminifera from the Panama Basin. Nature, 298, 841-844

FAIRBANKS, R.G., CHARLES, C.D., and WRIGHT, J.D., 1992. Origin of global meltwater pulses, in Radiocarbon After Four Decades, Springer-Verlag, New York, pp. 473-500.

FAUL, K.L., RAVELO, A.C., and DELANEY, M.L., 2000. Reconstructions of upwelling, productivity, and photic zone depth in the eastern equatorial pacific ocean using planktonic foraminiferal stable isotopes and abundance, Journal of Foraminiferal Research, 30, 110-125.

FLOWER, B.P., 1999. Warming without high CO_2? Nature, 399, 313-314.

FLOWER, B.P., KENNETT, J.P., 1993a. Middle Miocene ocean/climate transition: high-resolution oxygen and carbon isotopic records from DSDP Site 588A, Southwest Pacific. Paleoceanography, 8, 811-843.

FLOWER, B.P., KENNETT, J.P., 1993b. Relations between Monterey Formation deposition and Middle Miocene global cooling: Naples Beach section, California. Geology, 21, 877-880.

FLOWER, B.P., ZACHOS, J.C., PAUL, H., 1997a. Milankovitch-scale climate variability recorded near the Oligocene/Miocene boundary, In: Shackleton, N.J., Curry, W.B., Richter, C., Bralower, T.J. (Eds.), Proceedings of the Ocean Drilling Program, Scientific Results, 154: College Station, TX (Ocean Drilling Program), pp. 433-439.

FLOWER, B.P., ZACHOS, J.C., MARTIN, E., 1997b. Latest Oligocene through Early Miocene isotopic stratigraphy and deepwater paleoceanography of the western Equatorial Atlantic: Sites 926 and 929, In: Shackleton, N.J., Curry, W.B., Richter, C., Bralower, T.J. (Eds.), Proceedings of the Ocean Drilling Program, Scientific Results, 154: College Station, TX (Ocean Drilling Program), pp. 451-461.

KENNETT, J.P., SRINIVASAN, M.S., 1982. Oligocene Miocene boundary–correlation, biostratigraphy, and foraminiferal evolution, AAPG Bulletin, 66, 568-588.

HODELL, D.A., F WOODRUFF, 1994. Variations in the strontium isotopic ratio of seawater during the Miocene – stratigraphic and geochemical implications. Paleoceanography, 9, 405-426.

HOVAN, S.A., REA, D.K., 1992. The Cenozoic record of continental mineral deposition on Broken and Ninetyeast ridges, Indian Ocean: Southern African aridity and sediment delivery from the Himalayas. Paleoceanography, 7, 833-860.

KUMP, L., GARRELS, R.M., 1986. Modelling atmospheric O_2 in the global sedimentary redox cycle, Am. J. Sci., 286: 337-360.

LECKIE, R.M., 1989, An oceanographic model for the early evolutionary history of planktonic foraminifera, Palaeogeog, Palaeoclimatol., Palaeoecol., 73: 107-138.

LOHMAN, GP, 1995. A model for variation in the chemistry of planktonic foraminifera due to secondary calcification and selective dissolution, Paleoceanography, 10, 445-457.

MALLINSON, D.J., COMPTON, J.S., 1997. Linking phosphogenic episodes on the southeast US margin to marine delta C-13 and delta O-18 records. Geology, 25, 103-106.

MANABE, S., BROCCOLI, A.J., 1985. The influence of continental ice sheets on the climate of an ice age, J. Geophys. Res., 90, 2167-2190.

MCINTYRE, A., RUDDIMAN, W.F., KARLIN, K., MIX, A.C., 1989. Surface water response of the equatorial Atlantic Ocean to orbital forcing. Paleoceanography, 4, 19-55.

MERLE, J., 1980. Seasonal heat-budget in the equatorial Atlantic Ocean. J. Phys. Oceanography, 10, 464-469.

MILLER, K. G., M.D. FEIGENSON, J. D. WRIGHT, B.M. CLEMENT, 1991a. Miocene isotope reference section, Deep Sea Drilling Project site 608: An evaluation of isotope and biostratigraphic resolution, Paleoceanography, 6, 33-52.

MILLER, K. G., WRIGHT, J. D., FAIRBANKS, R.G., 1991b. Unlocking the ice house: Oligocene-Miocene oxygen isotopes, eustacy and margin erosion, Journal of Geophysical Research, 96, 6829-6848.

MOLFINO, B., MCINTYRE, A., 1990. Precessional forcing nutricline dynamics in the Equatorial Atlantic. Science, 249, 766-769.

NAJMAN, Y.M.R., PRINGLE, M.S., JOHNSON, M.R.W., ROBERTSON, A.H.F., WIJBRANS, J.R., 1997. Laser ^{40}Ar/^{39}Ar dating of single detrital muscovite grains from early foreland-basin sedimentary deposits in India: implications for early Himalayan evolution, Geology, 25, 535–538.

OSLICK, J.S., MILLER, K.G., FEIGENSON, M.D., WRIGHT, J.D., 1994. Oligocene-Miocene strontium isotopes: correlation to a glacioeustatic record. Paleoceanography, 9, 427-443.

PAGANI, M., M.A. ARTHUR, K.H. FREEMAN, 1999. Miocene evolution of atmospheric carbon dioxide. Paleoceanography, 14, 273-292.

PAILLARD, D., L. LABEYRIE, P. YIOU, 1996. Macintosh program performs time-series analysis, EOS, Trans. AGU, 77, (39), 379.

PAUL, H., ZACHOS, J.C., FLOWER, B.P., TRIPATI, A., 2000, Orbitally induced climate and geochemical variability across the Oligocene / Miocene boundary, Paleoceanography, 15, 471-485.

PEARSON P.N., PALMER M.R., 2000. Atmospheric carbon dioxide concentrations over the past 60 million years, Nature, 406, 695-699.

PEARSON, P.N., SHACKLETON, N.J., HALL, M.A., 1993. Stable isotope paleoecology of middle Eocene planktonic-foraminifera and multispecies isotope stratigraphy, DSDP Site 523, South Atlantic, J Foramin. Res. 23, 123-140.

PEARSON P.N., NORRIS R.D., EMPSON A.J., 2001. *Mutabella mirabilis* gen. et sp nov., a Miocene microperforate planktonic foraminifer with an extreme level of intraspecific variability, J Foramin. Res. 31, 120-132.

PEARSON, P.N., SHACKLETON, N.J., WEEDON, G.P., HALL, M.A., 1997, Multispecies planktonic foraminifer stable isotope stratigraphy through Oligocene/Miocene boundary climatic cycles, Site 926, In: Shackleton, N.J., Curry, W.B., Richter, C., and Bralower, T.J. (Eds.), Proceedings of the Ocean Drilling Program, Scientific Results, 154: College Station, TX (Ocean Drilling Program), pp. 441-449.

PETERSON, L.C., PRELL, W.L., 1985. Carbonate dissolution in Holocene sediments of the eastern equatorial Indian Ocean: Preservation patterns and carbonate loss above the lysocline. Marine Geology, 64, 259-290.

PHILANDER, S.G., PAKANOWSKI, R.C., 1986. A model of the seasonal cycle in the tropical Atlantic Ocean. Journal of Geophysical Research, 91, 14.212-14,220.

RAVELO, A.C., R.G. FAIRBANKS, 1992. Oxygen isotopic composition of multiple species of planktonic foraminifera: recorders of the modern photic zone temperature gradient. Paleoceanography, 7, 815-831.

RAVELO, A.C., FAIRBANKS, R.G. 1995. Carbon isotopic fractionation in multiple species of planktonic foraminifera from core-tops in the tropical Atlantic. J Foramin. Res., 25, 53-74.

RAVELO, A.C., FAIRBANKS, R.G., PHILANDER, S.G., 1990. Reconstructing tropical Atlantic hydrography using planktonic foraminifer and an ocean model. Paleoceanography, 5, 409-431.

RAVIZZA, G., 1993. Variations of the OS-187/OS-186 ratio of seawater over the past 28 million years as inferred from metalliferous carbonates; Earth and Planetary Science Letters, 118, 335-348.

RAYMO, M.E., 1994. The Himalayas, organic carbon burial, and climate in the Miocene. Paleoceanography, 9, 399-404.

REA, D.K., 1994. The paleoclimatic record provided by eolian deposition in the deep sea- The Geologic History of Wind. Reviews of Geophysics, 32, 159-195.

REUSCH, D.N., G. RAVIZZA, K.A., MAASCH, K.A., *et al.*, 1998. Miocene seawater Os187/Os188 ratios inferred from metalliferous carbonates. Earth Planet. Sci. Lett. 160, 163-178.

RICHTER, F.M., ROWLEY, D.B., DePAOLO, D.J., 1992. Sr Isotope evolution of seawater – the role of tectonics. Earth and Planetary Science Letters, 109, 11-23.

SHACKLETON, N.J., 1984. The carbon isotope record of the Cenozoic: History of organic carbon burial and of oxygen in the ocean and atmosphere, In: Marine Petroleum Source Rocks, Brooks, E., Fleet, A.J., (Eds.), Geol. Soc. Spec. Publ. London., 26, 423-434.

SHACKLETON, N.J., 1985. Oceanic carbon isotope constraints on oxygen and carbon dioxide in the Cenozoic atmosphere. In: The Carbon Cycle and Atmospheric CO_2: Natural Variations Archean to Present, Sundquist, E.T., Broecker, W. S., Geophysical Monograph 32, American Geophysical Union, Washington, D.C., pp. 412-417.

SHACKLETON, N.J., CROWHURST, S.J., WEEDON, G.P., LASKAR, J., 1999. Astronomical calibration of Oligocene-Miocene time. Philosophical Transactions of the Royal Society of London Series A-Mathematical Physical and Engineering Sciences, 357, 1907-1929.

SHACKLETON, N.J., HALL, M.A., RAFFI, I., TAUXE, L., ZACHOS, J., 2000. Astronomical calibration age for the Oligocene-Miocene boundary. Geology, 28, 447-450.

Shipboard Scientific Party, 1995. Site 926. In: Curry, W.B., Shackleton, N.J., Richter, C., *et al.*, Proc. ODP, Init. Repts., 154. College Station, TX (Ocean Drilling Program), pp. 153-232.

SPEZZAFERRI, S., 1995. Planktonic foraminiferal paleoclimatic implications across the Oligocene-Miocene transition in the ocean record (Atlantic, Indian and South-Pacific). Palaeogeog., Palaeoclimatol., Palaeoecol., 114, 43-74.

SRINIVASAN, M.S., KENNETT, J.P., 1983. The Oligocene-Miocene boundary in the South-Pacific. GSA Bulletin, 94, 798-812.

UDDIN, A. AND LUNDBERG, N., 1993. Unroofing history of the eastern Himalaya and the Indo–Burman ranges: heavy-mineral study of Cenozoic sediments from the Bengal Basin, associated with continent–continent collision, Geology 21, 885–888.

VINCENT, E., KILLINGLEY, J.S., 1985. Oxygen and carbon isotope record for the early and Middle Miocene in the central equatorial Pacific (Leg 85) and paleoceanographic implications, In: Mayer, L., Theyer, F., *et al.*, (Eds.), Initial Reports of the Deep Sea Drilling Project, 85, pp. 749-769.

VINCENT, E., BERGER, W.H., 1985. Carbon dioxide and polar cooling in the Miocene: The Monterey Hypothesis, In: Sundquist, E.T., Broecker, W.S., (Eds.), The Carbon Cycle and Atmospheric CO_2: Natural Variations Archean to Present. Geophysical Monograph 32, American Geophysical Union, Washington, D.C., pp. 455-468.

WEEDON, G.P., SHACKLETON, N.J., PEARSON, P.N., 1997. The Oligocene time scale and cyclostratigraphy on the Ceara Rise, Western Equatorial Atlantic. In: Shackleton, N.J., Curry, W.B., Richter, C., Bralower, T.J. (Eds.), Proceedings of the Ocean Drilling Program, Scientific Results, 154: College Station, TX (Ocean Drilling Program), pp. 101-114.

WEEDON, G.P., SHACKLETON, N.J., 1997. Inorganic geochemical composition of Oligocene to Miocene sediments and productivity variations in the western equatorial Atlantic: results from Site 926 and 929. In: Shackleton, N.J., Curry, W.B., Richter, C., Bralower, T.J. (Eds.), Proceedings of the Ocean Drilling Program, Scientific Results, 154: College Station, TX (Ocean Drilling Program), pp. 507-526.

WOODRUFF, F., SAVIN, S., 1991. Mid-Miocene isotope stratigraphy in the deep sea: high-resolution correlations, paleoclimatic cycles, and sediment preservation. Paleoceanography, 6, 755-806.

WRIGHT, J.D., MILLER, K.G., FAIRBANKS, R.G., 1992. Early and middle Miocene stable isotopes: implications for deepwater circulation and climate. Paleoceanography, 7, 357-389.

ZACHOS, J.C., FLOWER, B.P., PAUL, H., 1997. Orbitally paced climate oscillations across the Oligocene/Miocene boundary. Nature, 388, 567-570.

ZACHOS, J., PAGANI, M., SLOAN, L., THOMAS, E., BILLUPS, K., 2001a. Trends, rhythms, and aberrations in global climate 65 Ma to present. Science, 292, 686-693.

ZACHOS, J.C., SHACKLETON, N.J., REVENAUGH, J., PALIKE, H., FLOWER, B.P., 2001b, Climate response to orbital forcing across the Oligocene-Miocene boundary, Science, 292, 274-278.

ZACHOS J.C., ARTHUR, M.A., BRALOWER, T.J., *et al.*, 2002. Palaeoclimatology - Tropical temperatures in greenhouse episodes. Nature 419, 897-898.

Response of Miocene deep sea benthic foraminifera to paleoceanographic changes at DSDP site 237 (Mascerene Plateau), Northwest Indian Ocean

Ajai K. Rai[1], M.S. Srinivasan[2] and Virendra B. Singh[1]

[1] Department of Earth & Planetary Sciences, Nehru Science Center, Allahabad University, Allahabad 211 002, India.
[2] Department of Geology, Banaras Hindu University, Varanasi 221005, India

Introduction

Recent deep sea benthic foraminifera reflect the complex interaction of deep oceanic circulation, the character of surface water in the source regions, and local primary productivity (e.g. Douglas and Woodruff, 1981; Lutze and Coulbourn, 1984). The large influence of nutrient influx on the deep sea benthic foraminiferal fauna should be taken into account, while assessing the influence of deep water circulation on the faunal patterns (Gooday and Lambshead, 1989; Gooday, 1993, 1994). There is now growing realization that adaptations to nutrient levels rather than to water mass properties alone control the benthic foraminiferal distribution (Altenbach, 1992; Gooday 1993; Linke and Lutze, 1993; Altenbach *et al.*., 1999). The oxygen penetration depth and organic carbon content primarily control the deep sea benthic foraminiferal abundance and distribution (Miao and Thunnell, 1993). Their good response to the changes in oxygen content upto a threshold makes them excellent tools to reconstruct past changes in the oceanic circulation (Murray, 2001). In addition, their prompt response to the organic matter flux from the mixed layer is also important, although their value as proxy for flux is not unequivocal as for oxygenation. This is because at the same flux level different suites of species may occur dependent on the oxygenation of the environment (van der Zwaan *et al.*, 1999).

Significant changes in the physico-chemical character and circulation of the deep-sea have occurred since the beginning of the Cenozoic. However, detail $\delta^{18}O$ record reflects many intervals of global warming and cooling, and ice-sheet growth and decay (Miller *et al.*, 1987, Zachos *et al.*, 2001). The Antarctic ice-sheet formation started in the earliest Oligocene (or may be late Eocene) and permanently developed by the middle Miocene. From the early Miocene until the middle Miocene (ca. 15 Ma) the global ice volume remained low and bottom water temperature was relatively higher (Wright *et al.*., 1992). The early Miocene thermohaline circulation was largely influenced by the voluminous flow of relatively salty water from the Mediterranean region into the Indian Ocean (Woodruff and Savin, 1989). The early Miocene warm period was followed by a gradual cooling in the middle Miocene (started at ca. 14 Ma), when the flow from the Mediterranean region into the Indian Ocean probably terminated (Woodruff and Savin, 1989) and major ice-sheet on Antarctica was re-established by 10 Ma (Vincent *et al.*, 1985).

The present work is an attempt to quantitatively analyze Miocene deep-sea benthic foraminifera at DSDP Site 237 and to examine its relationships with the environmental parameters. The results of the benthic foraminiferal abundance data at this site earlier published by Rai and Srinivasan (1996) have been reinterpreted in the present study by employing multivariate methods (Q-mode factor analyses).

Oceanographic setting

DSDP Site 237 (Lat.7°04.99'S, Long. 58°07.48'E) is located on the Mascarene Plateau in the northwest Indian Ocean at a water depth of 1640m (Fig. 1.). During Miocene the site was situated at

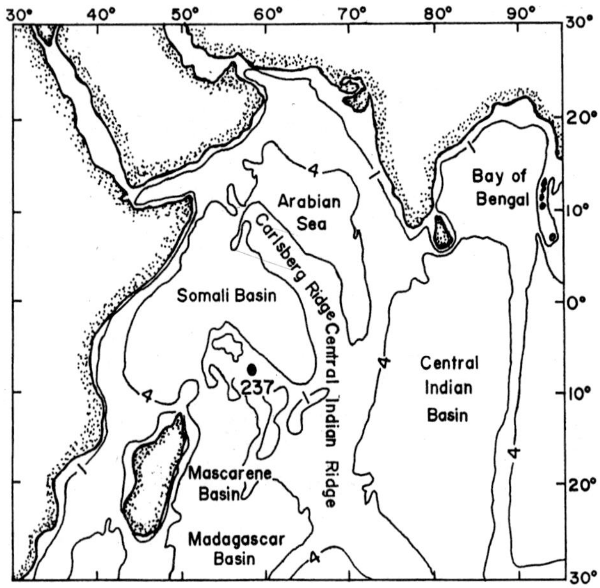

Fig. 1. Physiography of the Indian Ocean and location of DSDP site 237. Bathymetric contours are in km.

relatively shallower depth (1213m at 22 Ma; 1352m at 16 Ma; 1533m at 8 Ma) (Sclater *et al.*, 1985). The circulation pattern in the Indian Ocean is mainly characterized by the east-west flowing North Equatorial Current (NEC) between 10°N and 2-3°S and the South Equatorial Current (SEC) between 5°S and 15°S with a northward flow of the permanent anticyclonic circulation (Wyrtki, 1973; Tchernia, 1980). The Equatorial Counter Current (ECC) flows between 3°S and 10°S. Wyrtki (1971, 1973) described a hydrological front (near 10°S) separating high-nutrient, low-oxygen waters of the monsoon gyre in the northern Indian Ocean from low-nutrient, high-oxygen waters in the subtropical gyre. The hydrological front is a zone of very high activity of SEC and is marked by more surface productivity due to increased upwelling (high productivity equatorial belt). In the equatorial Indian Ocean, productivity is driven by the wind intensity of westerlies, which is related to the southern oscillations (Beaufort *et al.*., 1997). The surface water productivity in the northwest Indian Ocean is highly seasonal as a result of monsoon (Loubere, 1998). The southward flowing deep water at depths between 1200 and 3800m to the north of 30°S in the Indian Ocean is known as North Indian Deep

Water (NIDW). The NIDW of the northern Indian Ocean is relatively more saline, oxygen poor and nutrient rich and is formed due to sinking of surface waters from the Arabian Sea and from the Red Sea and the Persian Gulf outflow (Gupta, 1994). The intermediate waters of the Indian Ocean (500-1500m depth) are divided into Antarctic Intermediate Water (AAIW) in the south and Red Sea-Persian Gulf Intermediate Waters (RSPGIW) in the north (Emery and Meincke, 1986). The flow of AAIW to the north of 10°S is blocked by the hydrological front (Tchernia, 1980).

The Indian Ocean bottom water is primarily influenced by the Antarctic Bottom Water (AABW) which is mainly confined to the deep basins (>3800m) of the ocean (Kolla *et al.*, 1976). A major portion of AABW ascends to shallower depths at about 10°S and mixes with NIDW and AAIW (Vincent, 1974). AAIW mixes and sinks to greater depths after the blocking of its northward movement due to hydrological front. The above observations suggest that the water from the surface to the depth of about 3800m north of 10°S is of North Indian origin. Thus, site 237 is bathed by NIDW and situated under Equatorial Water (EQW) above the CCD which makes it ideal for carbonate deposition and preservation.

Previous work

It was after the commencement of the Deep Sea Drilling Project (DSDP) that systematic studies on the Neogene deep-sea benthic foraminifera of the Indian Ocean were more seriously undertaken. Prior to this, benthic foraminiferal studies were few as evident from a review by Bandy *et al.* (1971) on the history of foraminiferal research in the Indian Ocean. First systematic efforts to understand the benthic foraminiferal faunal-watermass associations in the Indian Ocean were made by Corliss (1979a, b, 1983) followed by Gupta and Srinivasan (1990, 1992a, b, 1996); Hermelin and Shimmield (1990, 1995); Nomura (1991, 1995); Rai and Srinivasan (1994, 1996, 2000).

The paleoceanographic significance of the Neogene deep-sea benthic foraminiferal diversity in the Indian Ocean were discussed by Gupta and Srinivasan (1992c; Ninetyeast Ridge), Rai and Srinivasan (1992; Arabian Sea and Central Indian Ocean) and Smart (1998; Somali Basin, northwestern Indian Ocean). Gooday *et al.*, (1998) presented a comparison of the deep-sea benthic foraminiferal diversity in the northeast Atlantic and northwest Arabian Sea. Variations in the late Neogene deep-sea benthic foraminiferal diversity in the eastern Indian Ocean are associated with the surface water productivity (Rai and Singh, 2001). Recently, Gupta *et al.* (2001) discussed the relation of Miocene benthic foraminiferal diversity with the water mass stratification and surface water productivity in the northeastern Indian Ocean. Studies on Neogene deep-sea benthic foraminifera in the northern Indian Ocean indicate that faunal distributions show consistent relationship to the influx of organic carbon in response to the changes in the surface water productivity (Gupta and Srinivasan, 1990, 1992a; Rai and Srinivasan, 1994, 1996, 2000; Nomura, 1995; Gupta, 1999; Gupta and Thomas, 1999). Den Dulk *et al.*. (1998 and 2000) discussed the response of benthic foraminifera to the organic matter flux and bottom water oxygenation in the northern Arabian Sea. Jannink *et al.*. (1998) and Gooday *et al.* (2000) have examined the nature of benthic foraminiferal distribution in the oxygen minimum zone of the Arabian Sea.

Material and methods

The present study is based on a total of 56 core samples from about 95.0m thick section at site 237. The Miocene section mainly comprises foraminifera-rich nannofossil ooze. Samples were dry sieved using a 149μm sieve. All the benthic foraminiferal specimens over 149μm size fraction were picked and counted as relative percentages of total assemblages which, consist of few dominant and a large proportion of rare species. In order to give more stress on dominant taxa only species with occurrence in at least two samples and with relative abundance of 3% or more in at least one sample were used for multivariate analysis (Tables. 1 and 2).

Table-1 Benthic foraminiferal species used for multivariate analyses

Vulvulina nicobarica (Schwager) = *Bigenerina nicobarica* Schwager, 1866
Textularia lythostrota (Schwager) = *Placanium lythostrotum* Schwager, 1866
Siphotextularia rolshauseni Phleger and Parker,1951
Eggerella bradyi (Cushman) = *Vemeuilina bradyi* Cushman, 1911
Martinottiella communis (d' Orbigny) = *Clavulina communis* d' Orbigny,1826
Sigmoilopsis schlumbergeri = *Sigmoilina schlumbergeri* Silvestri, 1904
Nodosaria spirostriolata Cushman, 1917
Robulus nicobarensis (Schwager) = *Cristellaria nicobarensis* Schwager, 1866
Robulus aft gibbus (dO Orbigny) = *Cristellaria gibba* dO Orbigny 1839
Orthomorphina modesta (Bermudez) = *Ellipsonodosaria modesta* Bermudez 1937
Sphaeroidina bulloides d' Orbigny, 1826
Favocassidulina australis Eade 1967
Favocassidulina favus (Brady) = *Pulvinulina fava* Brady, 1884
Stilostomella insecta (Schwager) = *Nodosaria insecta* Schwager, 1866
Stilostomella lepidula (Schwager) = *Nodosaria lepidula* Schwager, 1866
Bulimina alazanensis Cushman 1927
Bulimina macilenta (Cushman and Parker) = *Bulimina denticulata* Cushman and Parker, 1936
Bulimina jarvisi Cushman and Parker, 1936
Laticarinina pauperata (Parker and Jones) = *Pulvinulina repanda* var. *menardii* sub. var. pauperata Parker and Jones, 1865
Rectuvigerina royoi Bermudez and Fuenmayor, 1963
Uvigerina proboscidea Schwager, 1866
Cibicides bradyi (Trauth) = *Truncatulina bradyi* Trauth, 1918
Cibicides helferi Srinivasan and Shekhar, 1981
Cibicides kullenbergi Parker, 1953
Cibicides marialana gigas (Keijzer) = *Planulina marialana gigas* Keijzer, 1945
Cibicides wuellerstorfi (Schwager) = *Anomalina wuellerstorfi, 1866*
Planulina ariminensis d'Orbigny, 1826
Ehrenbergina hystrix Brady 1884
G/obocassidulina subglobosa (Brady) = *Cassidulina subglobosa* Brady, 1881
G/obocassidulina tumida (Heron-Allen and Earland) = *Cassidulina laevigata* d'Orbigny var *tumida* HeronAllen and Earland 1922
G/obocassidulina sp.
Pullenia bulloides (dO Orbigny) = *Nonionina bulloides* d' Orbigny, 1846
Oridorsalis umbonatus (Reuss) = *Rotalia umbonatus* Reuss, 1851
Osangularia culter (Parker and Jones) = *Planorbulina culter* Parker and Jones, 1865
Astrononion umbilicatulum Uchio,1952
Gyroidinoides nitidula (Schwager) = *Rotalia nitidula* Schwager, 1866
Anomalina globulosa Chapman and Parr, 1937

Table 2. Relative abundance of benthic foraminifera used in Q- mode factor analysis and factor loading of the first three factors alongwith the value of communnality at site 237. Total number of specimens counted in each sample and values of species diversity are also given

Core	Section	Depth (m)	*Vulvulina nicobarica*	*Textularia lythostrota*	*Siphotextularia roishauseni*	*Eggerella bradyi*	*Martinottiella communis*	*Sigmoilopsis schlumbergen*	*Nodosaria spirostriolata*	*Robulus nicobarensis*	*R. aff. gibbus*	*Orthomorphina modesta*	*Sphaeroidina bulloides*	*Favocassidulina australis*	*F. favus*	*Stilostomella insecta*	*S. lepidula*	*Bulimina alazanensis*	*B. macilenta*	*B. jarvisi*	*Laticarinina pauperata*	*Rectuvigerina royoi*	*Uvigerina proboscidea*	*Cibicides bradyi*	*C. helfen*	*C. kullenbergi*	*C. marialana gigas*	*C. wuellerstorfi*	*Planulina arminensis*	*Ehrenbergina hystrix*	*Globocassidulina subglobosa*	*Globocassidulina tumida*	*Globocassidulina sp*	*Pullenia bulloides*	*Oridorsalis umbonatus*	*Osangularia culter*	*Astrononion umbilicatulum*	*Gyroidinoides nitidula*	*Anomalina globulosa*	Specimens/10cc of sample	Communality	Factor 1	Factor 2	Factor 3	S_{50}
9	1	73.45	0	0.6	0	1.5	0	4.8	0	0	0.3	0	0	3	0	0.9	3.3	8.5	0	0	1.8	0	18	0.6	0	0.9	0	16	0	0	7.9	3	0	1.8	6.7	0.6	0	1.5	0.6	325	1	0.8	0.6	0.1	19
	2	74.95	0	0	0	1.2	0	4.4	0.2	0	0	0	0.5	3	0	1	2.2	7.6	0.2	0	1.5	0	22	2	0	0.7	0	14	0	0	7.6	1	0	1	9.4	1	1.5	2.7	1.5	406	1	0.8	0.5	0.1	19
	3	76.45	0	0	0	1.6	0	4.2	0.5	0	0	0	1.3	3.6	0	0	3.1	2.6	0	0	1.3	0	15	2.3	0	0	0	9.6	0	0	9.9	1.8	0	4.7	8.9	2.1	0.5	1.6	2.1	384	0.9	0.7	0.6	0.2	24
	4	77.88	0	0	0.3	3.8	0	4.7	1.2	0	0	1.8	0.6	0.6	0	1.2	3.2	1.2	0	0	2.9	0.9	24	0.9	0	0.3	0	11	0	0	2.6	3.2	0	2.3	2.9	0	1.8	1.2	3.5	341	0.7	0.7	0.5	-0	23
	5	79.45	0	0	1	1.6	0	7.8	0.6	0	0	0.6	0.6	5.2	0	0.3	1.6	6.2	1	0	1.9	0	18	1.3	0	1	0	13	0	0	2.9	4.2	0	1.9	7.5	1.3	1.3	2.3	1.9	308	0.9	0.8	0.5	0	21
	6	81.25	0	0	0.7	1.4	0	9	1.7	0	0	1	0	4.8	0	0.7	4.8	1.4	0	0	1.4	0.7	7.3	0	0	2.8	0	13	0	0	6.2	0	0	1.4	9	2.1	1	0	1.7	287	0.8	0.5	0.7	0.2	23
10	1	82.95	0	0.4	0.5	1.8	0.5	6.9	0.2	0	0.7	0	0	4	0	1.1	2.2	8	0.9	0	2.2	0.4	15	0.4	0	1.8	0	16	0	0	4.5	0.2	0	0	11	0.4	2.2	0.4	1.1	551	0.9	0.8	0.6	0.1	20
	2	84.45	0	0	0	3	0.8	4.5	0	0	0	0	0	2.3	0	0.4	1.1	9	0	0	1.5	0	20	1.5	0	3	0	11	0	0	4.5	0	0	0	6	0.8	5.3	0.8	0	266	0.9	0.9	0.4	0.1	22
	3	85.95	0	0	0	1.3	0	8.9	0	0	0	0	0.9	3.6	0	0	6.3	6.3	0	0	2.7	0	9.8	0	0	7.1	0	15	0	0	2.2	0	0	1.8	2.7	2.7	1.8	1.8	0	224	0.8	0.6	0.6	0.2	21
	4	87.45	0	0.9	0	1.9	0	1.9	0.9	0	0.6	2.2	1.3	2.5	0	0.3	3.5	1.3	0	0	1.3	1.3	15	0	0	2.5	0	18	0	0	5.7	0	0	1.3	7	0	5.7	0	0.6	316	0.9	0.5	0.8	0.2	22
	5	88.95	0	0	0	2.2	1.1	1.1	0.6	0	0	2.2	0	2.2	0	2.2	5.5	6.6	1.1	0	1.1	1.1	17	2.2	0	0	0	8.8	0	0	2.2	0	0	2.8	6.1	2.2	4.4	0	1.1	181	0.8	0.8	0.4	0.1	25
	6	90.75	0	0	0.8	1.7	0.6	1.7	0	0	0.4	0.6	0	1.3	0	0.4	2.3	1.9	1.1	0	2.1	0	24	1.7	0	1.3	0	19	0	0	3	0	0	1.3	4.6	1.7	2.1	0.2	3	474	0.9	0.6	0.7	0	22
11	1	92.41	0	0	0	2.7	0	4.1	0	0	0	0	1.4	0	0	0	3.4	5.4	0	0	2.7	1	14	0.7	0	2	0	22	0	0	5.1	0	0	2.7	6.8	0.7	0.7	1.4	3.4	294	0.9	0.6	0.8	0.1	21
	2	93.91	0	1.2	0	4.7	0.6	7.1	0	0	0	0	1.2	0	0	1.2	3	11	1.5	0	1.8	2.4	12	0	0	1.5	0	3.6	0	0	6.5	0	0	3.6	4.7	0	1.8	1.2	3	338	0.8	0.9	0.2	0.1	25
	3	95.41	0	1.3	0	5	0	9.4	0	0	0	0	1.3	0	0	1.3	3.8	2.2	1.6	0	0	0	14	3.1	0	0	0	11	0	0	10	0	0	5	5	0	0	4.4	1.3	319	0.8	0.6	0.6	0.1	21
	4	96.91	0	0	0.6	2.3	0	7	0	0	0	0	0	0	0	0.3	3.8	1.8	2.9	0	0.6	2.9	16	0.3	0	0.9	0	14	0	0	14	0	0	1.8	2.9	0.6	2.9	2.6	0	342	0.8	0.6	0.7	0.1	19
	5	98.41	0	0	0	0.9	0	7.2	0	0	0	1.4	2.7	0	0	0	0	7.2	0	0	0.9	0	17	0.9	0	0	0	14	0	0	9	0	0	1.4	9	3.6	0	0	0	222	0.9	0.8	0.6	0.1	18
	6	100.2	0	1.2	0	0	0	13	0	0	0	1.5	0	0	0	0.6	2.4	1.8	0	0	2.4	3.6	10	2.4	0	1.2	0	20	0	0	8.9	0	0	1.8	8.6	0	0.6	4.2	0.6	337	0.8	0.5	0.8	0.1	19
12	1	101.87	0	0	0	1.4	0.6	4.5	1.1	0	0.6	0.8	1.1	0	0	0.8	0.7	3.3	0.8	0	1.4	0.6	14	1.4	0	1.1	0.3	12	0	0	7.2	0	0	1.4	7.2	1.1	0.8	2.8	2.5	718	0.9	0.7	0.7	0.2	25
	2	103.37	0	0.5	0	0.4	0.4	6.6	1.5	0	0	0	0	0	0	0	2.2	1.3	1.8	0	0.4	1.5	17	1.1	0	2.2	0	12	0	0	7.5	0	0.4	1.1	5.9	2.6	1.5	1.5	2.6	547	0.9	0.6	0.7	0.2	22
	3	104.91	0.5	0	0	2.5	0	6.6	0.5	0	0	1	2	0	0	1.5	2	2.8	3.3	0	1	2.5	18	2.5	0	2	0.8	15	0	0	2.5	0	0	1.3	3	1.5	0	0.8	2.5	396	0.9	0.7	0.7	0	24
	4	106.41	0	0	0	3.3	0	6.7	0.7	0	0	0.7	0.7	1.5	0	0	6.7	0	0	0	1.5	1.5	15	0.7	0	1.5	0	12	0	0	6.7	0.7	0	1.5	8.1	0.7	0.7	3.7	1.5	270	0.9	0.6	0.7	0.3	23
	5	107.91	0	0.4	0	0.8	0.6	4.9	0	0	0.4	0.4	1.1	5.6	0	0.4	4.9	0.9	0.4	0	1.1	2.3	7.2	1.2	0	1.5	0	21	0	0	6.8	0	0	3	7.9	1.9	0.9	2.3	1.5	581	0.9	0.4	0.8	0.1	22
	6	109.41	0	0	0	1.6	0	13	0	0	0	0	0.8	3.9	0	0	1.6	31	0	0	6.2	0	3.9	3.1	0	4.7	0	16	0	0	13	0	0	2.3	6.2	0	0	0.8	0	256	0.6	0.4	0.7	0.2	18
13	1	110.39	0	0.8	0	1.9	0	4.3	0.6	0	0	0.8	0	5.8	0	0	5	7.4	1.6	0	1.2	1.2	13	0	0	1.6	0	10	0	1.2	11	0	0	1.4	5.2	0.8	0.8	1	0.8	515	0.9	0.8	0.5	0.2	23
	2	112	3.6	2.2	0	4	1.3	4	0	0	0	0	1.1	5	0	0	0.7	9.4	0.4	0	1.8	0.7	8.6	2.5	0	0.7	0	4.7	0	2.9	5.4	0	0	0	5.4	3.2	1.1	1.1	3.2	556	0.8	0.9	0.1	0.1	26
	3	113.91	1.6	0.5	0	1.3	0	4.4	0	0	0	0	0.6	2.5	0	0	0.6	28	0.3	0	0.3	1.3	7.8	0.3	0	2.2	0.2	8.1	0	0.9	6.3	0	0	0.8	8.4	1.6	0.3	1.9	0.7	640	0.7	0.8	-0	0.1	20
	4	115.79	1.1	0.6	0	0.6	0	6.1	0	0	0	0.6	0.6	6.1	0	0.3	3	24	0	0	2.8	0	5.5	0	0	0.8	0.6	10	0	0	9.4	0	0	0	7.5	2.2	0	0.6	0.6	361	0.6	0.8	0.1	0.1	19
	5	117.39	4.7	0.9	0	0.9	0	9.9	0	0	0	0	0.6	5.3	0	0	1.5	12	0.6	0	0.6	0	18	0	0	0	0	9.3	0.6	0	4.1	0	0	0	5.3	2.9	0	0.9	2.3	342	0.9	0.9	0.3	-0	20
	6	119.41	2	2.5	0	1	0	2.9	0	0	0.5	0.7	0	5.4	0	0.5	1.6	4.9	0	0	4.4	0	25	0	0	1.9	1	9.3	0	0.5	3.9	0	0	0	5.6	2.2	0	2.5	1.5	408	0.8	0.8	0.4	0.1	22
14	1	120.91	0.5	2.6	0	1	1	5.2	1	0	0	0.8	1.6	2.6	0	1	4.7	12	0.3	0	4.7	1	14	0.5	0	2.1	1	14	0.5	1.6	1.6	0	0	0	9.1	1.6	0	2.1	0.5	386	0.9	0.8	0.4	0.1	21
	2	122.41	2	1.5	0	1	0	8.3	0.5	0	0	1	0.5	4.1	0	0.5	5.2	8.8	1	0	0.5	1.5	18	1	0	6.2	0	9.3	0	2	4.7	0	1.5	0	4.1	0.5	0	3.1	0.5	382	0.9	0.8	0.4	0.2	22
	3	123.91	6.5	5.3	0	1.2	0	5.3	0	0	0	0	1.8	4.7	0	0	1.2	13	0	0	1.2	2.1	9.2	0	0	3	0	9.5	0	1.5	4.6	0	0.6	0	11	0.6	0	2.4	1.2	337	0.8	0.8	0.2	0.2	20
	4	125.41	1.1	0.5	0.5	1.1	0	6.9	0	0	0	0	1.1	1.9	1.3	0	12	19	1.1	0	1.6	3.2	11	0	0	2.7	0.5	10	0	1.6	3.7	0	0.5	0.5	3.2	0	1.1	0	0	376	0.7	0.8	0.2	0.1	20
	5	126.97	1.4	0	0	0.7	0	1.4	5.5	0	2.1	0	0.7	0	4.8	0.7	4.8	1.4	0	1.4	0	0	6.2	0	0	4.8	1.4	25	6.9	0.7	6.9	4.1	0	0	4.1	0.7	0	2.1	0	291	0.8	0.2	0.8	0.1	21

Table 2. contd.

Core	Section	Depth (m)	*Vulvulina nicobarica*	*Textularia lythostrota*	*Siphotextularia rolshauseni*	*Eggerella bradyi*	*Martinottiella communis*	*Sigmoilopsis schlumbergeri*	*Nodosaria spirostriolata*	*Robulus nicobarensis*	*R. aff. gibbus*	*Orthomorphina modesta*	*Sphaeroidina bulloides*	*Favocassidulina australis*	*F. favus*	*Stilostomella insecta*	*S. lepidula*	*Bulimina alazanensis*	*B. macilenta*	*B. jarvisi*
15	1	131.18	0	0	0	0	0	0	7.1	0	0	0	1.4	0	1.4	0	2.9	2.9	0	0
	2	132.68	0	0	0	0	0	0	1.5	0	2	1	1	0	1	0	3.9	2	0	2.9
	3	134.18	2.4	0	0	0.8	0	0	0	0	0.8	0	2.4	0	0	0	4.9	0.8	0	0.8
16	1	139.97	0	0	0	0.7	0.7	0.7	0.7	0.7	0	0	2.1	0	0	0	7.7	1.4	0	0.7
	2	141.47	2.3	0	0	0	1.1	0	0	0	0	1.1	4.5	0	1.1	2.3	3.4	0	0	0
	3	142.97	1.5	0	0	0.7	0.7	0	0	0.7	0	4.8	0	0	0.7	0	17	0	0	1.5
	4	144.47	1.7	0	0	0.9	0	0	0	0	0	2.6	0.9	0	0.9	0	1.7	1.7	1.3	2.1
	5	145.97	2.2	0	0	1.5	1.5	0	2.2	0	0	1.5	1.5	0	0.7	0	11	0.7	1.5	0
	6	147.87	2.4	0	0	1.2	0	0	0	0	2.4	1.8	1.2	0	0	0	16	1.2	0	0
17	1	149.97	0	0	0	1.9	0	0	1.9	0	1.9	1.9	0	0	0	3.7	5.6	0	0	0
	2	151.47	1	0	0	1.9	0	0	1.9	0	0	1.4	1	0	0	0	4.8	0	1	12
	3	153.47	0	0	0	0	3.4	0	0	0	0	1.7	3.4	0	0	0	1.7	0	0	0
	4	154.97	1.2	0	0	0	0	0	0	0	0	0	2.4	0	0	0	11	0	0	0
	5	156.47	2.8	0	0	0	0	0	1.4	4.9	2.1	0	2.8	0	0	0.7	3.5	0	1.4	4.2
	6	157.96	3.6	0	0	0	4.1	0	1.2	1.8	0.6	2.4	2.4	0	0	1.2	8.3	0	0	0
18	1	159.97	2.6	6	0	1.7	0	0	1.7	0.9	0.9	0	0.9	0	0	0.4	2.2	0	0	8.6
	2	161.47	2.3	3.4	0	0	1.2	0	2.3	0	4.6	2.3	0	0	0	0	7	0	0	5.7
	3	162.97	3	0	0	0	0	0	0	0	1.5	0	4.5	0	0	0	3	0	0	3
	4	164.47	2.4	0	4.8	0	2.4	0	0	0	4.8	0	2.4	0	0	2.4	7.1	0	4.8	0
	5	165.97	2.4	0	0	1.2	1.2	0	1.2	0	1.2	0	3.6	0	0	0	0	0	3.6	2.4
	6	167.47	3.9	0	0	1.9	0	0	0	0	0	0	5.8	0	0	0	1.9	0	0	1.9

Core	Section	Depth (m)	*Laticarinina pauperata*	*Rectuvigerina royoi*	*Uvigerina proboscidea*	*Cibicides bradyi*	*C. helleri*	*C. kullenbergi*	*C. marialana gigas*	*C. wuellerstorfi*	*Planulina ariminensis*	*Ehrenbergina hystrix*	*Globocassidulina subglobosa*	*Globocassidulina tumida*	*Globocassidulina* sp.	*Pullenia bulloides*	*Oridorsalis umbonatus*	*Osangularia culter*	*Astrononion umbilicatulum*	*Gyroidinoides nitidula*	*Anomalina globulosa*	Specimens/10cc of sample	Communality	Factor 1	Factor 2	Factor 3	S_{50}
15	1	131.18	2.9	5.7	1.4	2.9	0	10	1.4	17	1.4	2.9	11	0	0	0	2.9	4.3	1.4	0	0	140	0.6	0.1	0.7	0.3	22
	2	132.68	0	1	4.9	2	0	6.9	0	13	0	4.9	12	0	0	2.9	6.9	0	1	2	1	204	0.8	0.2	0.7	0.5	24
	3	134.18	0	0	2.4	0.8	0	12	1.6	16	0	3.3	7.3	0	2.4	1.6	5.7	4.9	1.6	2.4	3.3	245	0.8	0.1	0.7	0.6	23
16	1	139.97	0	1.4	7	2.1	0.7	7.7	0.7	4.2	0	0.7	9.8	0	3.5	2.1	12	2.1	0.7	1.4	2.1	286	0.9	0.3	0.3	0.8	25
	2	141.47	0	1.1	4.5	2.3	1.7	5.1	2.8	2.3	0	0	4.5	0	2.3	2.3	5.6	4.5	1.1	0	3.4	177	0.6	0.2	0.2	0.7	30
	3	142.97	0	3	5.9	1.9	0	6.3	0	2.2	0	0.7	0.7	0	0.7	1.5	4.5	3	0.7	3.7	2.2	269	0.4	0.1	0.2	0.6	26
	4	144.47	0	0	3.4	2.6	0	15	0.9	0	0	0	6	0	0	0.9	7.7	0.9	1.7	4.3	3.3	234	0.8	0.1	0	0.9	26
	5	145.97	0	0	6	0	0	11	0	6	0	0.7	6	0	1.5	4.9	3.7	1.5	0.7	1.9	0.7	267	0.7	0.2	0.4	0.7	27
	6	147.87	0	3.6	3.6	0	0	8.5	0	1.2	0	0	6.1	0	7.3	3.6	2.4	3.6	2.4	3.6	2.4	165	0.4	0	0.1	0.6	23
17	1	149.97	0	0	11	0	0	7.4	0	5.6	0	0	7.4	0	0	3.7	5.6	3.7	0	3.7	1.9	108	0.8	0.4	0.5	0.6	24
	2	151.47	0	1	4.8	3.8	1	5.7	0.5	4.3	0	0	5.7	0	1	4.8	9.6	4.8	1.9	2.9	0	209	0.5	0.1	0.3	0.7	24
	3	153.47	0	0	1.7	3.4	1.7	14	1.7	8.5	0	0	0	0	3.4	3.4	19	3.4	3.4	1.7	3.4	118	0.6	0.1	0.3	0.7	20
	4	154.97	0	2.4	1.2	1.2	0	9.7	0	3.6	0	0	4.8	0	1.2	3.6	16	1.2	0	6.1	1.2	165	0.8	0.1	0.2	0.9	24
	5	156.47	0	0	4.2	0	1.4	17	2.8	0	0	0	6.9	0	0	1.4	9.7	1.4	0	4.2	0	144	0.8	0	0	0.9	25
	6	157.96	0	0	11	2.4	0	7.1	2.4	0	0	0	7.7	0	1.2	1.2	8.3	4.7	2.4	2.4	0	169	0.7	0.3	0.2	0.7	24
18	1	159.97	0	4.3	0.9	1.7	2.6	13	0	0	0	0	3	0	2.2	0.9	11	4.3	0.9	2.6	4.3	232	0.7	-0	-0	0.8	25
	2	161.47	0	0	4.6	0	7	16	4.6	0	0	0	7	0	0	0	7	0	4.6	2.3	0	87	0.7	-0	0	0.8	21
	3	162.97	0	0	3	3	0	21	0	0	0	0	3	0	0	0	12	6.1	0	6.1	4.5	132	0.8	0	-0	0.9	20
	4	164.47	0	0	2.4	0	0	12	0	0	0	0	4.8	2.4	0	0	14	2.4	0	4.8	0	84	0.7	0.1	-0	0.9	22
	5	165.97	0	0	3.6	2.4	2.4	14	0	0	0	0	13	0	5.4	0	11	0	1.2	8.4	2.4	167	0.7	0.1	0.1	0.8	20
	6	167.47	0	1	2.9	0	0	12	0	0	0	0	6.8	0	0	0	19	0	1.9	5.8	3.9	103	0.8	0.2	0	0.9	22

Table 3. Important Miocene planktic foraminiferal events (FADs and LADs) at site 237 (after Singh and Srinivasan, 1995) and their estimated ages (after Berggren *et al.* 1995)

Planktic Foraminiferal Datums	Sample	Depth	Age (Ma)
Catapsydrax dissimilis LAD	18-4	164.47	17.3
Praeorbulina glomerosa curva FAD	18-2	161.47	16.1
Globorotalia peripheroacuta FAD	17-4	154.97	14.8
Globorotalia fohsi lobata FAD	16-4	144.47	12.5
Globorotalia (Fohsella) lineage LAD	15-2	132.68	11.9
Globorotalia plesiotumida FAD	13-6	119.41	8.3
Pulleniatina primalis FAD	11-1	92.41	6.4
Globorotalia tumida timida FAD	9-6	81.25	5.33

Q-mode factor analysis was performed on a reduced data set of 37 highest ranked species using commercially distributed statistical package (SPSS) in order to establish correlations between benthic foraminiferal assemblages and environmental conditions. This method involves a Principal Components Analysis followed by a VARIMAX rotation.

Species diversity is expressed as Hurlbert's diversity index (Hurlbert, 1971), because this index is used by many deep-sea workers who use variable sample sizes (Gage and Tyler, 1991). This method has an advantage over more commonly used diversity indices such as the Simpson index and Shannon-Wiener information index in that it eliminates differences in sample size (Malmgren and Sigaroodi, 1985). The Hurlbert's diversity index (S_m) is defined by the function;

$$Sm = \sum_{i=1}^{S} \left[1 - C(N - Ni, m) / C(N, m)\right]$$

Where, S_m is the expected number of species in a sample of m individuals selected at random from a collection containing N individuals (with $m \leq N$), and $[C(N\text{-}Ni,m) = (N\text{-}Ni)!/m!(N\text{-}Ni\text{-}m)!]$ for (N-Ni) $\geq$ m and is 0 for (N-Ni)< m, $(C(N,m) = N!/[m!(N\text{-}m)!]$, and Ni is the number of individuals of the ith species (Hurlbert, 1971). We set m at 50, well below the lowest numbers of specimens counted per sample.

The stratigraphic ranges of the most dominant benthic foraminiferal taxa have been plotted. The Miocene stable oxygen and carbon isotope values of benthic foraminiferal tests of site 237 (Vincent *et al.*, 1985) are also plotted against the quantitative benthic foraminiferal data to understand the relationship between faunal and isotopic changes. The carbon isotope values of planktic and benthic foraminiferal tests have been used to calculate planktic-benthic $\delta^{13}C$ gradients. The faunal and isotopic results are presented against the Miocene planktic foraminiferal biostratigraphic framework for site 237 (Singh and Srinivasan, 1995) with absolute chronologies of Berggren *et al.*. (1995) (Table. 3). Based on the abrupt replacement of stratigraphically important planktic foraminifera, Singh and Srinivasan (1995) recognized a hiatus across the middle/late Miocene transition at this site which corresponds to the hiatus NH4 of Barron and Keller (1982).

Results

Stratigraphic distribution of species

The stratigraphic distribution of the most common benthic foraminiferal species is shown in Fig. 2. Most of the taxa are distributed throughout the Miocene, either frequently or infrequently due to patchiness in distribution or as a result of environmental changes. Whereas, several other benthic foraminiferal species are characterized by lower and upper limits of their stratigraphic occurrences within the examined section (Table. 4). Large number of benthic foraminiferal events (first and last appearance datums) took place mainly during the middle Miocene. This broadly corresponds to the well-known middle Miocene benthic foraminiferal turnover (Woodruff and Douglas, 1981). From the late Miocene onwards the benthic foraminiferal composition did not show much variation and remained more or less unchanged showing close similarity with the recent benthic foraminiferal distribution at comparable depths in the Indian Ocean (Boltovskoy, 1977, 1979; Corliss, 1979b; Gupta, 1994).

Relative faunal abundance

The benthic foraminiferal population is represented by a total of 166 species in the Miocene section at site 237. The deep-sea benthic fauna is characterized by species rich and specimen poor assemblages in which only one or two individuals represent large numbers of species. However, *Uvigerina proboscidea, Bulimina alazanensis, Cibicides wuellerstorfi, Cibicides kullenbergi, Oridorsalis umbonatus, Sigmoilopsis schlumbergeri, Stilostomella lepidula* and *Globocassidulina subglobosa* are the most dominant species represented by 10% or more in abundance at least in one sample (Fig. 3). The late early Miocene to early middle Miocene interval (c. 18.0 – 12.5 Ma) is mainly dominated by the higher abundances of *C. kullenbergi* and *O. umbonatus. Cibicides wuellerstorfi* appeared in the early middle Miocene and increased in abundance during the late middle Miocene. The early late Miocene (upto c. 7.5 Ma) is characterized by higher abundances of *B. alazanensis* and *U. proboscidea* that correspond with low percentages of *C. wuellerstorfi*. Most of the late Miocene-early Pliocene time (c. 8.5 to 5 Ma) is dominated by *U. proboscidea* and *C. wuellerstorfi. Sigmoilopsis schlumbergeri* shows its distinct occurrences during most of the late Miocene.

Factor analysis

The results of Q-mode factor analysis with a VARIMAX rotation shows that the first three factors account for 78.3% of the total faunal variance (see Tables. 2, 5 and 6). The stratigraphic distribution of factor loading is shown in Fig. 4.

Uvigerina proboscidea – Bulimina alazanensis assemblage (Factor 1; 54.9% of the total variance) is characterized by high score of factor 1 in *U. proboscidea* (3.76) and *B. alazanensis* (3.58). This factor thus shows high loading in intervals where these two species are most abundant (Fig. 4). Subordinate forms of this assemblage are *Sigmoilopsis schlumbergeri* and *Oridorsalis umbonatus*. This assemblage is well developed during late middle Miocene to early Pliocene (c. 8.5 to c. 5.0 Ma).

Cibicides wuellerstorfi assemblage (Factor 2; 17.7% of the total variance) is dominantly linked to the relative abundance of *C. wuellerstorfi*, with a highest score of 4.93 and shows high loading where this species is abundant. This assemblage consists primarily of *C. wuellerstorfi*, and subsequently *Globocassidulina subglobosa* and *U. proboscidea*. It is dominantly distributed in the middle Miocene (c. 12.5-11 Ma) and later part of the late Miocene to early Pliocene (c. 8.5 to 5.0 Ma).

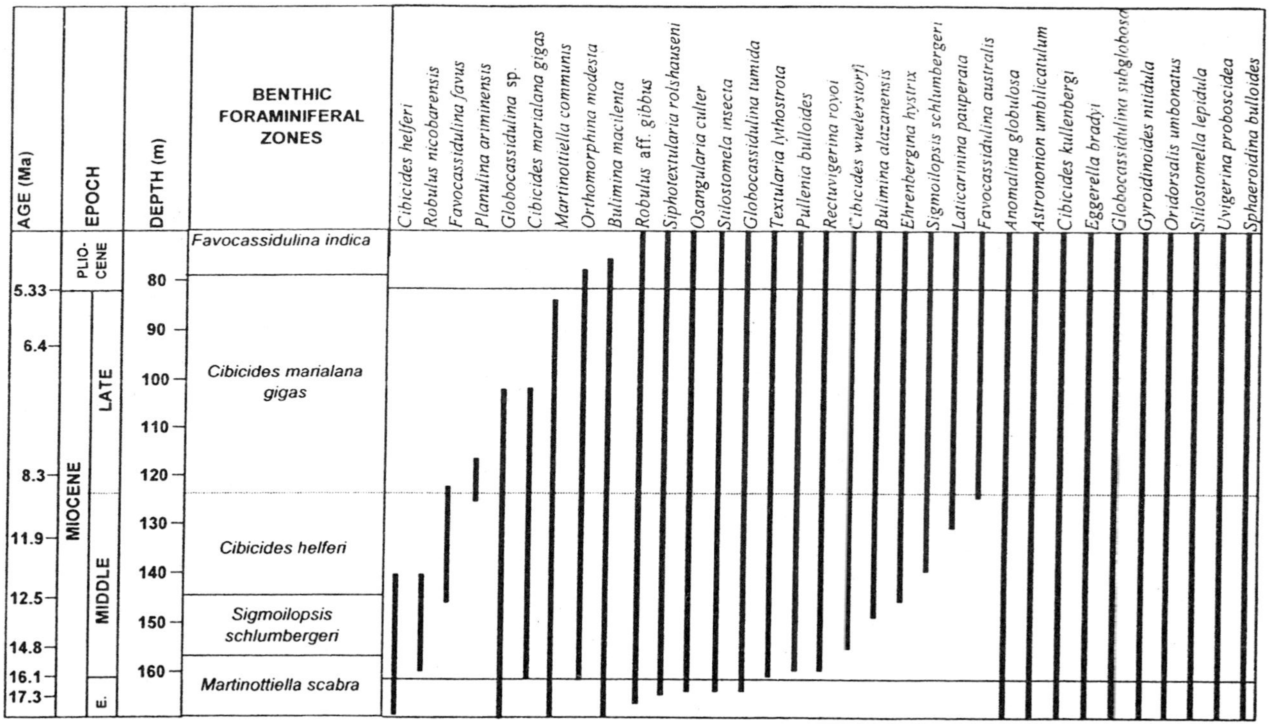

Fig. 2. Stratigraphic ranges of important benthic foraminiferal species during Miocene at site 237. Benthic foraminiferal zones are after Srinivasan *et al.*, 1993. Ages of planktic foraminiferal zones are after Singh and Srinivasan (1995) and Berggren *et al.*, (1995).

Table 4. First and Last Appearances (FADs and LADs) of taxa at site 237

Events		Sample	Depth (m)	Approx. Age (Ma)
LAD	*Bulimina macilenta*	9-2	74.95	4.9
LAD	*Martinottiella communis*	10-1	82.95	5.5
LAD	*Cibicides marialana gigas*	12-1	101.87	7.1
LAD	*Globocassidulina* sp.	12-2	103.37	7.2
LAD	*Planulina ariminensis*	13-5	117.39	8.2
FAD	*Favocassidulina australis*	14-4	125.41	9.9
FAD	*Planulina ariminensis*	15-1	131.18	11.5
FAD	*Laticarinina pauperata*	15-1	131.18	11.5
FAD	*Sigmoilopsis schlumbergeri*	16-1	139.97	12.3
LAD	*Cibicides helferi*	16-1	139.97	12.3
LAD	*Robulus nicobarensis*	16-1	139.97	12.3
FAD	*Favocassidulina favus*	16-5	145.97	12.7
FAD	*Ehrenbergina hystrix*	16-6	147.87	13.2
FAD	*Bulimina alazanensis*	16-6	147.87	13.2
FAD	*Cibicides wuellerstorfi*	17-4	154.97	14.8
FAD	*Rectuvigerina royoi*	18-1	159.97	15.8
FAD	*Robulus nicobarensis*	18-1	159.97	15.8
FAD	*Pullenia bulloides*	18-1	159.97	15.8
FAD	*Orthomorphina modesta*	18-2	161.47	16.1
FAD	*Textularia lythostrota*	18-2	161.47	16.1
FAD	*Cibicides marialana gigas*	18-2	161.47	16.1
FAD	*Globocassidulina tumida*	18-4	164.47	17.3
FAD	*Stilostomella insecta*	18-4	164.47	17.3
FAD	*Osangularia culter*	18-4	164.47	17.3

Cibicides kullenbergi – Oridorsalis umbonatus assemblage (Factor 3; 5.7% of the total variance) is distinguished by the higher scores of factor 3 in *C. kullenbergi* (3.89) and *O. umbonatus* (3.02). In general, intervals of higher relative abundances of these two species are characterized by higher loading of factor 3. This assemblage also has subordinate forms like *Stilostomella lepidula* and *G. subglobosa*. This assemblage shows higher loading during the late early Miocene to early middle Miocene (c. 18.0 – c. 12.5 Ma).

Faunal diversity

The values of Hurlbert's diversity index (S_{50}) vary from 18 to 30 during the Miocene at site 237 (Fig. 4). The late early Miocene (c. 18.0 – 16.0 Ma) interval is characterized by relatively low faunal diversity. Diversity started increasing gradually from the early/middle Miocene boundary onwards reaching its peak during the middle Miocene at about 12.5 Ma. A continuous decline in the values of diversity has been observed in the upper part of middle Miocene to middle part of late Miocene (c. 12.5 to c. 7.5 Ma). The rest of the late Miocene and early Pliocene (7.5 to 5.0 Ma) are characterized by relatively lower values of diversity with marked fluctuations.

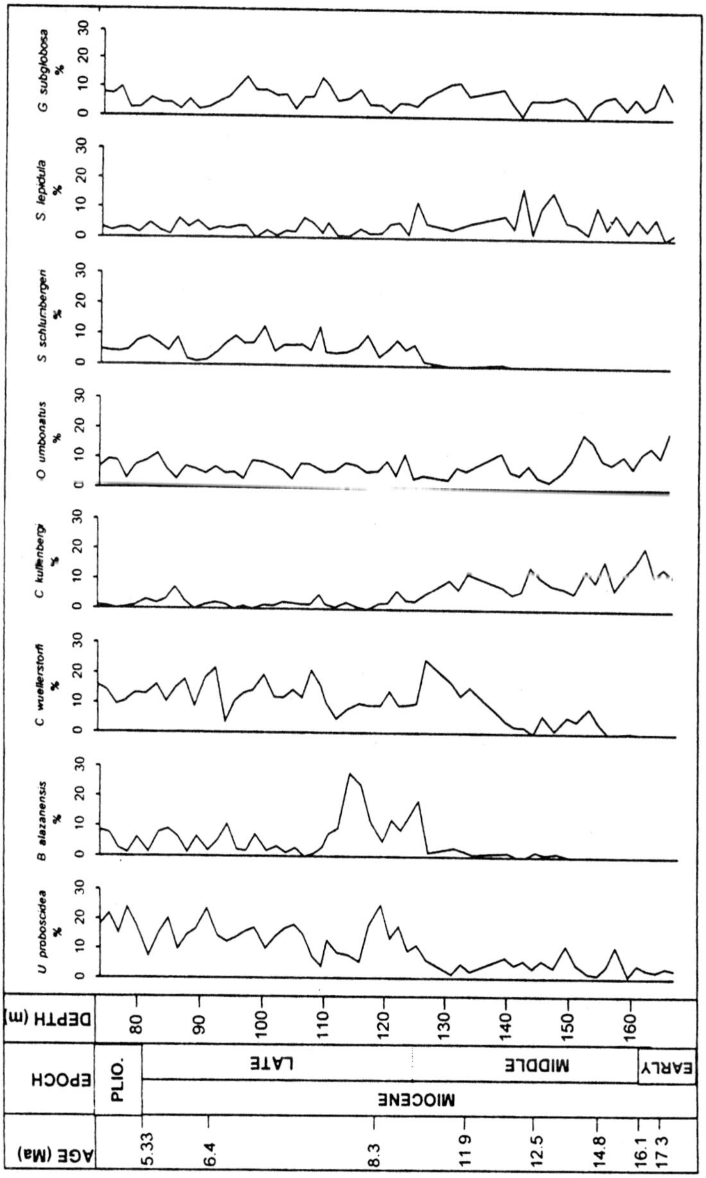

Fig. 3. Percent distribution of dominant benthic foraminiferal species at site 237.

Table 5. Factor scores of the first three factors from the Q – mode VARIMAX factor analysis of the benthic foraminiferal relative abundance data at 237.

Species	Factor 1	Factor 2	Factor 3
Vulvulina nicobarica	0.07224	-0.86269	0.18515
Textularia lythostrota	-0.03046	-0.62225	-0.42412
Siphotextularia rolshauseni	-0.45096	-0.29194	-0.56369
Eggerella bradyi	0.15658	-0.21773	-0.48771
Martinottiella communis	-0.35377	-0.38062	-0.32828
Sigmoilopsis schlumbergeri	1.17516	0.48155	-1.01398
Nodosaria spirostriolata	-0.75329	0.29697	-0.3985
Robulus nicobarensis	-0.4663	-0.35133	-0.45241
R. aff. gibbus	-0.57109	-0.20645	-0.22845
Orthomorphina modesta	-0.4342	-0.13692	-0.2719
Sphaeroidina bulloides	-0.32864	-0.30225	0.22466
Favocassidulina australis	0.60708	-0.41984	-0.7557
F. favus	-0.61099	-0.04875	-0.59799
Stilostomella insecta	-0.34426	-0.21156	-0.46732
S. lepidula	-0.13196	0.47919	1.48181
Bulimina alazanensis	3.57609	-2.32888	-0.60083
B. macilenta	-0.28271	-0.29543	-0.4012
B. jarvisi	-0.64208	-0.33037	0.29016
Laticarinina pauperata	-0.04081	0.00673	-0.77563
Rectuvigerina royoi	-0.3679	-0.01921	-0.32679
Uvigerina proboscidea	3.75976	1.15338	0.1326
Cibicides bradyi	-0.38101	-0.00062	-0.12679
C. helferi	-0.51768	-0.36425	-0.23841
C. kullenbergi	-0.76857	-0.09448	3.8861
C. marialana gigas	-0.51887	-0.23795	-0.26597
C. wuellerstorfi	0.42338	4.92707	-0.78091
Planulina ariminensis	-0.63073	0.00606	-0.70686
Ehrenbergina hystrix	-0.4107	-0.12238	-0.55052
Globocassidulina subglobosa	0.36543	1.31242	1.40859
Globocassidulina tumida	-0.44729	-0.09128	-0.67587
Globocassidulina sp.	-0.55105	-0.27244	-0.02421
Pullenia bulloides	-0.46912	0.36587	-0.0573
Oridorsalis umbonatus	1.16836	-0.16415	3.02063
Osangularia culter	-0.25892	-0.09577	0.34211
Astrononion umbilicatulum	-0.2643	-0.06604	-0.22291
Gyroidinoides nitidula	-0.18676	-0.23048	0.69221
Anomalina globulosa	-0.08968	-0.26318	0.08023

Table 6. Species composition of Miocene benthic foraminiferal Q- mode assemblages at Site 237 (Factor No., dominant and important associated species with factor scores are given.)

Q-mode assemblages	Dominant species	Factor scores	Important associated species	Factor scores	Variance %
Factor 1	*Uvigerina proboscidea* *Bulimina alazanensis*	3.76 3.58	*Sigmoilopsis schlumbergeri* *Oridorsalis umbonatus*	1.18 1.17	54.9
Factor 2	*Cibicides wuellerstorfi*	4.93	*Globocassidulina subglobosa* *Uvigerina proboscidea*	1.31 1.15	17.7
Factor 3	*Cibicides kullenbergi* *Oridorsalis umbonatus*	3.89 3.02	*Stilostomella lepidula* *Globocassidulina subglobosa*	1.48 1.41	5.7

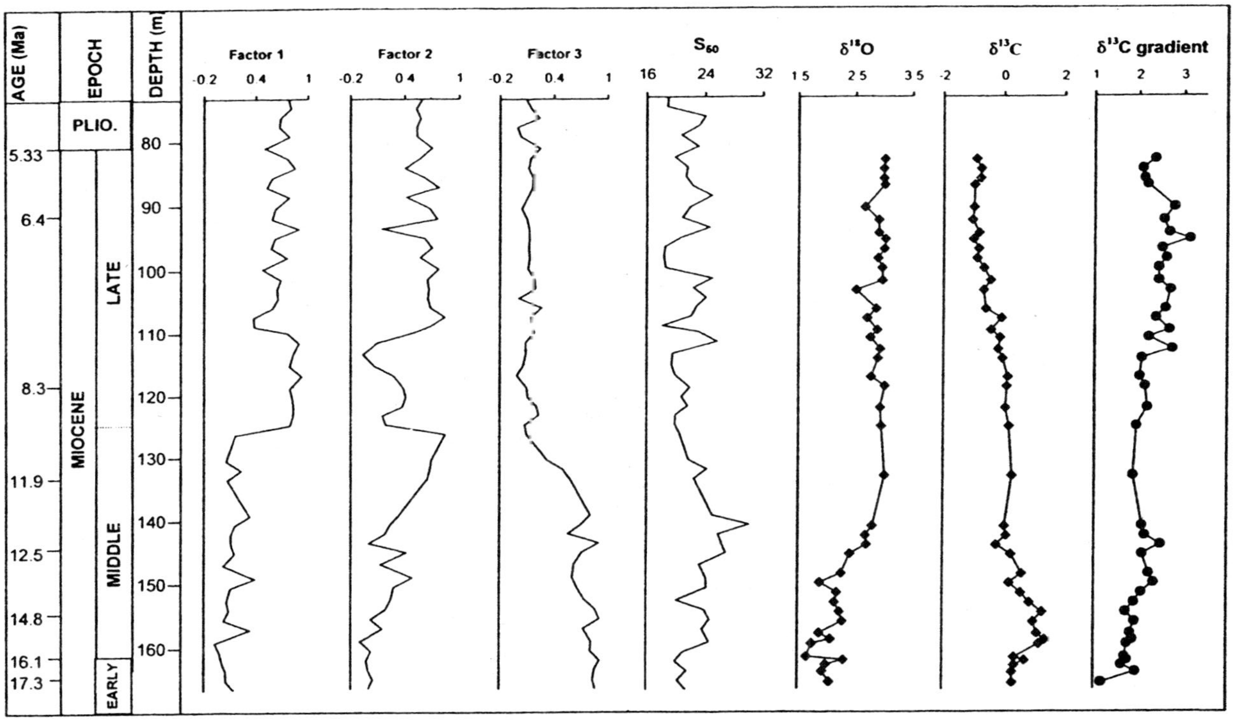

Fig. 4. Plots of factor loading of the first three factors extracted based on Q-mode VARIMAX factor analysis of the benthic foraminiferal relative abundance data and species diversity (Hurlbert's diversity index, S_{50}) of benthic foraminifera with benthic (O. umbonatus) stable isotope records (after Vincent et al. 1985) and planktic-benthic $\delta\ ^{3}C$ gradients at site 237.

Stable isotope records

The values of Miocene benthic stable oxygen and carbon isotopic records along with planktic-benthic $\delta^{13}C$ gradients at site 237 (after Vincent *et al.*, 1985) are plotted in Fig. 4. The early Miocene (c.18 to 16.0 Ma) is marked by relatively low values of $\delta^{18}O$. The $\delta^{18}O$ values gradually increased during the early middle Miocene (16.0-.12.5 Ma) which corresponds with the progressive decline in the values of $\delta^{13}C$. After 12.5 Ma, the rest of the younger Miocene interval does not show any marked change in $\delta^{18}O$ values. However, later part of late Miocene (c. 7.0 onwards) is characterized by a decrease in $\delta^{13}C$ and distinctly higher planktic-benthic $\delta^{13}C$ gradients. A shift toward lower $\delta^{13}C$ value in the latest Miocene (i.e. Chron-6 carbon shift) is also identified in the benthic record of this site (Fig. 4).

Discussion

Benthic foraminifera are considered as one of the valuable proxies to understand the marine paleoenvironment due to their wide and well-preserved occurrence. The benthic foraminiferal response to the changing oxygenation upto a critical threshold (Murray, 2001) makes them very useful to know the changes in the past oceanic circulation, which is supposed to be responsible for the changes in aeration of the benthic environment. In addition, benthic foraminifera also respond to changes in the organic matter flux from the mixed layer (e.g. Gooday, 1994; Murray, 2001). The higher rates of organic carbon flux at the sea floor can absorb a large amount of oxygen resulting into oxygen-poor benthic environment (Miao and Thunnell, 1993). Thus, the organic carbon flux rates are responsible for both the availability of food and oxygen content at various depths. The broad scale variations of the deep-sea oxygenation are dependent on the global thermohaline circulation pattern (Mantyala and Reid, 1983) however, changing magnitudes of nutrient supply and related biological metabolism are partially responsible for local oxygen depleting conditions (Sen Gupta and MaChain-Castillo, 1993).

The ecological preferences of various taxa associated with different benthic foraminiferal assemblages are broadly known which may be useful to explain the changes in the faunal composition through time. *Uvigerina proboscidea* is a well-known species in the organic–rich sediments below the areas of high surface water productivity (Gupta and Srinivasan, 1992a; Rai and Srinivasan, 1994, 2000; Wells *et al.*., 1994). *Bulimina (i.e. B. alazanensis)* occurs dominantly in bottom waters with decreased oxygen contents and higher trophic resource levels (Phleger and Soutar, 1973; Miller and Lohmann, 1982; van der Zwaan, 1982). *Cibicides kullenbergi* prefers a low organic carbon content substrate below low surface water productivity regions (Burke *et al.*., 1993; Nomura, 1995). The vertical distribution of *C. kullenbergi* is confined to oxygen-rich and nutrient poor NADW (Schmiedl *et al.*, 1997). *Bulimina alazanensis* and *C. kullenbergi* are associated with warm deep water in the northern Indian Ocean during the Miocene (Gupta and Srinivasan, 1992b). *Cibicides wuellerstorfi* is an attached epibenthic species which lives on the structures elevated above the sediment surface in high energy environment (Lutze and Thiel, 1989). As a suspension feeder and elevated epibiont, this species does not require high organic carbon levels and can withstand active bottom water currents (Linke and Lutze, 1993; Mackensen *et al.*., 1995; Jannink *et al.*, 1998; Gupta and Thomas, 1999; Ohkushi *et al.*, 2000 and others). The higher abundance of this species is attributed to decreased food supply associated with the regions of low surface productivity (Sarnathein and Altenbach, 1995; Kuhnt *et al.*., 1999). *Oridorsalis umbonatus* is a long-lived, cosmopolitan species and its environmental preferences are not clearly known. This species mainly occurs in the areas of low surface water productivity (Lutze and Coulbourn, 1984). It can survive in the sediments of lower oxygen contents and limited organic carbon flux (Rathburn and Corliss, 1994). However, Mackensen *et al.*. (1995) suggest that *O. umbonatus* occurs dominantly below the zones of seasonally high surface water productivity but requires well oxygenated pore waters. *Globocassidulina subglobosa* characterizes the influence of warm Antarctic Bottom Water in the Indian Ocean (Corliss, 1979a) and also prefers to live in the deep water with higher flux of organic carbon (Miao and Thunell, 1993; Nomura, 1995). However, in the eastern and western equatorial Pacific *G. subglobosa* may be related to the low surface water productivity and thereby low flux of organic matter to the sea floor (Loubere and Bannois, 1987; Burke *et al.*, 1993). Mackensen *et al..* (1995) observed that *G. subglobosa* dominated assemblage flourish in the areas of higher bottom current velocities with low flux and lying above carbonate corrosive environments. *Stilostomella lepidula* possibly shows preference to live in the sediments with higher organic carbon and low oxygen content (Boersma, 1990). *Sigmoilopsis schlumbergeri* is ecologically similar to the *Melonis barleeanum* (Hermelin and Shimmield, 1995) which prefers the bottom waters enriched with altered organic matters (Caralp, 1989).

The early Miocene and early middle Miocene interval (c.18.0 to c. 12.5 Ma) is characterized by the *C. kullenbergi – O. umbonatus* assemblage. We observed relatively low values of species diversity during the late early Miocene (c. 18.0 to 16.0 Ma) which also corresponds with the low $\delta^{18}O$ values reflecting warm and sluggish deep water circulation. This may possibly correspond to the lower trophic levels in the Indian Ocean deep water in response to low surface water productivity. Srinivasan and Singh (1991) observed distinct increase in the abundance of tropical planktic foraminiferal assemblages during the late early Miocene at site 237 indicating warming of surface water. This warm phase reached to its maximum during the middle Miocene climatic optimum (c. 17 to c. 15 Ma) and was followed by a gradual cooling and major Antarctic ice-sheet development by 10 Ma (Vincent *et al..*, 1985; Flower and Kennett, 1995). Smart and Ramsay (1995) and Ramsay *et al.*, (1998) observed increased abundance of smooth walled bolivinids between 19.5 and 16.5 Ma of the northwest Indian ocean and proposed a phase of reduced oxygen concentration within the Tethyan (= Mediterranean) Outflow Water (TOW) which did not affect the abundance of other benthic foraminiferal taxa.

Cibicides wuellerstorfi, a species with almost global distribution appeared at about 14.8 Ma at site 237 and exhibits increased abundance thereafter during the rest of the Miocene reflecting changed food resource level in the deep-sea and intensified bottom circulation. The Indian Ocean deep-sea fauna is characterized by lower food resource tolerant assemblages prior to the appearance of *C. wuellerstorfi* (Nomura, 1995 and present work). The widespread first appearance of this species in the early middle Miocene may be due to the closely similar trophic levels globally in the deep-sea and distinct development of thermohaline and active bottom water circulation (Nomura, 1995). However, there are slight differences in the first appearance of *C. wuellerstorfi* in different oceans (Pacific: c. 16-15 Ma; Subantarctic region: 16.1-15.7 Ma; North Atlantic: 14.8-14.1 Ma) (Nomura, 1995). Thus, based on the hypothesis that the change in food resource level is related to the appearance of this species, it can be suggested that the Indian Ocean bottom water changed its trophic level about one million year later than the Pacific. The early middle Miocene faunal changes correspond to the changes in water mass resulting from global warming and sluggish circulation followed by deep water cooling. The older benthic fauna of low diversity and less oxygenated conditions were gradually replaced by a new fauna characterizing higher species diversity and more oxygenated deep water in response to the gradual development of oxygen rich, cold deep water originating from Antarctica (Woodruff, 1991).

Major changes in the assemblages of deep-sea benthic foraminifera during the middle Miocene appear to be in response to the growth of permanent Antarctic ice sheet (Kennett and Shackleton, 1976; Woodruff, 1985; Kurihara and Kennett, 1992 and others). Originations and extinctions of several benthic foraminiferal species correspond closely to the well-known middle Miocene benthic foraminiferal turnover (Woodruff and Douglas, 1981). The middle Miocene (c. 14.5 Ma onwards) is also marked by the surface water cooling in the northwestern Indian Ocean as evidenced by the characteristic planktic foraminiferal assemblages at site 237 (Srinivasan and Singh, 1991). The significant middle Miocene enrichment in deep water $\delta^{18}O$ throughout the world ocean is generally interpreted as the combined effect of deep water cooling and growth of the Antarctic ice-sheet (Vincent *et al.*., 1991; Wright *et al.*, 1992). An atmospheric CO_2 draw down due to increased rates of chemical weathering and organic matter burial may have resulted in global cooling and the Antarctic ice sheet expansion by 15 Ma, culminating in the establishment of permanent continental ice sheet as indicated by higher $\delta^{18}O$ values between 15 and 13 Ma (Raymo, 1994). Recent workers do not support the middle Miocene atmospheric CO_2 draw down for the increased growth of Antarctic ice-sheet (Pagani *et al.*, 1999; Zachos *et al.*., 2001). However, Huybrechts (1993) and Bartek *et al.* (1996) suggested that moisture supply during the intervals of decreased green house gas levels was critical element in maintaining large polar ice sheets. A distinct change in the faunal assemblage occurred during late middle Miocene (c. 12.5 –11.8 Ma) as evidenced by significant drop in diversity and replacement of *Cibicides kullenbergi-Oridorsalis umbonatus* assemblage by *Uvigerina proboscidea-Bulimina alazanensis* and *Cibicides wuellerstorfi* assemblages reflecting an overall increase in the trophic level at the sea floor. These faunal changes occurred just after the major increase in $\delta^{18}O$ and decline in $\delta^{13}C$ values. Nomura (1995) observed the development of higher food resource tolerant faunal assemblage at the expense of lower food resource exploiting assemblages in the eastern Indian Ocean.

During the late middle Miocene (c. 11.5 Ma) onwards the benthic fauna at site 237 is dominated by *U. proboscidea-B. alazanensis* and *C. wuellerstorfi* assemblages reflecting increased trophic levels. However, distinctly high abundances of *B. alazanensis* alongwith low *C. wuellerstorfi* percentages and decreased species diversity during early late Miocene (upto c. 7.5 Ma) time indicate relatively less oxygenated deep-sea environment due to sluggish circulation in the northwest Indian Ocean. The distinct increase in the *U. proboscidea* abundances during the late Miocene (c. 8.5 Ma onwards) along with low values of $\delta^{13}C$ and higher planktic-benthic $\delta^{13}C$ gradients are suggested to be due to enhanced oceanic fertility and surface water productivity. Gupta and Srinivasan (1992a, b) and Nomura (1995) also observed the significant development of an *U. proboscidea* assemblage at bathyal depths in the eastern Indian Ocean interpreted as a response to increased surface water productivity. The higher rates of organic carbon burial between 8 to 5 Ma may be related to the enhanced chemical weathering and river nutrient fluxes stimulated by an increase in the height of the Himalayan range and associated intensification of the Asian monsoon around 8 Ma (Quade *et al.*., 1989; Prell and Kutzbach, 1992; Raymo, 1994). On the basis of carbon isotope and strontium isotope mass balance Raymo (1994) calculated a short-term maxima in organic carbon burial from 8.5 to 5 Ma with rates 16% higher than preceding values. Moody *et al.*., (1988) and Filippelli and Delancy (1994) also observed a distinct increase in the phosphorus accumulation rate to the ocean for this interval and assuming that this increase directly drives organic matter burial, yields an increase in organic carbon burial. A distinct interval of low species diversity at site 237 in the upper most Miocene coincides with the Chron-6 carbon shift (Vincent *et al.*., 1985 and 1991), expansion of western Antarctic ice-sheet (Kennett, 1977, 1986; Woodruff and Savin, 1985) and lowering of the sea level (Haq *et al.*., 1987). Gupta *et al.*. (2001) also observed distinctly low benthic foraminiferal diversity during this interval in the northeastern Indian Ocean. Srinivasan and Singh (1991) observed a major turnover in the planktic foraminifera at site 237 during the late Miocene which virtually coincides with Chron-6 carbon shift. Cerling *et al.*. (1993, 1997 and 1998) suggest that change in $\delta^{13}C$ values starting from about 7 Ma and continuing upto 5 Ma may possibly reflect a decrease in atmospheric CO_2 crossing a threshold that favoured C_3-photosynthetic plants near the end of Miocene. Pisias *et al.* (1995) discussed that increased chemical weathering in the Himalayan range in response to the intensified Asian monsoon was responsible for the 'biogenic bloom' event in the late Miocene. The 'biogenic bloom' was extended from the late

Miocene to the early Pliocene (c. 9 to 3.5 Ma) over large areas of the Indo-Pacific, which was most intense between 6 to 5 Ma (Peterson *et al..*, 1992; Dickens and Owen, 1994, 1999; Farrell *et al..*, 1995). Gupta and Srinivasan (1992) suggested vertical uplift of Andaman Islands during the late Miocene (7.7 to 5.0 Ma) which was also the period of enhanced monsoonal upwelling. The higher biological productivity during c. 8.5-5.0 Ma may be a fertilization pulse to the ocean from enhanced nutrient supply derived from intense weathering (Filippelli, 1997).

Conclusions

The variations in the Miocene deep-sea benthic foraminiferal assemblages and stable isotope records at DSDP site 237 are significant to understand paleoceanographic changes in the northwest Indian Ocean. The less diverse fauna and decrease in $\delta^{18}O$ values during early Miocene (c. 18-16 Ma) suggest warm bottom waters with low oxygen contents. Several benthic foraminiferal events (first and last appearances) occurred during the middle Miocene that corresponds closely to the well-known middle Miocene faunal turnover recorded from the equatorial Pacific. This was the period of large-scale climatic deterioration in response to the distinct growth of the Antarctic ice sheet. The first appearance of *C. wuellerstorfi* in different oceans during the middle Miocene (approximately around 15 Ma) indicates nearly similar trophic resource levels and significant development of thermohaline and intensified deep-sea circulation. The deep-sea benthic foraminifera are characterized by higher food resource exploiting assemblages after the first appearance of *C. wuellerstorfi*. After the main increase in the $\delta^{18}O$ values during middle Miocene at about 12.5-11.8 Ma species diversity dropped and *Cibicides kullenbergi-Oridorsalis umbonatus* assemblage is replaced by *Uvigerina proboscidea-Bulimina alazanensis* and *Cibicides wuellerstorfi* assemblages suggesting significant increase in the food supply to the ocean floor. The distinct occurrence of *U. proboscidea -B. alazanensis* assemblage alongwith low $\delta^{13}C$ values and high planktic-benthic $\delta^{13}C$ gradients during late Miocene (c. 8.5-5 Ma) reflect enhanced ocean fertility and surface water productivity. An increased erosion due to uplift of the Himalaya and associated intensification of the Asian monsoon was possibly responsible for the 'biogenic bloom' event in the late Miocene.

Acknowledgements

We thank Deep Sea Drilling Project (DSDP) for providing the core samples. We are thankful to J. W. Murray and E. Thomas for their valuable comments and suggestions on an early version of the manuscript. This work was partially supported by Council of Scientific and Industrial Research, New Delhi.

References

ALTENBACH, A. V., 1992. Short-term processes and patterns in the foraminiferal response to organic carbon flux rates. In: van der Zwaan *et al..* (Eds.), Approaches to Paleoproductivity Reconstructions. Marine Micropaleontology, 19: 119-129.

ALTENBACH, A. V., PFLAUMANN, U., SCHIEBEL, R., THIES, A., TIMM, S., TRAUTH, M., 1999. Scaling percentages and distributional patterns of benthic foraminifers with flux rates of organic carbon. Journal of Foraminiferal Research, 29: 173-185.

BANDY, O. L., LINDERBERG, H. G., VINCENT, E., 1971. History of Research, Indian Ocean Foraminifera. Journal of Marine Biological Association of India, 13: 86-105.

BARRON, A.J., KELLER, G., 1982. Widespread Miocene deep sea hiatuses: Coincidence with periods of global cooling. Micropaleontology, 10: 577-581.

BARTEK, L. R., HENRYS, S. A., ANDERSON, J. B., BARERTT, P. J., 1996. Seismic stratigraphy of McMurdo Sound, Antarctica: implications for glacially influenced early Cenozoic eustatic change? Marine Geology 130: 79-98.

BEAUFORT, L., LANCELOT, Y., CAMBERLIN, P., CAYRE, O., VINCENT, E., BASSINOT, F., LABEYRIE, L., 1997. Insolation cycles as a major control of equatorial Indian Ocean Primary Production. Science, 278: 1451-1454.

BERGGREN W.A., KENT, D.V., SWISHER III, C.C. AUBRY, M.P., 1995. A revised Cenozoic geochronology and chronostratigraphy: Geochronology, Time Scales and Global Stratigraphic Correlation. SEPM Spec. Pub. no. 54: 129-212.

BOERSMA, A., 1990. Late Oligocene to the Pliocene benthic foraminifers from depth traverses in the central Indian. Proc. ODP Sci. Results, 115: 315-319.

BOLTOVSKOY, E, 1977, Neogene deep water benthonic foraminifera of the Indian Ocean and biostratigraphy (ed Hertzler, J.R., Bolli, H.M., Davies, T.A., Saunders, J.B. and Sclater, J.G.) p.599-616. Am Geophy. Uni. Washington

BOLTOVSKOY, E, 1978, Late Cenozoic benthonic foraminifera of the Ninetyeast Ridge (Indian Ocean), marine Geology, vol. 26, p. 139-175, pl 1-8.

BURKE, S. C., BERGER, W. H., COULBOURN, W. T., VINCENT, E., 1993. Benthic foraminifera in box core ERDC 112, Ontong Java Plateau. Journal of Foraminiferal Research 23, 19-39.

CARALP, M. H., 1989. Abundance of Bulimina exilis and Melonis barleeanum: Relationship to the quality of marine organic matter. Geo-Marine Letters, 37: 37-43.

CERLING, T. E., WANG, Y., QUADE, J., 1993. Expansion of C4 ecosystems as an indicator of global ecological change in the late Miocene. Nature, 361: 344-345.

CERLING, T. E., HARRIS, J. M., MACFADDEN, B. J., LEAKEY, M. G., QUADE, J., EISENMANN, V., EHLERINGER, J. R., 1997. Global vegetation change change through the Miocene/Pliocene boundary. Nature, 389, 153-158.

CERLING, T. E., HARRIS, J. M., MACFADDEN, B. J., LEAKEY, M. G., QUADE, J.,EISENMANN, V., EHLERINGER, J. R., 1998. Reply (Scientific correspondence). Nature 393, 127.

CORLISS, B. H., 1979a. Quaternary Antarctic Bottom Water history: Deep-sea benthonic foraminiferal evidence from the Southeast Indian Ocean. Quaternary Research, 12: 271-279.

CORLISS, B. H., 1979b. Recent deep-sea benthonic foraminiferal distributions in the Southeast Indian Ocean: Inferred bottom-water routes and ecological implications. Marine Geology, 31: 115-138.

CORLISS, B. H. 1983. Distribution of Holocene deep-sea benthonic foraminifera in the Southwest Indian Ocean. Deep Sea Research, 30: 95-117.

DEN DULK, M., REICHART, G. J., MEMON, G. M., ROELOFS, E.M.P., ZACARIASSE, W. J., VAN DER ZAWN, G. J., 1998. Benthic foraminiferal response to variations in the surface water productivity and oxygenation in the northern Arabian Sea. Mar. Micropaleontol. 35, 43-66.

DEN DULK, M., REICHART, G. J., HEYST S. V., ZACARIASSE, W. J., VAN DER ZAWN, G. J., 2000. Benthic foraminifera as proxies of organic matter flux and bottom water oxygenation? A case history from the northern Arabian Sea. Palaeogeogr. Palaeoclimatol. Palaeoecolog. 161, 337-359.

DICKENS, G. R., OWENS, R. M., 1994. Late Miocene-early Pliocene manganese redirection in the central Indian Ocean: Expansion of the intermediate water oxygen minimum zone: Paleoceanography 9, 169-181.

DICKENS, G. R., OWEN, R. M., 1999. The latest Miocene-early Pliocene biogenic bloom: A revised Indian Ocean perspective. Marine Geology 161, 75-91.

DOUGLAS, R. G., WOODRUFF, F., 1981. Deep sea benthic foraminifera. In: Emiliani, C. (ed.): The Sea (The Oceanic Lithosphere), 7: 1233-1327.

EMERY, W. E., MEINCKE, J., 1986. Global water masses: summary and review. Oceanologia Acta, 9(4): 383-391.

FARRELL, J. W., RAFFI, I., JANECEK, T. R., MURRAY D. W., LEVITAN M., DADEY, K. A., EMRIS, K. C., LYLE, M., FLORES, J. A., HOVAN S., 1995. Late Neogene sedimentation patterns in the eastern equatorial Pacific Ocean, Proc. Ocean Drill. Program, Sci. Results, 138, 717-753.

FILIPPELLI, G. M., 1997. Intensification of the Asian monsoon and a chemical weathering event in the late Miocene-early Pliocene: Implications for late Neogene climate change. Geology, 25(1): 27-30.

FILIPPELLI, G. M., DELANCY, M. L., 1994. The oceanic phosphorus cycle and continental weathering during the Neogene. Paleoceanography 9, 643-652.

FLOWER, B. P., AND KENNETT, J. P., 1995. Middle Miocene deepwater paleoceanography in the southwest Pacific: Relations with East Antarctic Ice Sheet development. Paleoceanography 10, 1095-1112.

GAGE, J.D., TYLER, P.A., 1991. Deep-sea biology: a natural history of organisms at the deep-sea floor. Cambridge University Press, Cambridge, 504 pp.

GEOSECS, 1983. Indian Ocean Expedition, Hydrographic Data, 1977-1978, 5:1-48 and 6:1-63. Washington, D. C.: U. S. Government Printing Office.

GOODAY, A.J., 1993. Deep-sea benthic foraminiferal species which exploit phytodetritus: Characteristic features and controls on distribution. Mar. Micropal. 22: 187-205.

GOODAY, A. J., 1994. The biology of deep-sea foraminifera: A review of some advances and their applications in paleoceanography. Palaios 9: 14-31.

GOODAY, A.J., LAMBSHEAD, P. J. D., 1989. The influence of seasonally deposited phytodetritus on benthic foraminiferal populations in the bathyal northeast Atlantic. Marine Ecology Program Series, 58: 53-67.

GOODAY, A.J., BETT, B. J., SHIRES, R., LAMBSHEAD, P. J. D., 1998. Deep-sea benthic foraminiferal diversity in the NE Atlantic and NW Arabian area: a synthesis. Deep-Sea Research II, 45: 165-201.

GOODAY, A.J., BERNHARD, J. M., LEVIN, L. A., 2000. Foraminifera in the Arabian Sea oxygen minimum zone and other oxygen-deficient settings: taxonomic composition, diversity, and relation to metazoan faunas. Deep-Sea Deep-sea Research II, 47: 25-54.

GUPTA, A. K., 1994. Holocene deep-sea benthic foraminifera and water masses in the Indian Ocean and the Red Sea. Jour. of Geol. Soc. of India 43, 691-703.

GUPTA, A. K., 1999. Latest Pliocene through Holocene paleoceanography of the eastern Indian Ocean: benthic foraminiferal evidence. Marine Geology, 161, 63-73.

GUPTA, A. K., SRINIVASAN, M. S., 1990. Response of Northern Indian Ocean deep-sea benthic foraminifera to global climates during Pliocene-Pleistocene. Marine Micropaleontology, 16: 77-91.

GUPTA, A. K., SRINIVASAN, M. S., 1992a. Uvigerina proboscidea abundances and paleoceanography of the northern Indian Ocean DSDP site 214 during the Late Neogene. Mar. Micropaleontology 19, 355-367.

GUPTA, A. K., SRINIVASAN, M. S., 1992b. Deep-sea benthic foraminiferal changes and terminal Miocene Event at Tropical Indian Ocean DSDP Site 214. Jour. of Geol. Soc. of India 40: 262-278.

GUPTA, A. K., SRINIVASAN, M. S., 1992c. Species diversity of Neogene deep sea benthic foraminifera from northern Indian Ocean sites 214 and 216A. In: Takayanagi, Y. Saito, T.(Eds.), Studies in benthic foraminifera. Proc. of the Fourth Internat. Symp. on benthic foraminifera, Sendai, Tokai Univ. Press, pp. 249-254.

GUPTA, A. K., SRINIVASAN, M. S., 1996. Multivariate analysis of Pliocene-Pleistocene benthic foraminifera from DSDP site 214, eastern Indian Ocean: paleoceanic and paleoclimatic implications. In: Pandey, J., Azmi, R. J., Bhandari, A., Dave, A. (Eds.), Contr. Indian Colloq. Micropal. and Strat., Allied Printers, Dehra Dun, pp. 267-273.

GUPTA, A. K., THOMAS, E., 1999. Latest Miocene- Pleistocene productivity and deep-sea ventilation in the northwestern Indian ocean (Deep Sea Drilling Project Site 219). Paleoceanography, 14(1), 62-73.

GUPTA, A. K., JOSEPH, S., THOMAS, E., 2001. Species diversity of Miocene deep-sea benthic foraminifera and watermass stratification in the northeastern Indian Ocean. Micropaleontology, 47(2), 111-124.

Gupta, S. M., Srinivasan, M. S., 1992. Late Miocene radiolarian biostratigraphy and paleoceanography of Sawai Bay Formation, Neill Island, Andamans, India. Micropaleontology, 38(3), 209-235.

HAQ, B. U., HARDENBOL, J., VAIL, P. R., 1987. The chronology of fluctuating sea level since the Triassic. Science 235, 1156-1167.

HERMELIN, J. O. R., SHIMMIELD, G. B., 1990. The importance of the Oxygen Minimum Zone and sediment geochemistry in the distribution of Recent benthic foraminifera in the Northwest Indian Ocean. Marine Geology, 91: 1-29.

HERMELIN, J. O. R., SHIMMIELD, G. B., 1995. Impact of productivity on the benthic foraminiferal fauna in the Arabian Sea over the last 150,000 years. Paleoceanography 10(1), 85-116.

HURLBERT, S.H., 1971. The nonconcept of species diversity: a critique and alternative parameters. Ecol. 52, 577-586.

HUYBRECHTS, P., 1993. Glaciological modelling of the Late Cenozoic East Antarctic ice sheet: stability or dynamism, Geografiska Annaler 75A, 221-238.

JANNINK, N. T., ZACHARIASSE, W. J., VAN DER ZWAAN, G. J., 1998. Living (Rose Bengal stained) benthic foraminifera from the Pakistan continental margin (northern Arabian Sea) Deep-Sea Research I 45, 1483-1513.

KENNETT, J. P., 1977. Cenozoic evolution of Antarctic glaciations, the circum-Antarctic Ocean and their impact on global paleoceanography. Jour. of Geophy. Research 82, 3843-3860.

KENNETT, J. P., 1986. Miocene to early Pliocene oxygen and carbon isotope stratigraphy in the southwest Pacific, Deep Sea Drilling Project leg 90, Initial Rep. Deep Sea Drill. Proj., 90, 1383-1412.

KENNETT, J. P., SHACKLETON, N. J., 1976. Oxygen isotopic evidence for the development of the psychrosphere 38 m.y. ago, Nature, 260, 513-515.

KOLLA, V., SULLIVAN, L., STREETER, S. S., LANGSETH, M. G., 1976. Spreading of Antarctic Bottom Water and its effect on the floor of the Indian Ocean inferred from bottom-water potential temperature, turbidity and sea-floor photography. Mar. Geology, 21: 171-189.

KUHNT, W., HESS, S., JIAN, Z., 1999. Quantitative composition of benthic foraminiferal assemblages as a proxy indicator for organic carbon rates in the South China Sea. Mar. Geology, 156 : 123-157.

KURIHARA, K., KENNETT, J. P., 1992. Paleoceanographic significance of Neogene benthic foraminiferal changes in a southwest Pacific bathyal depth transect. Mar. Micropaleontol. 19, 181-199.

LINKE, P., LUTZE G. F., 1993. Microhabitat preferences of benthic foraminifera- a static concept or a dynamic adaptation to optimize food acquisition? Mar. Micropaleontol. 20, 215-234.

LOUBERE, P., 1998. The impact of seasonality on the benthos oas reflected in the assemblages of deep-sea foraminifera. Deep-Sea Research I 45, 409-432.

LOUBERE, P., BANONIS, G., 1987. Benthic foraminiferal assemblage response to the onset of northern hemisphere glaciation: paleoenvironmental changes and species trends in the northeast Atlantic. Mar. Micropaleontol. 12, 161-181.

LUTZE, G. F., COULBOURN ,W. T., 1984. Recent benthic foraminifera from the continental margin of northwest Africa: community structure and distribution. Marine Micropaleontol. 8, 361-401.

LUTZE, G. F., THIEL, H., 1989. Epibenthic foraminifera from elevated microhabitat: Cibicides wuellerstorfi and Planulina ariminensis. Jour. of Foram. Res. 19, 153-158.

MACKENSEN, A., SCHMIEDL, G., HARLOFF, J., GIESE, M., 1995. Deep-sea foraminiferain the South Atlantic Ocean: Ecology and assemblage generation. Micropaleontology 41, 242-258.

MALMGREN, B.A., SIGAROODI, M. M., 1985. Standardization of species counts – the usefulness of Hurlbert's diversity-index in paleontology. Bulletin of the Geological Institutions of the University of Uppsala, N.S. 10, 111-114.

MANTYALA, A. W., REID, J. L., 1983. Abyssal characteristics of the world's ocean waters. Deep-Sea Res. 30, 805-833.

MIAO, Q., THUNELL, R. C., 1993. Recent deep-sea benthic foraminiferal distributions in the South China and Sulu Seas. Marine Micropaleontol. 22, 1-32.

MILLER, K. G., LOHMANN, G. P., 1982. Environmental distribution of recent benthic foraminifera on the northeast United States continental slope. Geol. Soc. of Ame. Bull. 93, 200-206.

MILLER, K.G., Fairbanks, R.G. and Mountain, G.S. (1987) Tertiary oxygen isotope systhesis , sea level history and continental margin erosion Paleoceanography, v.2, p.1-9 *ET AL.*, 1987

MOODY, J. B., CHABOUDY, L. R., WORSLEY, T. R., 1988. Pacific pelagic phosphorus accumulation during the last 10 m.y.: Paleoceanography 3, 113-136.

MURRAY, J. W., 2001. The niche of benthic foraminifera, critical thresholds and proxies. Marine Micropaleontol. 41, 1-7.

NOMURA, R., 1991. Oligocene to Pleistocene benthic foraminifer assemblages at sites 754 and 756, eastern Indian Ocean. Proc. Ocean Drilling Program, Sci. Results, 121, 31-76.

NOMURA, R., 1995. Paleogene to Neogene deep-sea paleoceanography in the eastern Indian Ocean: benthic foraminifera from ODP Sites 747, 757 and 758. Micropaleontology. 41(3), 251-290.

OHKUSHI, K., THOMAS, E., KAWAHATA, H., 2000. Abyssal benthic foraminifera from the northwestern Pacific (Shatsky Rise) during the last 298 kyr. Mar Micropaleontol. 38, 119-147.

PAGANI, J., ARTHUR, M. A., FREEMAN, K. H., 1999. Miocene evolution of atmospheric carbon dioxide. Paleoceanography, 14, 273-292.

PETERSON, L. C., MURRAY, D. W., EHRMANN, W. U., HAMPEL P., 1992. Cenozoic carbonate accumulation and compensation depth changes in the Indian Ocean. Synthesis of Results from Scientific Drilling in the Indian Ocean, Geophys. Monogr. Vol. 70, edited by R. Duncan and D. Rea, 311-333, AGU, Washington D. C.

PHLEGER, F. B., SOUTAR, A., 1973. Production of benthic foraminifera in three east Pacific oxygen minima. Micropaleontology 19, 110-115.

PISIAS, N. G., LAYER, L. A., MIX, A. C., 1995. Paleoceanography of the eastern equatorial Pacific during the Neogene: Synthesis of Leg 138 drilling results. Proc. of ODP, Sci. Results 138, 5-21.

PRELL, W. L., KUTZBACH, J. E., 1992. Sensitivity of the Indian monsoon to forcing parameters and implications for its evolution. Nature, 360: 646-650.

QUADE, J., CERLING, T. E., BOWMAN, J. R., 1989. Development of Asian monsoon revealed by marked ecological shift during the latest Miocene in northern Pakistan. Nature 342, 163-166.

RAI A. K., SINGH, V. B., 2001. Late Neogene deep-sea benthic foraminifera at ODP Site 762B, eastern Indian Ocean: diversity trends and palaeoceanography. Palaeogeography, Palaeoclimatology, Palaeoecology, 173, 1-8.

RAI, A.K., SRINIVASAN, M.S., 1992. Neogene deep sea benthic foraminiferal diversity in the Indian Ocean: Paleoceanographic implications. . Proc. Ind. Acad. Sci. (Earth & Planet. Sci.), 101(3), 299-316.

RAI, A.K., SRINIVASAN, M.S., 1994. Pleistocene oceanographic changes indicated by deep sea benthic foraminifera in the Northern Indian Ocean. Proc. Ind. Acad. Sci. (Earth & Planet. Sci.) 103(4), 499-517.

RAI, A.K., SRINIVASAN, M.S., 1996. Miocene deep-sea benthic foraminifera and bottom waters at DSDP Site 237, central Indian Ocean. In: Pandey, J., Azmi, R. J., Bhandari, A., Dave, A. (Eds.), Contr. Indian Colloq. Micropaleontol . and Stratigr., Allied Printers, Dehra Dun, pp. 267-273.

RAI, A.K., SRINIVASAN, M.S., 2000. Deep sea benthic foraminiferal response to the Pliocene paleoenvironments of the northern Indian Ocean. GEOBIOS, 33(3), 301-308.

RAMSAY, A.T.S., SMART, C. W., ZACHOS, J. C., 1998. A model of early middle Miocene deep ocean circulation for the Atlantic and Indian Oceans. In: Cramp, A., *et al.*. (eds) Geological Evolution of Ocean Basins: Results from the ODP, Geol. Soc. London, Spec. Publ. 131, 55-70.

RATHBURN, A. E., CORLISS, B. H., 1994. The ecology of living (stained) deep-sea benthic foraminifera from the Sulu Sea. Paleoceanography 9(1), 87-150.

RAYMO, M. E., 1994. The Himalayas, organic carbon burial, and climate in the Miocne. Paleoceanography 9(3), 399-404.

SARNTHEIN, M., ALTENBACH, A. V., 1995. Late quaternary changes in surface water and deep water masses of Nordic Seas and north-eastern North Atlantic: a review. Geol. Rundsch. 84: 89-107.

SCHMIEDL, G., MACKENSEN, A., MÜLLER, P.J., 1997. Recent benthic foraminifera from the eastern South Atlantic Ocean : Dependence on food supply and water masses. Mar. Micropaleontol. 32, 249-287.

SCHWAGER,C, 1866. Fossile foraminiferen von Kar –Nicobar : Novara Exped. (1857-1859). Wien, Ed.2. Geol. theilpp.187-268

SCLATER, J. G., MEINKE, L., BENNETT, A., MURPHY, C., 1985. The depth of the ocean through the Neogene. Geological Society of America, Memoir 163, 1-19.

SEN GUPTA, B. K., MACHAIN-CASTILLO, M. L., 1993. Benthic foraminifera in oxygen-poor habitats. Mar. Micropaleontol. 20, 183-201.

SINGH, A. D. SRINIVASAN, M. S., 1995. Neogene planktic foraminiferal biochronology of the Central Indian Ocean DSDP sites 237 and 238. Jour. Geol. Soc. India 45, 445-462.

SMART, C. W., 1998. Diversity patterns of Miocene benthic foraminifera in the Somali Basin, northwestern Indian Ocean. Micropaleontology, 44(3), 256-264.

SMART, C. W., RAMSAY, A. T. S., 1995. Benthic foraminiferal evidence for the existence of an early Miocene oxygen-depleted oceanic water mass? Jour. Geol. Soc., London, 152, 735-738.

SRINIVASAN, M.S. AND SHARMA, V., 1980 Schwager's Car Nicobar foraminifera in the Reports of Novara Expedition-a revision; Today and Tomorrow's Printers and Publishers, New Delhi 83. pp. 1 1-8.

SRINIVASAN, M. S., SINGH, A. D. 1991. Planktic foraminiferal evidence for the Neogene paleoceanographic changes in the Indian Ocean. The Annual Technical Meeting 1989 and IGCP-246, 179-205.

SRINIVASAN, M. S., RAI, A. K., GUPTA, A. K., 1993. Neogene benthic foraminiferal biostratigraphy of the Indian Ocean deep sea sequences. Indian Journal of Geology, 65(4), 246-255.

TCHERNIA P., 1980. Descriptive Regional Oceanography. Pergamon Press, New York, 253p.

VAN DER ZWAAN G. J., 1982. Paleoecology of Miocene Mediterranean foraminifera. Utrecht Micropaleontol.Bull. 25, 1-202.

VAN DER ZWAAN G. J., DUIJNSTEE, I.A.P., DEN DULK, M., ERNST, S. R., JANNINK, N. T., KOUWENHOVEN, T. J., 1999. Benthic foraminifers: proxies or problems? A review of paleoecological concepts. Earth-Science Reviews 46, 213-236.

VINCENT, E., 1974. Cenozoic planktonic biostratigraphy and paleoceanography of the tropical western Indian Ocean. DSDP Init. Reports 24, 1111-1150.

VINCENT, E., KILLINGLEY, J. S., BERGER, W. H., 1985. Miocene oxygen and carbon isotope stratigraphy of the tropical Indian Ocean. In: Kennett, J. P. (ed.), The Miocene Ocean: Paleoceanography and Biogeography. Geol. Soc. of Amer. Memoir 163, 103-130.

VINCENT, E., SHACKLETON, N. J., HALL M.A., 1991. Miocene oxygen and carbon isotope stratigraphy of planktonic foraminifers at sites 709 and 758, tropical Indian Ocean. In: Weissel, J., Peirce, J., Taylor, E., Alt, J., *et al.*., Proc. Ocean Drilling Program, Scientific Results, vol. 121, 241-252.

WELLS P., WELLS G., CALI J., CHIVAS, A., 1994. Response of deep-sea benthic foraminifera to Late Quaternary climate changes, southeast Indian Ocean, offshore western Australia. Mar. Micropaleontol. 23, 185-229.

WOODRUFF, F., 1985. Changes in Miocene deep sea benthic foraminiferal distribution in the Pacific Ocean: Relationship to paleoceanography. Mem. Geol. Soc. Am., 163, 131-176.

WOODRUFF, F., 1991. Mid-Miocene isotope stratigraphy in the deep sea: High resolution correlations, paleoclimatic cycles, and sediment preservation. Paleoceanography 6, 755-806.

WOODRUFF, F., DOUGLAS, R. G., 1981. Response of deep-sea benthic foraminifera to Miocene paleoclimatic events: DSDP Site 289: Marine Micropaleontology 6, 617-632.

WOODRUFF, F., SAVIN, S. M., 1985. ^{13}C values of Miocene Pacific benthic foraminifera : correlations with sea level and biological productivity. Geology 13, 119-122.

WOODRUFF, F., SAVIN, S. M., 1989. Miocene deep water oceanography. Paleoceanography 4, 87-140.

WRIGHT, J. D., MILLER, K. G., FAIRBANKS, R. G., 1992. Early and middle Miocene stable isotopes: implications for deep water circulation and climate. Paleoceanography 7, 357-389.

WYRTKI, K., 1971. Oceanographic atlas of the International Indian Ocean Expedition: National Science Foundation, Washington, D. C., 531pp.

WYRTKI, K., 1973. Physical Oceanography of the Indian Ocean. In: ZEITZSCHEL B. (ed.): The Biology of the Indian Ocean. 18-36, Springer Verlag. New York.

ZACHOS, J., PAGANI, M., SLOAN, A., THOMAS, E., BILLUPS, K., 2001. Trends, rhythms, and aberrations in global climate 65 Ma to Present. Science, 292: 686-693.

Neogene planktic foraminiferal biostratigraphy on the Pacific coast of South America and its paleoceanographic implications

Masako Ibaraki
Department of Biology and Geosciences, Faculty of Science, Shizuoka University,
Shizuoka 422-8529, Japan

Introduction

The international Andean studies group of Shizuoka University has studied marine Neogene sequences on the Pacific coast of South America from 1985 to 1996 from a viewpoint of the trans-Pacific correlation (Fig. 1). The author had opportunities to obtain planktic foraminiferal samples from many of Neogene sequences in the studied area. Based on the planktic foraminiferal biostratigraphy of representative marine Neogene sequences on the Pacific coast of Ecuador, Peru, and Chile, a precise correlation chart was accomplished (Fig. 2) and various geologic events occurred during the Neogene were examined.

Geological setting and background

Representative sections of the Pacific coast of South America containing planktic foraminifera are, from north to south, those in areas of Esmeraldas, San Vicente, Jaramijo and Punta Canoa in Ecuador, those in areas Bayovar, Cerro Las Salinas, Loma Cuesta Chilcatay and Camana in Peru, those in areas of Costa Mejillones, Caleta Herradura de Mejillones, Cuenca del Tiburon, La Portada, Cuebrada Blanca, and Navidad in Chile.

Early Miocene sequences of Zones N4-N7 are observed in the Cerro Las Salinas section in Peru and Zones N5-N6 are found in the San Vicente section in Ecuador. Early Middle Miocene sequences of Zone N8 are frequently found in places of the Rio Esmeraldas, the San Vicente and the Jaramijo section in Ecuador, the Camana section in Peru, and the Caleta Herradura de Mejillones and the El Rincon section in Chile. Middle Miocene sequences of Zones N9-N10 and N14 are developed in the Esmeraldas area in Ecuador, and the Caleta Herradura de Mejillones area in Chile. Late Middle Miocene sequences of Zone N17 are extensively distributed in areas of the Esmeraldas in Ecuador, the Loma Questa Chilcatay section in Peru, ODP Site 684 off Peru, the Caleta Herradura de Mejillones section and the Navidad section in Chile. Pliocene sequences of Zones N19-N21 are found in the Esmeraldas area, although planktic foraminifea are rare in occurrence, because biosiliceous sediments become predominant in Pliocene and later ages in areas from central Ecuador to northern Chile (Tsuchi, 1988, 1990, 1992).

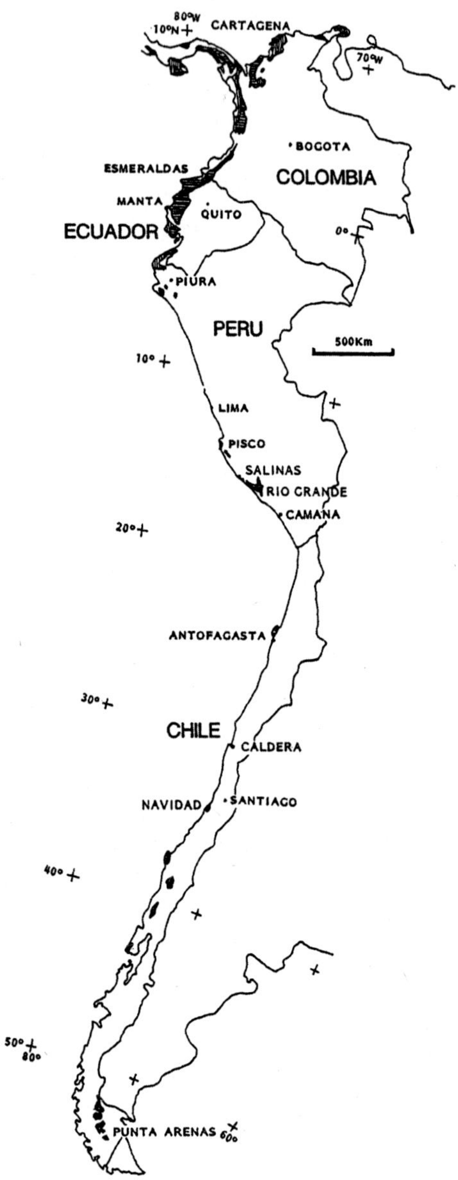

Fig. 1. Distribution of marine Cenozoic strata on the Pacific coast of South America and on the Caribbean coast of Colombia.

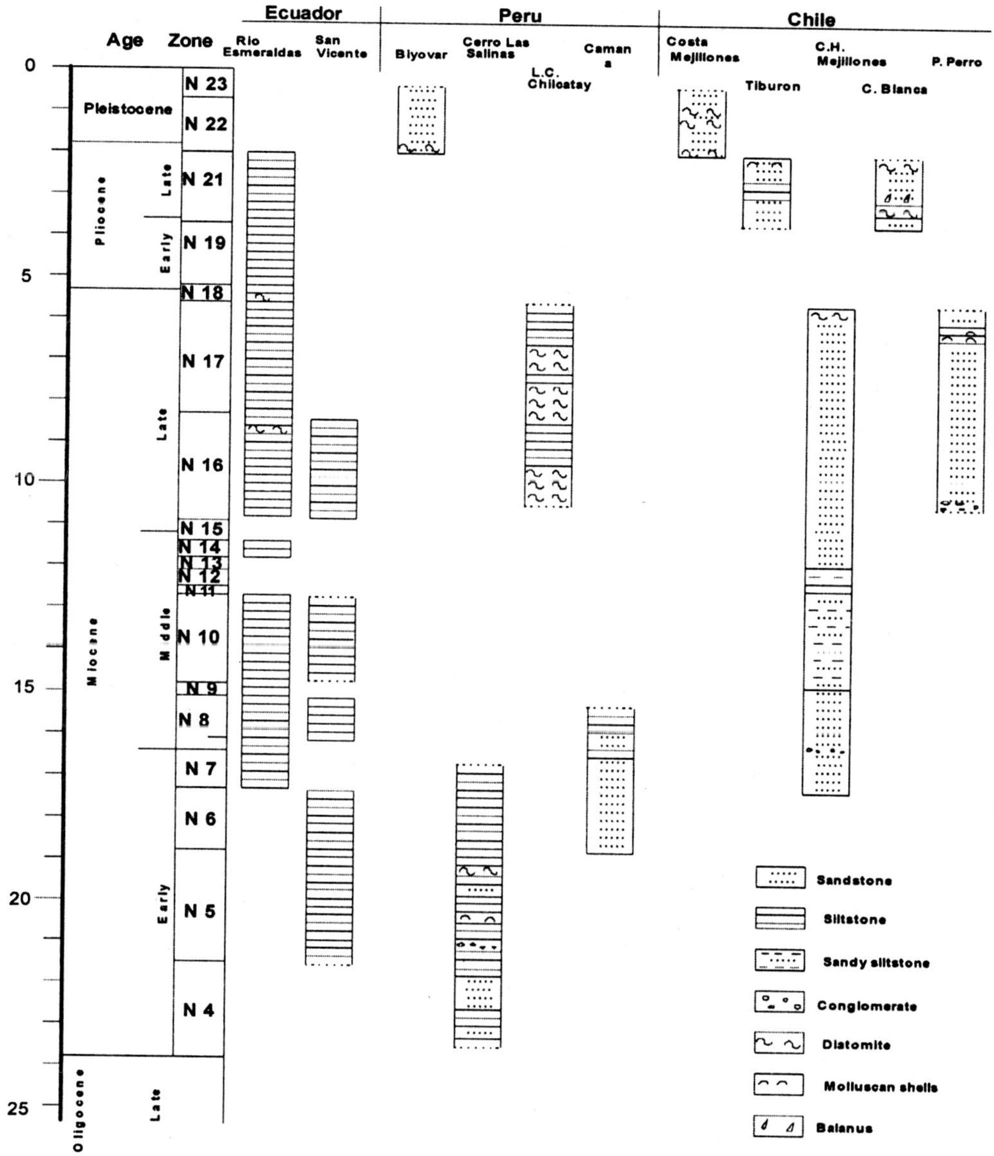

Fig. 2. Correlation chart of marine Neogene sequences on coastal areas of Ecuador, Peru and Chile based on planktic foraminifera.

Sampling and Method

During the course of the international Andean studies of Shizuoka University, the author had opportunities to obtain planktic foraminiferal samples from many of Neogene sequences on the Pacific coast of South America. Fieldwork was carried out in 1985, 1986, 1988, 1990, 1992 and 1996 in Ecuador, Peru, and Chile in cooperation with collaborators of these countries.

Dried rock samples of 100g each were washed through a 200-mesh screen and dried in an oven. Species of planktic foraminifera were picked up from washed residues, and the frequency of occurrence of each species was calculated. The ratio of warm-water and cold-water taxa to the total planktic foraminiferal population was calculated for each sample. Warm-water species recognized here are: *Catapsydrax* spp., *Dentoglobigerina* spp., *Globigeirnoides* spp., *Globoquadrina* spp., *Globorotalia cultrata, G. menardii, G. praefohsi, G. tumida, G. ungulata, Globoturborotalita decoraperta, G. druryi, G. nepenthes, G. rubescens, G. tenellus, Pulleniatina* spp., and *Sphaeroidinellopsis* spp. These species are regarded as characteristics of tropical and subtropical areas by Be (1977) and Kennnett and Srinivasan (1983). The planktic foraminiferal zonation used here is in accordance with Blow (1969), and age-assignments are those by Berggren *et al.* (1995a, b)

Neogene planktic foraminifera on the Pacific coast of South America

Biochronologic analyses of planktic foraminifera on the coastal area of Ecuador

In areas along the Pacific coast of Ecuador, marine Paleogene-Neogene sequences extensively develop, forming three sedimentary basins called, from north to south, Barbon, Manabi, and Progreso, respectively. Planktic foraminifera are obtained from the Esmeraldas area, the San Vicente area, the Jaramijo section of the Manta area and the Punta Canoa area (Ibaraki, 2002).

The Esmeraldas area located on the Pacific coast of northwestern Ecuador, 300km northwest of Quito, is known as a part of the Barbon Basin, one of Paleogene-Neogene sedimentary basins in Ecuador. The Neogene sequences are developing along the coast. In this area, planktic foraminifera were obtained from 44 localities (Ibaraki, 1994, 1988), including Zone P21 of Oligocene, Zone N7 of Early Miocene, Zones N8-N10 and N14 of Middle Miocene, Zone N16 and Zones N17-N18 of Late Miocene, and Zones N19-21 of Pliocene (Fig. 3).

The San Vicente area, located 380km southwest of Quito, belongs to the Manabi Basin of Neogene sequences. The Neogene sequences are exposed along the lower course of the Rio Chone and on the Pacific coast. In this area, planktic foraminifera were obtained from 16 localites, including Zones N5-N6 of Early Miocene, Subzone N8b of the latest Early Miocene, Zone N10 of Middle Miocene and Zones N16-N17 of Late Miocene (Fig. 4).

The Manta area is located 400km southwest of Quito. The Neogene sequences are exposed along the Pacific coast. In this area, planktic foraminifera were obtained from 12 localities, including Zones N8-N9 of the earliest Middle Miocene in the Jaramijo section, Zone N14 of Middle Miocene, Zones N16-19 of Middle Miocene to Pliocene, and Zone N21 of Pliocene in the Punta Canoa section (Fig. 5).

Biochronologic analyses of planktic foraminifera on the coastal area of Peru

The Peruvian coast on the Pacific side of South America is known as a typical upwelling region, where phytoplankton propagate due to a strong upwelling. Such a condition is considered to have continued from Early Miocene time up to the present. As a result, biosiliceous facies are well developed on the coastal area throughout most of the Neogene. In these sediments calcareous planktic

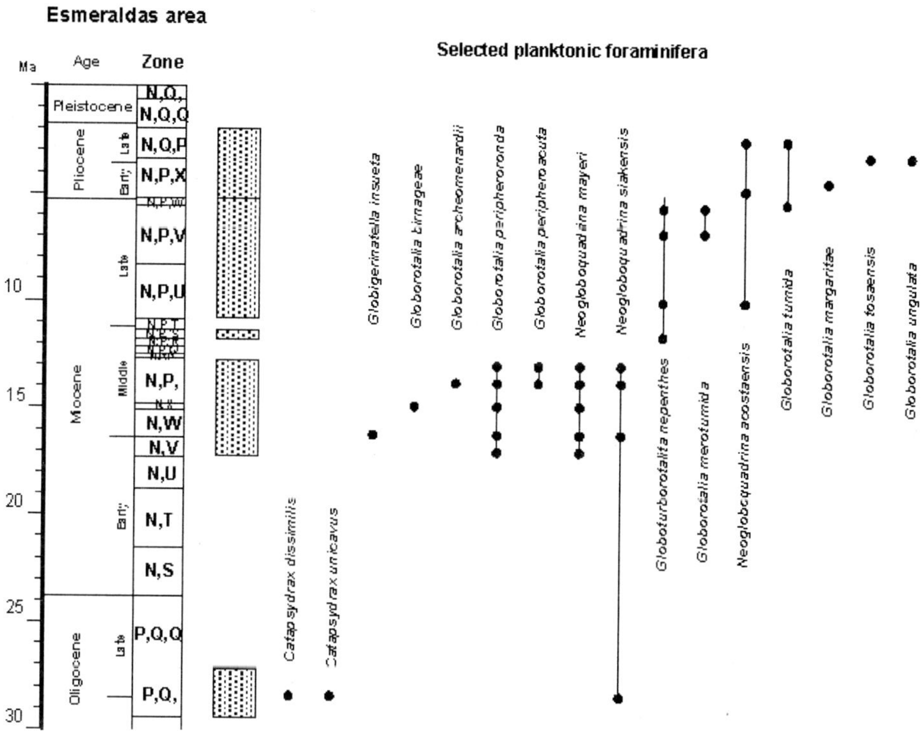

Fig. 3. Occurrences and stratigraphic ranges of selected planktic foraminiferal species and the biostratigraphy of the Esmeraldas area in Ecuador.

foraminifera are rare, however, we can find out some planktic foraminifera in occasional intercalations of calcareous facies.

The Bayovar area is located in northern Peru, 85km southwest of the Piura. Planktic foraminifera are obtained from an intercalated horizon among biosiliceous sediments. The horizon contains *Globigeina bulloides* and *Hastigerinella reideli* that is assignable to Zone N22-Recent (Fig. 6).

In the south of the Paracas area in central Peru, planktic foraminifera are obtained from the Cerro Las Salinas and the Loma Cuest Chilcatay sections (Ibaraki, 1992). In the Cerro Las Salinas section, planktic foraminifera were obtained from six horizons, including Zone N 4 to Zones N6-7 of Early Miocene (Fig. 7). Diatomaceous layers in this section are considered to be the oldest one in Neogene sequences on the Pacific coast of South America. The Loma Cuesta Chilcatay section is located in a desert on the coast of central Peru, west of the Ica. Planktic foraminifera obtained from two horizons indicate Zones N16-N17 of Late Miocene age.

The Camana area is located 840 km south of Lima in southern Peru. The Camana Formation is exposed in a basin along the coast, about 40 km from east to west, attaining a thickness of 400-500 meters and consisting of calcareous sandstone in the lower part and siltstone in the upper part with a

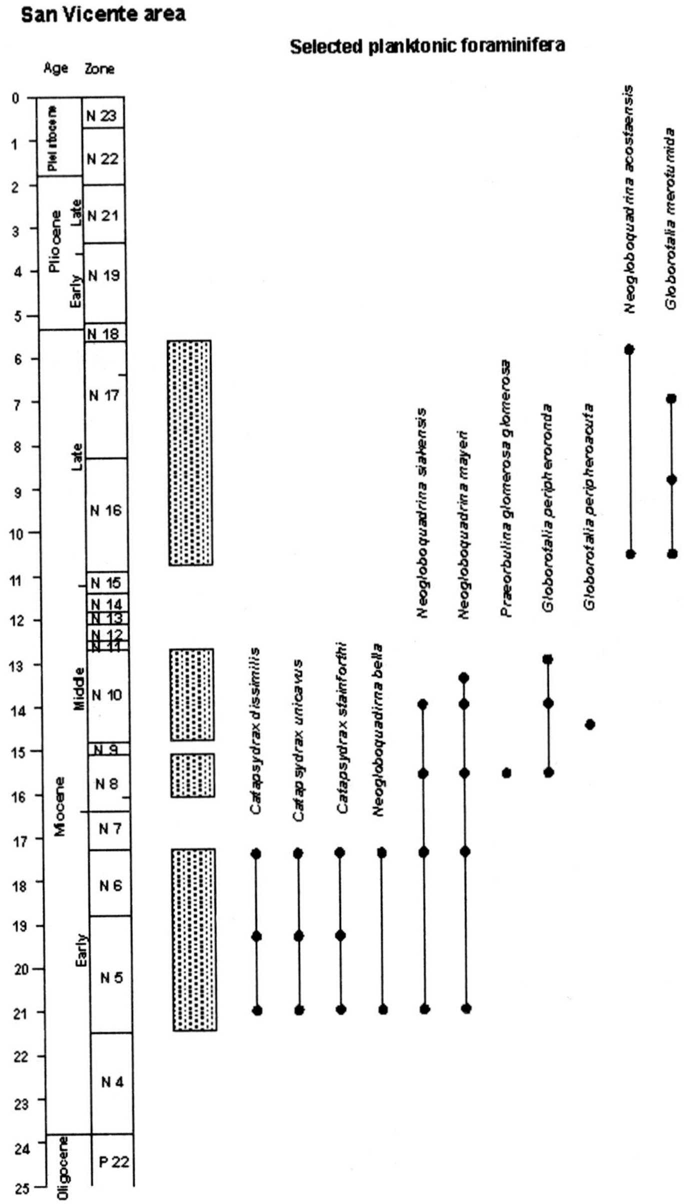

Fig. 4. Occurrences and stratigraphic ranges of selected planktic foraminiferal species and the biostratigraphy of the San Vicente area in Ecuador.

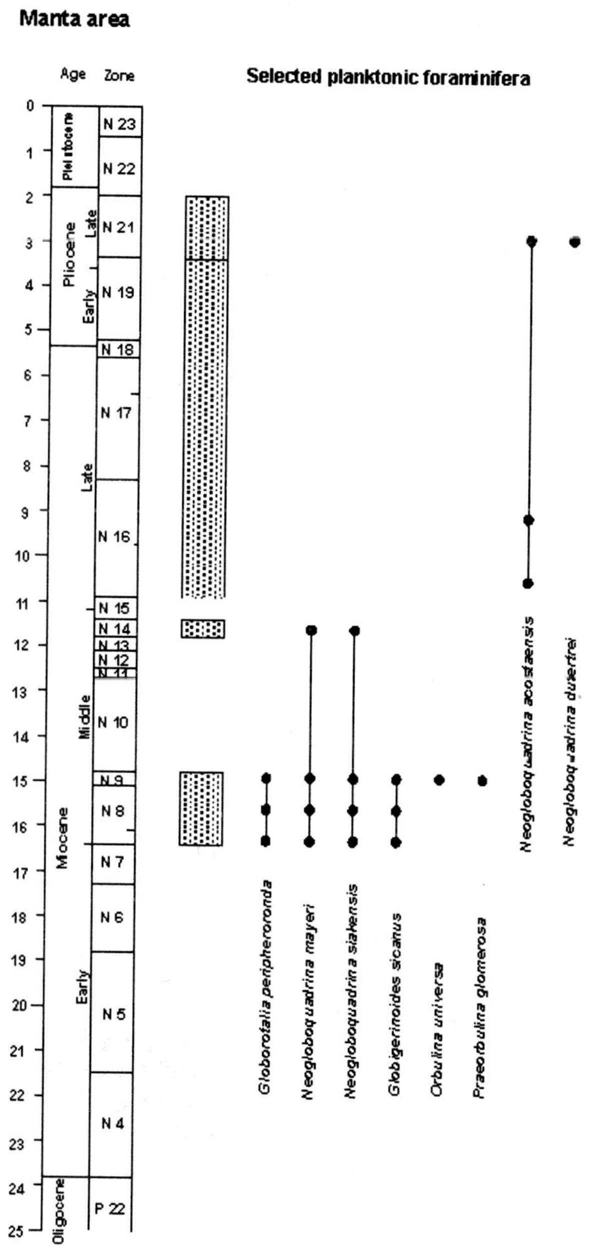

Fig. 5. Occurrences and stratigraphic ranges of selected planktic foraminiferal species and the biostratigraphy of the Manta area in Ecuador.

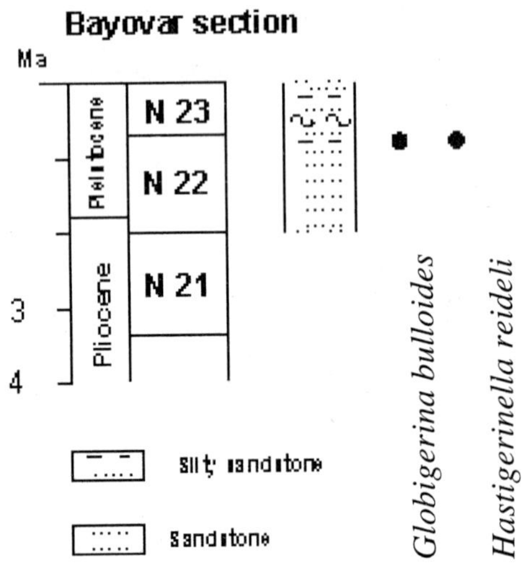

Fig. 6. Occurrences of selected planktic foraminiferal species and the biostratigraphy of the Bayovar section in Peru.

gently inclined structure to the coast. The formation yields rich miogypsinid tropical larger foraminifera. Fortyone planktic foraminiferal species were obtained from 12 localities. The geologic age of the Camana Formation is assigned to the interval from Zone N6 to Subzone N8b of the late Early Miocene to the earliest Middle Miocene in age (Fig. 8). Among them, the Miogypsina-bearing horizons are assignable to Zone N8 (Ibaraki, 1992a).

Biochronologic analyses of planktic foraminifera on the coastal area of Chile

Marine Neogene sequences are scattered on the coastal area of Chile. Planktic foraminifera were obtained from areas of the Mejillones, the Caldera, the La Serena, and the Navidad.

In northern areas of the Antofagasta, planktic foraminifera were obtained from sections of the Caleta Herradura de Mejillones, the Cuenca del Tiburon, the la Portada and the Costa Mejillones (Ibaraki, 2001, 2002).

The Caleta Herradura de Mejillones section is located about 50 km north of the Antofagasta, northern Chile. Marine Neogene sequences are continuously well exposed in a long lunate sea cliff on the coast, being unconformably underlain by the Jurassic-complex. The section is considered as one of the reference section for the Neogene planktic foraminiferal biostratigraphy on the Pacific coast of South America. Neogene sequences on the coast of the Caleta Herradura de Mejillones, attaining 280 m in total thickness and dipping gently southwestward, consisting mainly of sandstone with some intercalations of silty sandstone, calcareous layers, conglomeratic lenses, alternating sandstone and siltstone, and diatomaceous layers in the top part. Pliocene-Pleistocene marine diatomaceous sediments unconformably overlie the sequences. 81 Planktic foraminiferal species and subspecies in total, were

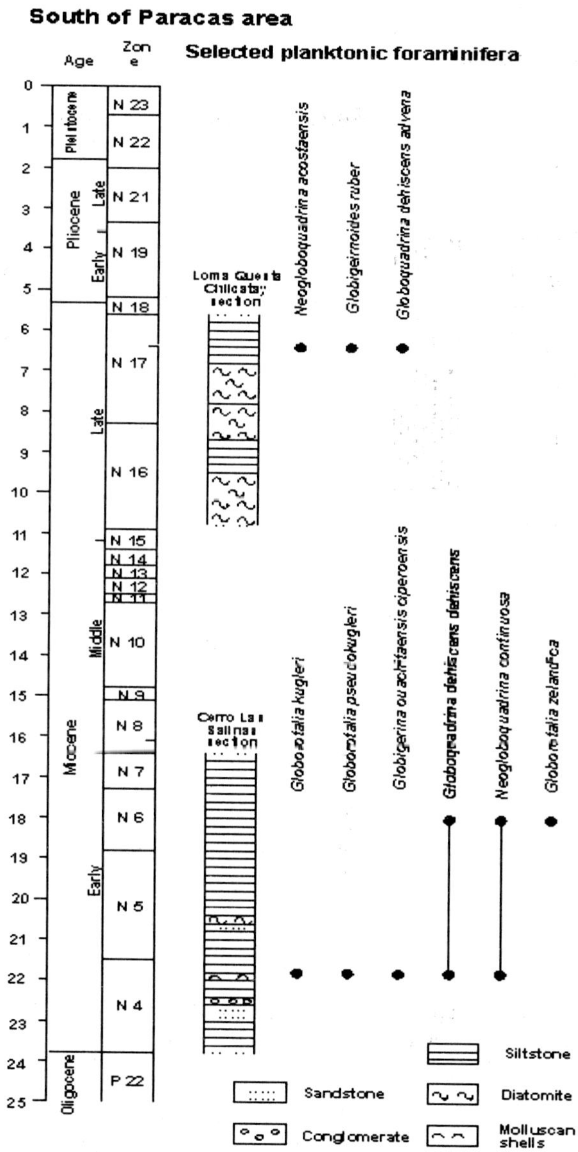

Fig. 7. Occurrences and stratigraphic ranges of selected planktic foraminiferal species and the biostratigraphy of the South of Paracas area in Peru.

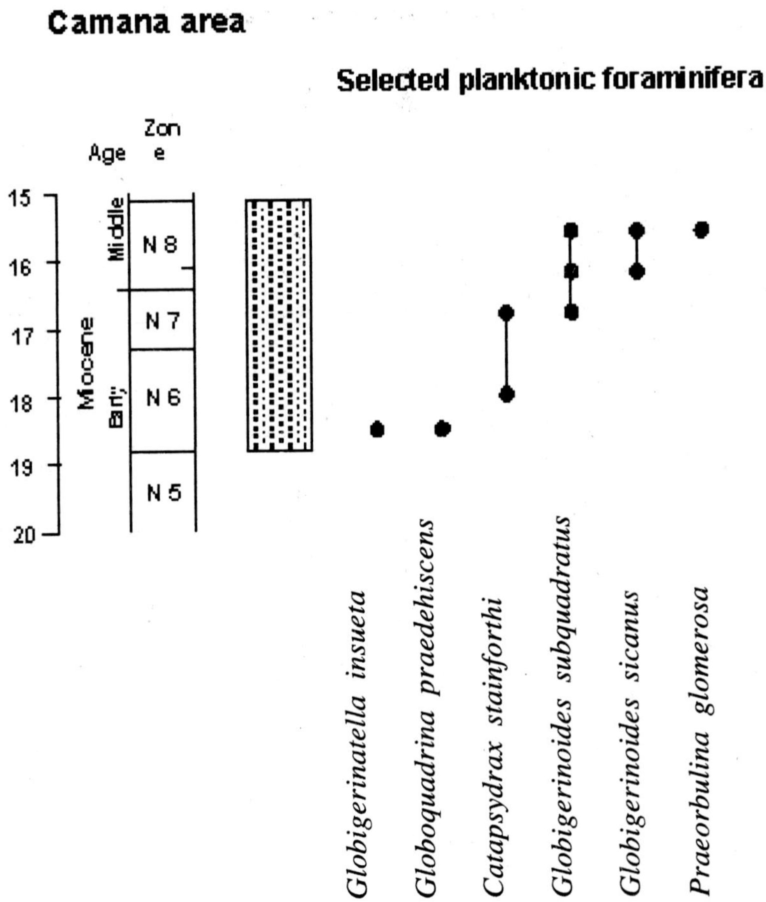

Fig. 8. Occurrences and stratigraphic ranges of selected planktic foraminiferal species and the biostratigraphy of the Camana area in Peru.

obtained from 42 horizons. Zones N7, N8a, N8b, N9, N10, N14, N17b of Early Miocene to late Miocene were recognized (Fig. 9).

The Cuenca del Tiburon section is located in the central part of the Mejillones Peninsula. The section is located on a steep slope in a hilly land in the basin of the Cuenca del Tiburon. The lower part consists of whitish diatomite siltstone with frequent intercalations of silty layers of a buff color, and the upper part consists of consolidated sandstone interbeded with s hell beds. Planktic foraminifera

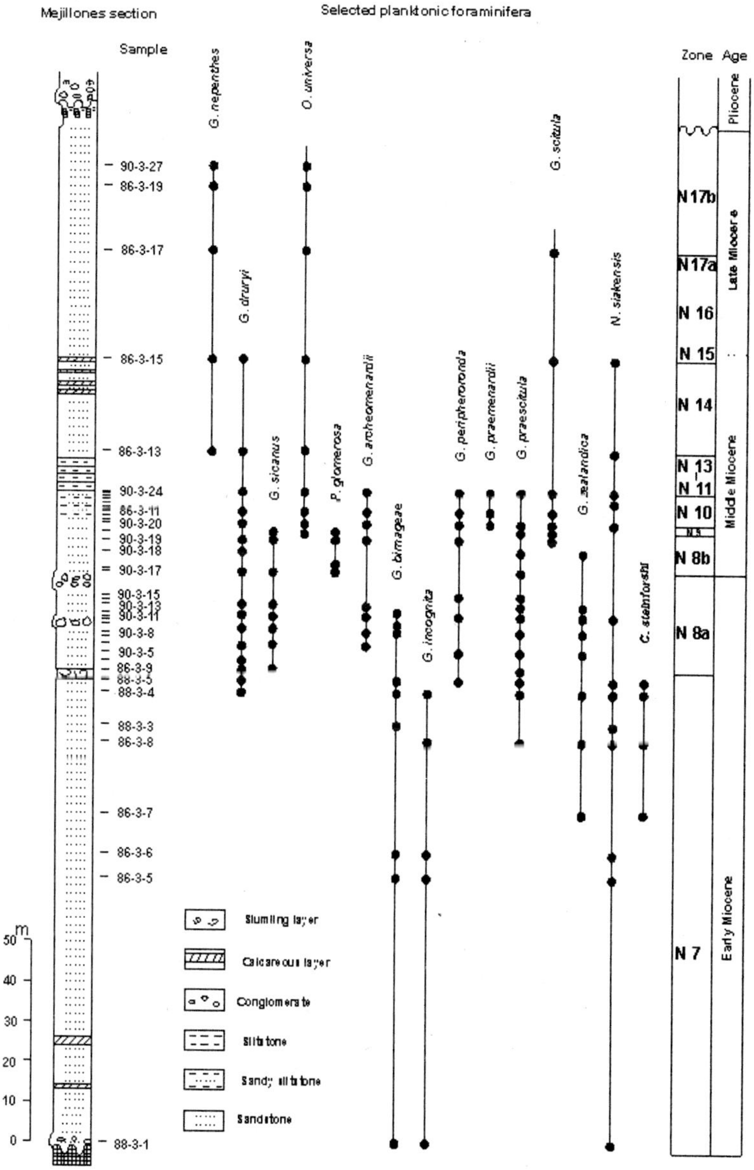

Fig. 9. Occurrences and stratigraphic ranges of selected planktic foraminiferal species and the biostratigraphy of the Caleta Herradura de Mejillones section in Chile.

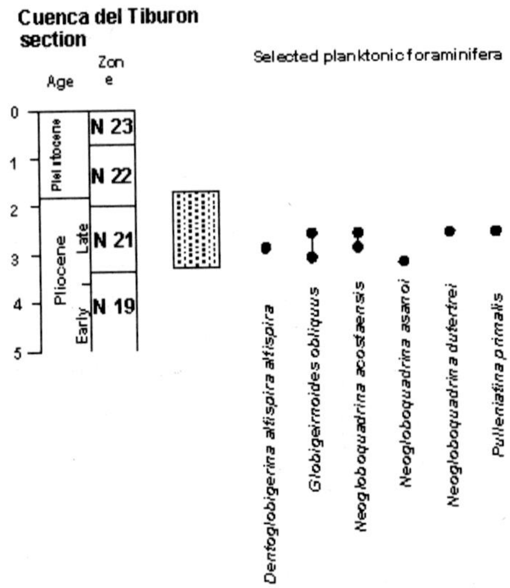

Fig. 10. Occurrences and stratigraphic ranges of selected planktic foraminiferal species and the biostratigraphy of the Cuenca del Tiburon section in Chile.

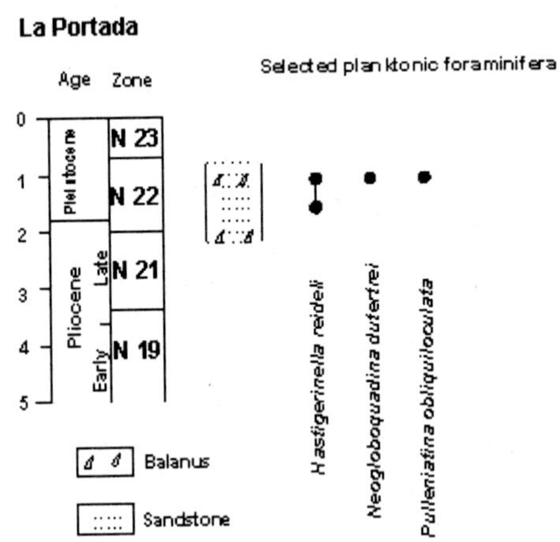

Fig. 11. Occurrences and stratigraphic ranges of selected planktic foraminiferal species and the biostratigraphy of the La Portada section in Chile.

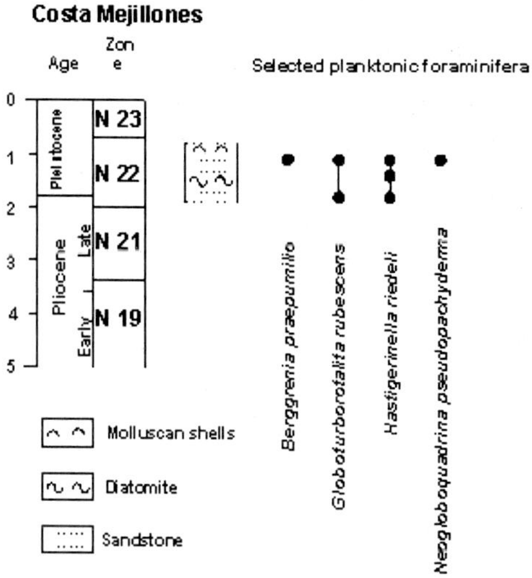

Fig. 12. Occurrences and stratigraphic ranges of selected planktic foraminiferal species and the biostratigraphy of the Costa Mejillones section in Chile.

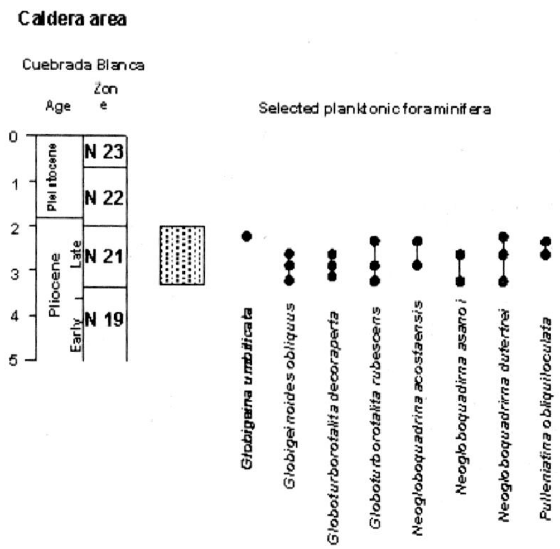

Fig. 13. Occurrences nad sratigraphic ranges of selected planktic foraminiferal species and the biostratigraphy of the Cuebrada Blanca section of the Caldera area in Chile.

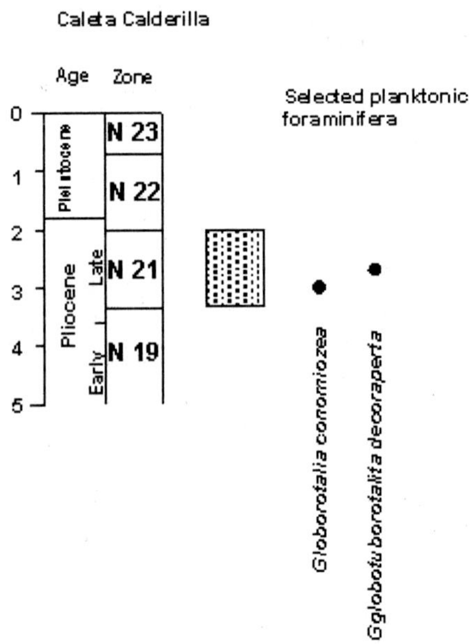

Fig. 14. Occurrences of selected planktic foraminiferal species and the biostratigaphy of the Caleta Calderilla section of the Caldera area in Chile.

were obtained from one sample of the lower sequences. Zones N21-N22 of Late Pliocene were recognized (Fig. 10).

The La Portada section is located 20 km north of Antofagasta. The section, a high sea-cliff at the La Portada, consists mainly of rarely stratified medium to coarse sandstone in the lower part, *Balanus*-bearing bioclastic sandstone in the middle and again clastic sand in the upper part. The top is unconformably covered by shell-bearing beach sand and dune sand. Planktic foraminifera were obtained from interbeded siltstone of the lowest part of the cliff. The section is assignable to Zones N21-N22 of late Pliocene (Fig. 11).

The Costa Mejillones section is located about 60 km north of Antofagasta, northern Chile. The section is a sea cliff at the Mejillones. Almost horizontal strata are divided into two units by an unconformity. The lower unit consists of weakly stratified silty clay, diatomite and well-sorted fine sandstone. The upper unit is gravelly sandstone with abundant molluscs. Planktic foraminifera were obtained from one sample. The section is assigned to Zone N22 of Pleistocene (Fig. 12).

In the Caldera area, planktic foraminifera were obtained from the Cuebrada Blanca and the Caleta Calderilla sections. The Caldera area is located about 490 km south of Antofagasta. In the Cuebrada Blanca section, a cliff consists of high, middle and lower level terrace deposits, and exposes finely laminated and non-laminated diatomaceous clay stone with intercalations of sand layers in the section. Those are unconformably overlain by gravelly sandstone with rich molluscs. Planktic foraminifera were obtained from lower, middle and high level deposits. Zone N21 of late Pliocene was recognized (Fig. 13).

In the Caleta Calderilla section, planktic foraminifera are very rare. The geologic age of the section is probably Zone N21 of Pliocene (Fig. 14).

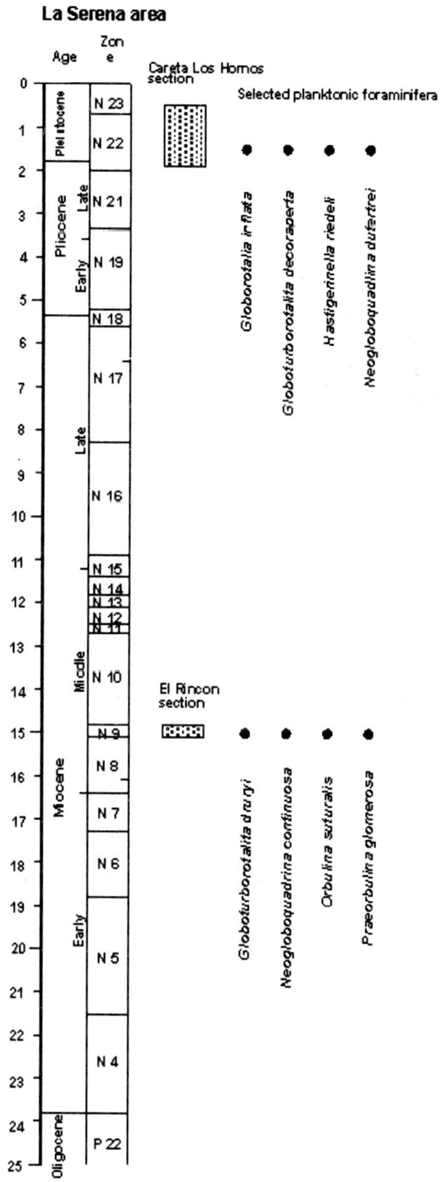

Fig. 15. Occurrences of selected planktic foraminiferal species and the biostratigraphy of the Careta Los Hornos and the El Rincon sections of the La Serena erea in Chile.

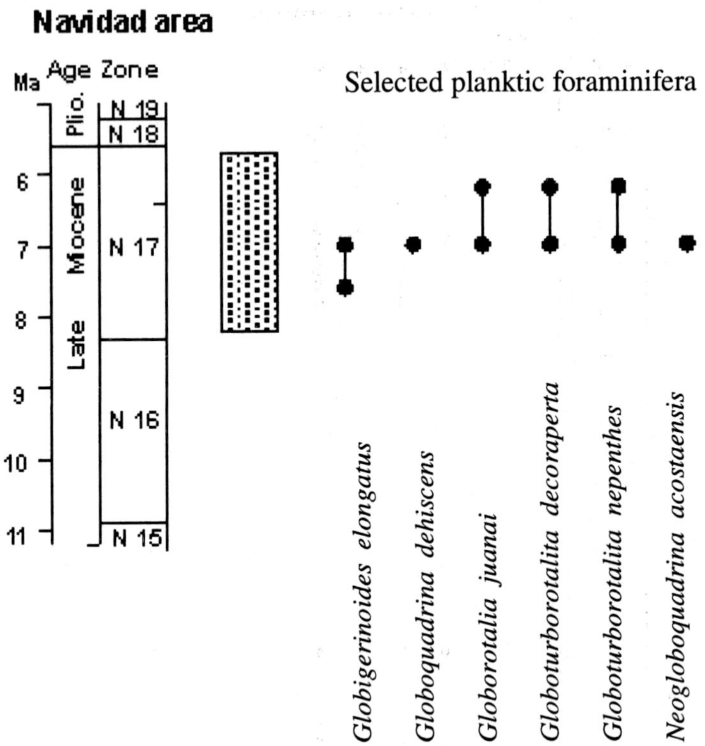

Fig. 16. Occurrences and stratigraphic ranges of selected planktic foraminiferal species and the biostratigraphy of the Navidad area in Chile.

In the La Serena area, planktic foraminifera were obtained from two sections of the Caleta Los Hornos and the El Rincon sections. The Caleta Los Hornos section is assigned to Zone N22 or younger age. In the El Rincon section, planktic foraminifera were obtained from two samples. The section is included in Zone N9 of early Middle Miocene (Fig. 15).

In the Navidad area, planktic foraminifera were examined on sections exposing along the coast of the Punta Perro and the Matanzas (Ibaraki, 1992b). The Navidad area is located 130 km southwest of Santiago. Planktic foraminifera occurred in three horizons in the Punta Perro section and two horizons in the Matanzas section. The Punta Perro section is assignable to Zone N17 of Late Miocene age, and the Matanzas section yields no marker planktic foraminifera (Fig. 16).

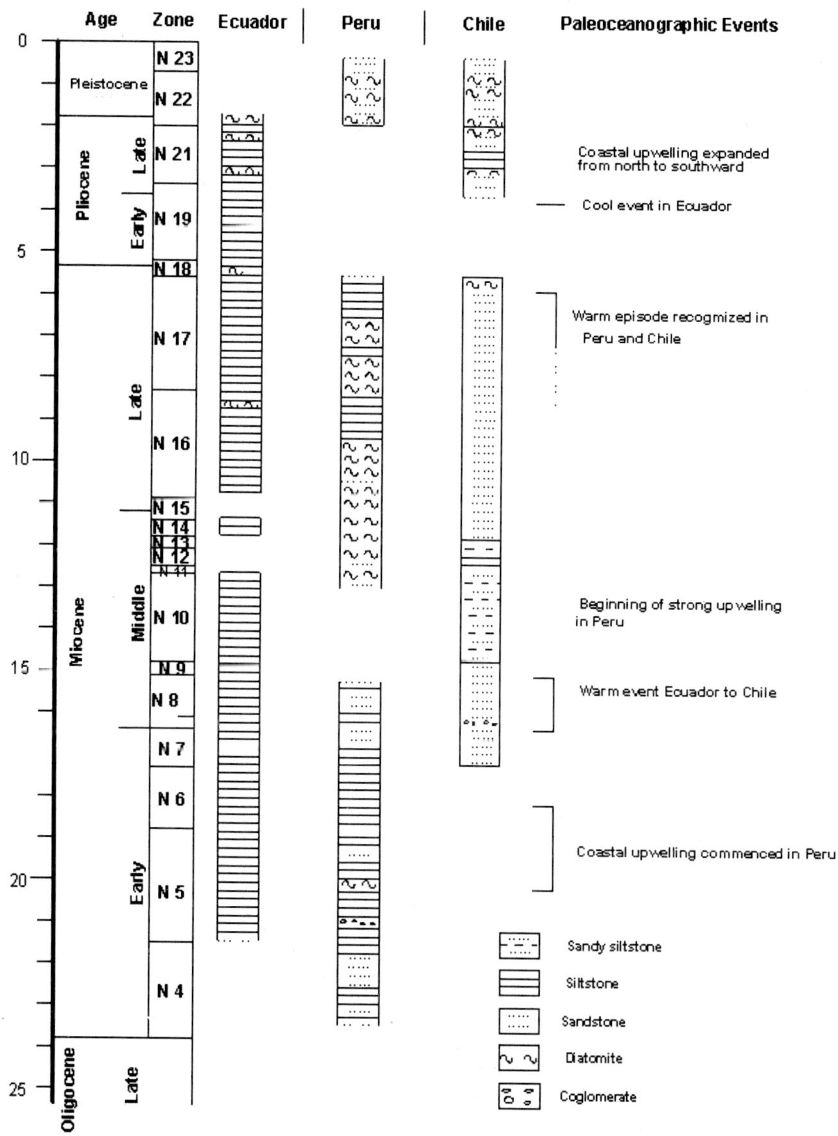

Fig. 17. Paleoceanographic events deduced from planktic foraminiferal assemblages during the Neogene on the Pacific coast of South America.

As mentioned above, the oldest sediments of the Neogene are found in the Cerro Las Salinas section in Peru. Diatomaceous layers are first intercalated in this section among Neogene sequences on the Pacific coast of South America. Early Middle Miocene planktic foraminifera of Zone N8 are frequently found in places, i.e., the Rio Esmeraldas, the San Vicente, and the Jaramijo sections in Ecuador, the Camana section in Peru, and the Calera Herradura de Mejillones and the El Rincon sections in Chile. Biosiliceous sediments are predominant from Middle Miocene time up to the present in Peru. The latest Late Miocene planktic foraminifera of Zone N17 occur in Ecuador, Peru and Chile. Pliocene-Pleistocene sediments of Ecuador to northern part of Chile consist mainly of biosiliceous facies.

Paleoceanographic Events recorded in Planktic foraminifera

A bio- and chronostratigraphic correlation chart of Neogene marine sequences on the coastal areas of Ecuador, Peru and Chile based on planktic foraminifera is presented in Fig. 2. Planktic foraminiferal communities are controlled by temperatures in the upper water column, and planktic foraminiferal assemblages in sediments are considered to indicate marine climatic changes. By means of this, some paleoceanographic events are recognized.

The Peruvian coast in the present time is known as a typical coastal upwelling region in the Pacific Ocean, where the cold Peruvian current flows from the south and phytoplankton propagate owing to the strong upwelling. The oldest Neogene diatomaceous sequences are found in the Cerro Las Salinas section in central Peru. The Cerro Las Salinas section is assignable to Zones N4 to N6-7 of Early Miocene. The beginning of diatom-rich facies deposition is also recorded in ODP Leg 112 Hole 682A off central Peru (Ibaraki, 1990a), where it is assignable to NN4 and Zone N7. Based on planktic foraminiferal biostratigraphy of offshore and onshore sections in Peru, the beginning of the coastal upwelling off Peru is considered to be Early Miocene in age (Fig. 17).

The Camana Formation in Peru is assignable to Zones N6-7 and Subzones N8a and N8b of Early Miocene to earliest Middle Miocene age, containing warm-water planktic foraminifera and miogypsinid tropical larger foraminifera, and none of upwelling feature. This suggests a warm event during the earliest Middle Miocene (Ibaraki, 1992a). This warm event is also recognized in the Caleta Herradula de Mejillones section in northern Chile. During the Neogene of the Pacific side of central Japan, the maximum seawater temperature is recognized during Subzone N8b of the earliest Middle Miocene age. This warm event is correlative with those recognized on the coastal area of the Pacific side of South America and the Pacific side of central Japan (Ibaraki, 1990b).

Middle Miocene and later sequences on the coast of Peru are comprised mostly of marine diatomaceous mudstone, containing few planktic foraminifera and no warm water taxa are present except in some sections. The Loma Cuesta Chilcatay section is assigned to the zonal interval N16-N17 of Late Miocene age. Planktic foraminiferal assemblage consists of warm-water and eurythermal dwellers and a few cold-water elements. The warm-water elements account for 43.3% of the total planktic foraminiferal fauna suggesting a warm event during the Late Miocene (Ibaraki, 1992). A warm event of a zonal interval N16-17 is recognized off Peru and in the Punta Perro section of the Navidad Formation in Chile. In ODP Hole 684A off Peru, a planktic foraminiferal assemblage, indicating Zones N16-17, includes warm-water elements as much as 33.8% of the total assemblage. Most of the Punta Perro section is assignable to Zone N16-17, characterized by warm-water planktic foraminifera (36.1-45.1%). A comparison of the planktic foraminiferal assemblage of the Punta Perro section with that of Recent sediments on the coast of the Manta in Ecuador shows that the sea water temperature during the Late Miocene time when the sequence of the Punta Perro section was deposited might have been much warmer than that of today off central Chile.

A warm episode of surface seawater temperature has been recognized in the latest Miocene on the Pacific side of Japan (Ibaraki, 1990), and in the northeast Pacific (Barron and Keller, 1983). The warm condition as recognized in the Navidad Formation in the southeast Pacific is probably correlative with the latest Miocene warm episode known from Japan and the Northeast Pacific.

The Esmeraldas section in Ecuador, being assignable to Zones N17 to N21 of Late Miocene to Pliocene, contains planktic foraminifera. The ratio of tropical-subtropical dwellers to the total assemblage decreases abruptly in the upper part of Zone N19, whereas eurythermal dwellers increase inversely. The abrupt decrease in surface water temperature and the abundant occurrence of *Neogloboquadina* spp. in the upper part of Zone N19 at about 3.95Ma also suggest an increased coastal upwelling in this region and a subsequent closing of the Central American Seaway (Ibaraki, 1997).

Biosiliceous sediments become dominant in Pliocene and Pleistocene sequences in central Ecuador and northern Chile. These depositional patterns suggest that the coastal upwelling expanded north and southward during the late Neogene, coincident with the closing of the Central American Seaway (Fig. 17).

Summary

Based on the biochoronologic analyses by means of planktic foraminiferal biostratigraphy, we can make a correlation chart of respective marine Neogene sections on the Pacific coast of South America. The oldest sequences of the Neogene are found in the Cerro Las Salinas in Peru and diatomaceous layers are first intercalated in this section among Neogene sequences on the Pacific coast of South America. The geologic age of the Cerro Las Salinas section includes horizons from Zone N4 to Zones N6-N7 of Early Miocene age. Early Middle Miocene planktic foraminifera of Zone N8 are frequently occurring in places, *i.e.*, the Rio Esmeraldas, the San Vicente, and the Jaramijo sections in Ecuador, the Camana section in Peru, and the Caleta Herradura de Mejillones and the El Rincon sections in Chile. In the Camana section, the lowest Middle Miocene sequences contain a dominantly warm-water planktic foraminiferal assemblage and rich assemblage of tropical miogypsinid larger foraminifera in calcareous facies. The early Middle Miocene sequence of Caleta Herradura de Mejillones section contains an abundant warm-water assemblage of diverse species. The Peruvian coast and northern Chile were under an influence of warm-water conditions in the earliest Middle Miocene. This warm event can be correlated with that of the Pacific side of central Japan.

Biosiliceous sediments are predominant from Middle Miocene time up to the present in Peru. This suggests that the coastal upwelling became strong in Peruvian coast. The latest Late Miocene planktic foraminifera of Zone N17 include warm-water dwellers. The warm episode of surface seawater temperature is correlative with the latest Miocene warm episode known from Japan and the northeast Pacific.

The abrupt decrease in surface temperature in an upper part of Zone N19 at about 3.95Ma in northwestern Ecuador, suggest that an increased coastal upwelling with closing of the Central American Seaway. Biosiliceous sediments become predominant in Pliocene and Pleistocene ages in areas from central Ecuador to northern Chile. These depositional patterns suggest that the coastal upwelling expanded from north to southward.

Acknowledgements

Fieldwork was carried out on the Pacific coast of South America in cooperation with international Andean studies group of Shizuoka University with collaborators of Ecuador, Peru and Chile. This work was supported by Ministry of Education, Science and Culture of Japan. I am grateful to Dr. R. Tsuchi, Professor Emeritus of Shizuoka University for his kind review of the manuscript, and Dr. Madhusoodhan Satish-Kumar of Shizuoka University for his beneficial advice on the English of the text. I also thank the staff of "Pacific Neogene events in Japan and South America" Project Party.

Appendix.

Planktic foraminiferal taxa cited in the text

The list includes all the taxa cited in the paper. The taxonomic concepts are those of Saito, Thompson and Breger (1981), Kennett and Srinivasan (1983), Bolli and Saunders (1985), and Hemleben et al. (1989).

Berggrenia praepumilio (Parker)=Gloanomalina praepumilio Parker 1967
Catapsydrax dissimilis (Cushman and Bermúdez) = Globigerina dissimilis Cushman and Bermudez, 1937
Catapsydrax stainforthi Bolli, Loeblish, and Tappan, 1957
Catapsydrax unicavas Bolli, Loeblich, and Tappam, 1957
Dentoglobigerina altispira altispira (Cushman and Jarvis)=Globigerina altispira Cushman and Jarvis, 1936
Globigerina bulloides d'Orbigny, 1826
Globigerina ouachitaensis ciperoensis Howe and Wallace, 1932
Globigerina umbilicata Orr and Zaitzeff, 1971
Globigerinatella insueta Cushman and Stainforth, 1945
Globigerinoides elongates (d'Orbigny) =Globigerina elongates d'Origny 1826
Globigerinoides obliquus Bolli, 1957
Globigerinoides ruber (d'Orbigny)= Globigeina rubra dOrigny 1839
Globigerinoides sicanus De Stefani, 1950
Globigeirnoides subquadratus Brönnimann, 1954
Globoturbototalita decoraperta (Takayanagi and Saito)=Globigerina druryi Akers decoraperta *Takayanagi and Saito*, 1962
Globoturborotalita druryi (Akers)= Globigerina druryi Akers, 1955
Globoturborotalita nepenthes (Todd) = Globigerina nepenthes Todd, 1957
Globoturborotalita rubescens (Hofker) = Globigeirna rubescens Hofker, 1956
Globoquadrina dehiscens advena Bermudez, 1959
Globoquadrina dehiscens dehiscens (Chapman, Parr, and Collins) = Globorotalia dehiscens Chapman, Parr, and Collins, 1934
Globoquadrina praedehiscens Blow and Banner 1962
Globorotalia archeomenardii Bolli, 1957
Globorotalia birnageae Blow, 1959
Globorotalia conomiozea Kennett, 1966
Globorotalia incognita Walters, 1965
Globorotalia inflata (d'Orbigny) = Globigeina inflata d'Orbigny, 1839
Globorotalia juanai Bermudez and Bolli, 1969
Globorotalia kugleri Bolli, 1957
Globorotalia margaritae Bolli and Bermudez, 1965
Globorotalia merotumida Blow and Banner, 1965
Globorotalia peripheroacuta Blow, and Banner, 1966
Globorotalia peripheroronda Blow, nd Banner, 1966
Globorotalia praemenardii Cushman and Stainforth, 1945
Globorotalia praescitula Blow, 1959
Globorotalia pseudokugleri Blow, 1969
Globorotalia scitula (Brady) =Pulvinulina scitula Brady, 1882

Globorotalia tosaensis Takayanagi and Saito, 1962
Globorotalia tumida tumida (Brady) = Pulvinulina menardii (d'Orbigny) var. tumida Brady, 1877
Globorotalia ungulata Bermudez, 1960
Globorotalia zealandica Hornibrook, 1958
Hastigerinell riedeli Rögl and Bolli, 1973
Neogloboquadrina acostaensis (Blow) Globorotalia acostaensis Blow, 1959
Neogloboquadrina asanoi (Maiya, Saito, and Sato) = Globoquadrina asanoi Maiya, Saito and Sato, 1976
Neogloboquadirna bellu (Jenkins)= Globorotalia bella Jenkins, 1967
Neogloboquadirna continuosa (Blow) = Globorotalia opima Boll subsp. continuosa Blow 1959
Neogloboquadirna dutertrei (d'Orbigny) = Globigerina dutertrei d'Orbigny, 1839
Neogloboquadrina mayeri (Cushman and Ellisor) = Globorotalia mayeri Cushman and Ellisor, 1939
Neogloboquadrina pseudopachyderma (Cita, Premoli-Silva and Rossi) = *Globorotalia pseudopachyderma* Cita, Premoli-Silva, and Rossi, 1965
Neogloboquadrina siakensis (LeRoy) = Globorotalia siakensis LeRoy, 1939
Orbulina suturalis Bronnimann, 1951
Orbulina universa d'Orbigny, 1839
Praeorbulina glomerosa glomerosa (Blow) = *Globigeirnoides glomerosa glomerosa* Blow, 1956
Pulleniatina obliquiloculata (Parker and Jones) = *Pullenia sphaeroides* (d'Orbigny) var. *obliquiloculata* Paker and Jones, 1865
Pulleniatina primalis Banner and Blow, 1967

References

BARRON, J. A. AND KELLER, G., 1983. Paleotemperature oscillations in the Middle and Late Miocene of the northeastern Pacific. Micropalcontol., 29, 150-181.

BE, A.W.H., 1977. An ecological zoogeographic and taxonomic review of Recentplanktic foraminifera. In: Ramsay A.T.S. (ed.), Oceanic Micropaleontology, 1. Academic Press, London, 1-100.

Berggren, W.A., Hilgen, F.J., Langereis, C.G. Kent, D.V., Ovradovich, J.D., Raffi, I., RAYMO, E., AND SHACKETON, N.J., 1995a. Late Neogene chronology new perspectivesin high-resolution stratigraphy: Geol. Soc. Am. Bull, . 107, 1272-1287.

BERGGREN, W.A., KENT, V.D., AND AUBRY, M-P, 1995b. A revised Cenozoic Geochronology and chronostratigraphy. SEPM Special Publication 54, Geochronology, time scale and global stratigraphic correlation. 129-212.

BLOW, W.H., 1969. Late middle Eocene to Recent planktic foraminiferal biostratigraphy. In BRONNIMANN, P. AND RENZ, H.H (eds), First International Conference of Planktic Microfossils, Geneve 1967, Proceedings 1: 199-422.

BOLLI, H.M. AND SAUNDERS, J.B. 1985. Oligocene to Holocene low latitude plannktic foraminifera. Bolli, H.M. and Saunders, J.B. and Perch-Nielsen, K., (eds), Plannkton Stratigraphy, Cambridge University Pres, 1032pp.

HEMLEBEN, CH., SPINDLER M. AND ANDERSON, O.R., 1989. Modern planktic foraminifera. SpringerVerlag New York Inc., 363pp.

IBARAKI, M., 1990a. Eocene through Pleistocene planktic foraminifera off Peru. Leg 112–biostratigraphy and paleoceanography. In: E. Suess, R. von Huene et al (eds), Proceedings of the Ocean Drilling Program. College Station, TX, 112, 239-262.

IBARAKI, M., 1990b. Neogene planktic foraminifera and events in JapanPalaeogeogr. Palaeoclimatol. Palaeoecol. 77, 335-343.

IBARAKI, M., 1992a. Neogene planktic foraminiferal biostratigraphy on the coast of Peru and its paleocanographic implications. In: R. Tsuchi and J.C. Ingle (eds), Pacific Neogene: Environment, Evolution, and Events. University of Tokyo Press, 71-81.

IBARAKI, M., 1992b. Planktic foraminifera from the Navidad Formation, Chile. Their geologic age and paleoceanographic implications. In K. Ishizaki and T. Saito (eds), Centenary of Japanese Micropaleontology. Terra Scientic Publishing Company, Tokyo, 91-95.

IBARAKI, M., 1994. Neogene planktic foraminifera from the Esmeraldas area on the Pacific coast of northwestern Ecuador. Repts. Fac. Sci., Shizuoka University, 28, 99-121.

IBARAKI, M., 1997. Closing of the Central American Seaway and Neogene coastal upwelling along the Pacific coast of South America. Tectonophysics, 281, 99-104.

IBARAKI, M. 1998. Planktic foraminifera from the Neogene sequences on the Pacific coast of northwestern Ecuador. Repts. Fac. Sci., Shizuoka University, 32, 75-97.

IBARAKI, M., 2001. Neogene planktic foraminifera of the Caleta herradura de Mejillones section in northern Chile: Biostratigraphgy and paleoceanographic implications. Micropaleontol., 47. 257-267.

IBARAKI, M., 2002. Neogene planktic foraminifera on the Pacific coast of South America– Biostratigraphy and Key-plates- Shizuoka University, 1-101.

KENNETT, J.P. AND SRINIVASAN, M.S., 1983. Neogene planktic foraminifera. A Phylogenetic Atlas. Hutchinson and Ross, Stroudsburg, Pa., 265pp.

SAITO, T. THOMPSON, P.R. AND BREGER D., 1981. Systematic index of Recent and Pleistocene planktic foraminifera.University of Tokyo Press, 190p.

TSUCHI, R. ed. 1988. Reports of Andean Studies, Shizuoka University, Special Volume 2: Trans-Pacific correlation of Cenozoic geohistry, 108p.

TSUCHI, R. ed., 1990. Reports of Andean Studies, Shizuoka University, Special Volume 3: Trans-Pacific correlation of Neogene geologic events, 77p.

TSUCHI, R. ed., 1992. Reports of Andean Studies, Shizuoka University, Special Volume 4: Pacific Neogene events and South America, their timing and nature, 50.

Episodes of weakening of Leeuwin Current (Southeast Indian Ocean) during Quaternary: Isotopic and planktic foraminiferal evidences

Devesh K Sinha and Ashutosh Kumar Singh

Department of Geology, Banaras Hindu University, Varanasi 221005, India

Introduction:

The Southeast Indian Ocean off Western Australia is the only eastern boundary current region undergoing no upwelling despite predominating equatorward winds as there is no continuous equatorward flow within 1000 km of the coast (Smith, 1992; Veeh *et al*, 2000). To the contrary, there is a strong poleward flow of warm water named Leeuwin current against the prevailing equatorward winds (Cresswell and Golding, 1980) whose principal forcing mechanism is the generation of meridional steric height gradients and associated zonal geostrophic flows (Godfrey and Weaver, 1991) linked to Western Pacific Warm Pool and Indonesian throughflow (Weaver, 1990). The current flows throughout the year but is strongest in the winters and during times when the El Nino southern Oscillation is not operating (Pearce and Phillips, 1988; Phillips *et al*, 1991). The current is associated with southeast trade winds which cause westward flowing south equatorial currents to drive tropical waters of the Pacific Ocean through the Indonesian Archipelago on to the northwestern shelf of Western Australia (Okada and Wells, 1997). Beneath and adjacent to the Leeuwin current the cold West Australian Current flows northward (Fig.1a) as a part of a major anticlockwise gyre in the Indian ocean (Tchernia, 1980, Okada and wells, 1997).

Previous studies on changing strengths of Leeuwin Current during Quaternary

The Leeuwin current turned on and off during the late Quaternary in concert with the interglacials and glacials at scales of 10^5 years (Mc Gowran *et al*, 1997). Contrasting generalizations were made regarding the weakening (Wilson and Allen, 1987) and strengthening (Kendrick *et al*, 1991) of the Leeuwin current during Pleistocene to explain changes in biogeographic pattern at southern Australian margin. Anomalously cold sea surface temperatures off western Australia during LGM were reported by Prell and Hutson (1979), Prell *et al.* (1979, 1980), CLIMAP (1984) and Prell (1985). During glacial times, the northward shifting of the subtropical convergence and associated strong offshore winds could have caused upwelling in the region (Webster and Streten, 1978, Prell *et al*, 1980). Major changes in ocean circulation caused by intensification in the West Australian Current and reduced strength or cessation in the flow of Leeuwin Current (Wells and Wells, 1994) have been attributed for the cold anomaly (Prell and Hutson, 1979; Prell *et al*, 1979, 1980; CLIMAP, 1984). Here we present a detailed quantitative planktic foraminiferal census data integrated with isotopic record from ODP site 763A,

Exmouth Plateau, off Western Australia and show that there have been five upwelling intervals during Quaternary when Leeuwin Current was weekend in response to ENSO induced changes in WPWP or due to Indonesian throughflow reduction in response to expansion of Antarctic ice sheet.

Relationship between Leeuwin current, planktic foraminiferal assemblage and stable isotopic record

Planktic foraminiferal assemblages preserved in oceanic sediments are excellent indicators of overlying surface water masses (Kennett *et al*, 1985). The relative abundance of water mass sensitive species groups has been widely utilized to infer surface water paleoceanography (Wei, Kuo Yen. 1998; Srinivasan and Sinha, 2000). Isotopic depth ranking studies of planktic foraminiferal species have further made them useful tool for inferring nature of upper water column (Gasperi, and Kennett, 1992). The ODP site 763 is influenced by poleward flowing warm Leeuwin Current and equatorward flowing cold West Australian Current (Fig.1a). The Quaternary interval at this site shows a mixture of tropical and temperate planktic foraminifera with tropical elements dominating most of the time. The tropical assemblages consist of *Globigerinoides conglobatus, Gs. sacculifer, Gs. trilobus, Gs. quadrilobatus, Pulleniatina obliquiloculata* and *Sphaeroidinella dehiscence*. The temperate (cool water) species group consists of mainly *Globorotalia inflata* and infrequent occurrence of *Neogloboquadrina pachyderma*, and *Globigerina woodi*. The site also shows episodic abundance of upwelling indicators *Gloigerinita glutinata* and *Globigerina bulloides* however out of these two the former dominates. We assume that intensification in the Leeuwin Current would result in a fossil record consisting of more warm water planktic foraminiferal assemblages like *Globigerinoides trilobus, Gs. sacculifer, Gs. quadrilobatus, Gs. conglobatus* groups at site 763A, while a weak Leeuwin Current would result in a fossil record showing abundance of upwelling indicator planktic foraminiferal species *Globigerinita glutinata* and *Globigerina bulloides* (Martinez *et al*, 1999). If such upwelling is associated with an increase in ice volume, the resultant fossil record would show increase in relative abundance of typical temperate water species *Globorotalia inflata* brought down by the West Australian Current as a result of the northward shifting of the subtropical convergence due to Antarctic ice cap expansion (Pell and Huston, 1979). As the upwelling in this region is prevented by the strong poleward flowing Leeuwin Current, we assume that intervals showing increased abundance of upwelling indicators *Globigerinita glutinata* and *Globigerina bulloides* are the periods of weakening or cessation of the Leeuwin Current in the region. These intervals would represent the dominance of equatorward wind driven surface circulation causing offshore Ekman transport and upwelling. Sea level changes induced by ENSO directly influence the strength of the Indonesian throughflow and consequently the strength of the Leeuwin Current (Godfrey, 1996). Thus if Leeuwin current can be weakened due to the causes lying in the south *i.e.* increase in Antarctic ice cap formation and the resultant lowering of the sea level allowing less water to flow across the Indonesian seaway it would result in deposition of temperate species *Globorotalia inflata* combined with increasing $\delta^{18}O$ at site 763A. On the other hand if Leeuwin current has been weekend due to decrease in WPWP possibly as a result of El Niño (without ice volume increase), then the resulting sediments would be devoid (or less) of *Globorotalia inflata* but will show upwelling indicator *Globigerina bulloides/Globigerinita glutinata* and increasing $\delta^{18}O$ values of surface waters. Here we show that such distinction is possible with the help of planktic foraminiferal census data.

During weakening of the Leeuwin current the upwelling intervals would definitely result in cooling of the surface waters at site 763A. This would be reflected in increase in $\delta^{18}O$ values of surface dwelling planktic foraminifera *Globigerinoides sacculifer*. If the upwelling is also associated with an ice volume expansion during glacial event, this would be reflected in more positive $\delta^{18}O$ values together with record of migration of the temperate species of planktic foraminifera.

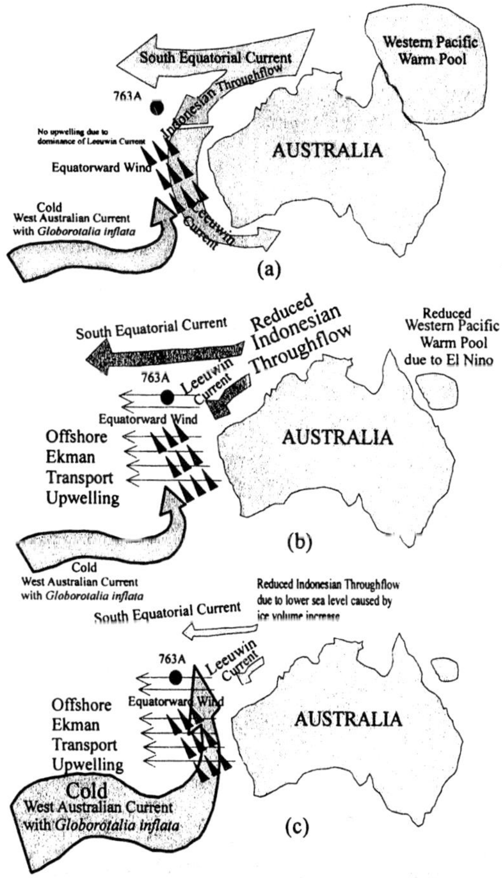

Fig. 1. Oceanographic setup of ODP site 763A, Southeast Indian Ocean. 1a. Modern setup when there is no upwelling due to suppressing of any equatorward flow of surface waters by southward flowing strong Leeuwin Current. 1b. Conditions during PL-1 and PL-2 events when reduced WPWP due to El Niño has caused decrease in steric height gradients and weak Leeuwin Current. The equatorward winds cause upwelling. 1c. Conditions during PL-3, PL-4 and PL-5 events when northward expansion of cooler waters, intensification of West Australian Current occurred as a consequence to ice volume increase. The resulting lowering of sea level caused Indonesian throughflow reduction resulting into weak Leeuwin Current. The equatorward winds cause upwelling. (After Sinha *et al*, 2006)

Material and Methodology

Samples

The material consists of 27 samples from the Quaternary interval of ODP site 763A, Exmouth plateau, southeast Indian Ocean. The samples are plugs sediment commonly taken at about 1.5 m intervals. The Quaternary section mainly comprises foram-nanno ooze. The site lies at $20^0 35.20'$ S and $112^0 12.50'$ E at a water depth of 1367.5 meters. The sampling interval is rather coarse with one sample per core per section. The preservation of planktic foraminifera is excellent and no evidence of dissolution is recorded.

Planktic foraminiferal census counts.

27 samples spanning 2.2 million years were studied for planktic foraminiferal census counts. The > 150 micron fraction was chosen for census counting as it gives maximum climatic information in the leaset time (Imbrie and Kipp, 1971) and is the size fraction now adopted by many major paleoclimatic studies (Wells *et al* 1994). A microsplitter was used to separate an aliquot of 300 or more planktic foraminiferal individuals. Each individual was identified up to species level following the taxonomy of Kennett and Srinivasan (1983) and mounted on assemblage slides. Percentage of each species was calculated and relative abundance curves were plotted for key species discussed in the previous section. The relative abundance curves are shown in Fig. 2.

Stable isotope data

The Oxygen isotopic analysis was carried out for a surface ocean dwelling planktic foraminifera *Globigerinoides sacculifer.* About 25 clean specimens of *Globigerinoides sacculifer* from > 150 micron size fraction with the same sized final chamber were picked up from each sample for which census counting was made. Specimens were cleaned further in an ultrasonic bath to remove the fine fraction contamination. The stable isotopic analyses was carried out on a Europa Scientific GEO 20-20 Stable isotope ratio mass spectrometer with an online "acid on individual carbonate" preparation device at thc Physical Research laboratory (PRL) Ahmedabad. Oxygen and Carbon isotopic data are reported in δ notation relative to the Vienna Peedee belemnite (V-PDB) standard (Fig. 2)

Dating of the paleocenographic events

The planktic foraminiferal species *Globigerinoides fistulosus* has its last appearance just above the Olduvai subchron which correlates with Pliocene/ Pleistocene boundary in the boundary stratotype section of Vrica in southern Italy (Tauxe *et al*, 1983).Our biostratigraphic record at site 763A shows last appearance of *Globigerinoides fistulosus* occurring in Core 4 section 5 at a depth of approximately 32 meters below sea floor. The paleomagnetic Stratigraphy (Cheng Tang, 1992) for the core shows top of the Olduvai Event (1.77 Ma) at the same level as our recorded LA of *Globigerinoides fistulosus.* Thus, based on Last Appearance Datum of *Globigerinoides fistulosus* (Srinivasan and Sinha, 1992) the Pliocene / Pleistocene boundary can be marked at 32m bsf at site 763A and its age can be fixed at 1.77 Ma (Berggren *et al*, 1995). Additional palomagnetic data with depth has yielded a good chronology for the core.

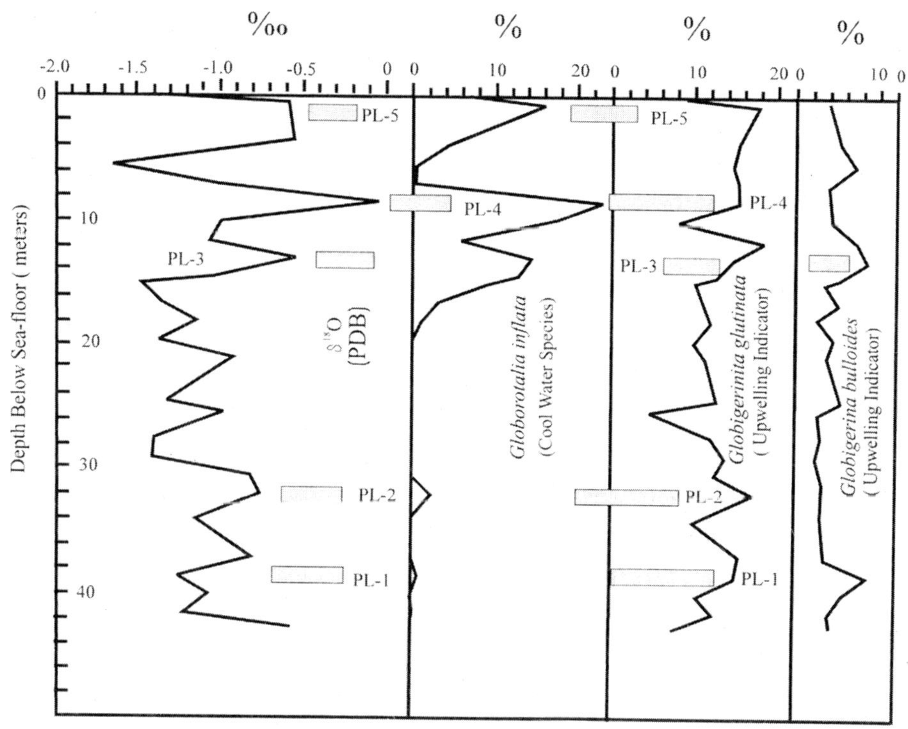

Fig. 2. Relative abundance of selected planktic foraminiferal species with $\delta^{18}O$ record of surface dweller, *Globigerinoides sacculifer* at Site 763A.

Observation

Variation in Planktic foraminiferal abundance during Quaternary at Site 763A

The relative abundance curve (Fig. 2) plotted for *Globorotalia inflata, Globigerina bulloides,* and *Globigerinita glutinata* show significant fluctuations during Quaternary. . The upwelling indicator *Globigerinita glutinata* shows major peak of abundance at 2.2Ma , 1.83 Ma, 0.68 Ma, 0.45 Ma and 0.04 Ma named as PL-1, PL-2,PL-3, PL-4 and PL-5 events Sinha et al, 2006). PL stands or Pleistocene. The percentage frequency curve of *Gr. inflata* shows three major peaks of abundance at 0.68 Ma, 0.45 Ma, and 0.04 Ma respectively. Another upwelling indicator *Globigerina bulloides* shows abundance peaks at 2.2 Ma and 0.45 Ma (PL-1 and PL-3 respectively).

Stable Isotope record

Our record shows five intervals (PL-1 to PL-5) during the Quaternary when there is an increase in the $\delta^{18}O$ values of a surface dwelling planktic foraminiferal species *Globigerinoides sacculifer*. The earlier two intervals (PL-1 and PL-2) show only an enrichment of ~5%₀ while the later three events (PL-3 to PL-5) document an enrichment of about 1‰ in $\delta^{18}O$ values (Fig. 2).

Discussion

ODP site 763A at present is influenced by poleward flowing warm Leeuwin Current from the north and equatorward flowing cold West Australian Current from the south. The by and large tropical warm planktic foraminiferal assemblages was intruded several times by temperate water species reflecting periods of enhanced equatorward transport of cool waters possibly via the West Australian Current (Zachariasse, 1992). The strong Leeuwin Current nullifies the wind dragged transport of surface waters towards north thus preventing any upwelling. Our planktic foraminiferal census counts reveals five intervals during Quaternary when upwelling indicator planktic foraminiferal species *Globigerinita glutinata* (Thiede and Junger 1992) shows major peaks of abundance, few of which are also accompanied by increase in percentage of *Globigerina bulloides*, another upwelling indicator (Fig. 2). Several studies have confirmed the abundance of *Globigerinita glutinata* as a proxy indicator of upwelling. Transmission Electron Microscopic investigations of *Globigerinita glutinata* points to a diet mainly consisting of diatoms (Hembleben, *et al*, 1989). Diatom blooms occur during upwelling conditions. *Ga glutinata* is commonly recorded in greater abundance in continental margins and high productivity areas (Be and Toldulund, 1971, Ottens, 1992). We attribute the abundance peaks of *Ga. glutinata* to upwelling and resultant productivity increase at site 763A. Further evidence of upwelling during these five intervals is gained through Oxygen isotopic data which shows higher $\delta^{18}O$ values during these five intervals possibly due to resultant cooling of surface waters. The magnitude of increase is less during PL-1 and PL-2 (~0.5‰) while during Pl-3, PL-4 and PL-5 the increase is higher (~1‰). The first two intervals (PL-1 and PL-2) of higher $\delta^{18}O$ values and concomitant increase in *Ga. glutinata* abundance indicate definite evidence of upwelling at site 763A. There is no record of increase in the abundance of temperate species *Globorotalia inflata* during these two intervals thus ruling out any possibility of northward transport of cool waters via West Australian Current. The upwelling thus can be attributed to either weakening or cessation of Leeuwin Current during PL-1 and PL-2 as it is the Leeuwin Current which prevents upwelling in modern times off Western Australia (Fig.1b) . Taking into consideration the possible causes for the weakening or cessation of Leeuwin Current one has to look at the principal forcing mechanism of Leeuwin Current which is generation of meridional steric height gradients and associated zonal geostrophic flows (Godfrey and Weaver, 1991) linked to Western Pacific Warm Pool and Indonesian throughflow (Weaver, 1990). Also the Leeuwin Current is associated to El Nino in a way that it is weak during El Nino conditions (Pearce and Phillips, 1988, Phillips *et al*, 1991). Thus the events PL1 and PL2 occurring at ~2.2Ma and ~1.83Ma respectively in the early Quaternary showing no northward transport of *Globorotalia inflata* (whose FAD is in Late Pliocene) but high $\delta^{18}O$ and increasing abundance of upwelling species *Globigerinita glutinata can be* attributed to weak Leeuwin Current. This indicates cooling of surface waters (high $\delta^{18}O$) induced by upwelling (Fig.1b). During these two intervals we believe that Leeuwin current diminished due to the causes lying in the Western Pacific probably due to ENSO induced lowering of sea level in the Western Pacific Warm Pool and resulting dominance of equatorward wind driven circulation resulting into offshore Ekman transport similar to the Peruvian coast. Numerical modeling studies by Godfrey and Weaver (1991) shows that when Pacific temperature profiles are replaced by the much colder ones observed in the eastern equatorial Pacific a typical eastern boundary flow regime is obtained with equatorward winds causing upwelling and shallow surface mixed layers off Western Australia. The decrease in WPWP is a reasonable explanation as it would result in decrease in streic height gradient causing reduced alongshore pressure gradient which is the principal mechanism forcing the Leeuwin Current poleward against the prevailing equatorward winds (Cresswell and Golding, 1980). Thus strength of the Leeuwin current we consider to be good monitor with caution for inferring past El Nino conditions. With our sampling interval rather coarse we can not comment as to what could have been duration of such long term El Nino like conditions. Thus the events PL-1 and PL-2 occurring at 2.2Ma and 1.83Ma represent weakening of Leeuwin Current due to WPWP reduction possibly induced by ENSO events.

During later part of the Quaternary at 0.68 Ma, 0.45 Ma and 0.04 Ma the peaks in *Ga. glutinata* are associated with remarkable invasion of typical temperate water species *Globorotalia inflata* at site 763A (Fig.2) which can be attributed to intensification and expansion of West Australian Current due to ice volume increase in Antarctica. Considering the modern oceanographic set up, the cooling of the surface waters at the region of ODP site 763A can occur by a northward shifting of the isotherms (subtropical convergence) (Prell and Hutson, 1979) due to episodic expansion in the Antarctic ice sheet and resulting increase in the influence of the West Australian Current. Supporting evidence for this conclusion is derived from increase in $\delta^{18}O$ values of surface dwelling planktic foraminifera by ~ 1‰. (Fig.2). Thus the cooling of surface waters due to upwelling and enhanced West Austrlian Current can explain planktic foraminiferal and isotopic record. As discussed earlier these intervals are the intervals of weakening or cessation of the Leeuwin Current. However the possible causes of reduced strength of Leeuwin Current during PL-3, PL-4 and Pl-5 (Fig.1c) can be a reduction in the Indonesian throughflow due to lowering of sea level in response to ice volume expansion. Thus, all these five intervals we show here are intervals of weakening of Leeuwin Current.

We consider these three events at 0.68 Ma, 0.45 Ma and 0.04 Ma to represent ice volume increase when sea levels must have been considerably lower reducing the strength of the Leeuwin current as a result of reduction in Indonesian throughflow (Srinivasan and Sinha, 1998, 2003) and increase in equatorward wind driven circulation and offshore Ekman transport (Fig. 1c). Major changes in the ocean circulation during the cooling intervals, involving the increased importance of West Australian current (CLIMAP, 1984; Prell and Hutson, 1979) and the cessation of the flow of Leeuwin current (Wells and Wells, 1994) has been shown earlier. Though we can not still answer the question about absence of an upwelling system in this region during LGM (Martinez *et al*, 1999), as our youngest sample showing glacial interval lies beyond the LGM level, we show here that during glacial advances earlier than LGM, there is a development of strong upwelling off Western Australia. The weakening of the Leeuwin current would have an effect on the Sea surface temperature in the Indian Ocean as the Indonesian through flow generally warms the Indian Ocean (CLIMAP, 1984). Thus these five intervals could probably be the intervals of considerable lowering of SST in the Indian Ocean and would definitely have influenced Indian Monsoonal intensity. At this juncture we would say that during the weak Leeuwin Current there is a possibility of reduced intensity of Indian Monsoon due to reduction in net heat input to the Indian Ocean through Indonesian throughflow. Thus a better understanding of the possible links between El Nino, Western Pacific Warm pool, Indonesian throughflow, wind pattern off Western Australia and ice volume expansion and contraction can be obtained by studying the behavior of Leeuwin current during Quaternary.

Conclusions

1. Based on planktic foraminiferal and isotopic record we have identified five intervals during Quaternary at 2.2Ma, 1.83Ma, 0.68 Ma, 0.45 Ma, and 0.40 Ma named here as PL-1, PL-2 Pl-3 PL-4 and PL-5 events respectively when the sea off western underwent intense upwelling contrary to modern Oceanographic conditions due to weakening or cessation of the flow of Leeuwin Current.
2. The PL-1 and PL-2 events of upwelling are attributed to EL Nino induced reduction in Western pacific warm Pool and weak Leeuwin Current while the PL-3, PL-4 and PL-5 events represent upwelling and weakening of Leeuwin Current due to reduced Indonesian throughflow due to lowering of sea level in response to ice volume increase.

3. These periods of weak Leeuwin Current are the intervals of reduced net heat inflow into the Indian Ocean from the Pacific via Indonesian Seaway and thus can be possibly the periods of reduced sea surface temperature in the Indian Ocean and reduced Summer Monsoonal intensity.

Acknowledgements

We thank Ocean Drilling Programme (ODP) and A.K. Rai, Allahabad University, for ODP samples. Department of Science and Technology (DST), New Delhi for financial support in the form of project P -7/293. Manish Tiwari carried out the isotopic analysis at PRL, Ahmedabad. We thank R. Ramesh (PRL) for providing stable isotope analytical facilities.

References

BE, A.W.H. and TOLDULUND, D.S. 1971. Distribution and ecology of living Planktonic foraminifera in surface waters of the Atlantic and Indian Ocean, In: The Micropaleontology of the Oceans, B.M.Funnel and W.R.Riedel (Eds) Cambridge Univ. Press London, 105-149.

BERGGREN, W.A. HILGEN, F.J., LANGEREIES, C.G., OBRADOVICH, J.D., RAFFI, I, RAYMO, M.E. and SHACKLETON, N.J., 1995. Late Neogene Chronology: New perspective in high resolution Stratigraphy, Geological Society of America Bull. 107 (11), 1271-1287.

CHENG TANG, 1992. Paleomagnetism of Cenozoic sediments in Holes 762B and 763A, Central Exmouth plateau, northwest Australia. Proceedings ODP Scientific results , 122, 717-733.

CLIMAP, 1984. The last interglacial ocean. Quat, Res. 21. 123-224 .

CRESSWELL G.R. and GOLDING, 1980. T.J. Observations of a south flowing current in the southeast Indian Ocean. Deep Sea Research 27, p.449-466.

GASPERI, J.T. and KENNETT, J.P., 1992. Isotopic evidence for depth stratification and paleoecology of Miocene Planktonic foraminifera : Western Equatorial Pacific DSDP site 289. In: Pacific Neogene Events: Tsuchi ,R and Ingle James Jr. (Eds.) Univ. of Tokyo press , 117-147.

GODFREY J. S. and WEAVER, A.J.., 1991. Is the Leeuwin Current driven by Pacific heating and winds? Progress in Oceangraphy. 27, 225-272.

GODFREY, J.S., 1996. The effect of the Indonesian throughflow on ocean circulation and heat exchange with the atmosphere: a review. Jour Geophysical Research 101 (C5), 12217-12237.

HIRST, A. C. and GODFREY, J.S., 1993. The role of Indonesian throughflow in a global ocean GCM. Journal of Physical Oceanography , 23, (6) , 1057-1086.

IMBRIE, J. and KIPP, 1971 A new micropaleontological method for quantitative paleoclimatology: Application to Late Pleistocene Caribbean core in The late Cenozoic glacial ages, K.K. Turekian (Ed), Yale University, New haven Conn. 71-182.

KENDRICK, G.W., WYRWOLL, K.H., SZABO, B.J., 1991. Pliocene Pleistocene costal events and history along the western margin of Australia. Quat. Sci. Rev. 10, 419-439.

KENNETT , J.P. KELLER, G. and SRINIVASAN, M.S., 1985., Miocene plakntonic foraminiferal biogeography and paleoceanographic development of the Indo-Pacific region. In: *The Miocene Ocean*. (Ed. Kennett, J.P.) GSA memoir 163, 197-236.

KENNETT, J.P. and SRINIVASAN, M.S.,1983 Neogene Planktonic foraminifera: A phylogenetic Atlas: Hutchinson Ross Publ Co. USA.

Mc GOWRAN, QIANYU LI, CANN J., PADLEY, D., MC KIRDY, D.M. and SHAFIK S., 1997 Biogeographic impact of the Leeuwin Current in southern Australia since the Late Eocene. Palaeogeogr. Palaeoclimatology, Palaeoecolo. 136, 19-40.

MRTINEZ J. IGNACIO, DECKKER, PATRICK DE AND BARRROWS, TIMOTHY, T., 1999. paleoceanography of the last glacial maximum in eastern Indian Ocean: Planktonic foraminiferal evidence Paleogeography, Paleoclimatology, Paleoecology 147, 73-99.

OKADA, H. and WELLS, P. 1997. Late Quaternary nannofossil indicators of climate change in two deep sea cores associated with the Leeuwin Current off western Australia. Palcogcogr. Paleoclimat. Paleoecolo. 131, 413-432.

OTTENS, J.J. 1992. Planktonic foraminiferal diversity as indicator of ocean environments, mar. Micropal., 19, 13-28.

PEARCE, A.F. and PHILLIPS, B.F., 1988. ENSO events, the Leeuwin Current , and larval recruitment of the western Rock lobster. Jour. Cons. Int. Explor. Mer., 45: 13-21.

PHILLIPS, B.F. PEARCE, A.F. LITCHFIELD, R.T., 1991. The Leeuwin current and the larval recruitment of the rock (spiny) lobster fishery off Western Australia. In: A.F. Pearce and D.I.Walkers (editors) The Leeuwin Current: An influence on the coastal climate and marine life of Western Australia. J.R.Soc. West Aust. . 74: 93-100.

PRELL, W.L. and HUTSON, W.H., 1979. Zonal temperature anomaly maps in the Indian Ocean surface waters, Modern and ice age patters. Science, 206, 454-456.

PRELL, W.L. and HUTSON, W.H., 1979. Zonal temperature maps of the Indian Ocean surface waters; modern and ice age patterns. Science 206, 454-456.

SMITH, R. L.,1992. Coastal upwelling in the modern ocean. In: Upwelling Systems: Evolution since the early Miocene. Geol. Soc. Lond. Spec. Publ. No. 64, Summerhayes, C.P., Prell, W.L. & Emeis, K.C. (Eds.) 9-28

SINHA, DEVESH K., SINGH, ASHUTOSH K. and TIWARI, M. (2006), Paleoceanographic and Paleoclimatic history of ODP site 763A (Exmouth Plateau). Southwest Indian Ocean: 2.2 Ma record of planktic foraminifera. Current Science, 90 (10), 25 May (in press)

SRINIVASAN M.S. and SINHA DEVESH K., 2000. Ocean circulation changes in the Indo-Pacifc during 5.6 to 4.2 Ma: Planktic foraminiferal and isotopic evidences. Earth and Planet Sci. Indian Acad. Sci.109 (3) , 315-328.

SRINIVASAN, M.S. and SINHA,DEVESH .K., 1998., Early Pliocene closing of the Indonesian Seaway: Evidence from northeast Indian Ocean and southwest Pacific deep sea cores. Journal of Southeast Asian Earth Sciences, Pergamon Press, U.K. ,16, 1, 29-44.

SRINIVASAN, M.S. and SINHA, D.K., 2003. Late Neogene ocean circulation changes in the tropical Indo-Pacific and evolution of Asian Monsoon. VIII RCPNS *Proc. VIII RCPNS,* Chaing Mai, p. 212-236.

TCHERNIA, P., 1980. Descriptive regional oceanography. Pergamon, Oxford, 252 pp.

THIEDE J, and JUNGER B.., 1992. Faunal and Floral indicators of coastal upwelling (NW African and Peruvian Continental margins). In Upwelling Systems. Evolution since the early Miocene. Geol. Soc. Lond. Spec. Publ. No. 64, Summerhayes, C.P., Prel, W.L. & Emeis, K.C. (Eds.). 47-76.

VEEH, H.H. McCORKLE, D.C. AND HEGGIE, D.T., 2000. Glacial interglacial variations of sedimentation on the west Australian continental margin : constraints from excess ^{230}Th. *Marine Geology*, 166, 11-30.

WEAVER, A. J., 1990. Ocean currents and climates, Nature 347, 432.

WEI, KUO YEN, 1998. Southward shifting of Tasman Front at 4.4 Ma (early Pliocene): paleobiogeographic and oxygen isotopic evidences. Journal of Asian Earth Sciences. 16, 97-106.

WELLS P. WELLS G. CALLI, J. and CHIVAS, A. 1994. Response of deep sea benthonic foraminifera to late Quaternary climate changes, SE Indian Ocean, offshore Western Australia. Mar. micropal, 23, 185-229.

WELLS, P.E and WELLS, G.M., 1994, Large scale reorganization of ocean currents offshore Western Australia during the late Quaternary. Mar. micropal. 24: 157-186.

WILSON, B.R. and ALLEN, G.R.., 1987., Major Components and distribution of marine fauna. In Dyne G.R., Walton, D.W. (Eds.) Fauna of Australia, General articles, vol. 1A, Aus. Govt. Publ. Service Canbera, pp 43-68.

ZACHARIASSE, W.J., 1992. Neogene Planktonic foraminiferas from site 761 and 762 off northwest Australia. Proceedings of the Ocean Drilling Programme Scientific results 122, 665-675.

Planktic foraminiferal evidence for Neogene deep sea hiatuses in the Northern Indian Ocean

A.D. Singh[1] and Rahul Mohan[2]

[1]Department of Geology, Banaras Hindu University, Varanasi 221 005, India
[2]National Centre for Antarctic and Ocean Research, Goa, India

Introduction

The ocean basin configuration, climatic conditions, surface productivity, circulation system and vertical water mass stratification collectively control the sedimentation pattern in the world oceans. With the advent of the Deep Sea Drilling Project (DSDP) and its successor Ocean Drilling Program (ODP) it is now possible to examine undisturbed deep sea sequences from equatorial to polar oceans ranging in age from the late Mesozoic to Recent. The studies conducted so far have provided a better insight into the development of major changes in deep sequences which are dominated by pelagic sediments (calcareous and siliceous). The sediment accumulation rate is mainly controlled by rate of supply of sediments (primarily biogenic productivity) and the rate at which they are removed. The rate of sediment removal depends on vigorous bottom water activity and dissolution within the water column and sediment-water interface. Thus, an increase in the carbonate accumulation may reflect high productivity and/or a deepening of carbonate lysocline. In pelagic regimes, however, the rate of accumulation is influenced by dissolution of sediments at various stages of sedimentation (Fig. 1). For instance, in some regions it occurs within the water column resulting into a solution resistant assemblage of Sidocoenosis. Further, dissolution at the sediment- water interface or sediment removal by vigorous bottom water currents may result into a gap in the geological record leading to deep sea hiatuses. Thus, non-deposition due to dissolution and/or physical removal of sediments by erosion are the two primary means of creating deep sea hiatuses. In general, the amount of missing sediment record per hiatus depends on the net difference between the rate of supply and rate of removal of sediments at the ocean floor. In high productivity regimes with high sediment flux, hiatuses tend to be of shorter duration as these breaks are caused by non- deposition or dissolution. Whereas, in regions with low sediment flux or/land active bottom currents attributing deep sea erosion, the hiatuses are likely to be of longer duration, often spanning several million years.

Detailed biostratigraphic investigations of the sediment cores from the world oceans have enabled to identify a number of hiatuses in the deep sea sequences, and much efforts have been made by marine geologists to understand their paleoceanographic implications (Berger, 1972; Kennett and Watkins 1975; Kennett, 1977; McGowran, 1978; Moore *et al.* 1978; Barron and Keller, 1982; Haq *et al.* 1987; Keller and Barron, 1983; Kennett *et al.* 1985; Ledbetter and Ciesiciski, 1986; Vincent *et al.* 1985;

Keller and Barron, 1987; Srinivasan and Chaturvedi, 1992; Srinivasan and Singh, 1992; Ramsay *et al.*, 1994; Singh and Srinivasan, 1995; Srinivasan, 1996 and Spencer- Cervato, 1998).

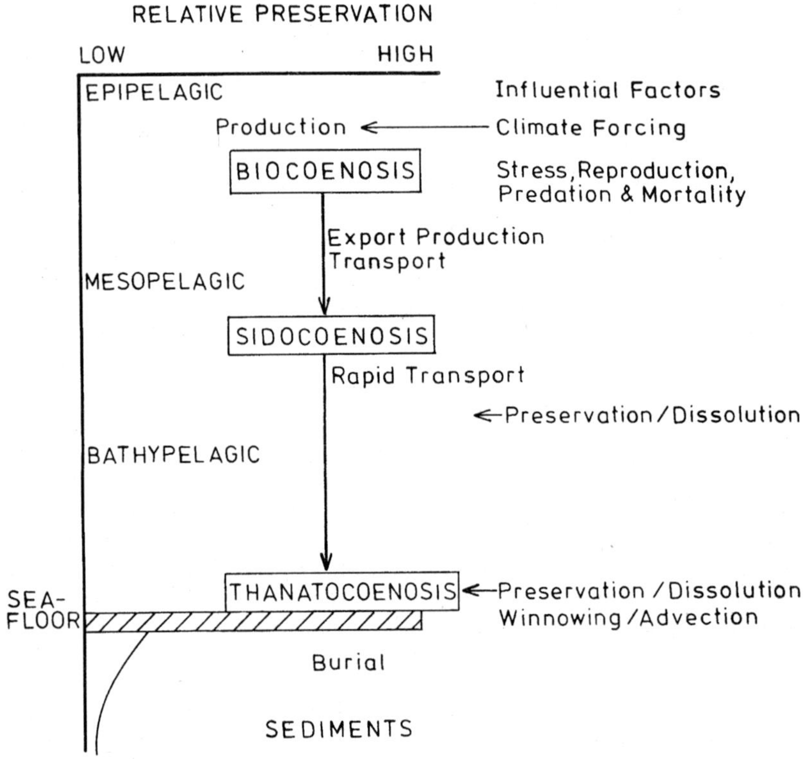

Fig. 1. Schematic illustration of various processes associated with production to burial of shell bearing plankton. (Modified after Takahashi 1995).

The temporal and spatial distribution of these hiatuses can be determined for the Cenozoic from the deep sea sections recovered by the Deep Sea Drilling Project and Ocean Drilling Programe. Kennett and Watkins (1975) while examining the late Cenozoic deep sea sequences of the southeast Indian Ocean observed that Paleogene sediments are less disturbed than the Neogene. In contrast, the Neogene sequences of southern ocean (eg. Tasman Sea and around) are found to be relatively continuous than the Paleogenes. This regional change in distribution patterns of hiatuses through Cenozoic has been interpreted as a result of major circulation change in the Circum- Antarctic Ocean during the late Paleogene (Kennett *et al.*, 1975). Studies by Moore *et al.* (1978) revealed that major deep sea hiatuses occurred in the late Eocene (across the Eocene/Oligocene boundary) and in middle to late Miocene, whereas, fewest hiatuses occurred during middle Eocene, early to middle Miocene and Quaternary. Fleisher (1974), Keller (1983) and Vincent *et al.* (1985) recorded hiatuses in the northern Indian Ocean deep sea sections and correlated these hiatuses with those identified from the Atlantic and Pacific deep sea sequences.

Because the deep sea hiatuses have a fundamental bearing on history of ocean circulation, effects of ice volume changes and their role in climatic evolution, an attempt has been made to identify these hiatuses in the selected Neogene deep sea sequences of the northern Indian Ocean. However, at certain sites it was difficult to differentiate between a geological hiatus and the information gap because of incomplete core (recovery. Hiatuses generally remove all or most of a particular microfossil zone. The recognition and magnitude of these hiatuses are determined on the basis of abrupt replacement of stratigraphically important planktic foraminifera. An integrated biostratigraphic, paleomagnetic and isotopic records have enabled to decipher precise timing of these hiatuses.

Location of DSDP Sites

The five DSDP sites (214, 219, 220, 237 and 238) discussed in this paper lie at present under equatorial water mass (Fig. 2). However, the paleolocations of these sites during the Miocene were further south near the transitional water mass. Sites 237 and 238 have moved to the north with lesser extent than the other three sites (Sclater *et al.* 1985). Sites 214, 237 and 238 are located presently in the tropical-subtropical transition zone whereas, sites 219 and 220 in the northern part of equatorial water (Sevrdrup *et al.* 1942). Core recovery was reasonably good at sites 214, 219, 237 and 238 with almost continuous late Neogene record, while early Neogene record was interrupted with hiatuses and dissolution intervals at different stratigraphic levels. Site 220, the deepest among the five sites examined appears to lie close to foranliniferal lysocline during the Miocene period as indicated by the assemblage dominated by solution resistant taxa and poor foraminiferal preservation (Fleisher, 1974 and Singh, 1990). Details of the studied sites are provided in Table I.

Methods

In the last two decades Neogene biostratigraphy has been greatly refined. Improved ties with paleomagnetic stratigraphy and radiometric scale have contributed to this refinement. In addition, applications of graphic correlation method and quantitative analyses have contributed significantly to improved biostratigraphic precisions. These refinements are applied in this paper to recognize the Neogene deep sea hiatuses.

The earlier reports on the Neogene deep sea hiatuses of northern Indian Ocean DSDP sites (Fleisher, 1974, Vincent, 1977 and Ramsay *et al.*, 1994) were based on wide sampling intervals; therefore, in the present study data generated from close stratigraphic intervals are used. The synchroncity of planktic foraminiferal datums used for delineating the zonal boundaries were checked with graphic correlation (Srinivasan and Sinha, 1992). These datums were also calibrated with magnetic polarity chrons (Berggren *et al.* 1985). Improved biostratigraphies thus enabled to identify the condensed planktic foraminiferal zones and also to update the record of Neogene deep sea hiatuses in these sites. In general, the existence of a hiatus is clearly defined by the absence of one or more planktic foraminiferal zones. Only these breaks in the stratigraphic record are considered as hiatus and are described here.

Further, the relative abundances of solution susceptible and solution resistant taxa were examined to demarcate the dissolution intervals within the Neogene sequences. In addition, quantitative planktic foraminiferal data from these sites were synthesized across the intervals of identified hiatuses with paleoceanographic climatic changes. The quantitative faunal data from sites 214, 237 and 238 were also compared with the foraminiferal oxygen stable isotope records published by Vincent *et al.* (1985), for these sites to gain a better insight into the linkages among isotopic switch, faunal turnover and climatic change. However, some discrepancies were unavoidable in the correlation of faunal data with major isotopic signals due to lack of one to one correlation. The hiatuses were also compared with the Neogene foraminiferal oxygen isotope records of the low- latitude Atlantic and Pacific Oceans (Miller *et al.*, 1987 and Keller and Barron, 1987) in order to assess the global nature of these erosional events.

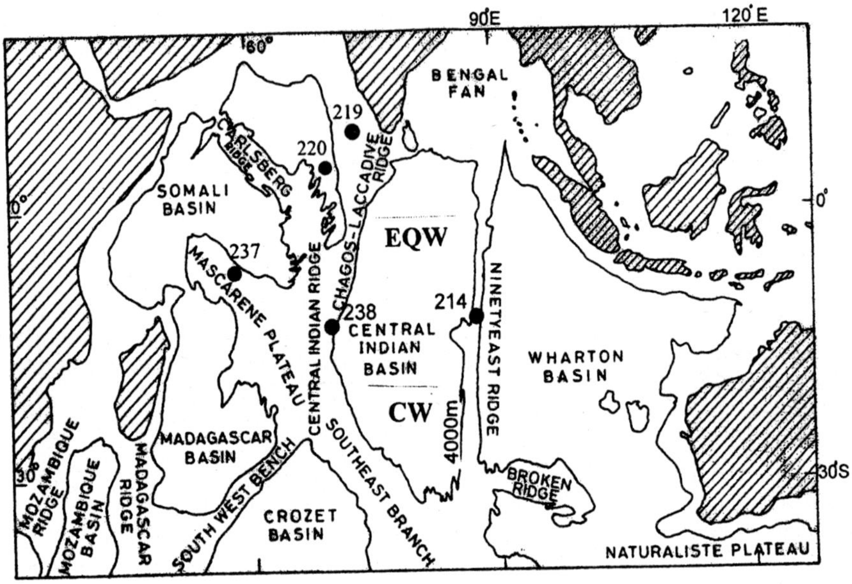

Fig. 2. Location of five DSDP sites examined in the northern Indian Ocean. EQW = equatorial water; CW=central water; TRANS = tropical-subtropical transition zone. (Modified after Vincent *et al.* 1985).

Stratigraphy of deep sea hiatuses

Moore *et al.* (1978) were the first to describe the Cenozoic hiatuses in pelagic sediments, and discussed their implications. Later, Barron and Keller (1982) and Keller and Barron (1987) identified a number of widespread deep sea hiatuses in the post-Oligocene deep sea sequences and labeled NH for Neogene Hiatus and numbered sequentially 1 to 8 from the base of the Miocene to the Middle Pliocene (Table 2). The greatly improved Neogene biostratigraphies established for DSDP sites 214, 219, 220, 237 and 238 provided an opportunity to identify four hiatuses. The temporal and spatial extent of the Neogene hiatuses identified at the examined sites are provided in Table 3, and the stratigraphic position of the hiatuses with respect to planktic foraminiferal zones are illustrated in figure 3.

The earliest Neogene hiatus is recognised in the middle Miocene between 16 Ma and 14 Ma, demonstrating a break in the planktic foraminiferal record, from the uppermost part of *Praeorbulina glomerosa* plus *Globorotalia peripheroronda*, *Globorotalia peripheroacuta* to the lower part of *Globorotalia praefohsi* Zones at sites 214 and 219. This hiatus spanning about 2 m.y. roughly corresponds to NH2 of Keller and Barron (1987). However, the hiatus NH2 recorded by Keller and Barron (1987) spans between 16.1 Ma and 15.1 Ma. Another hiatus in the middle Miocene has been detected by the absence of most of the *Globorotalia fohsi fohsi* Zone plus *Globorotalia fohsi lobata* Zone and lower part of the *Globorotalia siakensis* Zone at site 238 only (Fig. 3). This hiatus representing about 2 m.y. spans the interval between 13 Ma and 11Ma, and broadly correlates with the hiatus NH3 of Keller and Barron (1987). Immediately above the middle Miocene unconformity, a distinct hiatus is recognized encompassing the upper most part of planktic foraminiferal Zone *Globorotalia siakensis* plus *Globorotalia menardii* and the lower part of *Neogloboquadrina acostaensis* Zones. Within this stratigraphic interval Barron and Keller (1982) identified hiatus NH4 in the Pacific deep sea sequences. This hiatus spanning 10 – 10.5 Ma represents a shorter erosive event across the middle/late Miocene at Site 237. Vincent (1977) and Vincent *et al.* (1985) also observed this hiatus in other DSDP sites from the western Indian Ocean.

Table 1. Location and water depths of the examined DSDP Sites.

Site	Latitude	Longitude	Water Depth (m)
214	11°20.21'S	88°43.08'E	1,165
219	9°01.75'N	72°52.67'E	1,764
220	6°30.97'N	70°59.02'E	4,036
237	7°09.99'S	58°07.48'E	1,640
238	11°09.21'S	70°31.56'E	2,832

Table 2. Ages of Neogene hiatus events (after Keller and Barron, 1987).

Hiatus Events	Keller and Barron, 1987
NH8	3.7 – 3.2 Ma
NH7	5.2 – 4.7 Ma
NH6	7.0 – 6.0 Ma
NH5	8.6 – 8.0 Ma
NH4	10.5 – 9.2 Ma
NH3	16.1 – 15.1 Ma
NH2	16.1 – 15.1 Ma
NH1	18.2 – 17.2 Ma
NH1a	20.5 – 19.5 Ma

Table 3. Temporal and spatial extent of Neogene deep sea hiatuses at the examined DSDP Sites.

Hiatus	Site	Core & Section	Biostratigraphy	Separation Chronology (m.y.)
NH7	220	Core 4	*Pu. primalis* Zone/	~1
			Gr. tumida Zone	(4.5 Ma to 5.5 Ma)
NH4	237	Core 14	*Gr. siakensis* Zone/	~0.5
		Section 4	*N. acostaensis* Zone	(10 Ma to 10.5 Ma)
NH3	238	Core 32	*Gr. fohsi fohsi* Zone/	~2
		Section 3	*Gr. siakensis* Zone	(11 Ma to 13 Ma
NH2	214	Core 20	*Gr. glomerosa* Zone/	~2
		Section 6	*Gr. praefohsi* Zone	(14 Ma to 16 Ma)
NH2	219	Core 13	*Gr. glomerosa* Zone/	~2
		Section 1	*Gr. praefohsi* Zone	(14 Ma to 16 Ma)

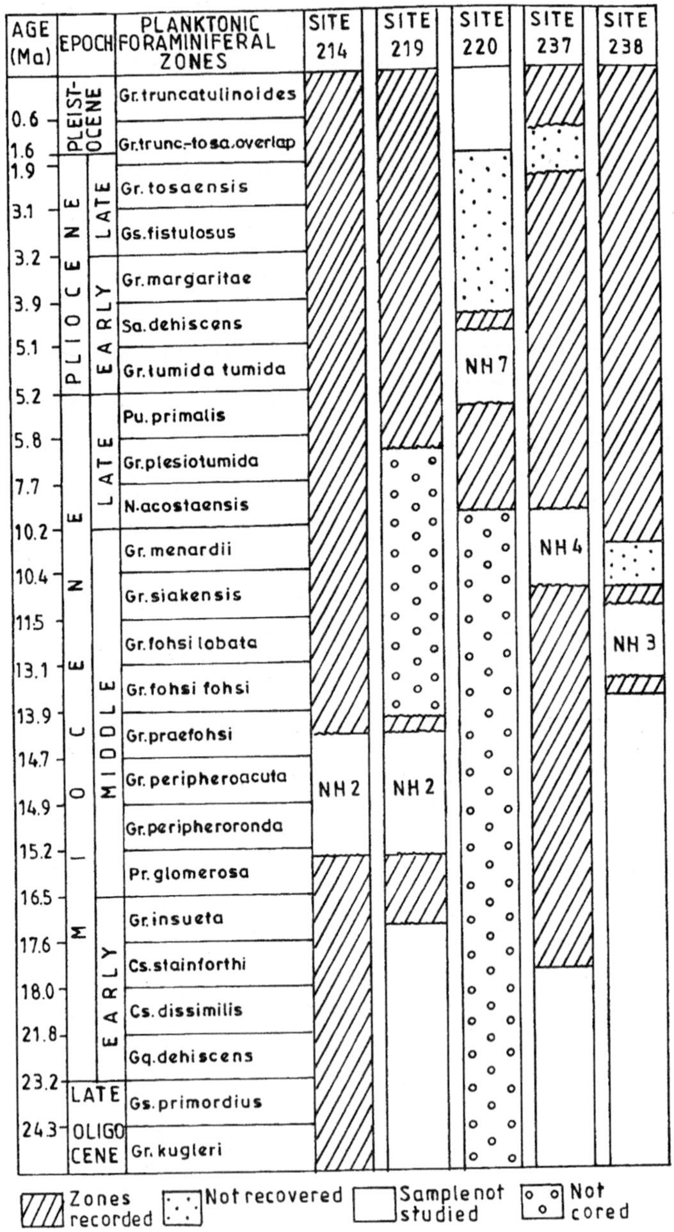

Fig. 3. Stratigraphic position of Neogene hiatuses (NH) at DSDP sites examined along with planktic foraminiferal biostratigraphy. Planktic foraminiferal Zones after Srinivasan and Chaturvedi (1992); Srinivasan and Singh (1992); and Singh and Srinivasan (1995).

The Pliocene sequences at the studied sites are fairly continuous except at site 220, where interval across the Miocene/Pliocene boundary is represented by a hiatus. This unconformity is identified by the absence of most of the *Pulleniatina primalis* Zone and the lower part of *Globorotalia tumida* Zone. This hiatus occurring between 5.5 Ma and 4.5 Ma appears to correspond with the NH7 of Keller and Barron (1987). Thus, the four hiatuses identified in the present work broadly correlates with NH2, NH3, NH4 and NH7 of Keller and Barron (1987). The hiatuses corresponding to NH4 and NH7 are of short duration, while those corresponding to NH2 and NH3 are of longer duration.

Dissolution and accumulation rate

Studies carried out on several Neogene deep sea sequences of world oceans revealed a number of dissolution intervals reflecting the chemistry of water column. Based on the fluctuations in percentage abundance of solution susceptible vs. solution resistant planktic foraminiferal taxa and the rate of accumulation, intervals of dissolution can be detected.

Based on the foraminiferal records (Srinivasan and Singh, 1991), two distinct dissolution intervals were detected in the sites examined. The first one during 17-20 Ma (early Miocene) at site 214 where a lowest accumulation rate of about 1 m/m.y. has been recorded pointing out strong dissolution (Fig. 4). The earliest Neogene hiatus NH1 of Keller and Barron (1987) is correlatable with this dissolution interval. The other notable dissolution interval characterizing a drastic decrease in abundance of solution susceptible taxa was recognised at site 237 between 16.5 Ma and 17 Ma. This period of strong dissolution during late Early Miocene is also indicated by an abrupt reduction in accumulation rate from 15.2 m/m.y. during 17.6–18 Ma to 1.4 m/m.y. between 16.5 Ma and 17 Ma. This late Early Miocene dissolution interval can be mistaken for a hiatus because of the lack of highly solution susceptible zonal marker *Globigerinatella insueta*, whose first appearance marks Zone N6. For instance, the early Miocene hiatus indicating the absence of Zone N6 in many Pacific and Atlantic cores may in fact represent strong dissolution interval. Therefore, differentiation of dissolution interval from a deep sea hiatus in pelagic sequences should be carried out with great caution. The site 220 which lies in water depth greater than 4000 m, dissolution of planktic foraminifera is prominent. As a result many index planktic foraminifera which are susceptible to solution are absent. Therefore, zonal assignment was not possible for the entire Neogene at this site.

In the deep sea sequences which are dominated by pelagic sediments, the sediment supply rate is controlled by a combination of various factors such as biogenic productivity, corrosiveness of water and bottom current's erosive intensity. In order to have a better understanding of these processes and to study spatial and temporal variations of sedimentation rate in the northern Indian Ocean, Age Vs Depth plots were made for the Neogene sequences at sites 214, 219, 237 and 238. The plots reveal broad similarity in sediment accumulation rate for all the sites (Fig. 5). In general, most of the early and middle Miocene is marked by low carbonate accumulation rate with its minima occurring around 17-20 Ma and 13-14 Ma. The maximum rate of accumulation in the middle Miocene occurred between 14.5 Ma and 16 Ma at site 237 with an average value of 9 m/m.y. The late Miocene sediments younger than 7.5 Ma accumulated at much higher rates with average values three to four times higher than those of the early and middle Miocene. For instance, 25 m/m.y. at site 219; 15 m/m.y. at site 237; about 44 m/m.y. at site 238 and about 40 m/m.y. just prior to the middle/late Miocene boundary at site 214 (Fig. 4).

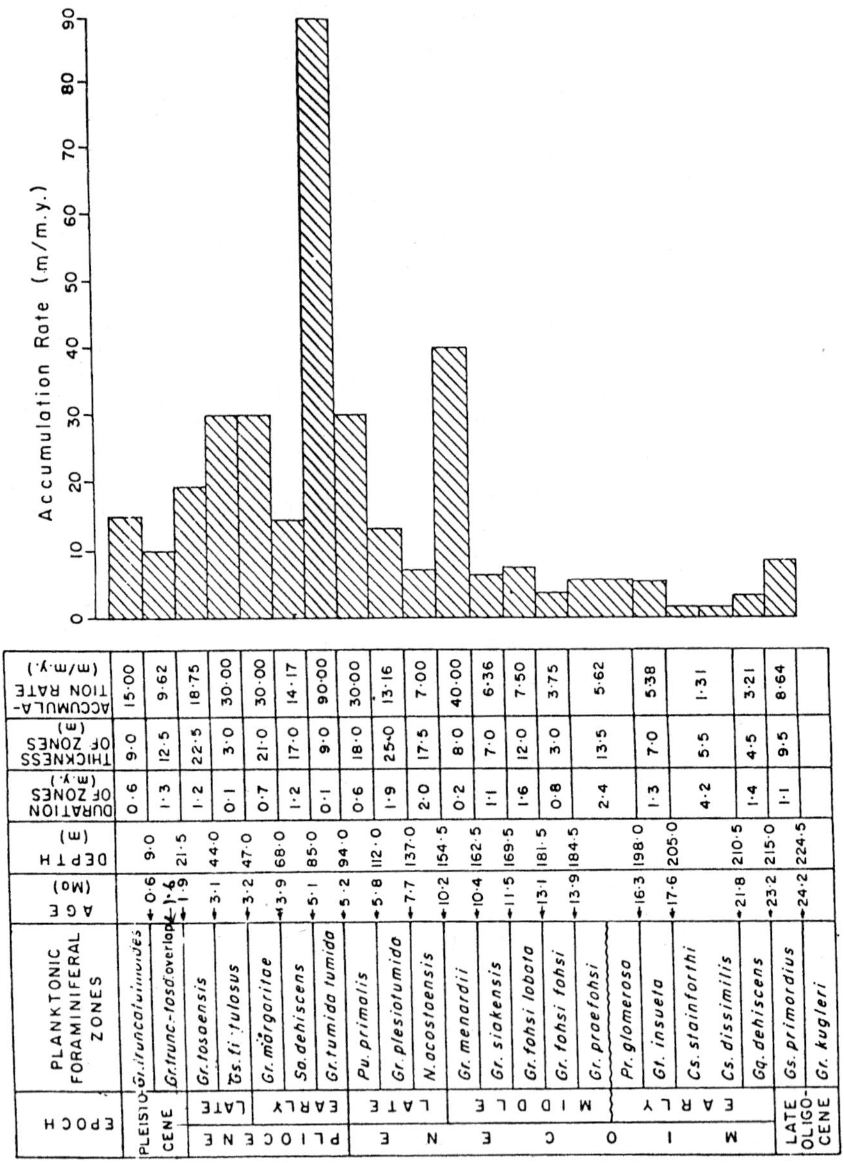

Fig. 4. Accumulation rates of Neoegene sediments at northern Indian Ocean Sites 214. Planktic foraminiferal zones after Srinivasan and Chaturvedi (1992) and ages after Saito (1984) and Berggren *et al.* (1985).

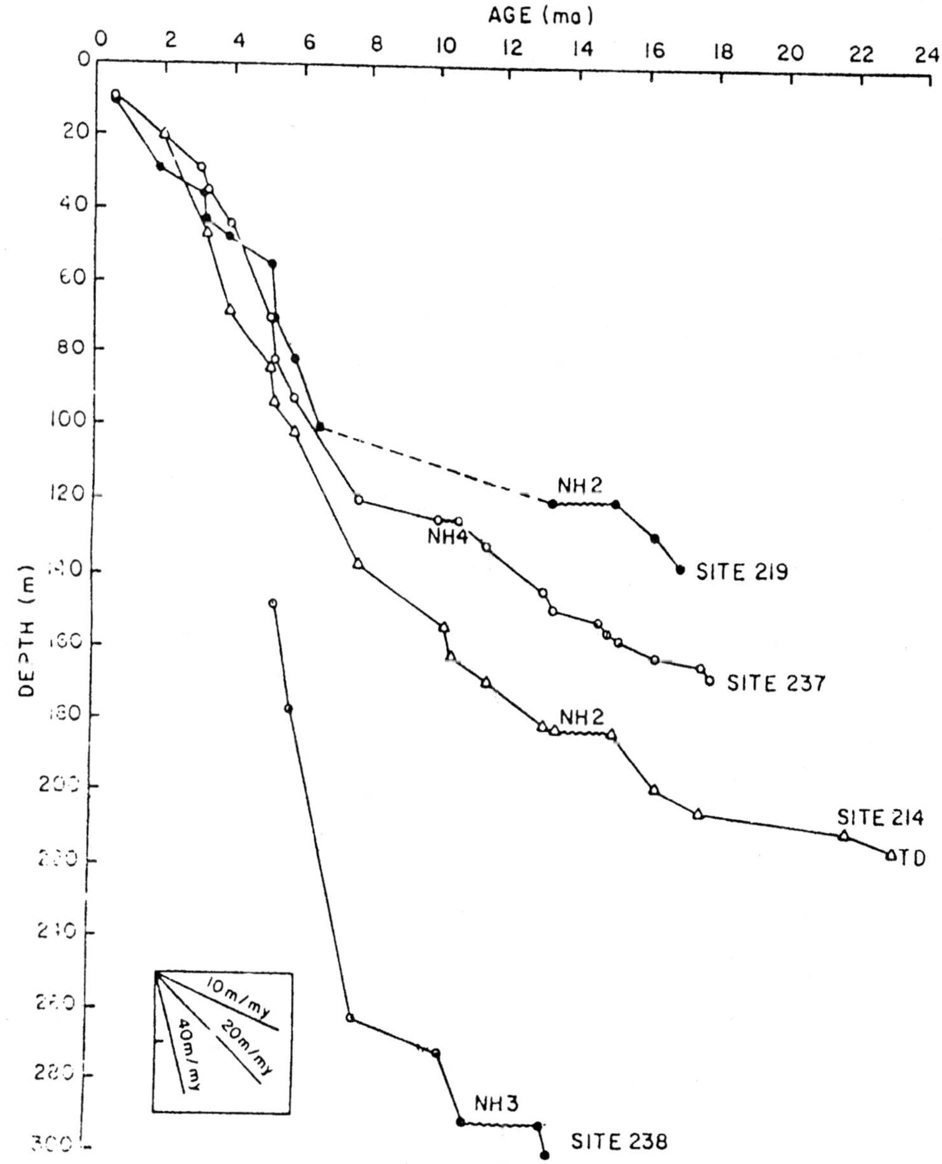

Fig. 5. Age Vs Depth Plot for the Neogene DSDP sites 214, 219, 237, 238. Ages after Saito *et al.* (1984) and Berggren *et al.* (1985), H = Hiatus; TD = Total Depth [Data on Sites 219, 237, 238 are from Srinivasan and Singh 1992 & Singh and Srinivasan 1995. Site 214 after Srinivasan and Chaturvedi 1988, 1992].

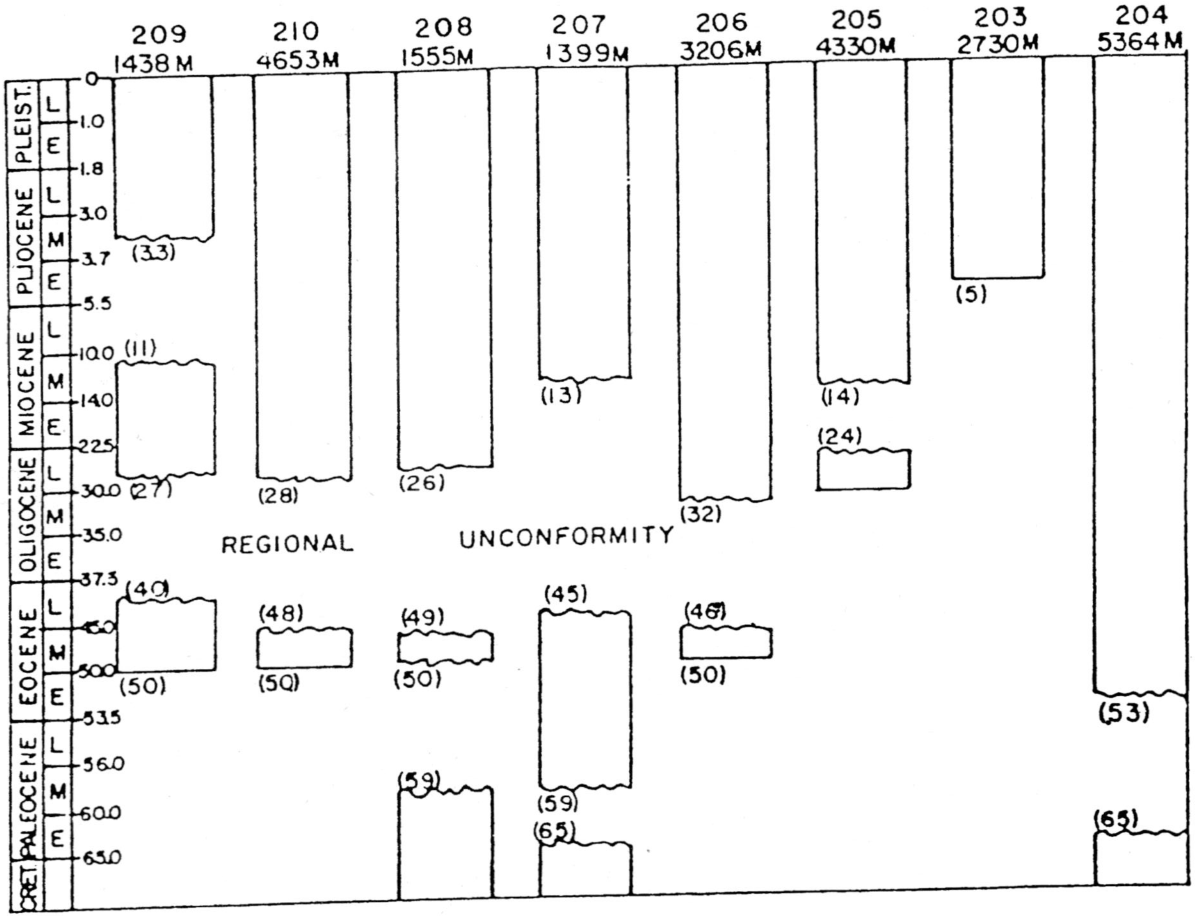

Fig. 6. Widespread deep sea hiatuses recorded in the Southwest Pacific Leg 21 DSDP Sites [after Kennett, 1973].

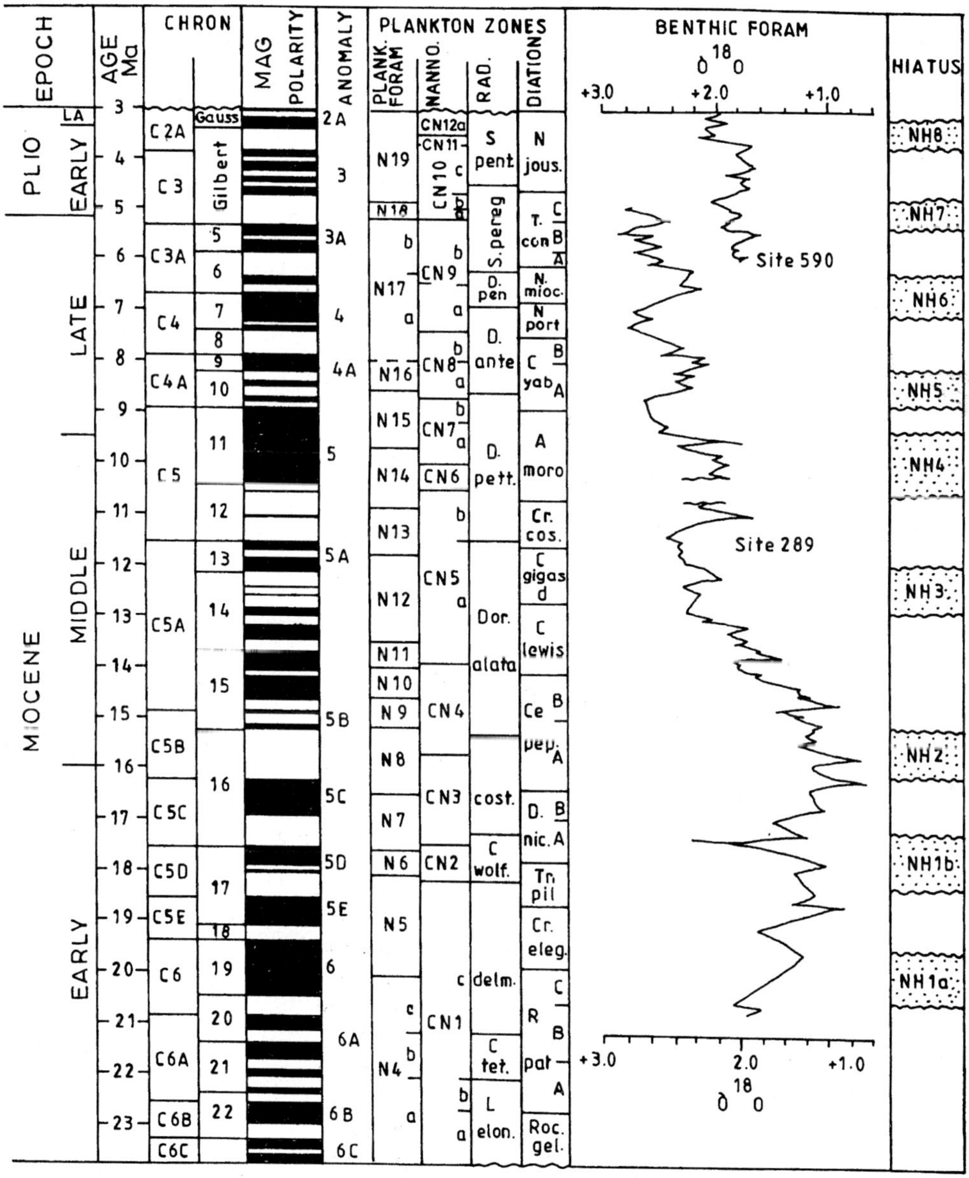

Fig. 7. Miocene to early Pliocene deep sea hiatuses and their correlation with paleomagnetic stratigraphy, plankton biostratigraphy and benthic foraminiferal oxygen isotope stratigraphy. [Modified after Keller and Barron 1987].

The Pliocene in general indicates a high rate of accumulation. The beginning of the Pliocene (5.2–5.1 Ma) is marked by an abrupt increase in accumulation rate in all the four sites excepting site 220 where the Miocene/Pliocene boundary is represented by a hiatus. Another peak of high accumulation rate occurs in all the sites between 3.2 Ma and 3.1 Ma. This observation suggests a broad similarity in sediment accumulation rate in the northern Indian Ocean during the Pliocene. The accumulation rate has been low at the beginning of the Quaternary followed by an increase in the late Pleistocene.

Causes of hiatus formation

The main objective of the present study is to identify the spatial and temporal distribution of Neogene hiatuses from the five sites examined and to understand the interrelationships of deep sea hiatuses and the tectonic, oceanographic/climatic evolution of the northern Indian ocean. Earlier studies have suggested that some hiatuses particularly those formed during the Paleogene are related to major tectonic events such as the opening and closing of ocean gateways, while others are primarily caused by intensified deep water circulation driven by increased bottom water production during intervals of polar cooling (Kennett, 1973; ref Fig. 6 and McGowran, 1978). It was also observed that Neogene hiatuses generally correlated to intervals containing positive increase in $\delta^{18}O$ values indicating polar cooling episodes (Keller and Barron, 1987).

Figure 7 shows correlation of the hiatuses identified in the present study with the Neogene deep sea hiatuses recorded earlier by Keller and Barron (1987) from the world oceans. An attempt has also been made to calibrate these hiatuses with paleomagnetic stratigraphy, planktic foraminiferal zonations and benthic foraminiferal oxygen isotopes. The increased carbonate dissolution during 17–20 Ma at site 214 and during 17.6–18 Ma at site 237 appears to have resulted due to intensified bottom water circulation accompanied with increased corrosiveness of bottom waters. This early Miocene dissolution interval has also been recorded from the Atlantic and Pacific and other regions of the Indian Oceans. The initiation of deep circum - Antarctic circulation and associated influx of Antarctic Bottom Water (AABW) into the major ocean basins is suggested to be a plausible cause for such widespread carbonate dissolution. As stated above, the early Miocene hiatus demonstrating the absence of Zone N6 in many deep sea cores corresponds to this dissolution interval.

The dissolution interval is succeeded by a distinct hiatus spanning about 2 m.y. at sites 214 and 219. This hiatus which roughly corresponds to hiatus NH2 of Keller and Barron (1987) appears to be linked with global climatic cooling and further intensification of bottom waters as indicated by quantitative planktic foraminiferal data and oxygen isotope record (Srinivasan and Singh, 1991 and Vincent *et al.*, 1985). A significant increase in benthic foraminiferal $\delta^{18}O$ during the early Middle Miocene has also been observed in the Atlantic and Pacific Oceans (Woodruff *et al*, 1981; Miller and Fairbanks, 1983 and Keller and Barron, 1987). According to Barron and Keller (1982) this hiatus is less widespread, identified only in the North Atlantic and the Caribbean. In the Indian Ocean this hiatus is probably resulted due to vigorous bottom water erosion linked with initiation of ice-sheet formation in Antarctica.

Significant drop in sediment accumulation rate and increase in foraminiferal $\delta^{18}O$ values at sites 214 and 237 marks a major middle Miocene hiatus at site 238 during 13–11 Ma. This hiatus corresponding broadly to the hiatus NH3, coincides with the major middle Miocene cooling event (Woodruff *et al.* 1981). This suggests a close linkage between formation of the hiatus NH3 and an increased intensity of circulation and corrosiveness of bottom currents during periods of increased polar glaciation. It is interesting to note that this hiatus occurs at the time of collision of Australian continent with southeast Asia resulting into the closure of the Indonesian Seaway (Kennett *et al.* 1985).

This resulted in bringing out profound changes in ocean circulation pattern both in the northern Indian Ocean and the equatorial Pacific (Srinivasan and Singh, 1991).

The hiatus identified at site 237 across the middle/late Miocene boundary is of only 0.5 m.y. duration and corresponds to hiatus NH4. Vincent (1977) also recognized this hiatus across the middle/late Miocene boundary at many DSDP sections from the western Indian Ocean and suggested regional tectonic activity and circulation readjustment following the opening of Gulf of Aden as plausible cause. However, a considerable increase in $\delta^{18}O$ values during this interval (Vincent *et al.*, 1985 and Miller *et al.*, 1987) appears to suggest this as an erosive event caused by intensified bottom water activity. The record of the characteristic temperate taxon *Globorotalia miozea* in samples preceding the NH4 event at sites 214, 237 and 238 (Srinivasan and Singh, 1991 and Srinivasan and Chaturvedi, 1992) provides evidence for the expansion of cool waters into the tropical northern Indian Ocean. However, the absence of NH3 and NH4 hiatuses at site 214 probably suggests relatively less corrosive bottom water activity prevailed in the eastern Indian Ocean region since middle Miocene.

The early Pliocene hiatus spanning 5.5-4.5 Ma identified at site 220 approximately correlates with the hiatus NH7 (5.2-4.7 Ma). The planktic foraminiferal record of site 220 exhibits a number of dissolution intervals resulted probably due to its greater depth (>4000 m). Therefore, intensified carbonate dissolution during early Pliocene appears to be a plausible cause for the formation of this hiatus. This hiatus is found to be synchronous with the interval of intense carbonate dissolution (5.2-4.7 Ma) identified by Dunn and Moore (1981) in the equatorial Pacific deep sea cores. A considerable higher accumulation rate of sediments during periods younger than 7.5 m.y. suggests increase in biogenic productivity, due to upwelling and increase in terrigenous input as a result of global lowering of sea level (Keller and Barron, 1987; Srinivasan and Singh, 1992).

Conclusions

Improved biostratigraphic refinements of Neogene deep sea sequences led to identify four major deep sea hiatuses and two strong dissolution intervals in the northern Indian Ocean. The deep sea hiatus occurs at 16.0-14.0 Ma, 13.0-1 1.0 Ma, 10.5-1 0.0 Ma and 5.5-4.5 Ma. These hiatuses broadly correlate with NH2, NH3, NH4 and NH7 respectively of Keller and Barron (1987). The hiatuses corresponding to NH2 and NH3 are of longer duration while NH3 and NH7 are of shorter duration. Two intervals of intense carbonate dissolution occur during 20.0-17.0 Ma and 14.0-13.0 Ma. The large hiatuses indicate physical erosion and transport of sediments by powerful currents, most likely related to Antarctic Bottom Waters (AABW).

Correlation of hiatuses with quantitative planktic foraminiferal data and oxygen stable isotope records reveals these erosive events correspond to intervals of positive excursion in $\delta^{18}O$, indicating global climatic cooling and consequent lowering of sea level.

Acknowledgements

We thank Professor M.S. Srinivasan for stimulating discussions, encouragement and review of the paper. The Deep Sea Drilling is thankfully acknowledged for providing the core samples. RM is thankful to the administration of the National Centre for Antarctic and Ocean Research Goa for providing the necessary facilities.

References

BARRON, J.A. and KELLER. G. 1982. Widespread Miocene deep-sea hiatuses: Coincidence with periods of global cooling. Geology, 10, p. 577-581.

BERGER, W.H. 1972. Deep sea carbonates: dissolution facies and age depth constancy. Nature, 236, 392-395.

BERGGREN, W.A., KENT, D.V. and VAN COUVERING, J.A. 1985. Neongene geochronology and chronostratigraphy. The Neogene: Part 2, In: The chronology of the Geological Record (Ed. N.J. Snelling), Geol. Soc. Mem. No. 10, Blackwell Scientific Publication, Oxford, London, 211-260.

DUNN, D.A. and MOORE, T.C. Jr. 1981. Late Miocene-Pliocene (Magnetic Epoch 9–Gilbert Magnetic Epoch) Calcium-carbonate stratigraphy of the equatorial Pacific Ocean. Geological Soc. America Bull., Pt. II, 91, 408-451.

FLEISHER, R.L. 1974; Cenozoic planktonic foraminifera and biostratigraphy, Arabian Sea, Deep Sea Drilling Project, Leg 23A. In: Initial Reports Deep Sea Drilling Project, 23, 1001-1072.

HAQ, B.U., HARDENBOL, J. and VAIL, P.R. 1987. Chronology of fluctuating sea levels since the Triassic (250 million years ago to present). Science, 235, 1158-1167.

KELLER, G. 1983. Biochronology and paleoclimatic implications of middle Eocene to Oligocene planktic foraminiferal fauna. Mar. Micropaleont., 7, 463-486.

KELLER, G. and BARRON, J.A. 1983. Paleoceanographic implications of Miocene deep-sea hiatuses. Geological Soc. America Bull., 94, 590-613.

KELLER, G. and BARRON, J.A. 1987. Paleodepth distribution of Neogene deep-sea hiatuses. Paleoceanography, 2, No. 6, 697-713.

KENNETT, J.P. 1973. Middle and Late Cenozoic planktonic foraminiferal biostratigraphy of the southwest Pacific-DSDP Leg 21. In: Initial Repts. Deep Sea Drilling Project, XXI, Washington (Eds.) Burns, R.E. *et al.*, 575-639.

KENNETT, J.P. 1977. Cenozoic Evolution of Antarctic Glaciation, the Circum-Antarctic Ocean, and their impact on Global Paleoceanography. Jour. Geophy. Res., 82, No. 27, 3843-3860.

KENNETT, J.P. and WATKINS, N.D. 1975. Late Micene planktonic foraminiferal biogeography and paleoceanographic development of the Indo-Pacific region. In: The Miocene Ocean (Ed.) Kennett, J.P., Mem. Geol. Soc. America, 163, 196-236.

KENNETT, J.P., KELLER, G. and SRINIVASAN, M.S. 1985. Miocene planktonic foraminiferal biogeography and paleoceanographic development of the Indo-Pacific region. Mem. Geol. So. America, 163, 197-236.

KENNETT, J.P., HOUTZ, R.E., ANDREWS, P.B., EDWARDS, A.R., GOSTIN, V.A., HAJOS, M., HAMPTON, M.A., JENKINS, D.G., MARGOLIS, S.V., OVENSHINE, A.T. and PERCH-NIELSEN, K. 1975. Cenozoic paleoceanography in southwest Pacific Ocean, Antarctic glaciation and the development of the circum-Antarctic current. In: Initial Reports of the Deep Sea Drilling Project, 29, 1155.

LEDBETTER, M.T. and CIESIELSKI, P.F. 1986. Post-Miocene disconformities and paleoceanography in the Atlantic sector of the southern Ocean. Palaeogeography, Palaeoclimatology and Palaeoecology, 52, 184-214.

McGOWRAN, B. 1978. Stratigraphic record of early Tertiary oceanic and continental events in the Indian Ocean region. Mar. Geol., 26, 1-39.

MOORE, T.C. Jr, VAN ANDEL T.H., SANCETTA, C., and PISIAS, N. 1978. Cenozoic hiatuses in pelagic sediments. Micropaleontology, 24, No. 2, 113-128.

MILLER, K.G. FAIRBANKS, R.G., and MOUNTAIN, G.S. 1987. Tertiary oxygen isotope synthesis, sea level history, and continental margin erosion. Paleoceanography, 2, no. 1, 1-19.

RAMSAY, A.T.S., SYKES, T.J.S. and KIDD, R.B. 1994. (Waxing and waning) lyrical on hiatuses: Eocene-Quaternary Indian Ocean hiatuses as proxy indicators of water mass production. Paleoceanography, 9, No. 6, 857-877.

SAITO, T. 1984. Planktonic foraminiferal datum planes for biostratigraphic correlation of Pacific Neogene sequences. In: Pacific Neogene Datum Planes (eds.) Ikebe, N. and Tsuchi, R., University of Tokyo Press, Japan, 3-10.

SCLATER, J.G., MEINKE, L., BENNETT, A. and MURPHY, C. 1985. The depth of the ocean through the Neogene. Mem. Geol. Soc. America, 163, 1-20.

SEVRDRUP, H.W., JOHNSON, M.S. and FLEMING, R.H. 1942. The Oceans, their physics, chemistry and general biology. Prentice Hall, New York, 1987.

SINGH, A.D., 1990. Neogene planktonic foraminiferal biochronology and paleoceanography of DSDP sites 219, 220 (Arabian Sea) and 237, 238 (central Indian Ocean). Ph.D. thesis (unpublished).

SINGH, A.D. and SRINIVASAN, M.S. 1995. Neogene planktic foraminiferal biochronology of the Central Indian Ocean DSDP sites 237 and 238. Jour. Geol. Soc. India, 45, 445-462.

SPENCER-CERVATO, C., 1998. Changing depth distribution of hiatuses during the Cenozoic. Paleoceanography, 13, No. 2, 178-182.

SRINIVASAN, M.S. and SINHA, D.K. 1992. Late Neogene planktonic foraminiferal events of the Southwest Pacific and Indian Ocean: A comparison. In: V – CPNS – IGCP – 246 Congress, 203-220.

SRINIVASAN, M.S. 1996. Oceanic Micropaleontology: Major advances since the beginning of DSDP, Presidential Address XV Indian Colloq. Micropal. Strat. Dehra Dun, 267-273.

SRINIVASAN, M.S. and CHATURVEDI, S.N. 1988. Neogene planktic foraminiferal biostratigraphy of DSDP Site along the crest of the Ninety east Ridge, Northern Indian Ocean. 75th Session, Indian Sci. Congress, Pune, Section of Geology and Geog. (Abst.), 10.

SRINIVASAN, M.S. and CHATURVEDI, S.N. 1992. Neogene planktonic foraminiferal biochronology of the DSDP sites along the Ninetyeast Ridge, Northern Indian Ocean. Cent. Japanese Micropal., Eds. K. Ishizaki and T. Saito, 175-188.

SRINIVASAN, M.S. and SINGH, A.D. 1991. Planktonic foraminiferal evidence for the Neogene paleoceanographic changes in the Indian Ocean. In: Ann. Tech. Meet. And IGCP-246, Chiang Mai Univ., Thailand, 179-205.

SRINIVASAN, M.S. and SINGH, A.D. 1992. Neogene planktonic foraminiferal biochronology of DSDP Site 219 (Chagos-Laccadive Ridge), Arabian Sea. Proc. Indian Natn. Sci. Acad., 58A, No. 4, 335-354.

TAKAHASHI, K. 1995. Opal particle flux in the subarctic Pacific and Bering Sea and Sidocoenosis preservation hypothesis. In: Global fluxes of Carbon and its related substances in the coastal Sea-Ocean-Atmosphere System, Proceedings of the 1994 Sapporo IGBP Symposium, S.Tsunogai *et al.*, (Eds.). M & International, Yokohama, Japan, 458-466.

VINVENT, E. 1977. Indian Ocean Neogene planktonic foraminiferal bistratigraphy and its paleoceanographic implications. In: J. Heirtzler (Ed). Indian Ocean Geology and Biostratigraphy: Am. Geophy. Union Mem., 469-584.

VINCENT, E., KILLINGLEY, J.S. and BERGER, W.H. 1985. Miocene oxygen and carbon isotope stratigraphy of the tropical Indian Ocean. In: Kennett, J.P. (Ed.) The Miocene Ocean. Geol. Soc. America Mem., 163, p. 103-130.

WOODRUFF, F. 1985. Changes in Miocene deep sea benthic foraminiferal distribution in the Pacific Ocean: relationship to paleoceanography. In: Kennett, J.P. (Ed.) The Miocene Ocean. Geol. Soc. America Mem., 163, p. 131-176.

WOODRUFF, F., SAVIN, S.M. and DUGLAS, R.G. 1981. Miocene stable isotope record: A detailed deep sea Pacific Ocean study and its paleoclimatic implications. Science, 21, p. 605-668.

Significance of stable oxygen ($\delta^{18}O$) and carbon ($\delta^{13}C$) isotopic compositions of individual foraminifera *O. universa* d'Orbigny in a sediment core from the Eastern Arabian Sea

R. Ramesh and Manish Tiwari

Physical Research Laboratory, Navrangpura, Ahmedabad 380009, India

Introduction

Stable Isotopes

Stable isotopes are called so because they do not decay during the known lifetime of the universe. A number of elements have stable isotopes, in addition to radioisotopes of various half-lives. For example, carbon has two stable isotopes ^{12}C and ^{13}C (Craig, 1953), in addition to the radioisotope ^{14}C. The radioisotope is well known to geologists, who use it to date Late Quaternary sediments. Stable isotopes on the other hand, are used for studying the physico-chemical processes that take place in nature and thus play a very useful role in paleoceanographic research (e.g., Srinivasan, 1998)..

Abundance of a stable isotope refers to the proportion in which it is present relative to the other stable isotopes of the same element. It is expressed either as a fraction or a percentage. For example, the abundances of ^{13}C and ^{12}C are 1.11% and 98.89% respectively (0.0111 and 0.9889, if expressed as fractions). In the case of light elements (mass number less than 50), the abundance of the heavy isotope is usually much less than that of the light isotope, as in the example above. Therefore reference is made to *isotope ratios*: $^{13}C/^{12}C$, $^{18}O/^{16}O$, $^{15}N/^{14}N$, etc. These ratios are small numbers normally, very much less than 1. In the above example, the ratio is 0.0112. This is the approximate mean terrestrial ratio. However, carbon is present in different forms in terrestrial reservoirs: as carbon dioxide gas in the atmosphere, dissolved inorganic (bicarbonate, carbonate and dissolved CO_2) and organic carbon in the hydrosphere (oceans, lakes and rivers), trapped CO_2 in the cryosphere (glaciers and ice sheets), sedimentary inorganic (carbonates) and organic carbon in the lithosphere (rocks) and as a fluid in the mantle. The isotope ratios in these different reservoirs are different depending on the how the reservoirs formed and the degree of *isotopic fractionation* that accompanied the process of formation.

During a physical or chemical process, the different isotopic species of carbon may not participate at the same rate. In the case of sublimation of dry ice, the lighter carbon dioxide ($^{12}CO_2$) will sublimate slightly faster than the heavier carbon dioxide ($^{13}CO_2$) because of the difference in their partial pressures, and therefore at any instant, there will be less of $^{13}CO_2$ in the gaseous phase. As they both have the same translational energy, i.e., $(3/2)k_BT$, (k_B, the Boltzmann's constant and T, the absolute temperature), their kinetic energies are the same ($0.5m_lv_l^2$ and $0.5\ m_hv_h^2$); therefore the heavier molecule of mass m_h has a lower speed v_h than the lighter molecule of mass m_l (of speed v_h). This process of modification of the isotope ratio during a process is called *isotopic fractionation*. The *isotopic fractionation factor* is defined as the ratio of the isotopic ratios: $\alpha = (^{13}C/^{12}C)_{gas}/(^{13}C/^{12}C)_{solid}$. This number is usually close to one, sometimes slightly less and sometimes slightly more than one, depending on how we define α (gas/solid or solid/gas). In the example under discussion, it is obvious that the value of α will be less that one, but close to one (e.g., 0.9990). To avoid dealing with such small numbers the *per mil* notation is used: unity is subtracted from α and the result multiplied by 1000. This is denoted by another symbol $\varepsilon = (\alpha - 1)X1000$. If $\alpha = 0.9990$, ε will be −1 per mil, denoted by the symbol ‰. E.g., 10‰ means 10 out of 1000 and is the same as 1%. In the above example, the minus sign in -1‰ signifies that the isotope ratio in the gas phase is 1‰ (equivalently, 0.1%) *less* than that in the solid phase. If $\alpha > 1$, then ε will be positive. e.g., if $\alpha = 1.005$, then $\varepsilon = 5$‰. This signifies that the phase corresponding to the numerator in our definition of α has an isotope ratio 5‰ (equivalently, 0.5%) *more* than the other phase. Likewise during chemical reactions, which involve breaking and making of bonds, the lighter isotopic species participate at a slightly greater rate than the heavier isotopic species, thus leading to isotopic fractionation (molecules containing the heavier isotopes have a lower zero point energy and are more difficult to dissociate). Here the fractionation factor is defined as the ratio of isotope ratios in the products and reactants, somewhat similar to the definition of chemical equilibrium constant. Just as the equilibrium constant is dependent on temperature, the isotopic fractionation factor is also *temperature dependent*. This forms the basis for calculating palaeotemperature from geological deposits (Urey, 1947, a Nobel Laureate).

In general, if two substances *a* and *b* (e.g., water and its vapour, calcium carbonate and the water from which it precipitates, such as foraminiferal shells and seawater, speleothems and drip water in caves, atmospheric carbon dioxide and the carbon dioxide dissolved in surface ocean etc.) are in *isotopic equilibrium* (i.e. each substance has acquired a constant isotopic ratio after exchanging isotopes for some time, this time being known as *isotopic equilibration time*), the fractionation factor α defined as R_a/R_b, is known as *equilibrium fractionation factor*, R being the isotope ratio of interest (e.g., $^{13}C/^{12}C$). For this the system must be closed, i.e., no addition/loss of isotopes from a third substance/reservoir. In the sublimation of dry ice, if it is kept in vacuum at a temperature just close to the sublimation temperature in a closed vessel, the sublimation will stop after some time because of the overhead pressure build-up. In such a case, both the dry ice and gaseous CO_2 will come to "isotopic equilibrium", i.e., the ratio of their isotopic ratios will attain a constant value (need not be equal to one). On the other hand, if the vessel is open, CO_2 moves away and is lost, the fractionation will be kinetic.

In the case of oxygen, there are three stable isotopes: ^{16}O, ^{17}O and ^{18}O. Their abundances are: 99.63%, 0.0375% and 0.1995% respectively. In earth sciences until recently, the minor isotope ^{17}O was not used because of the difficulty in measuring such a small abundance. The most widely used oxygen isotope ratio is $^{18}O/^{16}O$. This ratio has a mean value of ~0.002 (same as ~2000 ppm, meaning parts per million).

Another example of *kinetic fractionation* is the evaporation of water from the sea surface. Water vapour is depleted in the heavier isotope of oxygen (^{18}O). Winds take away the evaporated vapour and the remaining water gets progressively enriched in the heavier isotope. For e.g., during the ice age, ~21000 years ago, water was removed from the oceans by the hydrological cycle and stored as ice sheets on continents, therefore, the ocean water was enriched in ^{18}O.

Stable isotope ratios of light elements are usually measured using a gas source mass spectrometer. A solid or liquid sample must be converted into a suitable gas. For measuring the carbon isotope ratio of a carbonate (e.g., foraminifera), it is reacted in vacuum with 100% orthophosphoric acid in vacuum and the liberated CO_2 gas is collected quantitatively for the measurement of carbon and oxygen isotope ratios. Organic carbon is combusted with CuO in vacuum to obtain CO_2 gas. To measure the oxygen isotopic composition of a water sample, a CO_2 gas of known amount is isotopically equilibrated with the water and the CO_2 gas is then transferred to a mass spectrometer for isotope ratio measurement. In the stable isotope mass spectrometer there is a standard gas known as the 'reference'. The isotope ratios of the sample and the reference are measured in rapid succession repeatedly and the per mil difference in their isotope ratios is expressed as a *delta value*.

$$\delta^{13}C^{sam}_{std} = [\,(^{13}C/^{12}C)_{sam}/(^{13}C/^{12}C)_{std} - 1]X1000\ ‰ \quad (1)$$

$$\delta^{18}O^{sam}_{std} = [\,(^{18}O/^{16}O)_{sam}/(^{18}O/^{16}O)_{std} - 1]X1000\ ‰ \quad (2)$$

(sam = sample; std = standard)

The standard CO_2 gas used for measurement can be any arbitrary gas, provided its isotope ratio is within ~20‰ of the samples to be measured. These arbitrary laboratory standard gases are then calibrated with respect to international standards supplied by NBS (the National Bureau of Standards, USA) or IAEA (The International Atomic Energy Agency, Vienna, Austria). If the sample's δ value measured relative to the laboratory standard gas, is to be related to the international standard, the following relationship is useful:

$$\delta^{x}_{I} = \delta^{x}_{l} + \delta^{l}_{I} + 10^{-3}\delta^{x}_{l}\delta^{l}_{I} \quad (3)$$

where x = sample, I = International standard and l = laboratory standard.

As two different standards are used for oxygen isotopes (SMOW and PDB, in modern terminology, called V-PDB and V-SMOW, V standing for Vienna), they are mutually convertible using the formula:

$$\delta^{18}O^{x}_{SMOW} = 1.03086\ \delta^{18}O^{x}_{PDB} + 30.86 \text{ for carbonates, and} \quad (4)$$

$$\delta^{18}O^{x}_{SMOW} = \delta^{18}O^{x}_{PDB} + 0.2 \text{ for water samples.} \quad (5)$$

(the equations are different because of the different procedures followed for water and carbonates, see above).

It is important to note that in the mass spectrometer, the gases analysed yield the molecular abundance ratios and not the atomic abundance ratios. Therefore, one calculates the atomic abundance ratios from the measured molecular abundance ratios using the following procedure (Craig, 1957):

For example, to determine the $^{13}C/^{12}C$ ratio in organic carbon we combust the sample and produce CO_2 gas. The mass spectrometer essentially identifies ions of a particular [mass / charge] ratio. Assume that the ions are only singly charged (others filtered out). Then we have masses 44 (from $^{12}C^{16}O^{16}O$), 45 (from $^{13}C^{16}O^{16}O$ and also from $^{12}C^{17}O^{16}O$), 46 (from $^{13}C^{17}O^{16}O$, $^{12}C^{18}O^{16}O$ and $^{12}C^{17}O^{17}O$) and so on. We measure the ratios of abundances of masses: 45/44 and 46/44 (in a triple collector mass spectrometer). We are interested in getting the $^{13}C/^{12}C$ and $^{18}O/^{16}O$ ratios. If there is no contribution from $^{12}C^{17}O^{16}O$, we would get the $^{13}C/^{12}C$ ratio straight away as the 45/44 ratio measured. But there is always a finite contribution from $^{12}C^{17}O^{16}O$. This is known as *isobaric interference* (interference from another species of the same mass). Similarly we must correct for the isobaric interference form $^{12}C^{18}O^{16}O$ and $^{12}C^{17}O^{17}O$ for obtaining the $^{18}O/^{16}O$ ratio. These corrections were worked out by Craig (1957) and are given below for the triple collector mass spectrometer in a form that is directly usable:

$$\delta^{13}C = 1.0677\ \delta_{45} - 0.0338\ \delta_{46} \quad (6)$$

$$\delta^{18}O = 1.0011\ \delta_{46} - 0.0022\ \delta_{45} \quad (7)$$

where, δ_{45} and δ_{46} are defined as the measured δ values: $[(45/44)_{sam} / (45/44)_{std} - 1]$ X1000 and $[(46/44)_{sam} / (46/44)_{std} - 1]$ X1000 respectively. It is seen that when $\delta_{45} = 1‰$ and $\delta_{46} = -1‰$, then $\delta^{13}C = 1.1‰$ and $\delta^{18}O = -1.0‰$, i.e., the correction is small. However, when $\delta_{45} = -30‰$ and $\delta_{46} = 10‰$, then $\delta^{13}C = -32.37‰$ and $\delta^{18}O = 11.02‰$, i.e., the correction is not small.

Foraminifera

Foraminifera are eukaryotic (possessing a distinct nucleus), unicellular (single celled characterized by the absence of tissue or organs) organisms belonging to the phylum *Protozoa*. They secrete calcium carbonate shells of incredible beauty and structural complexity, and are preserved in sea sediments as fossils. Their size ranges from 0.01 mm (10µm to 10 mm but the average representative size range is 0.1 mm (100µm to 1.0 mm. The living organism consists of protoplasm encapsulated in the shell, which is further divided into different chambers. The wall separating one chamber from another is known as a *septum*. There is a hole in the septum through which the protoplasm extends throughout the shell. This hole is called as *foramen*, from which the name foraminifera (Latin, *foramen* = hole; *ferre* = to bear) is derived. The protoplasm extends outside the test through an opening known as *aperture* and surrounds the shell as a network of branching pseudopodia. The typical life spans of foraminifera are 2 to 4 weeks. Foraminifera are further classified into "spinose" and "non-spinose" forms. "spinose" forms are those that bear spines and "non-spinose" forms lack spines. Usually spines get broken and do not get fossilized with the shell.

Nearly all of the foraminifera are marine organisms and are found at all the latitudes. The highest diversity occurs in the tropics, which decreases with increasing latitude to a single species in the true Antarctic and Arctic waters. They are both planktic and benthic. Benthics are the bottom dwelling form and have existed since the Cambrian time (~570 Ma). They may be sessile i.e. constantly attached to the bottom or vagile, i.e. free bottom-dwelling organisms. They inhabit all depths ranging from the continental shelf to the abyssal plain. Planktic foraminifera float in the water column. They are more recent in origin than benthics and have existed since the Jurassic (~200 Ma). Most of the planktic species inhabit the top 150 m of the water column but their habitat can go upto 1000 m depth. On the basis of depth habitats, planktic foraminifera can be grouped into three categories *viz*., "shallow-water", "intermediate-water" and "deep-water" dwelling species. The "shallow-water" species live predominantly in the upper 50 m of the water column. They are mainly spinose forms and include all the species of genus *Globigerinoides* and some species of *Globigerina*. The "intermediate-water" species include both spinose forms and non-spinose forms that inhabit the upper 100 m but predominantly dwell in the 50-100 meters of the water column. The "deep-water" group consists of those species, which live in the euphotic zone as the juveniles, and predominantly below 100 m as adults: e.g. all the species of genus *Globorotalia*. Several planktic foraminifera called as "symbiont bearing" are known to host algae as symbionts within their shells. The species that host dinoflagellates (unicellular, 5-10 µm size, brownish-yellow colour) are *Globigerinoides.ruber*, *G. sacculifer*, *G. conglobatus* and *Orbulina universa*. The species hosting chrysophycophytes (yellow-green eukaryotic algae, 2-3 µm size) are *Globorotalia. siphonophera* and *G. humilis*. There are some species that are known to sometime possess or lack chrysophycophytes as symbionts. They are called as "symbionts facultative" and include the species *Globorotalia. inflata*, *Globorotalia. menardii*, *Neogloboquadrina dutertrei*, *Pulleniatina obliquiloculata* and *Globigerinita .glutinata*. On the other hand, species like *Hastigerina. pelagica*, *Neogloboquadrina pachyderma*, *Globigerina bulloides*, *Globorotalia truncatulinoides* and *G .hirusta* are not known to host symbionts and are called as "symbiont barren" (Hemleben et al, 1988). Studies of foraminifera are immensely useful in paleoceanography (e.g., Srinivasan, 1998, 2000, 2003).

Radiocarbon dating

The chronologies of the late Quaternary sediment cores are obtained using the Radiocarbon dating method (Libby, 1955, another Nobel Laureate). Radiocarbon (^{14}C) continuously produced in the atmosphere by the action of cosmic rays (energetic charged particles coming from outer space) on the atmospheric nitrogen nuclides is eventually oxidized to $^{14}CO_2$. The chemical cycle of the CO_2 in the ocean is governed by a series of equilibria shown below:

$$^{14}CO_2\,(g) \leftrightarrow {}^{14}CO_2\,(aq.) \leftrightarrow H_2{}^{14}CO_3 \leftrightarrow H^+ + H^{14}CO_3^- \leftrightarrow 2H^+ + {}^{14}CO_3^{2-}$$

(dissolved) (carbonic acid) (bicarbonate ion) (carbonate ion)

The bicarbonate ions, which are the dominant species in the oceans at the prevailing pH (~8), are removed by the microorganisms to secrete calcium carbonate shells:

$$2HCO_3^- + Ca^{2+} \Leftrightarrow H_2O + CaCO_3\downarrow + CO_2\uparrow$$

These calcitic shells of microorganisms contain the radioactive carbon, and are used for dating. Only planktic foraminifera are picked up for AMS (accelerator mass spectrometry facility is coming up at the Institute of Physics, Bhubaneswar) dating, *viz*., *Globigerinoides ruber, Globigerinoides sacculifer, Orbulina universa* and *Neogloboquandrina dutertrei*. This is because they inhabit the surface and near surface oceans (up to ~100 m) and therefore readily incorporate changes occurring in the surface ocean into their shells. For AMS radiocarbon dating, only ~10mg of foraminifera are required. This means approximately 200 individuals have to be handpicked for every date. (Linick et al, 1986; Somayajulu et al, 1999). The fundamental difference between conventional radiocarbon dating such as Liquid scintillation method (Yadava and Ramesh, 1999) and the AMS technique is that in the former, the rate of decay of the atom is counted whereas in the latter, the radioactive atoms themselves are counted. This allows us to date even small samples, and thus helps in improving the accuracy because various contaminants usually present, are not present here. AMS differs from the conventional MS in the energies to which the ions are accelerated. In normal mass spectrometers the energies are in thousands of eV (keV) whereas in the AMS they are in millions of eV (MeV). The practical consequence of having higher energies is that ambiguities in the identification of the atomic and molecular ions with the same mass (isobaric effects) are removed. Thus we can detect isotopes of specific elements to a level of 1 in 10^{15}.

The surface bicarbonate of the ocean today does not have zero age, because of mixing with deeper waters of lower radiocarbon content. This is known as the reservoir effect and has to be corrected. ΔR is the value that has to be added to the global mean reservoir age (400 yrs., Dutta et al., 2001) to obtain the age correction needed for the local reservoir effect in the Arabian Sea. The radiocarbon ages thus obtained are calibrated to calendar ages using the calibration program "CALIB 4.1 (INTCAL 98)" (Stuiver et al, 1998). Radiocarbon ages beyond 24 ka BP are also calculated using the "glacial polynomial" (Bard, E. 1998) but the age difference is only 2 years; so the CALIB 4.1 ages are retained.

Fig.1 presents the difference between bulk radiocarbon ages and AMS ages in a sediment core, SK-145-9 from the eastern Arabian Sea, very near to the site of the core 3268G5 studied by Sarkar et al (2000). It is clearly seen that bulk ages are older by a few thousand years, due probably to dead carbon contribution from detrital carbonate. This contribution also varies with time. For example, about 13 ka ago, there was no difference, which implies that the detrital contribution was absent; perhaps the monsoon was weak. But it is known that the monsoon intensified from the beginning of the Holocene (10 ka). Probably the detrital contribution also increased since then.

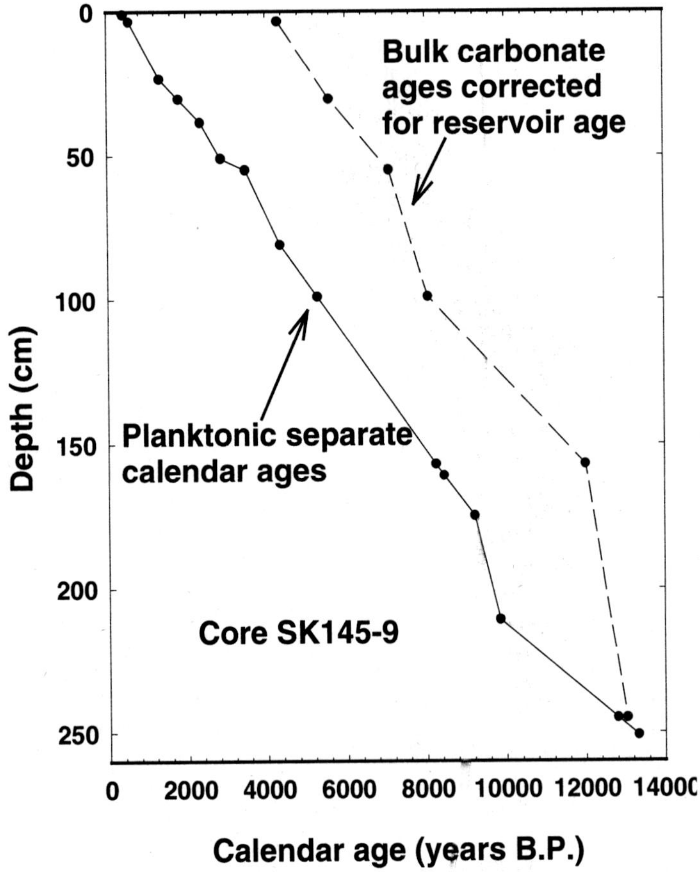

Fig. 1. Comparison of conventional and AMS ^{14}C ages in a sediment core from the eastern Arabian Sea. AMS dates courtesy: A.J.T. Jull and G.S. Burr of University of Arizona, Tucson. Bulk dates obtained at PRL radiocarbon laboratory; courtesy: B.L.K. Somayajulu.

Oxygen isotopes

In the case of foraminifera, calcite is precipitated from seawater and the following isotopic exchange process takes place:

$$(1/3)\ CaC^{16}O_3 + H_2{}^{18}O \leftrightarrow (1/3)\ CaC^{18}O_3 + H_2{}^{16}O$$

The equilibrium constant "K" for this reaction can be written as:

$$K = [CaC^{18}O_3]^{1/3}[H_2{}^{16}O] / [CaC^{16}O_3]^{1/3}[H_2{}^{18}O]$$

or, $K = ([CaC^{18}O_3] / [CaC^{16}O_3])^{1/3} / ([H_2{}^{18}O] / [H_2{}^{16}O])$

Thus equilibrium constant can be expressed as the ratio of $^{18}O/^{16}O$ in the carbonate phase to that in the water. In the $CaCO_3$-H_2O system the fractionation factor is defined as:

$$\alpha_{CaCO3-H2O} = R_{CaCO3} / R_{H2O} = 1.031 \text{ at } 25^{\circ}C$$

R_{CaCO3} is $^{18}O/^{16}O$ in calcite and R_{H2O} is $^{18}O/^{16}O$ in water. The fractionation factor is related to equilibrium constant by:

$$\alpha = K^{1/n}$$

where "n" is the number of atoms exchanged. For the sake of simplicity, isotope exchange reactions can be written in the way that only one atom is exchanged and then fractionation factor "α" becomes identical to the equilibrium constant "K", as shown above. The fractionation factor is temperature dependent in the way that calcites precipitated from water of constant isotopic composition but at different temperatures will possess different $^{18}O/^{16}O$ ratios. This is the basis for the paleotemperature scale, by which various empirical paleotemperature equations are deduced. The overall effect is that ~0.2 to ~0.25‰ depletion in carbonate $\delta^{18}O$ occurs for every $1^{0}C$ temperature increase. On the other hand, if temperature is held constant, calcites precipitated from two waters that differ in their $\delta^{18}O$ by some amount will have the same difference in their $\delta^{18}O$ values. For instance, it is easy to see that in the tropics, sea surface temperature is ~30°C and the planktic foraminifera have a $\delta^{18}O$ of ~ −2‰, whereas benthic forams which live in the deep sea at ~2°C, have $\delta^{18}O$ values close to +5‰.

For oxygen isotopic studies two international standards viz. SMOW and PDB are in use. SMOW (Standard Mean Ocean Water) is a hypothetical standard with $\delta^{18}O$ value close to the modern mean seawater value. The value of SMOW was defined by Craig (1961) with respect to an existing water standard NBS-1 (distilled Potomac river water) with $\delta^{18}O$ value of −7.94 ‰ vs. SMOW. As SMOW does not actually exist as a real water sample so it can't be used for calibrating laboratory measurements. Therefore a water sample with values identical to SMOW is distributed by IAEA, Vienna that is called as V-SMOW. Harmon Craig of the University of California (winner of the Belize prize, equivalent of Nobel Prize, as there is no earth science category in the latter) prepared it by mixing distilled ocean water (collected from the Pacific Ocean, 0^{0}, 180^{0} W in July1967) with small amounts of other waters to adjust the isotopic ratio to the required value (Gonfiantini R, 1981). It has a $\delta^{18}O$ value of 0 ‰ with respect to SMOW (by definition). This scale has been used for oxygen isotope analyses involving waters, silicates, phosphates, high temperature carbonate etc. The PDB standard is used in the low temperature carbonate studies. It is a carbonate obtained from fossilized rostrum (cigar shaped internal shell) of *Belemnitella americana*, a belemnite (an extinct cephalopod) from the PeeDee Formation of Cretaceous Period in South Carolina, USA. It was the laboratory working standard used by Urey's group in the University of Chicago in the 1950s for developing the initial paleotemperature scale (McCrea, 1950). Over the decades the original PDB standard has been exhausted and is no longer available. To overcome this problem IAEA introduced a hypothetical standard Vienna-PDB (V-PDB) with values nearly identical to PDB and it is defined by its relationship to carbonate materials NBS-19 and NBS-20. NBS-19 is calcium carbonate from a marble of geologically unidentified origin with $\delta^{18}O$ and $\delta^{13}C$ values of −2.20 ‰ and +1.95 ‰ respectively with respect to V-PDB. NBS-20 is a homogenized sample of Solnhofen limestone of Jurassic Period from Southern Germany with $\delta^{18}O$ and $\delta^{13}C$ values of −4.14 ‰ and −1.06 ‰ with respect to V-PDB (Hoefs, 1997). The conversion equations of $\delta^{18}O_{(PDB)}$ to $\delta^{18}O_{(SMOW)}$ and *vice versa* is (Coplen et al,1983):

$$\delta^{18}O_{(SMOW)} = 1.03091\ \delta^{18}O_{(PDB)} + 30.91 \qquad (8)$$

$$\delta^{18}O_{(PDB)} = 0.97002\ \delta^{18}O_{(SMOW)} - 29.98 \quad (9)$$

Erez and Luz (1983) determined the following empirical temperature equation by comparing isotopic composition of the planktic foraminifera *Globigerinoides sacculifer* with the actual growth temperature:

$$t\ (^{o}C) = 17.0 - 4.52\ (\delta^{18}O_{c} - \delta^{18}O_{w}) + 0.03\ (\delta^{18}O_{c} - \delta^{18}O_{w})^2 \quad (10)$$

Where t, $\delta^{18}O_c$, $\delta^{18}O_w$, are the estimated temperature and the isotopic compositions of the shell carbonate and the seawater respectively. But there are various complications that inhibit the widespread application of paleotemperature scale. To apply the paleotemperature equation we must know the $\delta^{18}O$ value of the seawater with which a given specimen of $CaCO_3$ had equilibrated. This is not easy to establish with certainty because the isotopic composition of seawater as a whole depends on the amount of ice stored on the continents, which gives rise to the so called "ice volume effect". When water evaporates from the ocean surface, the water vapour gets enriched in the lighter isotopes ($H_2{}^{16}O$), as vapour pressure of $H_2{}^{16}O$ is more than that of $H_2{}^{18}O$. Kinetic effects also take place that further depletes the vapour phase in the heavier isotope. As more and more evaporation takes place and as this isotopically lighter water gets locked in the form of continental ice sheets, the remaining ocean water gets more and more enriched in the heavier isotope. It is believed that during the Last Glacial Maximum the seawater $\delta^{18}O$ was 1.2 ‰ heavier than present (Labeyrie et al, 1987, Fairbanks, 1989).

The depletion of $\delta^{18}O$ in freshwater gives rise to the observed correlation of sea surface salinity with seawater $\delta^{18}O$. As evaporation takes place, the salinity as well as $\delta^{18}O$ of the seawater increases. Likewise when there is river runoff, there is reduction in both the surface salinity and the $\delta^{18}O$ of seawater. In the eastern Arabian Sea, ~1‰ decrease in surface salinity causes ~0.33 ‰ decrease in $\delta^{18}O$ of water (Duplessy et al., 1981, Sarkar et al, 2000). Fig. 2 presents the quantification of this effect (Ramesh, 2001), in terms of the excess of precipitation over evaporation (in mm).

Separately analyzing planktic and benthic foraminifera from the same site can help decouple the temperature and ice volume effects. The $\delta^{18}O$ values of the planktics reflect the changes in both the sea surface temperature and the $\delta^{18}O$ of surface seawater (the ice volume effect), whereas the bottom water is mostly believed to be unaffected by the temperature changes, hence the benthics are believed to record mostly the change in the isotopic composition of seawater (the ice volume effect). There are several effects associated with the biology of foraminifera that cause deviation from ratios that should be obtained if there were isotopic equilibrium. Such disequilibria effects can be classified as either metabolic or kinetic (McConnaughey, 1989 a, b). Metabolic effects result from the incorporation of ambient dissolved carbon and oxygen (produced due to respiration and photosynthesis, respectively) into the shell material. Kinetic effects are due to the preferential uptake of lighter isotopes during the hydration and hydroxylation of CO_2 and are generally associated with rapid calcification. These effects, collectively called "vital effects", include effects due to respiration, ontogeny, secretion of gametogenic calcite etc. The respiration products are depleted in the heavier isotopes (Lane and Doyle, 1956) and their utilization for shell secretion results in depleted $\delta^{18}O$ values. Progressive $\delta^{18}O$ enrichment from juvenile to mature chambers has been observed and is assigned to the incorporation of the respired CO_2 during calcification. Higher metabolic rates in the juvenile specimen cause stronger depletions, which then decrease in the adult specimen that has a reduced metabolism (Berger et al, 1978; Wefer and Berger, 1991). Furthermore, planktic foraminifera inhabit different depths at various stages of their ontogeny. In the later part of their lifecycles they tend to secrete shells at deeper

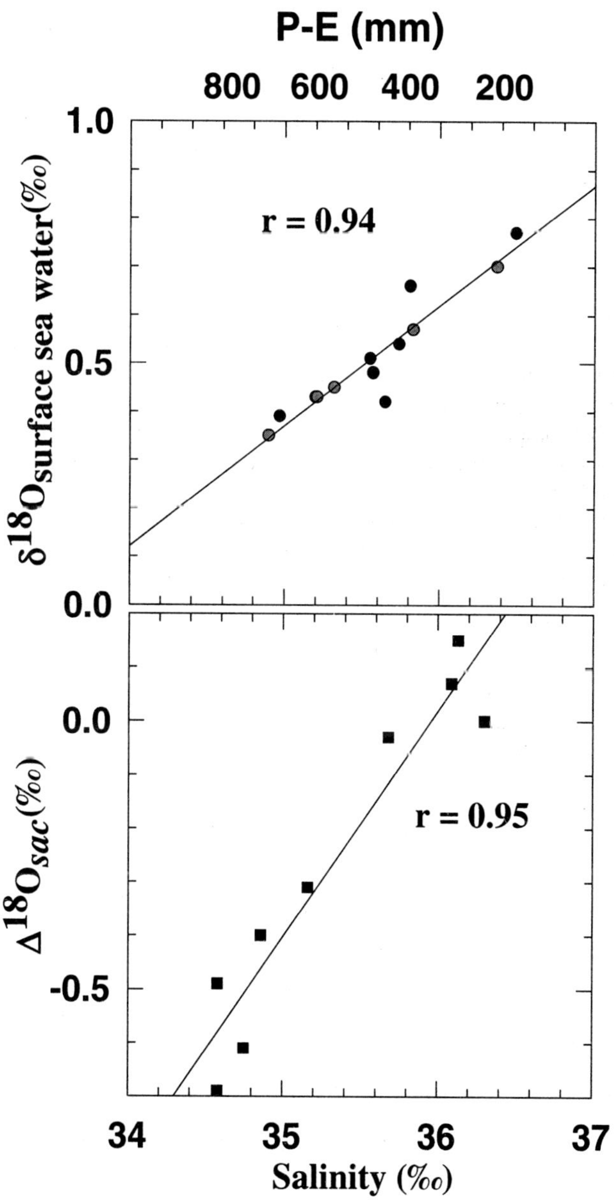

Fig. 2. Oxygen isotopic composition of surface sea water (top) and core top *G. sacculifer* (bottom) from the Eastern Arabian Sea, plotted against surface salinity. In the top P–E (excess of precipitation over evaporation) values in mm are marked (Data from Duplessy et al, 1981, and Sarkar et al, 2000).

and cooler waters resulting in enhanced $\delta^{18}O$ values (Bouvier-Soumagnaç and Duplessy, 1985; Emiliani, 1971).Towards the end of their life cycle foraminifera secrete gametogenic calcite. Before the gamete release, foraminifera move to deeper, cooler waters and secrete a layer of calcite that is enriched in $\delta^{18}O$ over its shell. This calcite layer can comprise 18 to 28% of the total shell mass of foraminifera (Be, 1980; Duplessy et al, 1981). Furthermore $\delta^{18}O$ has been found to vary with the varying light intensities (photosynthetic activities in the symbiont bearing foraminifera) and carbonate ion concentrations. $\delta^{18}O$ decreases with increasing irradiance that causes enhanced photosynthesis. Enhanced rate of skeletogenesis ensues increased photosynthesis that favours stronger kinetic fractionation resulting in more negative $\delta^{18}O$ (Spero and Lea, 1993; Wefer and Berger, 1991). $\delta^{18}O$ of foraminiferal carbonate decreases with increasing carbonate ion concentration, ascribed to abiological, kinetic fractionation effect (Spero et al, 1997) as it is also observed in rapidly precipitating inorganic $CaCO_3$ (McCrea, 1950). The effect of isotopic disequilibria due to 'vital effects' can be avoided by choosing some species known to precipitate their shells in isotopic equilibrium with seawater and by picking mature shells from a particular size range. Also, diagenetic alteration after the death of the organism can modify its isotopic composition significantly. The $\delta^{18}O$ values of planktic foraminifera from warm subtropical seas are particularly sensitive to alteration as the secondary calcite precipitates in equilibrium with the cold pore fluids at the sediment-water interface (Schrag, 1999). Shells may also recrystallize at the micron scale, yet preserving the shell features and ornamentation (Pearson et al, 2001). This obstacle can be overcome by selected pristine shells, which appear glassy under the microscope and are without any overgrowths. For a more detailed review see Rohling and Cooke (1999).

Carbon isotopes

Kinetic isotope effects during the photosynthesis cause a preferential uptake of ^{12}C in plant organic matter. On the other hand, isotope equilibrium exchange reactions within the inorganic carbon system "atmospheric CO_2 – dissolved bicarbonate – solid carbonate" cause enrichment in ^{13}C (Hoefs, 1997) in the carbonate. The carbonate chemical equilibria in the marine system can be represented by the following equations (Faure, 1986):

$$CO_2\ (g) \leftrightarrow CO_2\ (aq.)$$

$$CO_2\ (aq.) + H_2O \leftrightarrow H^+ + HCO_3^-\ (aq.)$$

$$CaCO_3\ (s) + H^+ \leftrightarrow Ca^{2+}\ (aq.) + HCO_3^-\ (aq.)$$

These equilibrium reactions govern the carbon and oxygen isotopic variations among various chemical species. Total dissolved inorganic carbon (DIC) comprises of HCO_3^- (aq.), CO_3^{2-} and CO_2 (aq.). The net equation for carbonate precipitation is:

$$Ca^{2+}\ (aq.) + 2HCO_3^-\ (aq.) \leftrightarrow H_2O + CaCO_3\ (s)\downarrow + CO_2\ (g)\uparrow$$

The equilibrium fractionation factors for carbon isotope in this system at 20°C are (Emrich et al, 1970):

$$\alpha_{CaCO3\ (s)-\ HCO3^-\ (aq.)} = 1.00185$$

$$\alpha_{HCO3^-\ (aq.)\ -\ CO2\ (g)} = 1.00838$$

$$\alpha_{CaCO3\ (s)\ -\ CO2\ (g)} = 1.01017$$

At low temperatures the largest fractionation occurs between the dissolved CO_2 and the bicarbonate ion. The bicarbonate ion is the dominant species at the prevailing pH of seawater (7.5 – 8.3) and DIC can be conveniently approximated by HCO_3^- (aq.). The isotopic enrichment factor for calcite vs. bicarbonate is (Grossman, 1984):

$$\varepsilon_{CaCO3-HCO3^-} (‰) = 10.51 - 2980/T, \text{ where T is temperature in Kelvin} \quad (11)$$

This relationship gives a temperature dependence of 0.035 ‰ increase in $\delta^{13}C$ per degree Celsius increase in temperature, which is same as that given earlier by Emrich et al (1970). Thus temperature has a negligible effect on $\delta^{13}C$ of calcite in comparison to the effect due to the organic productivity (the rate of fixation of carbon by photosynthesis) variations (Kroopnick, 1974). Phytoplanktons prefer lighter isotope (^{12}C) during photosynthesis enriching the ambient surface seawater in the heavier isotope (^{13}C). As the organisms die they settle to the deep, where they get oxidized releasing CO_2 depleted in ^{13}C to the water. Thus the DIC at the surface waters has the highest $\delta^{13}C$, while a minimum in $\delta^{13}C$ exists at base of the thermocline. The organisms secreting calcareous shells in isotopic equilibrium with the ambient water record such isotopic signatures. Thus a higher $\delta^{13}C$ value probably corresponds to an enhanced rate of photosynthesis in the euphotic layer, that indicates an increase in organic productivity in the surface ocean.

As $\delta^{13}C$ has a steep gradient in the water column, the species secreting their shells at different depths in isotopic equilibrium with the ambient water during their ontogeny will exhibit deviation from the surface seawater equilibrium values (Kroopnick et al, 1972). Isotopic disequilibria are also caused by "vital effect". Respiration products are depleted in heavier isotopes, the incorporation of which causes ^{13}C depletion relative to equilibrium values (Grossman, 1987; Weber and Woodhead, 1970; Vinot-Bertouille and Duplessy, 1973). Higher metabolic rates have been found to cause increased depletion in $\delta^{13}C$ values (Wefer and Berger, 1991, Oritz et al, 1996). Symbionts present in the foraminifera carry out photosynthesis in which they preferentially utilize ^{12}C. With increased irradiance, photosynthesis also increases, causing enhanced $H^{13}CO_3^-$ in the ambient microenvironment resulting in foraminiferal chambers enriched in ^{13}C (Spero and Lea, 1993). $\delta^{13}C$ in foraminiferal carbonate decreases with increasing carbonate ion concentration similar to the effect experienced by $\delta^{18}O$ that was ascribed to abiological, kinetic fractionation (Spero et al, 1997). Diagenetic alteration at the sediment-water interface or beneath the sediments lowers the $\delta^{13}C$ values, as the degradation of organic matter produces ^{13}C depleted CO_2, making the ambient water depleted in ^{13}C.

Experimental methods

Separation of foraminifera:

About 5-10 g of sediment samples are taken in a 400 ml beaker filled with ~200 ml of distilled water. Around ½ cc of 30% H_2O_2 is added to oxidize any organic matter that might be present in the sediments. Thereafter ½ spoonful of Calgon powder (sodium hexametaphosphate) is added. Calgon acts as a dispersing agent and helps in removing the agglutinated inorganic particles. They are then warmed at 60-80°C for an hour. Care is taken that the mixtures do not boil while heating. Thereafter the beakers are kept for 4-5 hours so that a total disaggregation of the sediments takes place. Subsequently wet sieving is carried out using sieves of the sizes 250 μm and 500 μm. The size fraction (500μm to 250 μm) is transferred to 50ml beakers containing ~25 ml distilled water and are ultrasonically cleaned to remove the remaining detrital particles sticking to the shells. The excess water is decanted and the shells were dried at 90°C. Thereafter they are kept in the furnace at 400°C for ~8 hours to remove any volatile organic matter still adhering to the shells. The dry bulk fraction thus obtained is stored in pre-cleaned plastic vials. The planktic species required are then handpicked under a stereoscopic microscope from the dry bulk fraction. While picking foraminifera care is taken to pick the shells falling approximately in the range 450 μm to 350 μm and all the chosen shells are clean, and more or less in the same size range.

$\delta^{18}O$ and $\delta\ ^{13}C$ measurements on the foraminifera

For measuring $\delta\ ^{18}O$ and $\delta\ ^{13}C$, CO_2 is fed into the mass spectrometer. CO2 preparation involves the acid decomposition of forams with 100% phosphoric acid at any chosen temperature between 25°C and 150°C (McCrea, 1950; Rosenbaum and Sheppard 1986; Swart et al, 1991) as shown below:

$$3CaCO_3 + 2H_3PO_4 \rightarrow 3\ CO_2 + 3H_2O + Ca_3\ (PO_4)_2$$

This reaction shows that only two-thirds of the carbonate oxygen is liberated in the product CO_2. The oxygen isotopic fraction factor for CO_2-calcite system during acid decomposition at 25°C is 1.01025 (Rosenbaum and Sheppard 1986). Many precautions have to be taken to follow the acid decomposition technique. The salt produced [$Ca_3(PO_4)_2$] should be soluble in the solution produced by the interaction of the acid with the carbonate, or else a protective covering will form on the latter, inhibiting further reaction. Also the acid should not produce radicals in the mass range 44-46, which excludes nitric acid (HNO_3) that decomposes to form nitrous oxide (NO_2). Most importantly, isotopic exchange between water and CO_2 is known to take place that necessitates an acid that has no free water (Wachter and Hays, 1985). For this reason 100% ortho-phosphoric acid is chosen as it has an extremely low vapour pressure and also the concentration of water is very low in comparison to other acids. The acid is prepared following Coplen et al (1983).

The mass spectrometer used in this study is the Europa–Scientific GEO 20 –20 stable isotope ratio mass spectrometer with an attached Carbonate Preparation System (CAPS). The schematic diagrams of CO_2 evolution using the CAPS and the mass spectrometer inlet system are shown in Fig. 3. CAPS system (top panel) had a factory fitted aluminum carousel that can hold upto 24 sample vials. The total CO_2 production assembly is maintained at 80°C for rapid acid digestion of the carbonates. At first, air is put into the acid reservoir by opening and closing of the valves V1 and V2. Thereafter sample vial is connected to the acid dosing assembly. Then the valves V4 and V3 are opened to evacuate the vial to medium vacuum of 10^{-3} Torr using a rotary pump. In the meantime, the rest of the preparation line is pumped to a high vacuum (~10^{-8} Torr) by keeping V6 closed and opening other valves that connect the lines to turbo pumps. When the total line is under vacuum and the temperature is 80°C, then the acid is dropped for 0.175 second (~0.5 ml) on the sample. V6 is closed as soon as the reaction is over, which takes around 1-2 minutes. The evolved gases are immediately removed from the reaction vial and are passed through the water trap between V6 and V7 that is kept at –100°C. Here H_2O molecules are frozen out and the remaining, moisture free CO_2 is frozen at the cold fingers (CF) using liquid nitrogen (-196°C). If the gas amount is very low then the CF near V9 is used or else CF near V10 is used. Any residual gases that may be present above the frozen CO_2 (at V9 or V10) are pumped out. This cryogenically cleaned CO_2 is then fed into the mass spectrometer inlet system.

The mass spectrometer sample inlet line (bottom panel, Fig. 3) is first brought to medium vacuum and then to high vacuum by opening V21 (connects to rotary pump) and V20 (connects to turbo pump) respectively. The V22 is closed and the sample gas is put into the variable sample reservoir by opening V11, V12 and V13. There the gas is left for 2 minutes for equilibration. Thereafter the V12 is closed and the gas is introduced into the mass spectrometer via a changeover valve by opening V14. The reference reservoir stores the reference gas, which is called as CD Standard. It is CO_2 prepared by reacting around 50g of a mixed foraminiferal assemblage from a sediment core (raised from the Arabian

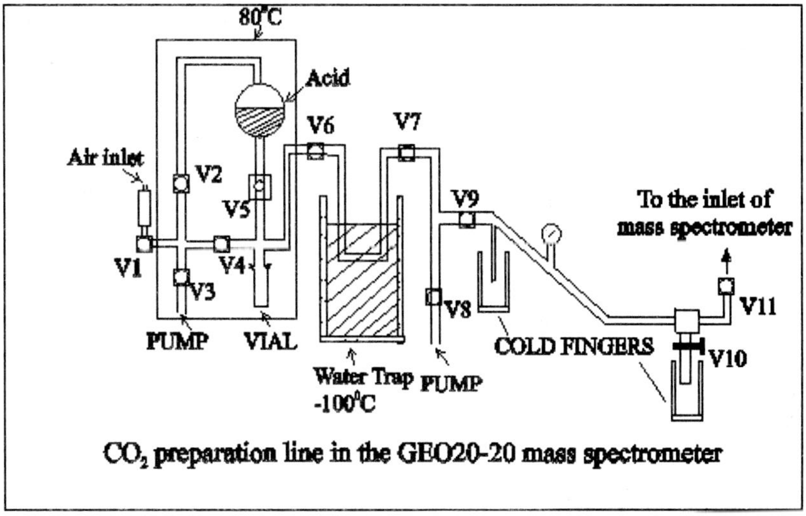

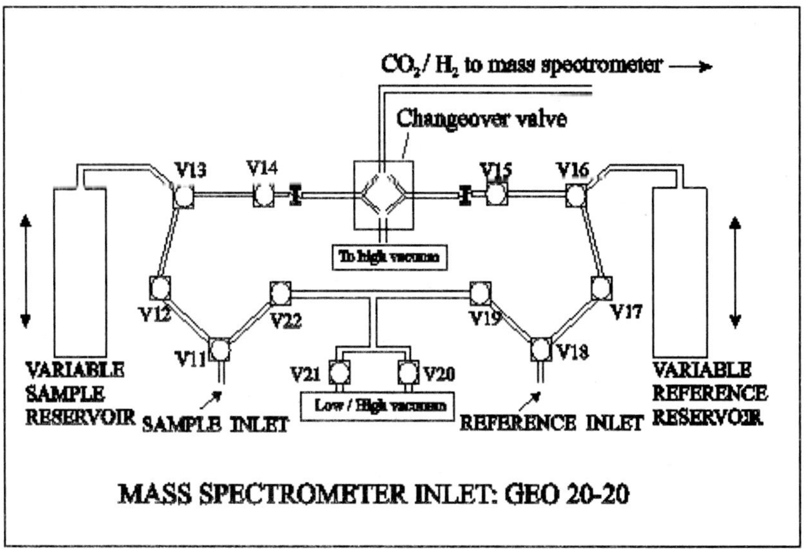

Fig. 3. Schematics: CO_2 preparation using CAPS (top), Mass Spectrometer inlet system (bottom).

Sea during a cruise of RV Charles Darwin in January 1997) with orthophosphoric acid at 25°C (Ghosh, 2000). Its $\delta\,^{18}O$ and $\delta\,^{13}C$ values with respect to PDB are 1.07‰ and 1.7‰ respectively. The changeover valve acts as dual inlet in the way that it allows the sample gas and reference gas to alternately enter the mass spectrometer at preset intervals and when one is connected to the mass spectrometer the other is connected to waste line so that both the reference and the sample gases face the same amount of depletion during the course of measurement. The variable volume reservoirs are adjusted to give a current of the major beam (mass 44) around 10 nanoamperes and the current is kept the same for both the sample and the reference by adjusting the volume of the variable reservoir.

GEO 20-20 mass spectrometer is a triple collector system having three faraday cups for collecting ions corresponding to three beams of masses 44, 45, and 46. Thus both $\delta\,^{18}O$ and $\delta^{13}C$ can be measured simultaneously under the same focusing conditions. The ion beams falling at the three collectors constitute currents and the ratio of appropriate currents are given as δ_{45} and δ_{46} :

$$\delta_{45} = [((I_{45}/I_{44})_{sam} / (I_{45}/I_{44})_{std}) - 1]\,10^3 \text{ and } \delta_{46} = [((I_{46}/I_{44})_{sam} / (I_{46}/I_{44})_{std}) - 1]\,10^3 \quad (12)$$

where I_{44}, I_{45} and I_{46} are the currents generated due the beams of masses 44, 45 and 46 respectively. δ_{45} and δ_{46} are then converted to $\delta\,^{13}C$ and $\delta\,^{18}O$ using the Craig correction (Craig, 1957).

In order to check the precision, a lab standard called as Z-Carrara was run at least three times daily viz. at the start of the measurement, in the middle and at the end of the day's measurement. It is a homogenized, fine carbonate powder (<63μm) from an Italian marble kindly provided by N.J. Shackleton. The $\delta^{13}C$ and $\delta^{18}O$ values of the Z-Carrara have been determined by repeated measurements and are 2.11‰ and -2.11‰ respectively. All the sample delta values are corrected with respect to Z-Carrara on a daily basis to account for any variation in the mass spectrometer and carbonate preparation line conditions. We measure sample (sam) as well as Z- Carrara (ZC) with respect to reference gas (rg) i.e. ${}_{rg}\delta^{sam}$ and ${}_{rg}\delta^{ZC}$. δ value for reference gas with respect to Z-Carrara can be calculated as follows:

$${}_{ZC}\delta^{rg} = \{(1/\,1 + {}_{rg}\delta^{ZC}\,10^{-3}) - 1\}\,10^3 \quad (13)$$

Now, sample δ values with respect to Z-Carrara can be calculated using the formula:

$${}_{ZC}\delta^{sam} = {}_{rg}\delta^{sam} + {}_{ZC}\delta^{rg} + {}_{rg}\delta^{sam}\,{}_{ZC}\delta^{rg}\,10^{-3} \quad (14)$$

Z-Carrara values with respect to PDB (${}_{PDB}\delta^{ZC}$) are known hence sample δ values with respect to PDB can be calculated as:

$${}_{PDB}\delta^{sam} = {}_{ZC}\delta^{sam} + {}_{PDB}\delta^{ZC} + {}_{ZC}\delta^{sam}\,{}_{PDB}\delta^{ZC}\,10^{-3} \quad (15)$$

Table 1 gives the $\delta^{13}C$ and $\delta^{18}O$ values of Z-Carrara standard measured from 19th June 2003 to 4th April 2004.

Table 1. Z-Carrara $\delta^{13}C$ and $\delta^{18}O$ values with respect to PDB in permil, measured between 19th July, 2003 and 4th April, 2004 (MN= Measurement No.)

MN	$\delta^{13}C$	$\delta^{18}O$	MN	$\delta^{13}C$	$\delta^{18}O$	MN	$\delta^{13}C$	$\delta^{18}O$	MN	$\delta^{13}C$	$\delta^{18}O$
1	1.98	-2.26	36	2.05	-2.07	71	1.94	-2.12	106	1.99	-2.11
2	2.04	-2.21	37	2.06	-2.03	72	1.92	-2.30	107	1.98	-2.14
3	1.99	-2.35	38	2.05	-2.02	73	2.10	-1.78	108	1.96	-2.17
4	1.90	-2.45	39	2.01	-2.21	74	1.98	-1.89	109	1.87	-2.17
5	1.92	-2.63	40	1.93	-2.44	75	1.97	-2.03	110	1.87	-2.25
6	1.84	-2.32	41	2.01	-2.14	76	2.02	-2.08	111	2.09	-1.79
7	1.95	-2.35	42	2.04	-2.00	77	1.96	-2.05	112	1.90	-2.15
8	1.96	-2.47	43	1.92	-2.32	78	1.96	-2.06	113	1.92	-1.87
9	1.92	-2.49	44	1.97	-2.29	79	2.03	-1.92	114	1.97	-2.25
10	1.99	-2.36	45	2.08	-1.98	80	2.22	-2.17	115	2.03	-1.97
11	1.92	-2.30	46	2.03	-1.96	81	2.18	-2.07	116	1.92	-2.16
12	2.03	-2.08	47	1.90	-2.24	82	2.25	-1.96	117	1.94	-2.11
13	2.10	-1.77	48	1.81	-2.51	83	2.07	-2.13	118	1.97	-2.00
14	2.03	-2.12	49	1.89	-2.26	84	1.94	-2.09	119	1.85	-2.19
15	2.06	-1.89	50	1.89	-2.32	85	2.04	-2.22	120	1.95	-2.17
16	1.99	-2.00	51	1.82	-2.34	86	1.99	-2.31	121	2.02	-2.00
17	2.11	-1.85	52	1.98	-2.22	87	2.12	-2.14	122	1.88	-2.22
18	1.98	2.13	53	2.09	-1.73	88	2.06	-2.35	123	1.94	-2.09
19	1.99	-2.09	54	2.04	-1.90	89	2.02	-2.43			
20	1.99	-1.93	55	2.04	-2.11	90	2.02	-2.21			
21	1.90	-2.35	56	2.02	-1.95	91	1.84	-2.36			
22	2.07	-1.76	57	1.99	-2.06	92	1.98	-2.12			
23	2.08	-1.95	58	1.96	-2.08	93	2.02	-2.22			
24	2.10	-1.82	59	2.03	-2.00	94	1.95	-2.24			
25	2.10	-1.88	60	2.00	-1.97	95	1.97	-2.19			
26	2.06	-2.09	61	2.09	-1.94	96	1.95	-2.31			
27	2.06	-1.88	62	1.93	-2.13	97	2.04	-1.87			
28	2.03	-1.94	63	1.99	-1.91	98	2.03	-2.01			
29	2.10	-1.70	64	2.01	-1.91	99	2.17	-1.96			
30	2.07	-1.86	65	2.01	-2.03	100	1.98	-2.12			
31	2.06	-1.81	66	1.99	-2.20	101	1.97	-2.15			
32	2.06	-1.75	67	2.06	-1.90	102	2.08	-1.76			
33	2.07	-1.87	68	2.00	-1.88	103	1.92	-2.08			
34	2.06	-1.82	69	1.93	-2.25	104	1.98	-1.94			
35	2.07	-1.83	70	1.79	-1.89	105	1.89	-2.28			

Thus the external precision on $\delta^{13}C$ measurement is ± 0.1‰ while that for $\delta^{18}O$ is better than ± 0.2 ‰. Uncertainties are given as one standard deviation (1σ). Another standard gas that is measured occasionally to check the stability of the machine is a "check standard" with $\delta^{13}C$ and $\delta^{18}O$ values of 0.33‰ and 2.33‰, respectively with respect to PDB.

Results and Discussion

Spero and Williams (1988), based on a laboratory culture study, proposed that stable isotope measurements on individual shells of *O. universa* can be useful, as their oxygen isotopic composition is affected only by surface salinity and temperature, whereas, the carbon isotopic ratio is not affected by these, but by solar irradiance. We wished to study the isotopic compositions of *O. universa* shells

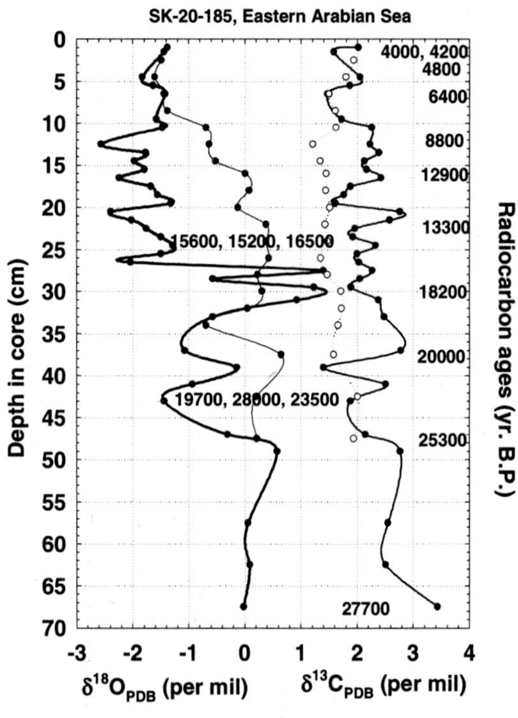

Fig. 4. Carbon and oxygen isotopic variations of *O. universa* in a sediment core SK-20-185 from the Eastern Arabian Sea. Thick lines denote variations in individual and thin lines, bulk (~30 forams of the same species) analyses. Conventional, uncalibrated, radiocarbon ages are shown on the right; these could be smaller than the true ages by a few thousand years (e.g., see Fig. 1). Multiple ages for a given depth interval indicate replicate measurements made on the same layer (bulk sediments repeatedly sampled sand processed).

from the Arabian Sea sediments to determine whether they can be used as a paleomonsoon proxy, by using the fact that during monsoons, the light level decreases because of cloudiness and $\delta^{13}C$ is expected to decrease due to reduced light levels (higher light levels enhance organic productivity that depletes the surface ocean of ^{12}C). Individual foraminifera of the species *O. universa* of size 400 to 500μm were picked and analysed from a sediment core in the eastern Arabian Sea (300 cm long gravity core SK-20-185, 10°N and 71° 50'E, water depth 2523m). Sarkar (1989) and Sarkar et al (1990a,b; 2000) had studied this core earlier. About 30 individuals of *O. universa* (size: 250 to 400μm) from each depth interval were analyzed. The average sedimentation rate based on bulk radiocarbon dates (shown on the right in Fig.4) was ~2.2 cm/ka. It must also be kept in mind that the isotope ratios vary with size: higher sizes are depleted in $\delta^{18}O$ and enriched in $\delta^{13}C$ (Sarkar et al., 1990b). Higher size foraminifer are expected to be more buoyant and hence closer to surface, where, the temperature is higher (hence lower $\delta^{18}O$) and productivity is higher due to higher sunlight (hence higher $\delta^{13}C$). This is reflected in Fig. 4, where data for individual shells are compared with data for the bulk.

Interpretation of $\delta^{13}C$ measurements of planktic forams made on bulk samples from the eastern Arabian Sea is complex because of the following factors. Upwelling during the monsoon season may bring up to the surface ^{13}C depleted waters from below. However, enhanced organic productivity due to nutrients coming from below and also from the continental runoff may increase the surface water $\delta^{13}C$. Decrease of solar radiation due to cloudiness may decrease the ^{13}C content of the surface water HCO_3^-. Thus the value measured on bulk foraminifera depends on which of these effects dominates.

It is seen from Table 1 that the experimental precision (external precision, σ_m) in the measurement of oxygen isotope ratio is about 0.2 ‰. But the natural variability in the foraminifera of the same species from a given sediment layer can be greater than this. The total variance can be expressed as: $\sigma^2 = \sigma_i^2/n + \sigma_m^2$, where the subscript *m* refers to bulk measurement, the subscript *i* to individual shell measurements and *n* the number of foram specimens in the bulk sample (Killingley, et al., 1981). Thus, measurement of single foraminifera (i.e., n = 1) a higher variance is expected. This is clearly seen in Fig.4. At 18ka, the single foram data show a sudden jump in the $\delta^{18}O$ values, whereas the bulk data show a more gradual transition. The negative excursion of ~1‰ reported earlier by Sarkar et al., (1999a) is also seen in the single foram $\delta^{18}O$ data. There is an overall decreasing trend in the $\delta^{18}O$ during the last 30ka, as a result of global warming, accompanied by monsoon intensification. With intensifying monsoons, the cloudiness should have enhanced; this is seen as a long term decreasing trend in the $\delta^{13}C$ data. Thus it appears that $\delta^{13}C$ of *O. universa* can serve as a good paleomonsoon proxy.

Conclusions

We have provided a brief overview of the stable isotope methodology and concepts as applicable to paleoceanography to non-specialist readers. We have also given a calibration curve to convert planktic foraminiferal oxygen isotopes to the excess of precipitation over evaporation in the Eastern Arabian Sea. Preliminary results from a sediment core here show that the $\delta^{13}C$ of individual shells of *O. universa* can be a potential monsoon proxy; however, this needs to be confirmed by more measurements on additional samples.

References

BARD, E. 1998. Geochemical and geophysical implications of the radiocarbon calibration. Geochimica Cosmochimica Acta, 62, 2025-2038.

BE', A.W.H. 1980. Gametogenic calcification in a spinose planktic foraminifer, Globigerinoides sacculifer (Brady). Marine Micropaleontology, 5, 283-310.

BERGER, W.H., KILLINGLEY, J.S. and VINCENT, E. 1978 Stable isotopes in deep carbonates: box core ERDC-92, West equatorial Pacific. Oceanologica Acta, 1, 203-216.

BOUVIER-SOUMAGNAC, Y. and DUPLESSY, J.C. 1985 Carbon and isotopic composition of planktic oraminifera from the laboratory culture, plankton tows and recent sediment: implications for the reconstruction of paleoclimatic conditions and of the global carbon cycle. Journal of Foraminiferal Research, 15 (4), 302-320.

COPLEN, T.B., KENDALL, C. and HOPPLE, J. 1983. Comparison of stable isotope reference samples. Nature, 302, 236-238.

CRAIG, H. 1953. The geochemistry of the stable carbon isotopes. Geochimica et Cosmochimica Acta, 3, 53-92.

CRAIG, H. 1957. Isotopic standards for carbon and oxygen and correction factors for mass spectrometric analysis of carbon dioxide. Geochimica et Cosmochimica Acta, 12, 133-149.

CRAIG, H. 1961. Standard for reporting concentrations of deuterium and oxygen-18 in natural waters, Science, 133, 1833-1834.

DUPLESSY, J.C, BE', A.W.H. and BLANC, P.L. 1981. Oxygen and carbon isotopic composition of planktic foraminifera in the Indian Ocean. Palaeogeogr. Palaeoclimat. Palaeoeco., 33, 9-46.

DUTTA, K, BHUSHAN, R. and SOMAYAJULU, B.L.K. 2001. □R correction values for the northern Indian Ocean. Radiocarbon, 43, 483-488.

EMILIANI, C, 1971. Depth habitats of the growth stages of pelagic foraminifera. Science, 173, 1122-1124.

EMRICH, K., EHHALT, D.H., and VOGEL J.C., 1970. Carbon isotope fractionation during the precipitation of calcium carbonate. Earth and Planetary Science Letters, 8, 363-371.

EREZ, J. and LUZ, B. 1983. Experimental paleotemperature equation for planktic foraminifera. Geochimica Cosmochimica Acta, 47, 1025-1031.

FAIRBANKS, R.G. 1989. A 17,000-year glacio-eustatic sea level record: influence of glacial melting rates on the Younger Dryas event and deep ocean circulation. Nature, 342, 637-642.

FAURE, G. 1986. Principles of Isotope Geology. John Wiley and Sons publishers, 2nd ed., 497p.

GHOSH, P. 2000. Isotopic and Elemental Studies of Gondwana Carbonates and their Implications. Unpub. PhD thesis, Devi Ahilya Vishwa Vidhyalaya, Indore, p47.

GONFIANTINI, R. 1981. The □notation and the mass spectrometric measurement techniques. In "Stable isotope Hydrology: Deuterium and Oxygen-18 in the water Cycle". Technical Reports Series no. 210. Gat JR & Gonfiantini R, Eds.), IAEA, Vienna, 35-84.

GROSSMAN, E.L. 1984.: Carbon isotopic fractionation in live benthic foraminifera: comparison with inorganic precipitate studies. Geochimica Cosmochimica Acta, 48, 1505-1512.

GROSSMAN, E.L. 1987. Stable isotopes in modern benthic foraminifera: a study of vital effect. Journal of Foraminiferal Research, 17 (1), 48-61.

HEMLEBEN, C., SPINDLER, M. and ANDERSON, O.R. 1988. Modern Planktic Foraminifera. Springer-Verlag publishers, 88-111.

HOEFS, J. 1997. Stable Isotope Geochemistry. Springer-Verlag Pub., 4th ed., pp 27-63.

KILLINGLEY, J.S., JOHNSON, R.F. and BERGER, W. H. 1981. Oxygen and carbon isotopes of individual shells odfplanktic foraminifera from Ontong Java Plateau, equatorial Pacific., Paleogeo. Paleoclim. Peloecol., 33, 193-204.

KROOPNICK, P, WEISS R.F and CRAIG, H. 1972. Total CO_2, ^{13}C, dissolved oxygen – ^{18}O at GEOSECS II in the north Atlantic. Earth and Planetary Science Letters, 16, 103-110.

KROOPNICK ,P. 1974. The dissolved $O_2 - CO_2 - {}^{13}C$ system in the eastern equatorial Pacific, Deep sea Research, 21, 211.

LABEYRIE, L.D, DUPLESSY, J.C and BLANC, P.L. 1987. Deep water formation and temperature variations over the last 125,000 years. Nature, 327, 477-482.

LANE, G.A. and DOYLE, M. 1956. Fractionation of oxygen isotopes during respiration. Science, 123, 574-576.

LIBBY, W.F. 1955. Radiocarbon Dating. Univ. of Chicago Press, 2nd ed., pp 175

LINICK, T.W., JULL ,A.J.T, TOOLIN, L.J. and DONAHUE, D.J 1986. Operation of the NSF – Arizona facility for radioisotope analysis and results from selected collaborative research projects. Radiocarbon, 28, 522 – 533.

McCONNAUGHEY, T. 1989a. ^{13}C and ^{18}O disequilibrium in biological carbonates.1. Patterns. Geochimica et Cosmochimica Acta, 53, 151-162.

McCONNAUGHEY, T. 1989b. ^{13}C and ^{18}O disequilibrium in biological carbonates.2. In vitro simulation of kinetic isotope effects. Geochim. Cosmochim. Acta, 53, 163-171.

McCREA, J.M. 1950. The isotopic chemistry of carbonates and a paleotemperature scale. J. Chem. Phys., 18, 849-857.

ORITZ, J.D, MIX A.C, RUGH, W. WATKINS, J.M. and COLLIER R.W 1996. Deep dwelling planktic foraminifera of the northeastern Pacific Ocean reveal environmental control of oxygen and carbon disequilibria. Geochimica et Cosmochimica Acta, 60 22., 4509-4523.

PEARSON, P.N, DITCHFIELD, P.W, SINGANO, J., HARCOURT-BROWN, K.G, NICHOLAS, J.C, OLSSON, R.K, SHACKLETON, N.J and HALL, M.A 2001. Warm tropical sea surface temperatures in the Late Cretaceous and Eocene epochs. Nature, 413, 481-487.

RAMESH, R. 2001. High resolution Holocene monsoon records from different proxies: An assessment of their consistency, Curr. Sci., 81, 101-105.

ROHLING, E. J. and COOKE, S. 1999. Stable oxygen and carbon isotopes in foraminiferal carbonate shells, in: Modern Foraminifera (ed. Sen Gupta, B.K.), Kluwer Academic Publications, 239-258.

ROSENBAUM, J. and SHEPPARD S.M.F. 1986. An isotopic study of siderites, dolomites and ankerites at high temperature. Geochim. Cosmochim. Acta, 50, 1147-1150.

SARKAR, A. 1989. Oxygen and carbon isotopes in Indian Ocean Sediments and their paleoclimatic implications, unpubl. Ph.D. Thesis, Gujarat University, 166p.

SARKAR, A., RAMESH, R., BHATTACHARYA, S.K. and RAJAGOPALAN, G. 1990a. $\delta^{18}O$ evidence for increased north east monsoon current during the last glaciation, Nature, 343, 549-551

Sarkar, A., Ramesh, R .and Bhattacharya, S. K. 1990b. Effect of sample pretreatment and size fraction on the $\delta^{18}O$ and $\delta^{13}C$ values of foraminifera in Arabian Sea sediments, Terra Nova, 2, 489-493

SARKAR ,A, RAMESH. R, SOMAYAJULU B.L.K., AGNIHOTRI, R, JULL A.J.T. and BURR GS 2000.: High resolution Holocene monsoon record from the eastern Arabian Sea, Earth and Planetary Science Letters, 177, 209-218.

SARKAR, A, RAMESH, R., BHATTACHARYA, S. K. and PRICE, N. B. 2000. Paleomonsoon and paleoproductivity records in $\delta^{18}O$, $\delta^{13}C$ and $CaCO_3$ variations in the northern Indian Ocean sediments Proc. Ind. Acad. Eci. (Earth Planet. Sci.) 109, 1, 157-169.

SCHRAG, D.P. 1999.: Effects of digenesis on the isotopic record of late Paleogene tropical sea surface temperatures. Chemical Geology, 161, 215-224.

SOMAYAJULU B.L.K., BHUSHAN, R., SARKAR, A., BURR, G.S and JULL, A.J.T. 1999.: Sediment deposition rates in the continental margins of the eastern Arabian Sea using ^{210}Pb, ^{137}Cs, ^{14}C, The Science of Total Environment, 237/238, 429-439.

SPERO, H. J. and WILLIAMS, D.F. 1988. Extracting environmental information from palnktonic foraminiferal $\delta^{13}C$ data, Nature, 335, 717-719.

SPERO, H.J and LEA, D.W. 1993.: Intraspecific stable isotope variability in the planktic foraminifera Globigerinoides sacculifer: results from laboratory experiments. Marine Micropaleontology, 22, 221-234.

SPERO, H.J., BIJMA, J., LEA, D.W. and BEMIS, B.E. 1997.: Effect of seawater carbonate concentration on foraminiferal carbon and oxygen isotopes. Nature, 390, 497-500.

SRINIVASAN, M. S. 2003. Cenozoic evolution of ocean gateways, impact of global ocean circulation and climate, 32nd Prof. Birbal Sahni Memorial Lecture, Published by the Dept. Of Geology, Banaras Hindu University, Varanasi, India, 1-28.

SRINIVASAN, M. S. 1998. Ocean Science Trends and future directions (ed. B. L. K. Somayajulu) Academia Books International, New Delhi, 113-128.

SRINIVASAN, M. S. 2000. Significant contributions to Geoscience research in India during the nineties, a status report, Indian National Science Academy, New Delhi, 158p.

STUIVER, M, REIMER, P.J, BARD E, BECK, J.W, BURR, G.S., HUGHEN, K.A., KROMER, B, MCCORMAC, G., VAN DER PLICHT, J. and SPURK, M. 1998.: INTCAL98 Radiocarbon age calibration, 24,000-0 cal BP. Radiocarbon, 40 (3), 1041-1083.

SWART, P.K, BURNS S.J and LEDER J.J. 1991. Fractionation of the stable isotopes of oxygen and carbon in carbon dioxide during the reaction of calcite with phosphoric acid as a function of temperature and technique. Chemical Geology, 86, 89-96.

UREY , H.C 1947. The thermodynamic properties of isotopic substances. J. Chem. Soc., 1947, 562-581.

VINOT-BERTOUILLE A.C and DUPLESSY, J.C. 1973.: Individual isotopic fractionation of carbon and oxygen in benthic foraminifera. Earth and Planetary Science Lett., 18, 247-252.

WACHTER EA and HAYES J.M. 1985. Exchange of oxygen isotopes in carbon dioxide–phosphoric acid systems. Chemical Geology, 52, 365-374.

WEBER JN and WOODHEAD PJJ. 1970. Carbon and oxygen isotope fractionation in the reef building corals. Chemical Geology, 6, 93-117.

WEFER, G. and BERGER, W.H. 1991. Isotope paleontology: growth and composition of extant calcareous species. Marine Geology, 100, 207-248.

YADAVA, M. G. and RAMESH, R. 1999. Speleothems – useful proxies for past monsoon rainfall, J. Sci. Ind. Res., 58, 339-348.

Acknowledgements

I consider it a great privilege and honor to contribute this article for the special volume dedicated in honour of Prof. M. S. Srinivasan, a pioneering and eminent paleoceanographer of India. I take this occasion to thank my former/present colleagues K. Gopalan, B. L. K. Somayajulu, S. Krishnaswami, S. K. Bhattacharya, R. A. Jani, J. T. Padia, Ashish Sarkar and former/current Ph.D. students A. Sarkar, S. Chakraborty, Geeta Rajagopalan, J. S. Ray, D. Jagadheesha, M. G. Yadava, R. Bhutani, S. Rajesh, Sanjeev Kumar, Manish Tiwari, Satya Prakash and S. Ramaswami other long/short term collaborators in stable isotope/climate/ocean research over the past 25 years. I thank Devesh Sinha (BHU) for the kind invitation to contribute this paper. I also thank ISRO-GBP for continued funding for Paleoceanographic research.

Symbiosis, Calcification and Stable Isotopes in Foraminifera: A Review

Pratul Kumar Saraswati

Department of Earth Sciences, Indian Institute of Technology, Bombay, Mumbai 400 076, India

Introduction

Among the 150 families of foraminifera living today, twelve families harbor algal symbionts. Both the benthic and planktonic foraminifera are hosts for symbiotic algae of diverse types, comprising of diatoms, dinoflagellates, chlorophytes, rhodophytes and chrysophytes. It is suggested that foraminifera have the advantages of obtaining energy and enhanced calcification from algal-photosynthesis (Hallock, 2000) while the algae are benefited by special adaptations of the host for their establishment and maintenance (Lee and Anderson, 1991). Hallock (1981) proposed mathematical models to demonstrate that the energetic advantage of algal symbiosis to foraminifera-algae system is several orders of magnitude more than the competing non-symbiotic foraminifera in nutrient-deficient environments. It has been argued that similar symbiotic relationships occurred in the geologic past (Hottinger, 1983) and the enhanced calcification aided by symbiont photosynthesis caused some benthic foraminifera to reach gigantic size in Permian, Cretaceous and Eocene times. There is no direct evidence though to infer photosymbiosis in fossil taxa. The experimental results on photosynthesis, the mechanisms of calcification and stable isotopic analysis of the living symbiont-bearing foraminifera have greatly improved our understanding of foraminifera-algae symbiosis. The results obtained so far suggest a possibility of finding distinct signatures of photosymbiosis in the skeletal carbonates of foraminifera.

Symbiont-diversity in foraminifera

The well-known example of host-symbiont system is coral, and coral-reef appears to be the most favourable habitat of symbiont bearing foraminifera. But in contrast to corals, which are host to practically a single species of dinoflagellate, *Symbiodinium microadriaticum,* the foraminifera host a diverse group of algal symbionts. Seven living families of benthic and five living families of planktonic foraminifera are known to host algal symbionts (Table-1). The rotaliid benthic foraminifera are host to diatoms while the miliolids are host to diverse types of algae including red alga, chlorophyte, dinoflagellate and diatoms. The planktonic foraminifera contain dinoflagellate and chrysophyte. There appears to be a fundamental difference between benthic and planktonic species in their adaptation to symbiosis. There is an architectural adaptation in benthic species and behavioral adaptation in planktonics (Lee and Anderson, 1991). The morphological complexity in large benthic foraminifera and generally thin, glassy tests of symbiont bearing species are considered to be natural greenhouse for the endosymbiotic algae (Haynes, 1965). Larger surface area of the shell and the presence of glassy pustules and pits in certain genera are also considered functionally significant in enhancing light for photosymbiosis. The planktonic foraminifera, in stead of adapting morphologically, adapt behaviorally by bringing their symbionts into rhizopodial network during the day time and retracting them into the shell cytoplasm in the night when the photosynthetic activity ceases.

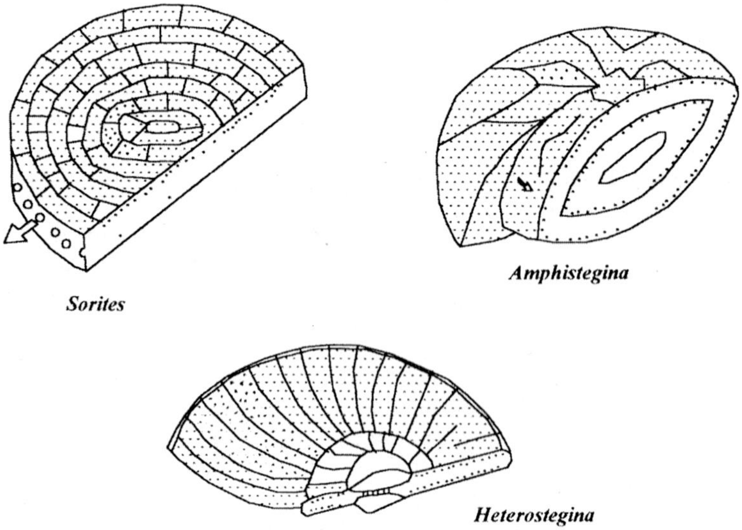

Fig. 1. The position of symbionts in *Sorites, Heterostegina* and *Amphistegina* (after Leutenegger, 1984).

The distribution of symbionts in the host is another important aspect especially from the view point of a hospitable microhabitat to the algal taxa. The species of *Amphistegina, Nummulites* and *Peneroplis* are seen to lift their shells by 45° from the surface and thereby both the lateral sides get exposed for the bilaterally distributed symbonts (Leutenegger1984). The disc shaped soritines, however, lie flat on the surface and sometimes one side is permanently fixed to substratum (Saraswati, 2002) and therefore their symbionts are arranged only below the side exposed to the sunlight (Fig.1). In *Sorites marginalis* and *Amphisorus hemprichii* the dinoflagellate symbionts are concentrated mostly in the inner three-fourth part from the embryonic apparatus (Lee and Anderson, 1991). The density of symbionts also varies. In a planktonic foraminifer, *Orbulina universa,* Spero and Parker (1985) estimated a symbiont density of 23,000 algae per foraminifer.

Leutenegger (1984) suggested that symbiotic association between foraminifera and algae are hereditarily fixed and possibly established in Palaeogene to Neogene times. In the superfamily Soritacea, the three families Peneroplidae, Soritidae and Archaiadae harbor different kinds of symbionts (Table-1). There are exceptions, however, to this host-symbiont specificity. The family Alveolinidae, unlike the other miliolids, is host to diatoms – the same symbionts which occur in unrelated families of rotaliid foraminifera. Leutenegger (op.cit.) cites *Peneroplis* as another exception. *Peneroplis proteus* contains chlorophycean symbionts unlike the other two species, *P. pertusus* and *P. planatus,* which host rhodophycean symbionts. The foraminiferal species may, thus, differ from each other in the type of symbiotic algae in their cytoplasm and in the distribution and density of symbionts in the shell.

Table 1. Types of symbionts in foraminifera (after Lee and Anderson, 1991).

Family	Species	Symbiont Type
Peneroplidae	*Peneroplis pertusus, P. planatus, P. arietina*	Unicellular red alga
Archaiadae	*Archaias angulatus, Cyclorbiculina compressa, Androsina lucasi*	Chlorophyte
Soritidae	*Sorites marginalis, S. orbiculus, Amphisorus hemprichii, Marginopora vertebralis, M. kudakajimaensis*	Dinoflagellate
Alveolinidae	*Alveolinella quoii, Borelis schlumbergeri*	Diatom
Amphisteginidae	*Amphistegina lobifera, A. lessonii, A. radiata, A. papilosa*	Diatom
Calcarinidae	*Calcarina calcar, C. gaudichaudi, C. hispida, C. spenglieri, Baculogypsina sphaerulata*	Diatom
Nummulitidae	*Nummulites cumingii, Operculina ammonoides, Heterostegina depressa, Cycloclypeus carpenteri*	Diatom
Globigerinidae	*Globigerinoides rubber* *G. saculifer* *Globigerina bulloides* *G. conglobatus* *G. cristata* *Orbulina universa,*	Dinoflagellate Dinoflagellate Chrysophyte Dinoflagellate Chrysophyte Dinoflagellate
Candeinidae	*Candeina nitida*	Chrysophyte
Pulleniatinidae	*Pulleniatina obliqueloculata*	Chrysophyte
Hastegerinidae	*Hastigerina pelagica*	Chrysophyte
Globorotalidae	*Globorotalia menardii, Neogloboquadrina dutertrei*	Chrysophyte

Calcification

The miliolid and rotaliid foraminifera are found to have two different mechanisms of calcification. Based on experimental data ter Kuile (1991) has presented generalized schemes for calcification in foraminifera. In a rotaliid species, *Amphistegina lobifera*, the inorganic carbon for calcification is derived from the cytoplasm, sea-water and photosynthetically assimilated metabolic source and first concentrated in an inorganic carbon pool. In *Amphisorus hemprichii*, a miliolid larger foraminifer, there is no internal inorganic carbon pool and it obtains carbon for calcification directly from the sea-water. Erez and Bentov (2002) have further observed the biomineralization process in *Amphistegina lobifera* using fluoroscent imaging confocal microscopy. According to them, primary calcite is precipitated as microspherules with organic matrix at the boundary between ectoplasm and endoplasm. This is Mg-rich calcite, the source of ion for which is endoplasmic birefringent granules. This is followed by the deposition of low-Mg secondary calcite, forming 90% of the total mass. The main source of this calcite is sea-water vacuoles in the endoplasm. It is said that the diffusion of metabolic CO_2 into the sea-water vacuoles concentrates inorganic carbon to form the carbon pool. And it is due to this process that there are deviations from equilibrium for carbon and oxygen isotopes in foraminiferal carbonates.

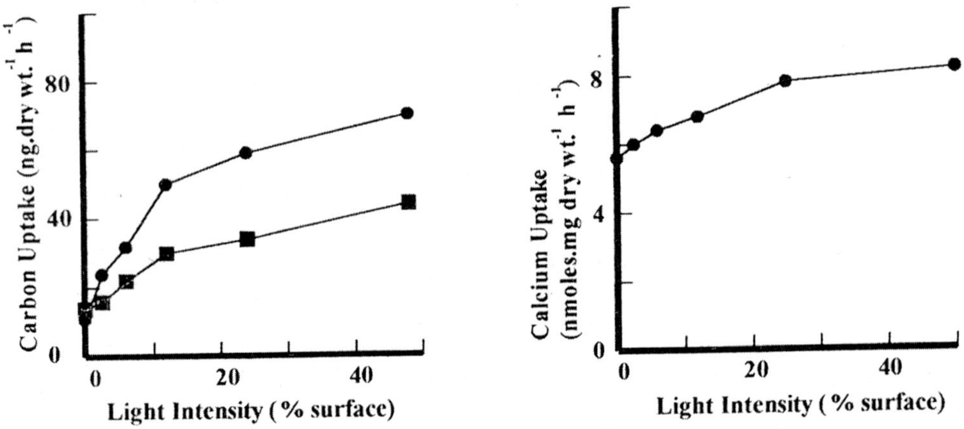

Fig. 2. The effect of light intensity on carbon and calcium uptake by *Archaias angulatus*. The organic carbon uptake is shown by filled-in circle and inorganic carbon uptake by filled-in rectangles. (after Duguay, 1983).

Several experimental studies have shown that in symbiont-bearing foraminifera the photosynthetic activity of the symbionts play an important role in calcification. Rottger and others (1980) measured growth rate of *Heterostegina depressa* and *Amphistegina lessonii* in clone cultures at different light intensities and darkness. The growth rate of the two species increased with light intensity and the optimal light intensity was 400 – 800 lux. No growth occurred in dark but growth continued without food when incubated in light. A detailed experimental study was carried out by Duguay (1983) in which he measured calcium incorporation and photosynthetic carbon fixation by microalgal symbionts in *Archaias angulatus, Sorites marginalis* and *Cyclorbiculina compressa* (Fig.2). In all the three species, calcification rates increased with the enhancement of the intensity of light and algal symbionts appeared to stimulate carbonate production.

Oxygen and Carbon Isotopes in symbiont-bearing foraminifera

During the process of calcification there is partitioning of oxygen and carbon isotopes between the ambient water and the carbonate species, known as isotopic fractionation. There are various factors due to which the isotopic composition of foraminiferal carbonates may deviate from the equilibrium isotope fractionation. It may be due to growth and ontogeny, symbiont photosynthesis, respiration, gametogenic calcite and variation in carbonate ion concentration (Rohling and Cooke, 1999). It may not be possible to distinguish the effects of these factors because the underlying processes may be related. For example, photosynthesis stimulates growth. A number of studies have shown that symbiont photosynthesis has a significant role in the disequilibrium fractionation in foraminiferal carbonates, and the effect is distinctly high in carbon isotopes.

Erez (1978) is among the most notable early workers to experimentally demonstrate the effect of symbiont photosynthesis on the rate of calcification and the stable isotopic composition of the foraminiferal shells. According to him the skeletal carbonate consists of two components, the seawater bicarbonate (incorporated into the skeleton in isotopic equilibrium process) and metabolic CO_2 (enriched in ^{12}C and ^{16}O, causing deviation from isotopic equilibrium). The symbiotic algae enhance the overall metabolic activity of the host-algae complex and increase the amount of metabolic CO_2 in the organisms' internal CO_2 pool. This would cause more negative isotopic composition in species with higher rates of photosynthesis, as seen in the rotaliine species *Amphistegina* and *Heterostegina* compared with the milioline species *Amphisorus* and *Borelis* (Fig. 3). The effect of photosymbiosis is found in both oxygen and carbon isotopic ratios. The $\delta^{18}O$ value in *Amphistegina* is 1.2‰ lighter than the expected equilibrium value. There have been several other reports of non-equilibrium oxygen isotopic fractionation in larger foraminifera (Hansen and Buchardt, 1977; William and others, 1981; Zimmermann and others, 1983). Wefer and others (1981) found both depletion and enrichment in ^{18}O in the species of larger foraminifera from different localities. Thus, as regards to oxygen isotopic composition of symbiont-bearing species there are reports of near-equilibrium to disequilibrium fractionation (with both depletion and enrichment of heavier isotope of oxygen). One of the possible reasons for reporting enrichment by some workers and depletion by others could be that the data is not corrected for $MgCO_3$ content of the shell as suggested by Tarutani and others (1969). In a recently concluded study on several species of larger foraminifera, near equilibrium to slight enrichment in oxygen isotopic composition has been observed when the data is corrected for magnesium carbonate (Saraswati and others, Commn.). Luz and others (1983) found oxygen isotopic ratios in *Amphisorus hemprichii* to be close to expected equilibrium values. Earlier, Wefer and Berger (1980) also reported depletion in *Marginopora vertebralis* and enrichment in *Cyclorbiculina compressa*, and stated that if it is corrected for $MgCO_3$ both the species may actually be depleted in ^{18}O.

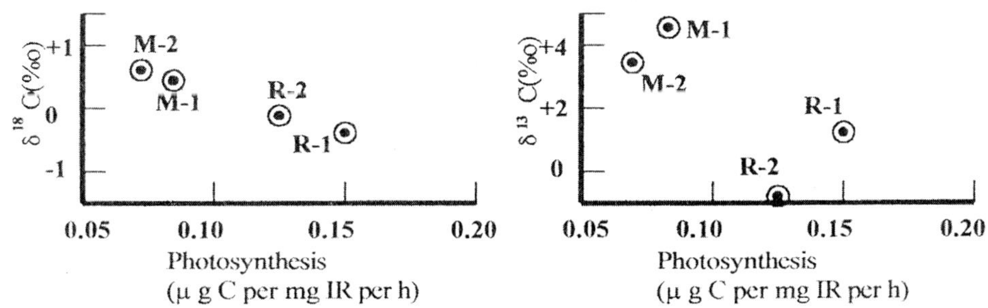

Fig. 3. Relationship between rate of photosynthesis and $\delta^{18}O$ and $\delta^{13}C$ values of some of the photosymbiotic species of foraminifera (after Erez, 1978). The miliolid species are *Borelis sp.* (M-1) and *Amphisorus hemprichii* (M-2) and the rotaliid species are *Amphistegina lobifera* (R-1) and *Heterostegina sp.* (R-2).

There have been differences of opinion about isotopic fractionation in planktonic foraminifera too, and both equilibrium and non-equilibrium oxygen isotopic values have been reported. Erez and Honzo (1981) carried out a detailed investigation of planktonic foraminifera from plankton tows, sediment traps and sediments. They observed non-equilibrium isotopic values in the symbiont-bearing species of *Globigerinoides* and equilibrium to slight positive deviation from equilibrium in symbiont-free globorotalids. The conflicting reports of non-equilibrium values in plankton tow samples and equilibrium values in sediment samples is explained as a result of continued skeletal growth below the photic zone. Isotopically heavier skeleton is added as they sink down the cool bottom water and the activity of the symbiotic algae also ceases below the photic zone. Spero and Deniro (1987) analyzed laboratory grown, symbiont-bearing *Orbulina universa* and found that symbiont photosynthesis does not affect the oxygen isotopic composition of shell calcite.

While explaining the general nature of oxygen and carbon isotopic disequilibrium in biogenic carbonates, McConnaughey (1989) suggests that photosynthesis and respiration are unlikely to have any direct effect on ^{18}O depletion. The same is not true for carbon isotopes. The carbon isotopic disequilibrium may often be large and it is supposed to reflect changes in $\delta^{13}C$ of dissolved inorganic carbon, caused by photosynthesis and respiration. Due to preferential removal of ^{12}C during photosynthesis, it is argued that symbiont-bearing taxa should generally have more positive $\delta^{13}C$ than asymbiotic taxa. The reported values of $\delta^{13}C$ of both symbiotic benthic and planktonic foraminifera indicate either depletion or slight enrichment from the equilibrium values (Erez and Honzo, 1981; Wefer and others, 1981, Saraswati and others, Commn.). As discussed in the later part of this section, microenvironment of calcification in symbiont-bearing foraminiferal shells are significantly different due to respiration and photosynthesis. The enrichment or depletion of $\delta^{13}C$ of shell carbonate will depend on the relative contributions of respired and photosynthetic flux of carbon.

Spero and Deniro (1987) found a large variation in the $\delta^{13}C$ values between pre-gametogenic and post gametogenic groups as well as high light and low light groups of in laboratory cultured *Orbulina universa*. They regarded the relative role of photosynthesis and respiration and the timing of calcification as causes of equilibrium or deviation from equilibrium of the $\delta^{13}C$ of foraminiferal carbonates. Further, Spero and others (1991) presented a quantitative model of carbon isotopes based on two planktonic foraminiferal species *Orbulina universa* and *Globigerinoides sacculifer*. Notwithstanding certain limitations, variations in symbiont density and photosynthetic rate are primary physiological control on intraspecific carbon isotope variability. In *Globigerinoides sacculifer* Spero and Lea (1993) found increase in $\delta^{13}C$ with irradiance and, therefore, with symbiont photosynthetic rates. They hypothesized that increased $\delta^{13}C$ under high light level is due to elevated utilization of $^{12}CO_2$ by the symbiotic algae at higher photosynthetic rates and consequently the calcifying microenvironment gets enriched in $H^{13}CO_3^-$ producing ^{13}C enriched chambers. At low light levels photosynthetic activity is subdued, and possibly due to incorporation of respired CO_2 , the chamber $\delta^{13}C$ is low.

The mechanism and timing of calcification are important in understanding the isotopic behavior of foraminiferal carbonates. As stated elsewhere the rotaliine species have a large internal inorganic carbon pool which serves mainly for calcification and possibly also for photosynthesis while the milioline species take up carbon for calcification indirectly from seawater or have a very small inorganic carbon pool. The timing of calcification is equally significant. In photosynthetic corals it is found that they calcify mainly during the day time when the photosynthetic CO_2 uptake is several times faster than respiratory CO_2 release. As a result, photosynthetic effect on skeletal $\delta^{13}C$ is more than the respiratory effect (McConnaughey *et al.*, 1997). Laboratory observations on relationship of light to growth and calcification are made only for a few foraminiferal species (see Lee and Anderson, 1991 for details). Broadly, foraminifera differ in their life cycles, growth, timing of calcification and in the kind and location of symbionts. All these factors are likely to influence the stable isotopic composition of the symbiont-bearing foraminifera.

In spite of a large number of isotopic studies, the mechanism of isotopic fractionation, especially of the carbon isotopes, in foraminiferal carbonates is poorly understood. A systematic approach in the last few years in the areas of foraminiferal biology, particularly the measurements on photosynthesis and calcification, and stable isotopic studies of laboratory grown foraminifera, are promisng. Another significant development is the recognition of symbiont-bearing foraminifera as a distinct microenvironment where calcification, respiration and symbiont photosynthesis change the pH, and shift the concentrations of CO_2, CO_3 and HCO_3 (Fig. 4) and consequently the carbonate chemistry in its vicinity is quite different from the sea-water (Wolf-Gladrow and others, 1999). In view of this, a numerical model has been proposed by Zeebe and others (1999) to calculate $\delta^{13}C$ of foraminiferal carbonates as a function of sea-water chemistry and vital effects. The vital effects in symbiont-bearing foraminifera includes (i) respiration of foraminifer (ii) respiration of symbiotic algae, and (iii) photosynthesis of the symbiotic algae. Respiration produces CO_2 which is enriched in ^{12}C. Photosynthesis preferentially removes ^{12}C and therefore the foraminiferal microenvironment gets enriched in ^{13}C. The change in carbon isotopic composition of this environment will depend on the relative contributions of respired and photosynthetically produced carbon flux. The effects of respiration, photosynthesis and calcite precipitation on shell $\delta^{13}C$ are shown qualitatively (Fig. 5) by Zeebe and others (*op. cit.*). In this model spherical symmetry of the foraminifer is assumed and it is compared with the results of *Orbulina universa*, cultured in the laboratory. In the opinion of the authors, the length and direction of the three vector components might vary significantly for different foraminiferal species. The microsensor study of the benthic species, *Amphistegina lobifera* and *Amphisorus hemprichii*, by Kohler-Rink and Kuhl (2001) corroborates the above view that O_2, CO_2 and pH in the vicinity of symbiont-bearing foraminifera are strongly affected by the combined processes of respiration, photosynthesis and calcification.

Isotopic evidence of Photosymbiosis in Fossil Taxa

Symbiosis is a driving force in the evolution of larger foraminifera (Lee and Hallock, 1987) and in planktonic foraminifera it is said to be active for the last 75 million years (D'Hondt and Zachos, 1995 cited in Norris, 1996). In benthic foraminifera the morphological features believed to be advantageous to symbionts are well documented in living species. The fossil species with similar morphological adaptations are recognized as symbiont bearing. The planktonic foraminifera having spines are, similarly, believed to be symbiont bearing because most modern species with algal symbiosis are characterized by spines, used for supporting dinoflagellate symbionts during photosynthesis in the day time (Anderson and Be, 1976). The morphological feature alone, however, may not be sufficient to infer photosymbiosis in fossil taxa. Spines, a typical adaptive feature of most symbiotic planktonic foraminifera, are also present in *Hastigerina pelagica* and *Globigerina bulloides* which otherwise lack obligate symbiosis (D'Hondt and others, 1994). In recent years attempts are made to use stable isotopic patterns to infer photosymbiosis in fossil species. The isotopic trends of Neogene planktonic foraminifera known or believed to harbor algal symbionts are found to be different from those of asymbiotic taxa. The investigations by Norris (1996), Houston and Huber (1998) and Houston and others (1999) have suggested the following isotopic features of photosymbiosis: (i) $\delta^{18}O$ values more negative than co-existing asymbiotic taxa (ii) relatively small size-related changes in $\delta^{18}O$ values (iii) a strong positive size-related shift in $\delta^{13}C$ values of 0.5‰ or more (iv) a poor co-variance between size-related changes in $\delta^{18}O$ and $\delta^{13}C$ values and (v) significant intra-specific variability in $\delta^{13}C$ values. On the basis of the above criteria, Houston and others (1999) found the Maastrichtian taxa *Planoglobulina acervulinoides* and *Racemiguembelina fructicosa* to be photosymbiotic. Another species in the same assemblage, *Planoglobulina multicamerata,* showing most positive adult $\delta^{18}O$ values and a strong co-variance between size-related $\delta^{18}O$ and $\delta^{13}C$ values is interpreted to have lacked photosymbionts. The Palaeocene taxa *Acarinina* and *Morozovella* were also interpreted to be symbiont-bearing on stable isotopic criteria (D'Hondt and others 1994).

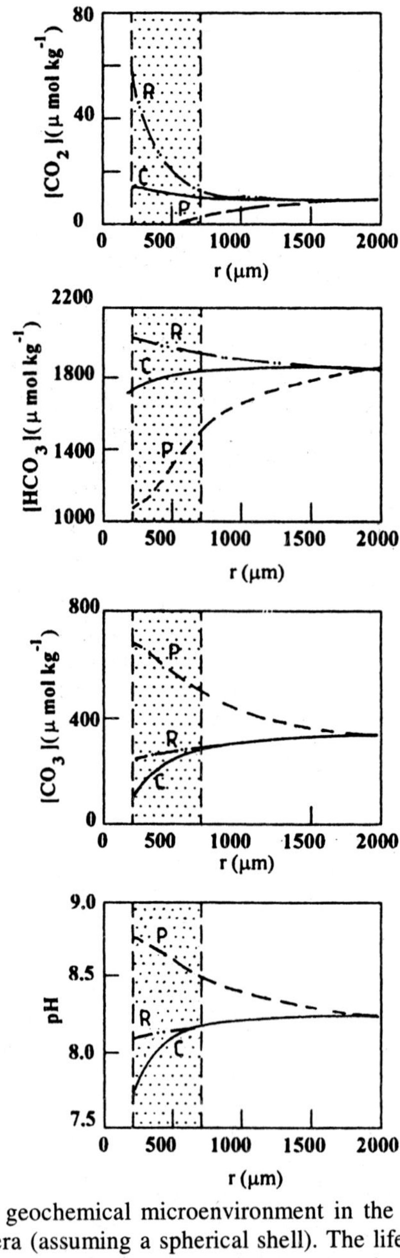

Fig. 4. Modelled profiles of geochemical microenvironment in the vicinity of symbiont-bearing planktonic foraminifera (assuming a spherical shell). The life processes, calcification (C), photosynthesis (P) and respiration (R), influence the characteristics of the carbonate microenvironment and the effect is the function of the distance from the center of the shell. (after Wolf-Gladrow and others, 1999).

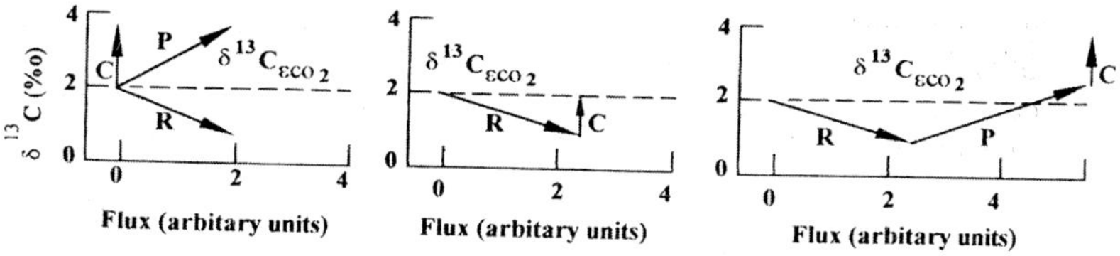

Fig. 5. A qualitative representation of the influence of photosynthesis (P) and respiration (R) on $\delta^{13}C$ value of the shell calcite. The effect of calcification (C) is constant. The $\delta^{13}C$ values of shell calcite is increased by photosynthesis and decreased by respiration (left); in dark condition, where there is no photosynthesis, the calcification effect may cancel the respiration effect (center); in light condition photosynthesis increases $\delta^{13}C$ of shell calcite (right). (after Zeebe and others, 1999).

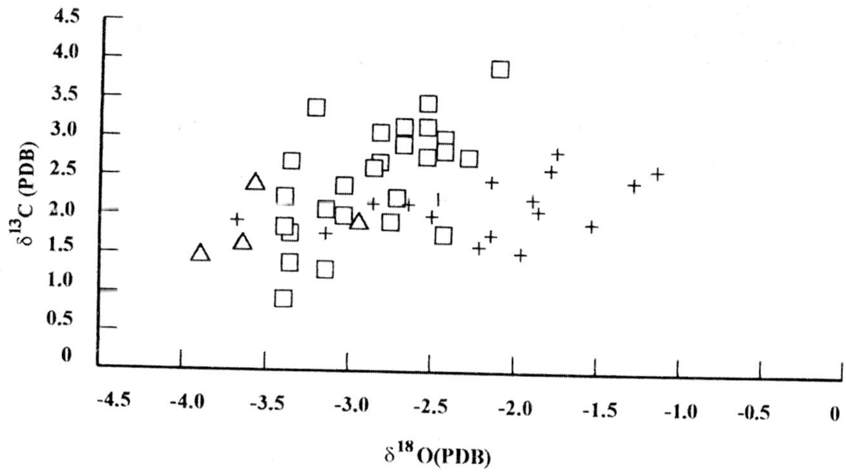

Fig. 6. $\delta^{18}O$ - $\delta^{13}C$ plot of Indo-Pacific *Marginopora* from different localities showing no correlation between the two values, a characteristic feature of photosymbiotic taxa. The data are from Philippines (open square; after Wefer and others, 1981), Papua New Guinea (open triangle; after Langer, 1995) and Ryukyu (cross; Saraswati, unpub.).

The above criteria of photosymbiosis in planktonic foraminifera may broadly be applied to larger foraminiferal species too, except for an important difference. The size-related change in $\delta^{18}O$ is reported to be small in planktonic foraminifera but in larger foraminifera the trend is different. On carefully dissected shells through different ontogenetic stages it is found that in shallow water species, such as *M. kudakajimaensis,* it is large but in deep water species, such as *Cycloclypeus carpenteri,* it is small (Saraswati and others, Commn.). In contrast to planktonic foraminifera, the larger benthic foraminifera have a longer life-span, living through complete range of seasonal change, and therefore show a significant size-related change in $\delta^{18}O$. *Cycloclypeus carpenteri*, a deeper water species, however, is subjected to smaller variation in temperature and therefore shows a minor variation in $\delta^{18}O$ with size. The $\delta^{18}O$ and $\delta^{13}C$ do not correlate in the species of *Marginopora* from different localities (Fig. 6) and, according to McConnaughey (1989), this is characteristic of photosymbiotic taxa unlike a positive correlation due to kinetic effect found in asymbiotic taxa.

Conclusions

Foraminifera – algal symbiosis has a long geological history and it has played a significant role in the evolutionary history of the host foraminifera. Foraminifera is unique in hosting diverse group of symbionts in contrast to the widely known coral-algal symbiosis. There are certain morphological features typical of living symbiont-bearing foraminifera and by analogy algal symbiosis has been presumed in certain fossil taxa. The morphological analogy needs to be supported by other proxies in skeletal carbonates. The oxygen and carbon isotopic composition of the modern symbiont-bearing foraminiferal species show particular trends different from the symbiont-free species. The experimental studies on calcification and photosynthesis have further improved our understanding of photosymbiosis and isotopic fractionation in foraminifera. Some recently proposed models of chemical micro-environment around symbiont-bearing foraminiferal shells and its effect on isotope fractionation will have significant implications in palaeobiology and palaeoceanography. At this stage there is a need to integrate the microsensor studies of cultured foraminifera with their ontogenic isotopic analysis.

References

ANDERSON, O.R. and BE, A.W.H., 1976. The ultrastructure of a planktonic foraminifer, *Globigerinoides sacculifer* (Brady), and its symbiotic dinoflagellates. Jour. Foram. Res., 6, 1-21.

D'HONDT, S. and ZACHOS, J.C.,1995. 75 million years of photosymbiosis in planktonic Foraminifera. Geol. Soc. Am. Abs. Prog., 27, A-244.

D'HONDT, S., ZACHOS, J.C. and SCHULTZ, G., 1994. Stable isotopic signals and photo-symbiosis in the Paleocene planktic foraminifera. Paleobiology, 20, 391-406.

DUGUAY, L.E., 1983. Comparative laboratory and field studies on calcification and carbonfixation in foraminiferal-algal associations. Jour. Foram. Res., 13, 252-261.

EREZ, J., 1978. Vital effect on stable-isotope composition seen in foraminifera and coralskeletons. Nature, 273, 199-202.

EREZ, J. and BENTOV, S., 2002. Calcification processes in foraminifera and their paleo-oceanographic implications. Abs. Forams 2002, Int. Symp. Foram., Perth, 32-33.

EREZ, J. and HONZO, S., 1981. Comparison of isotopic composition of planktonicforaminifera in plankton tows, sediment traps and sediments. Palaeogeog. Palaeoclimat. Palaeoecol. 33, 129-156.

HALLOCK, P., 1981. Algal symbiosis, a mathematical analysis. Mar. Biol., 62, 249-255.

HALLOCK, P., 2000. Symbiont-bearing foraminifera: harbingers of global change.Micropal., 46, suppl. 1, 95-104.

HANSEN, H.J. and BUCHARDT, B., 1977. Depth distribution of Amphistegina in the Gulf of Elat, Israel. Utrecht Micropal. Bull., 15, 205-224.

HAYNES, J.R., 1965. Symbiosis, wall structure and habitat in foraminifera. Contr.Cushman Fon. Foram. Res., 16, 40-43.

HOTTINGER, L., 1983. Processes determining the distribution of larger foraminifera in spaceand time. Utrecht Micropal. Bull., 30, 239-253.

HOUSTON, R.M. and HUBER, B.T., 1998. Evidence of photosymbiosis in fossil taxa?Ontogenetic stable isotope trends in some Late Cretaceous planktonic foraminifera. Mar. Micropal., 34, 29-46.

HOUSTON, R.M., HUBER, B.T. and SPERO, H.J., 1999. Size-related isotopic trends in someMaastrichtian planktic foraminifera: methodological comparisons, intra-specific variability and evidence for photosymbiosis. Mar. Micropal., 36, 169-188.

KOHLER-RINK, S. and KUHL, M., 2001. Microsensor studies of photosynthesis andrespiration in the larger symbiont bearing foraminifera *Amphistegina lobifera* and *Amphisorus hemprichii*. Ophelia, 55, 111-122.

TER KUILE, B., 1991. Mechanisms for calcification and carbon cycling in algal symbiont-bearing foraminifera. *In* Lee, J.J. and Anderson, O.R. (eds) Biology of Foraminifera, 73-90.

LANGER, M., 1995. Oxygen and carbon isotopic composition of Recent larger and smallerForaminifera from the Madang lagoon (Papua New Guinea). Mar. Micropal., 26,215-221.

LEE, J.J. and ANDERSON, O.R., 1991. Symbiosis in foraminifera. *In* Lee, J.J. and Anderson, O.R.(eds) Biology of Foraminifera, Academic Press. 73-90.

LEE, J.J. and HALLOCK, P., 1987. Algal symbiosis as a driving force in the evolution of larger foraminifera. Ann. New York Acad. Sc. 503, 330-347.

LEUTENEGGER, S., 1984. Symbiosis in benthic foraminifera: specificity and hostadaptations. Jour. Foram. Res., 14, 16-35.

LUZ, B., REISS, Z. and SEGEV, T.Z., 1983. Seasonal variation of stable isotopes in*Amphisorus* from the Gulf of Aqaba (Elat), Red Sea *in* Meulenkamp, J.E. (Ed) Reconstruction of marine paleoenvironments. Utrecht Micropal. Bull., 30, 229-237.

McCONNAUGHEY, T. A., 1989. ^{13}C and ^{18}O isotopic disequilibrium in biological carbonates I :Patterns. Geochim. Cosmochim. Acta, 53, 151-162.

McCONNAUGHEY, T. A., Burdett, J., Whelan, J.F. and Paull, C.K., 1997. Carbon isotopesin biological carbonates: Respiration and photosynthesis. Geochim. Cosmochim.Acta, 61, 611-622.

NORRIS, R.D., 1996. Symbiosis as an evolutionary innovation in the radiation of planktic foraminifera. Paleobiology, 22, 461-480.

ROHLING, E.J. and COOKE, S., 1999. Stable oxygen and carbon isotopes in foraminiferalCarbonate shells. *In* Sen Gupta, B.K. (ed) Modern Foraminifera, Kluwer Academic, 239-258.

ROTTGER, R., IRWAN, A., SCHMALJOHANN, R. and FRANZISKET, L., 1980. Growth of theSymbiont bearing foraminifera *Amphistegina lessonii* d'Orbigny and *Heterostegina depressa* d'Orbigny (Protozoa). *In* W. Schwemmler and H.E.A.Schenk (eds) Endocytobiology I, 125-132.

SARASWATI. P.K., 2002. Growth and habitat of some recent miliolid foraminifera: Palaeoecological implications. Current Science, 82, 81-84.

SARASWATI, P.K., SETO, K. and NOMURA, R., Commn. Oxygen and carbon isotopic variationin co-existing larger foraminifera from a reef flat of Akajima, Okinawa, Japan,Marine Micropal.

SPERO, H.J. and DENIRO, M.J., 1987. The influence of symbiont photosynthesis on the$\delta^{18}O$ and $\delta^{13}C$ values of planktonic foraminiferal shell calcite. Symbiosis, 4, 213-228.

SPERO, H.J. and LEA, D.W., 1993. Intraspecific stable isotope variability in planktonic Foraminifera *Globigerinoides sacculifer*: Results from laboratory experiments.Marine Micropal., 22, 221-234.

SPERO, H.J., LERCHE, I. and WILLIAMS, D.F., 1991. Opening of the carbon isotope "vitaleffect" black box, 2, Quantitative model for interpreting foraminiferal carbonisotope data. Paleoceanography, 6, 639-655.

SPERO, H.J. and PARKER, S.L., 1985. Photosynthesis in the symbiotic planktonicforaminifer *Orbulina universa* and its potential contribution to oceanic primary productivity. Jour. Foram. Res., 15, 273-281.

TARUTANI, T., CLAYTON, R.N. and MAYEDA, T.K., 1969. The effect of polymorphism and magnesium substitution on oxygen isotope fractionation between calcium carbonate and water. Geochim. Cosmochim. Acta, 33, 987-996.

WEFER, G. and BERGER, W.H., 1980. Stable isotopes in benthic foraminifera: Seasonalvariation in large tropical species. Science, 209, 803-805.

WEFER, G., KILLINGLEY, J.S. and LUTZE, G.F., 1981. Stable isotopes in recent largerforaminifera. Palaeogeog. Palaeoclimat. Palaeoecol., 33, 253-270.

WILLIAMS, D.F., ROTTGER, R., SCHMALJOHANN, R. and KEIGWIN, L., 1981. Oxygen and carbon isotopic fractionation and algal symbiosis in the benthic foraminiferan *Heterostegina depressa*. Palaeogeog. Palaeoclimat. Palaeoecol., 33, 231-251.

WOLF-GLADROW D.A., BIJMA, J. and ZEEBE, R.E., 1999. Model simulation of the carbonateChemistry in the microenvironment of symbiont bearing foraminifera. Mar.Chem., 64, 181-198.

ZEEBE, R., BIJMA, J and WOLF-GLADROW, D.A., 1999. A diffusion-reaction model of carbonisotope fractionation in foraminifera. Mar. Chem., 64, 199-227.

ZIMMERMANN, M.A., WILLIAMS, D.F. and ROTTGER, R., 1983. Symbiont-influenced isotope disequilibrium in *Heterostegina depressa*. Jour. Foram. Res., 13, 115-121.

Early Neogene nalcareous nannofossils from Lacam Point Section, Havelock Island, Ritchie's Archipelago, Andaman Sea

Om Prakash Singh

Department of Geology, Banaras Hindu University, Varanasi 221 005, India

Introduction

The Andaman-Nicobar Islands in the northeastern Indian Ocean represents a part of a larger geotectonic unit of Southeast Asia stretching from Arakan-Yoma in the north to Mentwai group of Islands in the south. The Andaman-Nicobar group of Islands are situated between the latitudes 6° 45'N and 14° N and between the longitudes 92° 15' E and 94° E separating the Bay of Bengal from the Andaman Sea (Fig. 1). It is geologically one of the most important regions in India, comprising an almost continuous, deep sea marine sequences ranging from Late Mesozoic to Quarternary age (Srinivasan, 1978, 1984, 1986). The Late Mesozoic and Paleogene sediments displaying highly deformed, weathered and poorly fossiliferous nature are exposed on the main Islands, whereas Neogene sediments displaying less deformed and richly fossiliferous nature are exposed in Ritchie's Archipelago and southern group of Islands. The Neogene sediments are characterized by abundant and well preserved pelagic microfossils such as foraminifera, calcareous nannofossils, radiolaria, silicoflagellates and diatoms, which provide basis for high resolution biostratigraphy and enable better understanding of the paleoceanographic evolution of the region (Srinivasan and Azmi, 1976a,; Srinivasan *et al.*, 1983; Wei and Srinivasan, 1984; Srinivasan, 1986; Sharma and Sharma, 1988, 1989; Gupta and Srinivasan, 1992; Singh and Jafar, 1995; Jafar and Singh, 1996; Singh *et al.*, 2000; Singh, 2002).

Havelock Island (Latitude 12° N, Longitude 93° E), a member of Ritchie's Archipelago in the Andaman Sea, is the study area for the present work (Figs. 1B, 2A). It is the largest among the Archipelago group of Islands. It is about 17 km long and 8 km in width. Greater part of the Island is covered with thick vegetation but outcrops stand out at several places forming steep cliffs particularly in the northern part of the Island.

Geology of Havelock Island

Havelock Island predominantly consists of grey silty mudstone and nanno-foram chalk belonging to Inglis Formation. The chalk beds are well exposed in the northwest promontory forming steep cliffs around the coast. Exposures of silty mudstone and limestone occur on the western part of the Island whereas major part of the south coast consists of nanno-foram chalk. Neogene microfossils viz. calcareous nannofossils (Pant and Bandopadhyay, 1972; Mathur, 1973; 1987), foraminifera (Srinivasan and Azmi, 1976a,b), diatoms and silicoflagellates (Mathur, 1987) are well known from this Island.

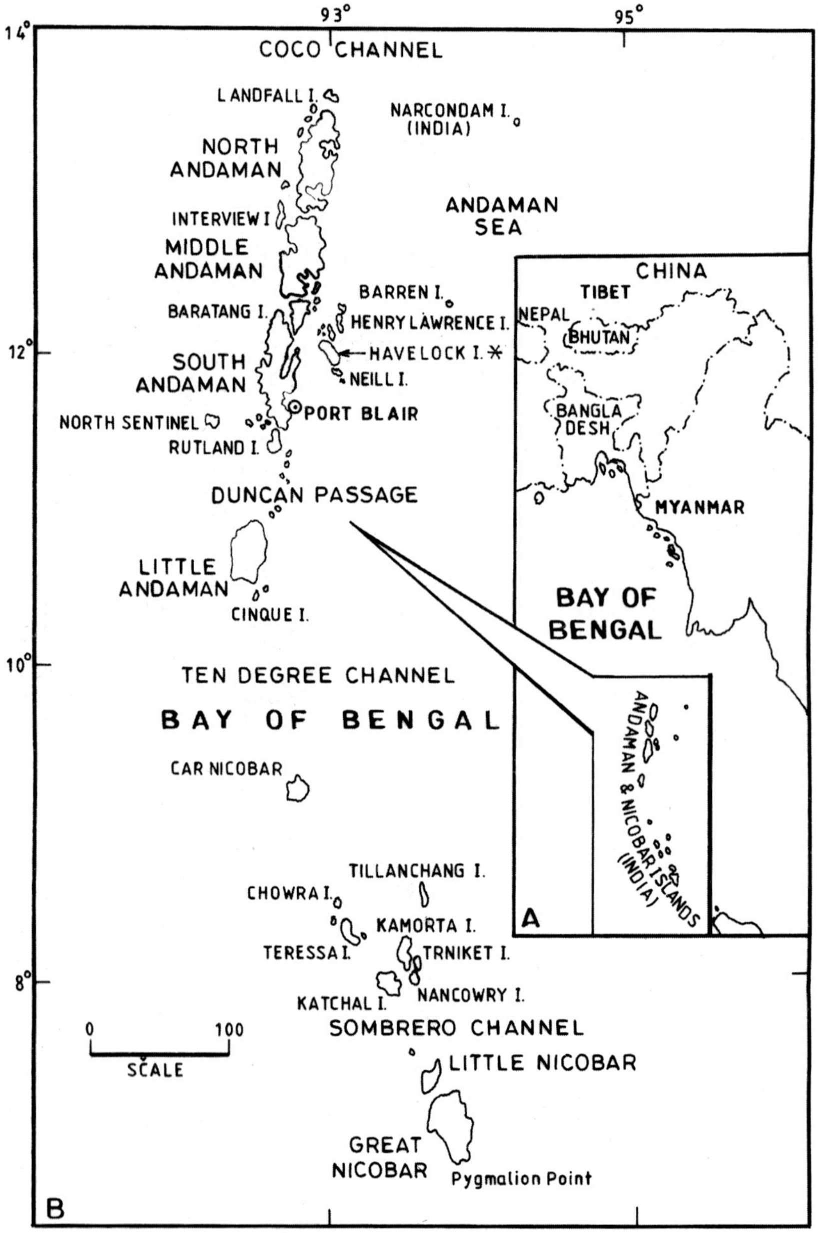

Fig. 1. (A) Map Showing the Andaman-Nicobar Islands in the Bay of Bengal. (B) Map showing the study area marked by asterisk.

Lacam Point Section

The Lacam Point Section, consisting of moderately hard, creamish-white nanno-foram chalk representing the Inglis Formation is well exposed at the northwest point, near Lacam Harbour of Havelock Island (Figs. 2A,B). This section is about 16 metre thick. Eight samples (H/Mf. 21 to H/Mf. 28) were collected from this sequence (Figs. 2B, 3). The base of the section is obscured and the top of the section is covered with vegetation. Samples are rich in pelagic microfossils such as foraminifera, calcareous nannofossils, radiolaria, diatoms and silicoflagellates.

Material and methods

Eight samples (H/Mf. 21 – H/Mf. 28) from the Lacam Point Section of Havelock Island were made available through the kind courtesy of Prof. M.S. Srinivasan, Department of Geology, Banaras Hindu University, Varanasi. Stratigraphic position of the samples bearing prefix H/Mf. is shown in the litholog (Figs. 2B, 3). For the preparation of smear slides, a small fraction of the sample was taken out by means of a pointed knife in a porcelain crucible. Turbid suspension was made by addition of distilled and neutral water. Little suspension was spread over a clean glass slide kept on a hot plate. After drying the suspension, permanent mounting was performed by covering it with a coverslip using canada balsam as mounting medium. Two sets of conventional slides were prepared for each sample. One set contains the finer fraction whereas another set contains the coarser fraction. Calcareous nannofossils were studied by using a Leitz Polarizing Research Microscope and an 100/1.32 oil immersion objective. The study was carried out both under the normal (Single polarizer) and crossed polarized illuminations.

Scanning electron microscopic studies were carried out with the SEM 505 Philips model installed at BSIP, Lucknow with accelerating voltage of 30.0 KV. Scanning electron micrographs are of variable magnifications illustrated in Plate 1. The smear slides and SEM negatives have been deposited in Micropaleontology Laboratory, Department of Geology, Banaras Hindu University, Varanasi.

Results and discussion

The examined samples from Lacam Point Section yielded nineteen species of calcareous nannofossils. The assemblage is represented by the genera *Calcidiscus, Coccolithus, Discoaster, Helicosphaera, Pontosphaera, Cyclicargolithus, Dictyococcites, Reticulofenestra, Sphenolithus* and *Thoracosphaera* (Fig. 3).

Presence of zonal marker *Helicosphaera ampliaperta* along with *Sphenolithus heteromorphus* and the absence of *Sphenolithus belemnos* throughout the samples of Lacam Point Section suggest that the entire section can be assigned to the *Helicosphaera ampliaperta* Zone (NN4) of Martini (1971; Fig. 3) corresponding to Zone CN3 of Okada and Bukry (1980) which is of late Early Miocene to early Middle Miocene age (Fig. 3) and corresponds to Jarawaian Stage of Andman-Nicobar Islands (Fig. 2B).

Srinivasan and Azmi (1976a) carried out detailed studies on planktic foraminifera from the Lacam Point Section and assigned it to late Early Miocene to early Middle Miocene (*Globigerinatella Insueta* Zone to *Globorotalia fohsi peripheroronda* Zone). Thus, the age of Lacam Point Section exposed at Havelock Island inferred on the basis of calcareous nannofossil data (present study) is in conformity with the age indicated by planktic foraminifera (Srinivasan and Azmi, 1976a).

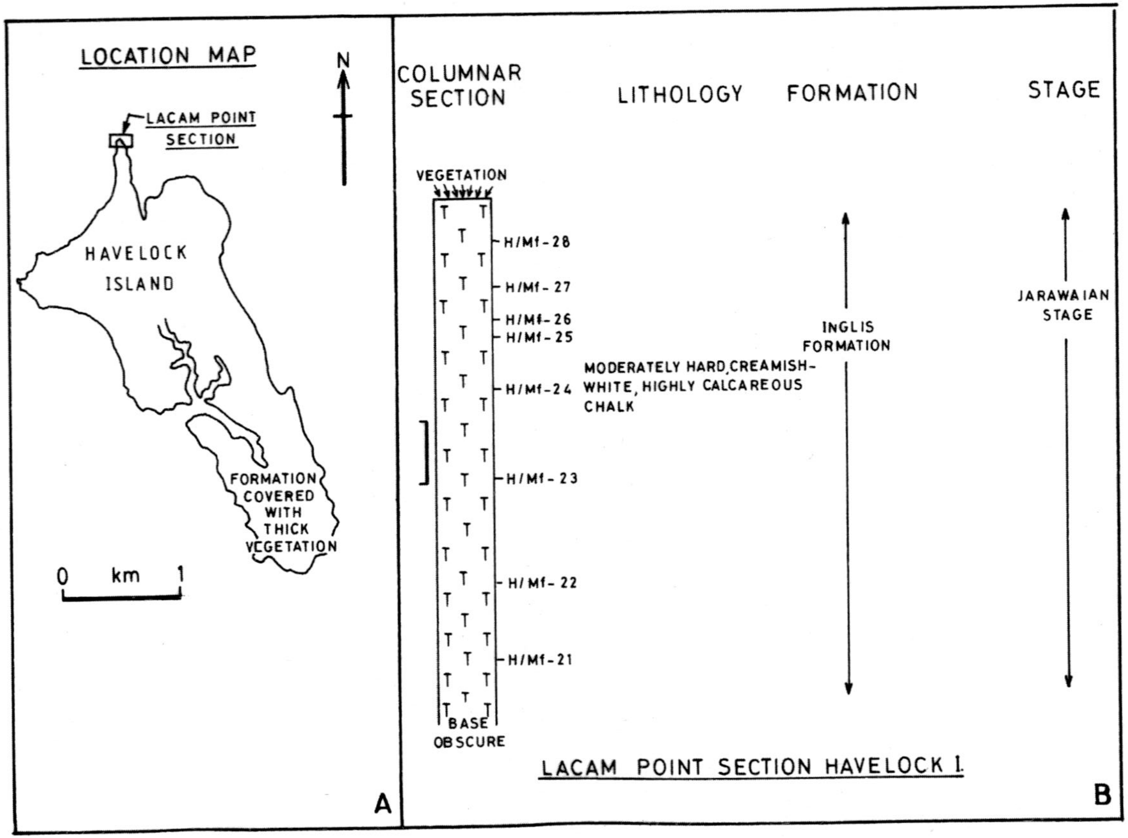

Fig. 2. (A) Map showing the location of the Lacam Point Section on Havelock Island, (B) Columnar section showing the lithology and sample locations at Lacam Point Section (after Srinivasan and Azmi, 1976a).

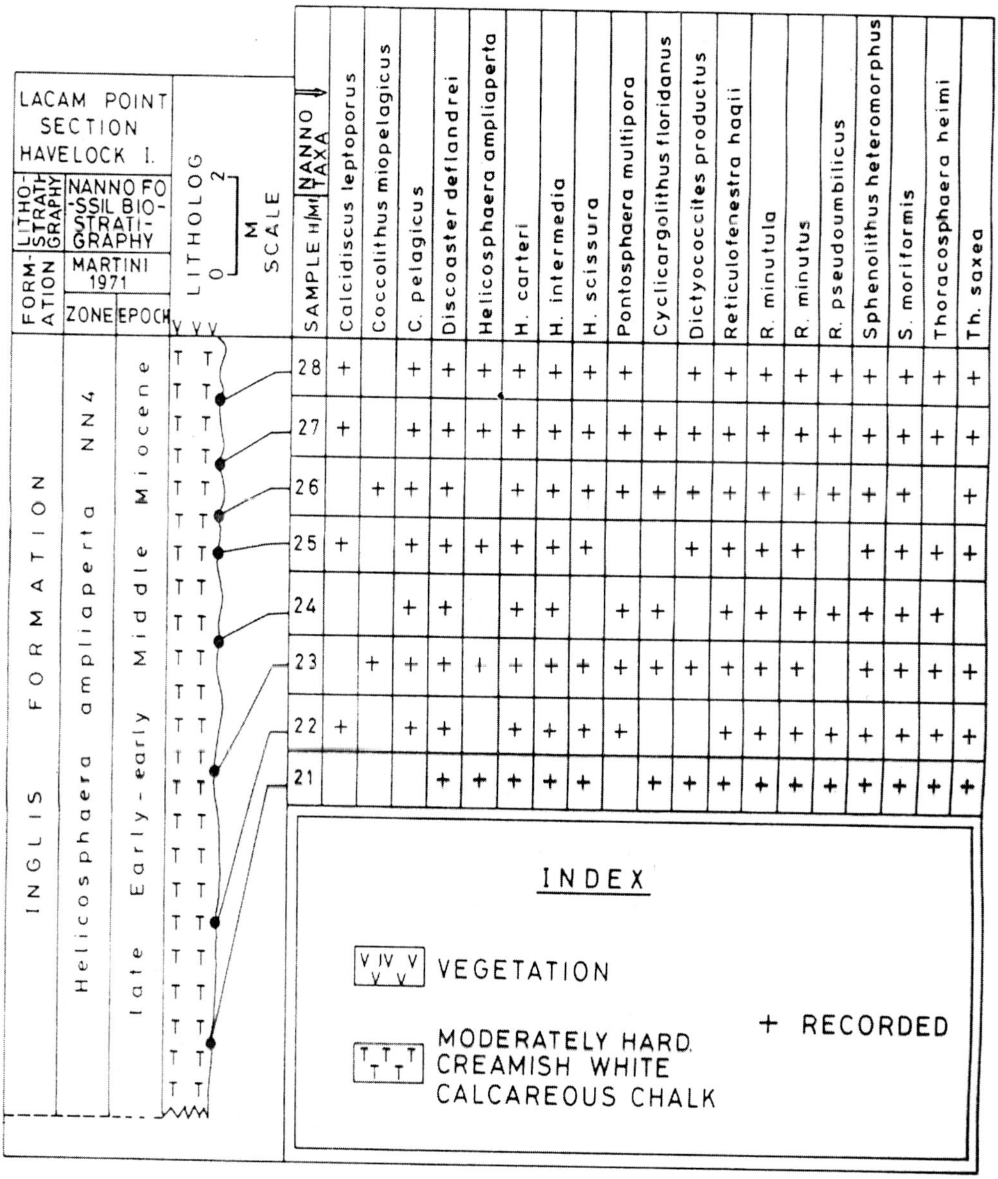

SAMPLE H/MI	Calcidiscus leptoporus	Coccolithus miopelagicus	C. pelagicus	Discoaster deflandrei	Helicosphaera ampliaperta	H. carteri	H. intermedia	H. scissura	Pontosphaera multipora	Cyclicargolithus floridanus	Dictyococcites productus	Reticulofenestra haqii	R. minutula	R. minutus	R. pseudoumbilicus	Sphenolithus heteromorphus	S. moriformis	Thoracosphaera heimi	Th. saxea
28	+		+	+	+	+	+	+	+		+	+	+	+	+	+	+	+	+
27	+		+	+	+	+	+	+	+	+	+	+	+	+	+	+	+	+	+
26		+	+	+		+	+	+	+	+	+	+	+	+	+	+	+		+
25	+		+	+	+	+	+	+			+	+	+	+		+	+	+	+
24			+	+		+	+		+	+		+	+	+	+	+	+	+	
23		+	+	+	+	+	+	+	+	+	+	+	+	+		+	+	+	+
22	+		+	+		+	+	+	+			+	+	+	+	+	+	+	+
21				+	+	+	+	+		+	+	+	+	+	+	+	+	+	+

Fig. 3. Range chart showing the sample positions in the litholog, biostratigraphic zone and distribution of calcareous nannofossils.

Plate 1.

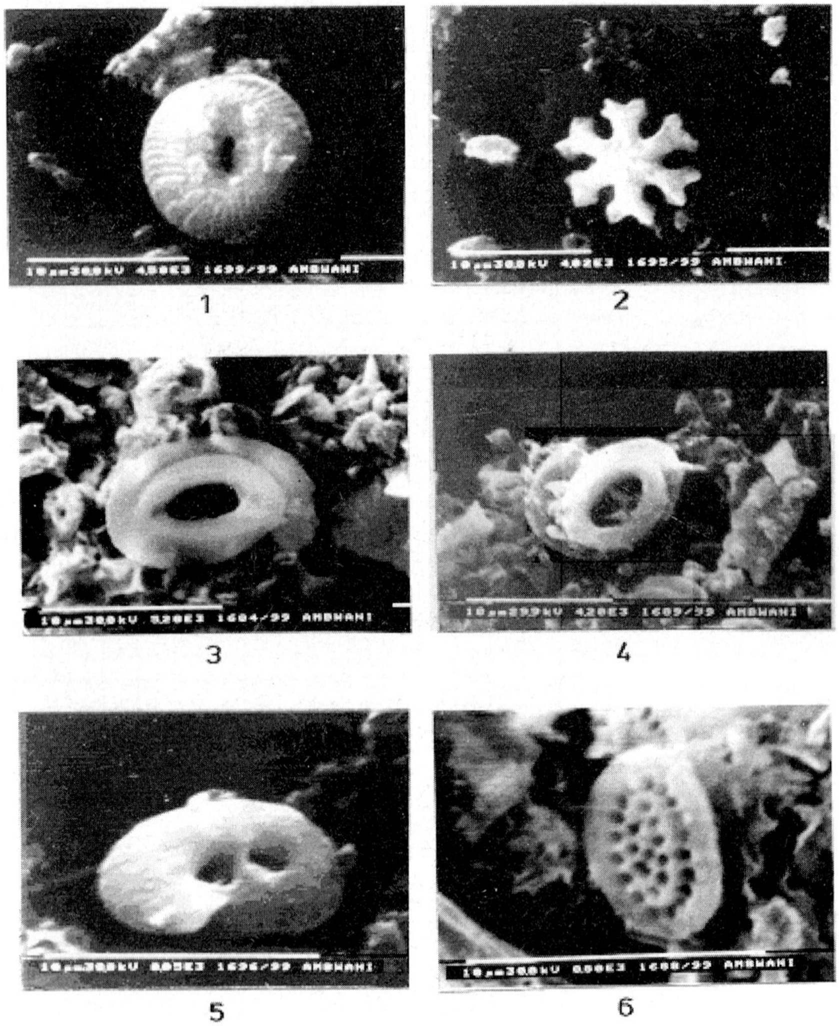

Explanation:

All scanning electron micrographs are of variable magnifications shown by scale bar =10μm.

1. *Coccolithus pelagicus* (Wallich) Schiller (1930)
2. *Discoaster deflandrei* Bramlette and Riedel (1954)

3,4. *Helicosphaera ampliaperta* Bramlette and Wilcoxon (1967).

5. *Helicosphaera carteri* (Wallich) Kamptner (1954)
6. *Pontosphaera multipora* (Kamptner) Roth (1970)

Conclusions

1. A rich assemblage containing nineteen species of calcareous nannofossils is recorded from the Lacam Point Section of Havelock Island.
2. Lacam Point Section of Havelock Island can be referred to the *Helicosphaera ampliaperta* Zone (NN4) of Martini (1971) corresponding to Zone CN3 of Okada and Bukry (1980).
3. Present study on calcareous nannofossils reveals that the age of Lacam Point Section of Havelock Island is range from late Early Miocene to early Middle Miocene. This is in conformity with the age indicated by planktic foraminifera (Srinivasan and Azmi, 1976a).

Acknowlwdgements

The author is extremely grateful to Prof. M.S. Srinivasan, Department of Geology, Banaras Hindu University, Varanasi, for providing the rock samples and for going through the manuscript.

References

BRAMLETTE, M.N. and RIEDEL, W.R. 1954. Stratigraphic value of discoasters and some other microfossils related to Recent coccolithophores. Jour. Paleontol, 28, 385-403.

BRAMLETTE, M.N. and WILCOXON, J.A. 1967. Middle Tertiary calcareous nannoplankton of the Cipero section, Trinidad. W.I. Tulane Stud. Geol, 5, pp. 93-131.

GUPTA, S.M. and SRINIVASAN, M.S. 1992. Late Miocene radiolarian biostratigraphy and paleoceanography of Sawai Bay Formation, Neill Island, Andamans, India. Micropaleontol., 38, no. 3, 209-235.

JAFAR, S.A. and SINGH, O.P. 1996. Late Miocene calcareous nannofossils from Sawai Bay Formation, Neill Island, Andaman Sea, India. Contrib. XV Indian Colloq. Micropaleont. Strat., 733-749.

KAMPTNER, E. 1954. Untersuchungen uber den Feinbau der Coccolithen. Arch. Protistenk, 100, 1-90.

MARTINI, E. 1971. Standard Tertiary and Quaternay calcareous nannoplankton zonation. In: A. Farinacci (Ed.). Proc. II Planktonic Conference, Roma, 1970, v. 2, 739-785.

MATHUR, K. 1973. Fossil nannoplanktons from Havelock Island, Andamans. Bulletin of ONGC, 10, no. 1-2, 89-92.

MATHUR, K. 1987. Studies on the Neogene diatoms, silicoflagellates and nannoplankton of Havelock Island, Andaman Sea, India. Int. Symp. Phycology, Madras (Abst.), 41.

OKADA, H. and BUKRY, D. 1980. Supplementary modification and introduction of code numbers to the low-latitude coccolith biostratigraphic zonation. Mari. Micropaleontol., v. 5, no. 3, 321-325.

PANT, S.C. and BANDOPADHYAYA, S. 1972. Calcareous nannoplankton from Ritchie's Archipelago, Andaman. Indian Minerals, 26, 2, 73-76.

ROTH, P.H. 1970. Oligocene calcareous nannoplankton biostratigraphy. Eclog. Geol. Helv., 63. 799-881.

SCHILLER, J. 1930. Coccolithineae. In: kryptogamen-Flora von Deutschland, Osterreich und der Schweiz, Akad. Verlagsges., Leipzig, Rabenhorst, L. (Ed.), 10, 2, 89-267.

SHARMA, V. and SHARMA, G.K. 1988. Radiolaria from Neill Island, Andaman Sea, and their distributional characteristics. Jour. Paleontol. Soc. India, 33, 7-19.

SHARMA, V. and SHARMA, G.K. 1989. Late Miocene to Early Pliocene radiolarian biostratigraphy of Neill Island, Andaman Sea. Jour. Geol. Soc. India, . 34, 76-82.

SINGH, O.P. 2002. Early Neogene calcareous nannofossil biostratigraphy of Henry Lawrence Island, Ritchie's Archipelago, Andaman Sea. Jour. Geol. Soc. India, . 60, 193-198.

SINGH, O.P. and JAFAR, S.A. 1995. Late Miocene discoasters from Swai Bay Formation, Neill Island, Andaman Sea, India. Palaeobotanist, . 44, 189-206.

SINGH, O.P., SRINIVASAN, M.S. and SHARMA, V. 2000. Early Neogene Multiple Microfossil Biostratigraphy, John Lawrence Island, Andaman Sea. Micropaleontol., . 46, . 4,343-352.

SRINIVASAN, M.S. 1978. New chronostratigraphic divisions of the Andaman-Nicobar Late Cenozoic. Recent Res. Geol., 4, . 22-36. Hindustan Publ. Corp. Delhi.

SRINIVASAN, M.S. 1984. The Neogene of Andaman-Nicobar. In: Ikebe, N., and Tsuchi, R. (Eds). Contributions to biostratigraphy and chronology, . 203-207. Univ. Tokyo Press, Japan.

SRINIVASAN, M.S. 1986. Geology of Andaman and Nicobar Islands. Jour. Andaman Sci. Assoc., 2, 1, 1-12.

SRINIVASAN, M.S. and AZMI, R.J. 1976a. New Developments in the Late Cenozoic lithostratigraphy of Andaman-Nicobar Islands. Proc. VI Indian Colloq. Micropaleontol. Strat., 302-327.

SRINIVASAN, M.S. and AZMI, R.J. 1976b. Paleobathymetric trends of the Late Cenozoic foraminiferal assemblages of Ritchie's Archipelago, Andaman Sea, Proc. VI Indian Colloq. Micropaleontol. Strat, 328-354.

SRINIVASAN, M.S., LOMBARI, G. and DAVE, A. 1983. Early Miocene planktonic foraminiferal and radiolarian zonation, Colebrook Island, Andaman Sea. Jour. Geol. Soc. India, 24, 1-18.

WEI, K.Y. and SRINIVASAN, M.S. 1984. Miocene calcareous nannofossils from Colebrook, North Passage and Great Nicobar Islands, Northeastern Indian Ocean. Rev. Esp. Micropaleontol., 16, 1-3, 345-366.

Distribution pattern of benthic foraminifera off Krishna-Godawari Delta and their environmental significance

D.Bhattacharjee[1], P.Vijoy Kumar [2] and B.R.Rao[2]

[1]OpEC-I, MW, GSI, Kolkata 700 032, India
[2] OpEC-II, MW, GSI, Visakhapatanam 530017, India

Introduction

Benthic foraminiferal studies carried out by several marine geologists during the last established that foraminifera is an important ecological indicator and one of the primary means of interpreting the depositional environments. Oil exploration off Krishna-Godavari delta area in the Bay of Bengal in the last few years has shown the oceanographic significance of foraminifera from the continental shelf, slope and abyssal domains of the region. The foraminiferal data base generated by marine wing of GSI from the *R. V.Samudra Manthan* Cruises of SM-150 &155 reveals the behavior of benthic foraminifera off Krishna-Godavari delta and inters deltaic regions. This study suggests the down slope sediment displacement from different physiographic provenances and their environmental significance. It also helps to identify the dissolved oxygen content of the marine domain which may in turn indicate the favorable environmental set up for the search of oil and natural gas in the Bay of Bengal.

Previous work

The occurrence of benthic foraminifera from the territorial water, continental shelf: slope and' abyssal' plain off Andhra Pradesh coast in the Bay of Bengal, has been studied by various workers over the last few decades. However, no comprehensive foraminiferal biofacies map is published till date. Ganapati & Sarojini (1959) studied ecology of foraminifera off Visakhapatanam coast. Bhalla (1968) studied the benthic taxa from the Andhra coast and correlated their relation with geographic provinces. Vedantam & Rao (1970) described foraminifera off Pentakota coast. Rao & Subba Rao (1971) studied living foraminifera from tidal creek at Pudimadaka. Subba Rao et. al.(1979) studied the distribution and ecology of benthic foraminifera of Visakhapatanam shelf. Bhalla & Nigam (1988) carried out cluster analysis of foraminiferal fauna from beaches of the east coast with reference to foraminiferal provinces. Anantharaman (1989) demarcated the foraminiferal biofacies in the sediments of Nizaamapetanam Bay. Madabhusi et. al. (1994) studied the foraminifera from the outer continental shelf off Pudimadaka within the lime mud sediments of Holocene time. Bhattacharjee et.al.(2002) studied the foraminiferal biofacies and examined their environmental implication in the northern parts of Bay of Bengal.

Present Work

In the present work benthic foraminiferal biofacies map has been prepared on the basis of the relative abundance of different benthic taxa in the surficial sediments of thirty gravity cores collected during the Cruise SM-155 from an area of about 6000 sq. km between the latitudes 15° 05'N to16015'N and longitudes 81006'E to 82° 15'E within the water depths 30m and 1766m. Location of different samples and sample particulars are shown in Fig.l (Location map) and Table.l respectively. The above seabed samples are studied under binocular microscope to examine the relation of different groups of microorganisms to the physiographic domain and the bottom sediments. Quantitative study of benthic foraminifera is carried out (Table.3) and their environmental interpretation is discussed.

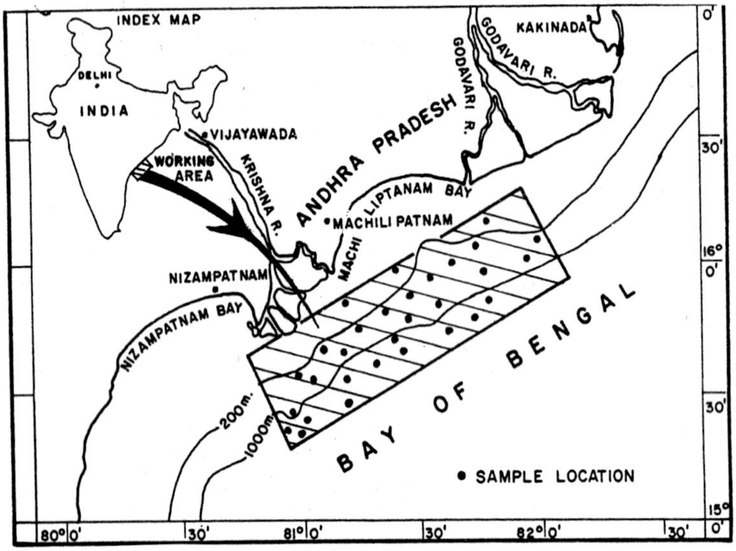

Fig. 1. Location map of the study area (SM-155).

Methodology

Seabed sediments are processed following the standard laboratory techniques suggested by Boltovoskoy & Wright (1976). The processed samples having good number of foraminifera are split with an Otto-microsplitter in order to obtain a suitable aliquot that contains about 300 specimens. The individual specimen was hand picked and kept in separate assemblage trays for taxonomic identification. The biogenic materials of the coarser fraction (+230 ASTM) are mainly comprise of foraminifera and radiolaria with some amount of spines, diatom, ostracod, pteropods, bivalves, gastropods, and scaphopods. Quantitative study of foraminifera has been carried out due to their abundance and their ecological significance. A total of 14 planktic and 68 benthic foraminiferal taxa have been identified from the gravity core samples of this region. Planktic foraminifera are identified as per the taxonomic classification of Kennett & Srinivasan (1983) and Bolli et.al. (1985) whereas benthic foraminifera are identified following the systematic works of Loeblich & Tappan (1988), Barker (1960), Ingle et.al. (1980), Gupta & Srinivasan (1990) and Wells et.al. (1994). Benthic foraminiferal biofacies has been prepared after the method adopted by Walton (1964), Patterson (1993) as well as Patterson & Cameroon (1991).

Results

The abundance of major groups of microorganisms in the seabed sediment samples is shown in Table-I. The relation of the dominant biota with the bathymetry and bottom sediments is also presented in Table-I. Quantitative study of the benthic taxa in the surficial seabed sediments of different gravity cores is given in Table.3. Spatial distribution pattern of biofacies is illustrated in Fig.2 and the dominant taxa of each biofacies are shown in Table-2. A brief description of the each biofacies is as follows.

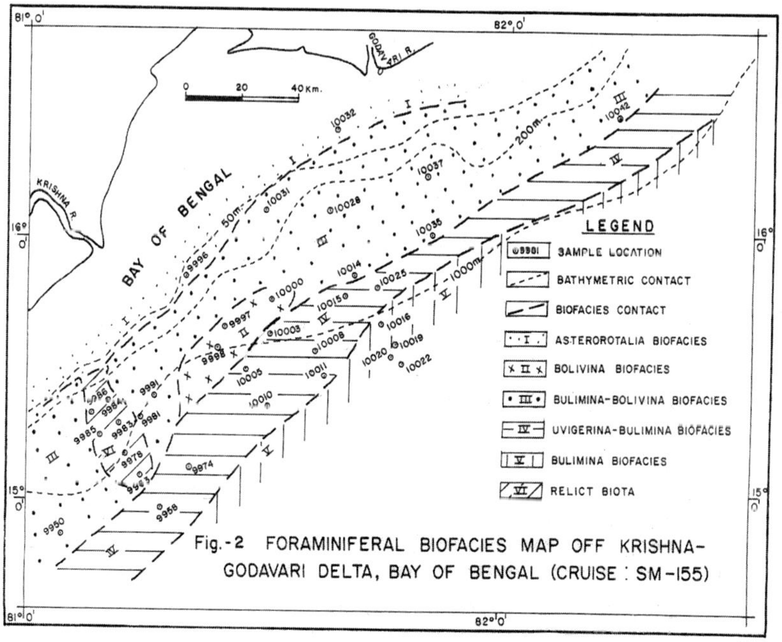

Fig. 2. Foraminiferal biofacies map off Krishna- Godavari delta,Bay of Bengal (SM-155).

Astrorotalia biofacies

This biofacies occurs in the northern part of the area, occupying the shallower region within the water depths between 30m and 70m (Sample no.GC-9996&10032). Bottom sediment of this facies is mainly sandy and contains a number of benthic foraminifera, bivalves and gastropod. Important species of this biofacies are *Ammonia beccarii, Bolivina striata, Florilus scapha, F.asterizans, Lenticulina calcar, Quinqueloculina parkari, Q. vensta* & *Elphidium crispum*. Planktic foraminifera are rare in this biofacies.

Bolivina biofacies

Bolivina biofacies occurs as a lenticular patch in the continental slope region within the silty clay sediments (Fig.2). The depth range of this biofacies varies from 550m to 1200m as observed in sample nos. GC-9993, 9995, 9997,9998 and 10000 . Important taxa of this facies are *Chrysalidinella dimorpha, ,Reophax dentaliniformis, Hyalina baltica, Bulimina pyrula, Bolivina spatulata, Bolivinata quadrilateral, Uvigerinaperegrine, cassidulina carinata, Cibicides brady, Haplophragmoides canariensis, Discammina compressa, Lenticulina thalmanni & L.calcar.* Planktic foraminifera is very common in this biofacies and Planktic /benthic ratio varies from 8 to 14.

Bolivina - Bulimina biofacies

This biofacies shows a wide range of depth variation from shelf region (135m) to middle bathyal zone(1575m) and occurs in the silty to clayey sediments (Sample nos.GC9950,9981,9991,10014,10031 &10042). Important benthic members of this facies are mainly made up of both non-agglutinated and agglutinated families. They are represented by *Cassidulina teretis, C.carinata, Lenticulina convergens, Bulimina marginata, Virgulina rotunda, Loxostomum karrerianun, Discamina compressa, Cibicides wullerstrofi, Siphotextularia concave, Textularia conica, Ammobaculites agglutinaus, Quinqueloculina seminulum, Lagina sp. & Fissurina radiata.* Planktic/ benthic ratio of this unit varies from 8 tol6.

Bulimina- Uvigerina biofacies

Bulimina~ Uvigerina biofacies occurs between 1033m and 1766m water depth in the clayey'bottoin sediments (Sample nos. GC-9958, 9975, 10005, 10010, lOOll, 10015, & 10043). Percentage of infaunal fauna is more than the epifaunal taxa in this unit. Important taxa. of this biofacies are *Globobulimina pacifica , Chillostomella oolina, Uvigerina asperula, Bulimina marginata, Cassidulina globolusa, Globocassidulina subglobosa, Dentalina inornata, Gyrodina neosoldani,Bolivina alata, Gavelinopsis labatulas, Virgulina brady, D.communis, Hauglandina elegan & Hyalina baltica.* Planktic / Benthic ratio (P.B.) of foraminifera is more in the northeastern side but considerably decreases towards the southwestern end of the biofacies due to influx of some shallow water benthic species from the northern direction.

Bulimina biofacies

This biofacies occurs within 1208m to 1622m water depth in clayey sediments (Sample nos. GC-I0016, 10018, 10020, 10022, 10024).Diversity of benthic population increases in this unit but with a reduction of their sizes. Occasionally the infaunal and epifaunal benthic taxa show blackening of their tests. Important members of this facies are *Uvigerina peregrine, U.asperula, Hyalina baltica, Epistomonella exigua,Bulimina exilies, Bolivina robust a, B.spatulata, Cassidulina subglobosa, Pullania bulloides, Pyrgo murrhyna, Gyrodina neosoldani, Oridorsalis umbonatus, Virgulina bradyi, Oolina seminula & Fissurina orbignyana.* P:B ratio of the foraminifera varies from 15 to 22.

Relict zone

A lensoid patch of sandy silty sediment lying between 929m to 1735m water depth, is observed in the southwestern side of the area contains considerable amount of skeletal and non skeletal grains (Sample nos.GC-9963,9978,9983,9984&9985). The non skeletal grains are made up of rounded to

Table 1 . Particulars of seabed sediments off Krishna-Godawari delta for paleontological studies in the Bay of Bengal (SM-155)

S.No	SmpleNo.	Latitude°N		Longitude°E		Depth	Lithology	Dominant Biota
		Deg	Min	Deg	Min	(m)		
1	GC-9950	15	24.66	81	6.46	1575	Silty clay	Foraminifera+Radiolaria
2	GC-9958	15	27.85	81	16.72	1766	Clay Silt	Foraminifera+Radiolaria
3	GC-9963	15	31.61	81	14.32	1735	Silt	Foraminifera+Radiolaria
4	GC-9974	15	34.6	81	20.06	1751	Clay	Foraminifera+Radiolaria
5	GC-9978	15	35.99	81	12.92	1473	Silt	Foram+Radiolaria+Diatoms
6	GC-9981	15	38.3	81	14.29	1295	Silty clay	Foram+Radiolaria+Diatoms
7	GC-9984	15	39.22	81	10.76	1098	Silt	Foraminifera
8	GC-9992	15	44.52	81	17.06	977	Silty clay	Foraminifera
9	GC-9996	15	55.65	81	20.25	70	Sand	Benthic Forams+Molluscus
10	GC-9998	15	47.27	81	26.79	890	Silt	Forams +pteropod
11	GC-9986	15	39.72	81	8.82	929	Sandy silt	Forams +pteropod
12	GC-9991	15	41.23	81	16.92	1185	Silty clay	Foraminifera
13	GC-9995	15	46.7	81	23.45	1127	Clayey silt	Foraminifera
14	GC-9997	15	49.78	81	25.13	566	Sandy silt	Forams +pteropod
15	GC-10000	15	51.85	81	29.4	556	Sandy silt	Forams +pteropod
16	GC-10003	15	48.96	81	30.89	817	Silt	Forams +pteropod
17	GC-10005	15	45.01	81	27.93	1033	Clay	Foraminifera
18	GC-10007	15	46.81	81	32.36	998	Clay	Foraminifera
19	GC-10010	15	40.76	81	30.85	1375	Clay	Foraminifera+Radiolaria
20	GC-10011	15	44.14	81	37.87	1430	Clay	Foraminifera+Radiolaria
21	GC-10014	15	55.58	81	41.4	1032	Silty clay	Foraminifera
22	GC-10016	15	49.95	81	45.46	1332	Clay	Foraminifera+Radiolaria
23	GC-10019	15	47.07	81	43.65	1479	Clay	Foraminifera+Radiolaria
24	GC-10020	15	47.55	81	43.68	1543	Clay	Foraminifera+Radiolaria
25	GC-10022	15	45.56	81	47.42	1622	Clay	Foraminifera+Radiolaria
26	GC-10028	16	5.67	82	42.1	531	Sandy silt	Foraminifera+pteropod
27	GC-10031	16	3.33	81	30.17	135	Sandy silt	Foraminifera+pteropod
28	GC-10032	16	10.35	81	40.25	30	Sand	Benthic Forams+Molluscus
29	GC-10037	16	6.04	82	46.45	881	Silty	Foraminifera+pteropod
30	GC-10042	16	12.47	82	15.44	800	Silty clay	Foraminifera+pteropod

Table 2. Distribution pattern of benthic foraminiferal biofacies in seabed sediments off Krishna-Godawari delta, Bay of Bengal

Distribution pattern of Benthic Foraminiferal Biofacies in seabed sediments off Krishna-Godavari Delta, Bay of Bengal					
S.No.	**Samp.No.**	**Depth (m)**	**Physiographic Domain**	**Dominent Taxa**	**Biofacies unit**
1	GC-10032	30	Inner shelf	*A.trispinasa,A.beccarii*	*Astrorotalia*
2	GC-9996	70	Middle shelf	*A.trispinasa,A.beccarii*	*Astrorotalia*
3	GC-10031	135	Outer shelf	*B.robusta,B.popoide,B.pyrula*	*Bolivina-Bulimina*
4	GC-10028	531	Middle bathyal off Godavari	*B. striatula,B.spatulata*	*do*
5	GC-10000	556	Upper bathyal	*B.alata, B.robusta*	*Bolivina*
6	GC-9997	566	do	*B.alata,B.robusta,*	*Bolivina*
7	GC-10042	800	Middle bathyal off Godavari	*B.pyrula, B. alata, B.spatulata*	*Bolivina-Bulimina*
8	GC-10003	817	Middle bathyal off Krishna	*U.peregrina,B.pyrula,B.marginata*	*Bulimina-Uvigerina*
9	GC-10037	881	Middle bathyal off Godavari	*B.alata,B.robusta,B.marginata*	*Bolivina-Bulimina*
10	GC-9998	890	Middle bathyal off Krishna	*B.alata,B.striatula*	*Bolivina*
11	GC-9986	929	do	*T.conica,A.trispinosa*	Relict biota
12	GC-9992	977	do	*B.robusta,B.pyrula B.alata*	*Bolivina-Bulimina*
13	GC-10007	998	do	*U.peregrina,B.popoides,B.margin*	*Bulimina-Uvigerina*
14	GC-10014	1032	Middle bathyal off Godavari	*B.pyrrula,B.popoides,B.spatulata*	*Bolivina-Bulimina*
15	GC-10005	1033	do	*U.peregrina,B.pyrula,B.marginata*	*Bulimina-Uvigerina*
16	GC-9984	1098	Middle bathyal off Krishna	*Amphistegina sp,E.crispun,C.calcar*	Relict biota
17	GC-9995	1127	do	*B.alata,B.robusta,B.striatula*	*Bolivina*
18	GC-9991	1185	do	*B.pyrula,B.alata.*	*Bolivina-Bulimina*
19	GC-9981	1295	do	*do*	*Bolivina-Bulimina*
20	GC-10016	1332	Middle bathyal off Godavari	*B.marginata,B.popoide,B.pyrula,*	*Bulimina*
21	GC-10010	1375	do	*U.peregrina,B.pyrula,B.marginata*	*Bulimina-Uvigerina*
22	GC-10011	1430	do	*U.peregrina,B.pyrula,B.marginata*	*Bulimina-Uvigerina*
23	GC-9978	1473	Middle bathyal off Krishna	*Amphistegina sp, A. beccarii, Quinqueloculina sp.*	Relict biota
24	GC-10019	1479	do	*B.marginata, B.popoide,*	*Bulimina*
25	GC-10020	1543	Lower bathyal off Godavari	*B.popoide, B.pyrula,*	*Bulimina*
26	GC-9950	1575	Lower bathyal off Krishna	*B.alata,B.pyrula,B.striatula*	*Bolivina-Bulimina*
27	GC-10022	1622	Lower bathyal off Godavari	*B.pyrula,B.marginata,B.aculata*	*Bulimina*
28	GC-9963	1735	Lower bathyal off Krishna	*A.beccarii,H.baltica,H.elegan*	Relict biota
29	GC-9974	1751	do	*U.peregrina,B.pyrula,*	*Bulimina-Uvigerina*
30	GC-9958	1766	do	*U.peregrina,B.popoides,B.margin*	*Bulimina-Uvigerina*

Table 3. Check list of benthic foraminifera off Krishna –Godawari delta , Bay of Bengal (SM-155)

Sp.no.GC-	9950	9958	9963	9978	9986	9991	9996	10000	10003	10010	10011	10016	10020	10031	10032	10042
Depth(m)	1575	1766	1735	1473	929	1183	70	556	817	1375	1430	1332	1543	135	30	800
Name of the Taxa	1	2	3	4	5	6	7	8	9	10	11	12	13	14	15	16
Ammonia beccarii (Linnaeus)			D	D	P		D							F	D	
Ammobaculites agglutinous (d'orbgny)									R	R	F	R	F			P
A.americanus cushman			F							R		F	R			R
Alveolofragmium subglobosum (Sars)	R	F			P				F		R					F
Alveolinella quoyi (d'orbigny)			P	F	R											
Amphistegina lessoni. d'Orbigny		R	F	D	P											
Astrorotalia trispinosa (Thalmann)					D		D							F	D	
Bolivina alata sequenza	D	C		F	P	D		D	P	P	F	R	R	P		D
B. striatula Cushman	D	P		R	R	P		F	R							R
B. robusta Brady			C	P				D				P	R	D		P
B. spathulata (williamson)				R	F			C	P			P		F		D
Bolivinata quadrilatera (schwager)			F	P	F			F	R					F		R
Bulimina exilies Brady						P			F	P	P	P	P			F
Bulimina popoides d'Orbigny		D		R					P	P	P	D	D	D		C
B.Pyrula d'orbigeny	D	C				D			D	D	D		D	D		D
B.marginata d'orbigeny	F	D							D	D	D	D	P			F
Cassidulina carinata Silvestri					P	F	F	P	F	R				P		
C.teretis Tappan					P	R	R	P	R	F				R		
Calcarina Calcar d'orbigeny		F	P	P	F									F		
Cancris Oblongus (Williamson)			P				F	P		R		R				
Ceratobulimina pacifica Cushman and Harris	R	F			R			F		R		F				R
Cibicides psudoungerianus (Cushman)		F		R		P			F		R		R			
C.wullerstrofi (Schwager)	P	p	R		F				R			F		R		
C. bradyi (Trouth)					F			P	P							F
C.margaritiferous (Bredy)		R	P	F	F	F						R				
C.praecinctus (Karrer)		F	P		F	F										
Chilostomella oolina (Schwager)	P	P	P	F	R		F			P	P	R				
Chrysalidinella dimorpha (Brady)			F	P	P			P	F							
Discammina Compressa (Goes)	P	P							F	R			R	R		
Dentalina inornata d'orbigny			F	F	R								F			
Elphidium Crispum (Linnacus)			C	C			P							F	P	
Epistominella exigua (Bredy)	P								P	P			F	F		
Eponoides umbonatus (Reuss)			P	F	R				F							
Florilus asterizans (Fitchel and Moll)							P		F					P	R	F

Table 3 (Contd.)

	1	2	3	4	5	6	7	8	9	10	11	12	13	14	15	16
F.scapha (Fichel &Mole)			F		R		P		F					P	F	R
Fissurina radiata Sequenza	P		F							R	F		R			
Glomospira Charoides (Jones and Parker)	F		R							R	F					
Gyriodina broeckhiana (Karrer)	F	P	R								P		R			
G. neosoldani Brotzen			F	F		R				P	P	F				
Gavelinopsis Lobatulus (Parr)	F	P									F	P	R			F
Hoglundina elegans(Schwager)		P	D	F	F			R	R	F			P			
Hyalinea baltica (schroeter)	P	F	D		C	C		P	P	F	F		F			
Haplophragmoides canariensis (d'orbigeny)	R		F						F	R		F	R	F		
Hormosina globulifera Brady	R		P	F							F		R			
Involutina calcar (Linnaeus)			F	F					P		F	F				
Lenticulina calcar (Linnaeus)			P	F				R				P		F		
L. gibba (d'orbigny)				F	F		F	F	R							
L. convergens(Bornemann)			P	P		F	P	R				R		F		R
Loxostomum karrerianum(Brady)			P				P		P	F		R	R	P	F	F
Osangularia bengalensis(Schwager)	F	R	R							F						R
Oridorsalis umbonetus(Reuss)	P	F	R			F				P	F					
Pyrgo depressa(d'orbigny)			F	F		R		F				R				
P. murrhyna(Schwager)	R	F								R	R					
Pullenia bulloides(d'orbigny)		R	R							R		F				
Pseudorotalia schroeterina Carpenter,Parker&Jones		R					P							F	C	
Quinqueloculina venusta Karrer		R	P	C			P			R				R	F	
Q. seminulam(williamson)			F	P			F	F				R			F	
Q. parkari Brady		R	F	F			R								R	
Reophax rodulosa Brady						P		F	R		F				R	
R.diffugiformis Brady				F		R		F			R				R	
Spiroloculina depressa d'orbigny			F	F			F						R	R		
S.communis Cushman and Todd			F	R			R							R		
Textularia conica d'orbigny		P	P	F	D				R					P	P	
T. pseudogramen chapman and parr				R			R								R	
Triloculina tricarinata d'orbigny			P	F	R			F							R	
Trochammina inflata(Montagu)		F	F	R	F							P			F	
Uvigerina asperula czizek		F	R	R					F	R	R		F			
U.peregrina Cushman	C	D	C	C	C	P		C	D	D	D		P		F	R
Virgulina rotundata Parr	F				F	P		R	R					R		R
V. bradyi Cushman	F					F				F	R		F			

R:Rare(0-5) ; F:Few (6-10) ; P:Present(11-15) ; C:Common (16-20) & D:Dominant (>20/10gm dry sample)

well rounded detrital materials, pellets, and ooides. Their shape varies from sphaeroidal to ellipsoidal but some irregular grains with smooth, polished and pitted surface are also present. The 'worn type' and abrated skeletal constituents are represented by shallow water benthic foraminifera, pteropod, microgastropod, bivalves, ostracod, scaphopod, coral and bryozoa fragments. The benthic foraminiferal assemblage contains *Amphistegina lessoni, A.radiata, Alveolinella quoyi, Elphedium crispum, Ammonia beccarii, Chrysalidinella dimorpha, Trochammina inflate, Quinquelina parkari, Q.lamarcana, Tri/oculina tricarinata, Spiroloculina sp., Borolis sp and Calcarina calcar*

The nature of skeletal and non skeletal grains as well as the benthic foraminiferal assemblage of this sediment patches points towards their relict origin. This zone shows antipathetic (incongruous) relation with the bathymetric contour.

Interpretation and discussion

Five benthic foraminiferal biofacies *(Astrorotalia, Bolivina, Bulimina-Bolivina, Bulimina Uvigerina* and *Bulimina)* have been identified in the study area. These biofacies show sympathetic relation with bathymetric contours. The nature of bottom sediments appears to favour the productivity of *A. beccarii* and *A.trispinosa* in the shelf region off Krishna- Godavari delta region. The width and extent of the biofacies are being controlled by the physiographic set up of the region.

The presence of 'worn type' and abraded skeletal taxa of shallow water facies as a lensiod patch of the present deep water environment reveal the occurrence of relict sediment to the southwestern side of the area. The benthic foraminiferal_assemblage *(Amphistegina lessoni, A.radiata, Elphidium crispum, Ammonia becarii, Trilocu/ina tricarinata* and *Calcarina calcar)* of the relict sediments indicates that their deposition in the inner shelf environment probably during Late Pleistocene low sea stand and subsequently during the Holocene time transported to lower bathyal zone.

Occurrence of different non skeletal grains (ooides, peloides and rounded to sub rounded detrital materials) within the relict sediment patch also supports the evidence of sediment transport from shelf region to present day deeper environment.

The characteristic benthic foraminiferal assemblage, in the major part of the survey area, indicates an environment of low oxygen and high organic carbon content of the sediment. These biota are again characterized by thin test and higher surface area: volume ratio suggesting a reducing environment. The P:B ratio varies between 8 and 20 within the depth range of 500m to 1700m isobaths and shows lower values than the normal marine environment. The P:B value of the survey area does not corresponds with the same by Van Marle (1987) and DiesterHass (1983) who worked on foraminifera in the Bay of Bengal and the Andaman Sea respectively. High fragmentation of planktic foraminifera and considerable low P: B ratio of foraminifera suggest poor mode of preservation of planktic population. This episode also corroborates a reducing environment. Aragonite Compensation Depth (ACD) has been demarcated on the basis of surficial distribution pattern of pteropod and their mode of preservation /fragmentation in the seabed sediments. The average ACD level is observed at about 1000m bathymetric level.

Conclusion

Dispersal pattern of the characteristic taxa reveals that transportation of shallow water biota in the vicinity of Godavari river system is confined within the upper bathyal zone due to higher shelf width and very gentle gradient in the northeastern part of the area. The slope is wide in the northern part off Godavary delta due to sediment build up on the shelf as well as on the slope

On the other hand transportation of very shallow water and high energy benthic species from Krishna river mouth to the lower bathyal zone are observed in the southwestern side of the area. Large scale sediment transport in the latter region is possible due to less shelf width, more gradient and number of channels present in the continental shelf region. Number of skeletal and non skeletal materials of shelf facies has been transported down through the steep gradient of Krishna channel system and might-have channelised their ways into the middle and lower bathyal zones.Hemipelagic clay followed down depth by layers of mica rich coarser sediments mixed with other minor detritalas (Quartz,garnet, pyroxene,amphibole and opaques),shallow benthics and ooides represent typical assemblage of shelf facies sediment of relict nature. The occurrence of the relict sediment patch and its antipathic relation with the isobath lines off Krishna delta region reveals the different environmental condition . from Godavari delta and interdelta domains off the Andhra coast in the Bay of Bengal.

Acknowledgement

The authors are grateful to sri B.K. Saha, Dy.Director General, Marine Wing, GSI, for providing all possible facilities to carryout the work. They sincerely thank Sri R.Sengupta, Director (CT)/OPEC-I, for his guidance and encouragement during the work. They also thank all cruise participants of SM-155 who extended their kind help during the cruise and laboratory studies.

References

ANANTHARAMAN, S. 1989: Foraminiferal biofacies in the sediments off Nizaampatnam Bay, XII Col. Micro. Stra. Abs. p.11.

BARKER, R. W. 1960: Taxonomic notes on the species figured by H.R Brady in his report on the foraminifera during the year 1873-1875. Soc. Eco. Pal. Min. Spl.Vol. pp. 1-238.

BHALLA,S.N. 1968: Recent foraminifera from beach sands and its relation to the known foramgeographical provinces of the Indian Ocean .In: Symp. Indian Ocean, Bull Nat. Inst. Sci. India. V.38(1), pp. 376-392.

BHALLA, S.N. & NIGAM., R. 1988: Cluster analysis of foraminifer fauna from beaches of the east coast of India with reference to the foramgeographical provinces of the Indian Ocean. Jour.Geol. Soc. India., V. 32, pp. 516-521.

BOLLI,H.N; SAUNDER, J.B &NIELSON, K.P.1987:Planktic Stratigraphy, Cambridge University Press, Cambridge, pp. 158-265.

BOLTOVOSKOY, B & WRIGHT, R. 1976: Recent Foraminifera, (Ed:W. Junk), RV. Publisher, The Haquye, 515p.

BHATTACHARJEE, D; GHOSH, S.K; BANDYOPADHYAY, R.R. 2002: Foraminiferal biofacies in-the northern parts of Bay of Bengal .and their environmental implication. Four decades of Marine Geosci. Ind. GSI,MW, pp. 80-85.

CORLIS, B.H & EMERSON, S. 1990:Distribution of Rose Bengal stained deep-sea benthic folaminifera from Nova Scotian Continental margin and Gulf of Maxico.Deep sea Res. V. 37, no. 3A, pp. 381-400.

DIESTER-HASS, L. 1985: Late Quaternary sedimentation on the eastern Walvis Ridge, SE Atlantic., Mar. Geo. V. 65, pp. 145-189.

GANAPATI, P.P.N & SAROJINI, D. 1959: Ecology of foraminifera off Vishahkapatnam coast. All Indian Cong. Zool. Proc. V. 2, pp. 311-315.

GRAZZINI, C.V & PIERRE, C. 1992: The carbon Isotope distribution in deep Co_2 and benthic foraminifera of the Alboran basin, Western Mediterraean, Mar. Micopal, V. 19, pp. 147-161.

GUPTA, A.K & SRINIVASAN, M.S. 1990: Quaternary bottom water circulation and benthic foraminiferal changes in the north Indian Ocean. Indian Jour. Geol. 62(2), pp.102-116.

HERMELIN, J.O.R & SHIMMIELD, G.D.1990: The importance of oxygen minima zone and sediment geochemistry in the distribution of recent benthic foraminifera in the northwestern Indian ocean, Mar. Geo. V.91, pp. 1-29.

INGLE, J.C., KELLER, G & KALPACK, R.L., 1980: Benthic foraminiferal biofacies, sediments and water masses of the southern Peru- Chile trench. Mar.Micropal, V .26 (2), pp.I13-150.

KENNETT, J. P. & SRINIVASAN, M.S., 1983: Neogene planktic foraminifera, A phylogenetic Atlas: Hutchinson Ross Pub..USA, pp. 163.

LEOBLICH, A.R. & TAPPAN, H., 1987: Foraminiferal genera and their classification, Van Nostrand Reinhold Company, pp. 889.

MADABHUSHI, S., 1991: Anomalies in foraminiferal distribution in some parts of Bay of Bengal. Jour.Geol.Soc. Ind., V.38, pp.312-318.

MADABHUSHI, S., RAO, B.L. & NEELKANTAM, S., 1994: Is the lime mud off Pudimadaka Holocene? News Letter, Marine Wing, GSI, V.X, no.l, pp.16-17.

PATTERSON, R. T., 1993: Late Quaternary benthic foraminiferal biofacies and paleoceanography of Queen Charlotte sound and southern hecate strait, British Columbia, Jour. For.Res.,V.23, no.l, pp.I-18.

PATTERSON, R. T. & CAMERON, B. E. B., 1991: Paleo environmental significance of the foraminiferal- biofacies succession in the late quaternary sediments of Fraser river delta, British Columbia, Jour. For.Res.V.21, pp.228-243.

RAO, T.V. & SUBBA RAO, M., 1974: Living foraminifera from the tidal creek at Pucijmadaka, near Vishakhapatnam shelf, east coast of India. In: Indian Ocean Adj. Seas, Mar. Bio. Asso. India, V.6, pp.32-35.

SUBBA RAO, M, VEDANTAM, M. & RAO, J.N., 1979: Distribution and ecology of benthic foraminifera in the sediments of the Vishakhapatnam shelf, east coast of India. Paleo.Paleo.Paleo, V.27, pp.349-369.

VAN MARLE, L.J., VAN HINTE, J.E. & NEDERBRAGT, A. J., 1987: Planktic percentage of the foraminiferal fauna in seafloor Samples from the Australian-irian Jaya continental margin, Eastern Indonesia, Mar.Geol., V.77, pp.151-156.

VEDANTAM, D. & SUBBA RAO, M.1970: Recent foraminifera from off Pentakota, east coast of India. Micropal.V.16 (3), pp.325-344.

WALTON, W.R., 1964: Recent foraminiferal ecology and paleoecology .In: J. Imprai & D. Newell (Eds.).Approaches to Paleoecology, Wiley, New .Approaches to Paleoecology, Wiley, New York, pp.151-171.

WELLS, P., WELLS, G., CALI, J. & CHIVAS, A., 1994: Response of deep-sea benthic foraminifera to Late Quaternary climate change, SE Indian Ocean, off shore western Australia. Mar. Micropal. V.23, pp.185-229.